Recent Titles in This Series

(Continued in the back of this publication)

CONTEMPORARY MATHEMATICS

180

Domain Decomposition Methods in Scientific and Engineering Computing

Proceedings of the Seventh
International Conference
on Domain Decomposition
October 27–30, 1993
The Pennsylvania State University

David E. Keyes
Jinchao Xu
Editors

American Mathematical Society
Providence, Rhode Island

The Seventh International Conference on Domain Decomposition was held at the Pennsylvania State University, from October 27–30, 1993, with support from the National Science Foundation, Grant No. DMS-9301980; Department of Energy, Grant No. DE-FG0293ER25184; Institute for Mathematics and its Applications (Minneapolis); GAMNI (France); IBM Corporation; Cray Research Corporation; and Pennsylvania State University.

1991 *Mathematics Subject Classification.* Primary 65M55, 65N55; Secondary 65F10, 65N30, 65N35, 65Y05.

Library of Congress Cataloging-in-Publication Data

International Conference on Domain Decomposition (7th : 1993 : Pennsylvania State University)
  Domain decomposition methods in scientific and engineering computing: proceedings of the Seventh International Conference on Domain Decomposition, October 27–30, 1993, the Pennsylvania State University / David E. Keyes, Jinchao Xu, editors.
    p. cm. — (Contemporary mathematics, ISSN 0271-4132; v. 180)
  Includes bibliographical references.
  ISBN 0-8218-5171-3 (alk. paper)
  1. Decomposition method—Congresses. 2. Differential equations, Parital—Congresses.
I. Keyes, David E.   II. Xu, Jinchao, 1961– .   III. Title.   IV. Series: Contemporary mathematics (American Mathematical Society); v. 180.
QA402.2.I55   1993
515′.353–dc20

94-41503
CIP

10 9 8 7 6 5 4 3 2 1   99 98 97 96 95 94

Contents

CONTENTS

Part III. Parallelism

Part IV. Applications

PREFACE

This volume captures the main part of the proceedings of the Seventh International Conference on Domain Decomposition Methods, which was hosted by The Pennsylvania State University, October 27–30, 1993. Over one hundred and fifty mathematicians, engineers, physical scientists, and computer scientists came to this nearly annual gathering – nearly half of them for the first time. Those attending from outside the United States accounted for about one-third of the registrants and came from 18 countries.

Since parallel sessions were employed at the conference in order to accommodate as many presenters as possible, attendees and non-attendees alike may turn to this volume for the latest developments. Most of the authors are to be commended for their efforts to balance the conflicting demands of writing for a diverse audience and staying within limits of twelve pages for invited lecturers and six pages otherwise. Enforcing page quotas was essential in accommodating the largest title count in the seven-volume history of the conference — selected from an even larger number of submissions.

After seven meetings, it would be natural to expect that domain decomposition methods would have moved into the mainstream, and therefore ceased to justify the special focus of their own conference. The interest of authors from many fields in entrusting final versions of their latest work to these proceedings supports this premise — while at the same time contradicting its conclusion! "Divide and conquer" domain decomposition may be the most basic of algorithmic paradigms, but theoreticians and practitioners alike are still seeking — and finding — incrementally more effective forms, and enjoy an interdisciplinary forum for them.

We comment briefly on the the term "domain decomposition" that has for nearly a decade been associated with this meeting and its proceedings. In the past few years, "domain decomposition" has become a synonym for "data parallelism" in the parlance of computer science, where it stands in contrast to "task parallelism." In a generic sense, any algorithm that achieves concurrency by applying *all* of the operations independently to *some* of the data, as opposed to *some* of the operations independently to *all* of the data, may properly be called "domain decomposition," but casting the net this broadly almost ceases to be useful. The PDE-motivated subject matter of this meeting has traditionally revolved around two foci within this very broadly defined class of algorithms: iterative subspace correction methods, and block elimination methods. In the former, which we may for convenience call "Schwarz methods" (though Schwarz's recorded perspective was narrower), a domain solver is used as a subdomain solver inside an iteration over subdomains. In the latter, which have been classified "iterative substructuring," and which we may in the same spirit call "Schur methods," an operator equation for a lower-dimensional interface between subdomains is derived. Schwarz and Schur methods may be unified in certain cases by regarding the Schwarz iteration as a map between iterates restricted to the interfaces. Both Schwarz and Schur approaches are now customarily used inside of a Krylov iteration (such as conjugate gradients), leading to what may be called "Krylov-Schwarz" and "Krylov-Schur" methods. These are terms that computer scientists are less likely to borrow.

The notion of a geometrical domain decomposition leads immediately to the notion of a function space decomposition, in which the subspaces are associated with subregions of support, and, in turn, to an operator decomposition, in which the relevant operators are restrictions of the original operator to the subspaces. This association between geometry and subspaces and operators allows many domain decomposition methods to be analyzed as iterative subspace correction methods, along with their relatives from classical iterative methods and multilevel methods, in which the decomposition of the corresponding function spaces is motivated by factors other than geometry.

Organizing the contents of an interdisciplinary proceedings is an interesting job, and our decisions will inevitably surprise a few authors, though we hope without causing offense. It is increasingly artificial to assign papers to one of the four categories of theoretical foundations, algorithmic development, parallel implementation, and applications, that are traditional for this proceedings series. Readers are encouraged not to take the primary divisions very seriously, but to trace all the connections.

Browsers turning to a preface expect a few words of context, particularly about what's new. The volume-wide subject classifications, viz.,

> **65N55:** Multigrid methods; domain decomposition for BVPs
> **65N30:** Finite elements, Rayleigh-Ritz and Galerkin methods, finite methods
> **65F10:** Iterative methods for linear systems
> **65Y05:** Parallel computation
> **65M55:** Multigrid methods; domain decomposition for IVPs
> **65N35:** Spectral, collocation and related methods

give only a majority-weighted impression of the contents. Specific noteworthy trends in the Seventh International Conference on Domain Decomposition Methods are highlighted below.

- *elliptic PDE problems and progress in dealing with so-called "bad parameters":*

For smooth problems, algorithms guaranteeing convergence rates that are asymptotically only weakly dependent of the size of the subdomains (H) into which a domain is cut and the finest resolved length scale (h) have been known for nearly a decade. The Poisson problem has inspired further theoretical work, primarily in establishing links to multilevel theory [Bank–Xu, Bornemann, Griebel]. The framework of "stable splittings" for iterative subspace correction methods is described in [Oswald], and permits discretization error estimates to be obtained as a by-product of algebraic convergence monitoring [Ruede]. The same benefit obtains from the cascade principle for the solution of general elliptic BVPs [Deuflhard]. The dependence of convergence rate on physical parameters such as jumps in the coefficients [Bakhvalov–Knyazev, Dryja, Le Tallec–Mandel–Vidrascu, Nepomnyaschikh] and irregular domain geometry not resolvable by a coarse grid [Kornhuber–Yserentent] have also come under study herein. Obstacle problems have been extended to include first-order terms [Kuznetsov–Neittaanmaki–Tarvainen].

- *PDE developments outside of the elliptically-dominated framework:*

The problem of large first-order terms, which manifests itself in both discretization and solution phases of a PDE, has traditionally been handled by some form of elliptic domination. This is achieved by considering h sufficiently small, so that second-order derivative contributions to the stiffness matrix dominate first-order, or by artificially diffusive discretizations of the first-order terms in the operator to be inverted. (In this case, more accurate discretizations are typically used in the computation of the true residual). Recently, investigators have been working in from the other end of the Reynolds number spectrum, starting from methods that become all the more accurate as the elliptic terms become vanishingly small [Katzer, Layton-Maubach-Rabier]. (The approach in [Katzer] is related to frequency decomposition methods.) Preconditioners so derived may be much more efficient to apply than preconditioners coming from all of the terms of the discretization [Ashby–Kelley–Saylor–Scroggs]. Generalized Schwarz splittings, involving mixed (Robin-type) or tangential boundary conditions on artificial boundaries, with parameters dependent upon the advection seem promising [Tan–Borsboom], especially in the context of inexact subdomain solves, where parameters can be found that mitigate the loss of coupling in an overlapped block ILU technique [de Sturler]. "Outflow" boundary conditions are preferred in [Nataf–Rogier]. Meanwhile, the Schwarz alternating method has been generalized in another way, namely, overdetermined matching conditions within a layer of finite thickness [Sun–Tang] instead of well determined conditions along an edge. In addition to dealing with operator nonsymmetry and boundary conditions, research has continued into operator-splitting approaches for advection-diffusion. Such semi-implicit, semi-Lagrangian methods are able to exploit the method of characteristics for the pure advection part and the symmetric elliptic theory for the pure diffusion part [Chefter–Chu–Keyes, Wang–Dahle–Ewing–Lin–Vag]. Nonlinear problems, whose theory greatly lags practice, have come in for more theoretical attention during the past year [Cai–Dryja, Dawson–Wheeler, Tai]. (The ability to treat the full nonlinearity on the coarse grid only is studied in [Dawson–Wheeler].) Wave-Helmholtz problems are also considered [Kim].

- *non-traditional discretizations:*

Earlier volumes of this proceedings series were devoted almost entirely to low order discretizations based on conforming finite elements, finite differences, or finite volumes. This volume contains new convergence and/or complexity results for several other discretizations, including nonmatching grids [Le Tallec–Sassi–Vidrascu], nonconforming elements [Brenner, Sarkis], spectral multidomain [Azaiez–Quarteroni, Pavarino], sinc functions [Lybeck–Bowers], $h-p$ finite elements [Oden–Patra–Feng], and the multiresolution-like "sparse grids" approach [Bungartz–Griebel–Roeschke–Zenger]. In several of these methods, the discretization and solution processes are intertwined. Related to developments in the discretizations themselves are adaptive grid refinements [Mishev, Shih–Liem–Lu–Zhou].

- *coarsened operators:*

With the non-traditional discretizations come new challenges for the derivation of appropriate operators for one or more coarsened spaces. Coarsened spaces play at least two critical roles as far as the mathematical analysis of domain decomposition methods are concerned (and perhaps others from a computer architecture point of view). They may be used to weaken or remove the dependence of the convergence rate on the number (or size) of the subdomains, and on jumps in the coefficients of the differential equation. A coarsened unstructured grid will, in general, not be nested in the fine unstructured grid for which it is created. Therefore, the corresponding function spaces are not nested. This situation has been dealt with in multigrid theory, and is practically addressed in [Bank–Xu, Chan–Smith]. [Kornhuber–Yserentent] consider the case in which the domain itself is not resolved by the coarse grid, illustrating with a fractal-like domain. Fundamental requirements for a good coarse space and some practical examples, including for higher-order fine spaces, are given in [Widlund].

• *other preconditioner developments:*

Regarding preconditioners, as has been noted in a different context, anyone "has a perfect chance to find a better one."[1] Regarding parallel preconditioners, the field is even more wide open, since serial suboptimality may be tolerated in trade-offs that favor computer architectural considerations. Preconditioner candidates appearing in these proceedings include fast summation techniques (essentially $O(n \log n)$ dense convolutions) for the coarse grid solver required in many elliptic preconditioners [Scott] and preconditioning a Krylov method by another Krylov method, applied independently within subdomains [Pernice]. A "one shot" method combines domain decomposition on the base grid with the fictitious domain (or domain imbedding) method for potential problems with irregular geometry [Glowinski–Pan–Periaux]. The boundary element formulation of the potential problem is employed in conjunction with a two-level BPS-type preconditioner with optimal results for the model problem in [Steinbach]. The rich theory of wire-basket preconditioners continues to be undergirded, most recently by [Pavarino, Shao]. In addition, "probing" for interface preconditioner blocks has received new experimental attention in the context of large coefficient variations [Giraud–Tuminaro].

• *non-PDE problems:*

One-dimensional problems under shooting are interpreted as domain decomposition methods in [Lai]. Calling the partitioning of a search space for the roots of an algebraic problem for a system of nonlinear equations a form of domain decomposition, [Mejzlik] proposes a generalized bisection root finder.

• *implementations and architectural considerations:*

Domain decomposition leads to a truncated form of nested dissection ordering in [Lin, Mehrabi–Brown]. This purely Schur form of parallelism on a distributed memory system turns out to be competitive with conventional finite element software on vector supercomputers. Implementations of capacitance matrix and box- and strip-based domain decomposition preconditioners are compared on shared memory parallel and superscalar architectures in [Ciarlet]. Parallel implementations of Krylov-Schwarz domain decomposition algorithms on networks of high-performance workstations are introduced in [Bjørstad–Coughran–Grosse]. Their practical discussion of the capabilities and limitations of networks will help orient researchers who are contemplating this seemingly cost-effective environment. The coarse grid problem, though key to the optimal convergence rates achievable by domain decomposition methods, is a bottleneck to parallelism on most realizable architectures, as

---

[1] G. Strang, *Introduction to Applied Mathematics*, Wellesley-Cambridge Press, 1986.

it also is in multigrid. Interesting attempts to overcome this manifestation of the "conservation of tsuris"[2], are featured in [Farhat–Chen, Roux–Tromeur-Dervout], which also consider the algorithmically related problem of multiple right-hand sides in a Krylov method.

- *partitioning tools and environments:*

As solvers mature, and the dependence of their convergence rates and parallel efficiencies on the partitioning is exposed, the partitioning itself is becoming a significant research interest. A mathematically elegant solution to the partitioning problem is recursive spectral bisection [Pothen], in which a small number of eigenvectors of the Laplacian matrix of the graph of the grid are used to find the partitioning cuts. A compiler for parallel finite element methods with domain-decomposed unstructured meshes is demonstrated in [Shewchuk–Ghattas]. The paper [Mu–Rice] argues for domain decomposition as a framework for the design of object-oriented software, while [Chrisochoides–Fox–Thompson] present a problem-solving environment for mapping subdomains and generating grids in parallel.

- *applications:*

The technologically important subject of semiconductor device simulation had not taken its rightful place alongside computational fluid dynamics and computational structural mechanics in domain decomposition proceedings until the current developments in [Bjørstad–Coughran–Grosse, Coomer–Graham, Giraud–Tuminaro, Micheletti–Quarteroni–Sacco]. The stiff problems of structural mechanics, caused by aspect ratios and anisotropic material properties that are extreme even in model problems have steadily driven domain decomposition theory, with the latest installment in [Le Tallec–Mandel–Vidrascu]. The physically smooth blending between the domains of applicability of Euler and Navier-Stokes models that occurs in the physical world is mimicked in the "χ" approach of [Arina–Canuto]. The primarily internal flow realm of incompressible Navier-Stokes is considered in [Jacobs–Mousseau–McHugh–Knoll, Raspo–Ouazzani–Peyret, Vozovoi–Israeli–Averbuch]. Specific applications of an expanding jet [Ku–Gilreath–Raul–Sommerer], detonation combustion [Cai], external aerodynamics [Cai–Gropp–Keyes–Tidriri, Hsiao–Marcozzi–Zhang] and the shallow water equations in geophysical contexts [Cai–Navon, Chefter–Chu–Keyes] are also taken up herein. The elliptically-dominated nonlinear Poisson-Boltzmann equation is solved by a variety of multigrid and domain decomposition-based methods in [Holst–Saied], one of the few settings apart from the Poisson problem in which these two families of methods have been carefully compared. Perhaps the most novel application area for domain decomposition relative to previous proceedings is to the Bellman equations [Camilli–Falcone–Lanucara–Seghini], where the state space of the relevant PDE may have dimension much larger than three.

For the convenience of readers coming recently into the subject of domain decomposition methods, a bibliography of previous proceedings is provided below, along with some major recent review articles and related special interest volumes. This list is about twice as large as could have been offered last year, and yet, will probably be embarrassingly incomplete by the time it is published. (No attempt

---

[2]from the Yiddish for "trouble"

has been made to supplement this list with the larger and closely related literature of multigrid and general iterative methods, except for the book by Hackbusch,
which has a significant domain decomposition component.)

(1) T. F. Chan and T. P. Mathew, *Domain Decomposition Algorithms*, Acta
    Numerica, 1994, pp. 61-143.

(2) T. F. Chan, R. Glowinski, J. Périaux and O. B. Widlund, eds., *Proc. Second
    Int. Symp. on Domain Decomposition Methods for Partial Differential
    Equations* (Los Angeles, 1988), SIAM, Philadelphia, 1989.

(3) T. F. Chan, R. Glowinski, J. Périaux, O. B. Widlund, eds., *Proc. Third
    Int. Symp. on Domain Decomposition Methods for Partial Differential
    Equations* (Houston, 1989), SIAM, Philadelphia, 1990.

(4) C. Farhat and F.-X. Roux, *Implicit Parallel Processing in Structural Me-
    chanics*, Computational Mechanics Advances **2**, 1994, pp. 1–124.

(5) R. Glowinski, G. H. Golub, G. A. Meurant and J. Périaux, eds., *Proc. First
    Int. Symp. on Domain Decomposition Methods for Partial Differential
    Equations* (Paris, 1987), SIAM, Philadelphia, 1988.

(6) R. Glowinski, Yu. A. Kuznetsov, G. A. Meurant, J. Périaux and O. B.
    Widlund, eds., *Proc. Fourth Int. Symp. on Domain Decomposition Meth-
    ods for Partial Differential Equations* (Moscow, 1990), SIAM, Philadelphia,
    1991.

(7) W. Hackbusch, *Iterative Methods for Large Sparse Linear Systems*, Springer,
    Heidelberg, 1993.

(8) D. E. Keyes, T. F. Chan, G. A. Meurant, J. S. Scroggs and R. G. Voigt,
    *Proc. Fifth Int. Conf. on Domain Decomposition Methods for Partial
    Differential Equations* (Norfolk, 1991), SIAM, Philadelphia, 1992.

(9) D. E. Keyes, Y. Saad and D. G. Truhlar, eds. *Domain-based Parallelism
    and Problem Decomposition Methods in Science and Engineering*, SIAM,
    Philadelphia, 1995 (to appear).

(10) P. Le Tallec, *Domain Decomposition Methods in Computational Mechanics*,
     Computational Mechanics Advances **2**, 1994, pp. 121–220.

(11) A. Quarteroni, J. Périaux, Yu. A. Kuznetsov and O. B. Widlund, eds.,
     *Proc. Sixth Int. Conf. on Domain Decomposition Methods in Science and
     Engineering* (Como, 1992), AMS, Providence, 1994.

(12) B. F. Smith, P. E. Bjørstad and W. D. Gropp, *Domain Decomposition:
     Parallel Multilevel Algorithms for Elliptic Partial Differential Equations*,
     Cambridge Univ. Press, Cambridge, 1995 (to appear).

(13) J. Xu, *Iterative Methods by Space Decomposition and Subspace Correction*,
     SIAM Review **34**, 1991, pp. 581-613.

The technical direction of the Seventh International Conference on Domain Decomposition Methods in Scientific and Engineering Computing was provided by
a scientific committee consisting of: James H. Bramble, Tony F. Chan, Ray-
mond C. Y. Chin, Peter J. Deuflhard, Roland Glowinski, Gene H. Golub, David E.
Keyes, Yuri A. Kuznetsov, Jacques Périaux, Alfio Quarteroni, Olof B. Widlund, and
Jinchao Xu. Local organization was undertaken by the following members of the
faculty of The Pennsylvania State University: Douglas N. Arnold, Jerry L. Bona,
Min Chen, Ali Haghihat, Jie Shen, Simon J. Tavener, and Jinchao Xu (Chair).
The scientific and organizing committees, together with all attendees, are grateful

to the following agencies, organizations, corporations, and departments for their financial and logistical support of the conference: the National Science Foundation (DMS-9301980), U. S. Department of Energy (DE-FG0293ER25184), the Institute for Mathematics and its Applications at the University of Minnesota, France's GAMNI, the IBM Corporation, Cray Research Corporation, and the Applied Research Lab, the College of Engineering, the Department of Mathematics, and the Eberly College of Science at The Pennsylvania State University.

The American Mathematical Society is publishing the proceedings of the International Conference on Domain Decomposition Methods for the second consecutive time. The editors are very grateful to Donna L. Harmon of AMS for her pacing and her patience, and to Ling Shen, a doctoral candidate in the Department of Mathematics at The Pennsylvania State University for her assistance with reformatting several papers received in nonconforming styles. A single thorough round of editing, followed by the authors assuming final responsibility for the revised camera ready copy, seems to permit a good balance between the quality of the proceedings and the promptness of its publication.

Our families graciously forsook many weekends together for this collection and are trusting, as are we, in a useful shelf life.

David Keyes
Hampton, Virginia

Jinchao Xu
University Park, Pennsylvania

September 1994

LIST OF PARTICIPANTS

I.K. Abu-Shumays
Bettis Atomic Power Lab
P.O. Box 79, Zap 34DD
W. Mifflin, PA 15122-0079

Douglas N. Arnold
Dept. of Mathematics
319 McAllister Bldg.
Penn State University
University Park, PA 16802
dna@math.psu.edu

Randolph E. Bank
Dept. of Mathematics
University of California at San Diego
LaJolla, CA 92093-0112
reb@sdna2.ucsd.edu

Jerry Bona
Dept. of Mathematics
316 McAllister Bldg.
Penn State University
University Park, PA 16802
bona@math.psu.edu

Mart Borsboom
Delft Hydraulics
P.O. Box 152
8300 AD Emmerloord, The Netherlands
Borsboom@wldelft.nl

Susanne Brenner
Dept. of Mathematics
University of South Carolina
Columbia, SC 29208
brenner@math.scarolina.edu

Robert A. Brown
Dept. of Chemical Engineering
66-342, MIT
77 Massachusetts Ave.
Cambridge, MA 02139
rab@mit.edu

Wei Cai
Dept. of Mathematics
University of North Carolina
Charlotte, NC 28223
wcai@mosaic.uncc.edu

Renzo Arina
Aerospace Dept., Politecnico di Torina
24 Corso Duca Delgi Abruzzi
Torino, 10129 Italy
arina@athena.polito.it

Majdi Azaiez
Universite Paul Sabatier
U.F.R. M.E.G.
118 route de Narbonne
31062 Toulouse Cedex, France
azaiez@soleil.ups-tlse.fr

Petter E. Bjorstad
Institute for Informatik
University of Bergen
Hoyteknologisenteret
Bergen, N-5020 Norway
Petter@ii.uib.no

Folkmar Bornemann
Konrad-Zuse-Zentrum
Heilbronner Strasse 10
Berlin, D-10711 Germany
bornemann@sc.ZIB-Berlin.DE

Ken L. Bowers
Dept. of Math Sciences
Montana State University
Bozeman, MT 59717

Franco Brezzi
IAN-CNR
Universita di Pavia
Via Abbiategrasso 209
27100 Pavia, Italy
brezzi@ipvian.bitnet

Hans J. Bungartz
University Muenchen
Institut fuer Informatik
University Muenchen, Postfach
Muenchen, D-80290 Germany
bungartz@informatik.tu-muenchen.de

Xiao-Chuan Cai
Dept. of Computer Science
University of Colorado
Boulder, CO 80309
cai@cs.colorado.edu

Yihong Cai
Dept. of Chemical Engineering
MIT
77 Massachusetts Ave.
Cambridge, MA 02139
ycai@mit.edu

Maria Morandi Cecchi
Dipartimento di Matematica
Pura ed Applicata
Universita di Padova
via Belzoni 7, 35131 Padova, Italy
mcecchi@pdmat1.unipd.it

Tony F. Chan
Dept. of Mathematics
U.C.L.A.
Los Angeles, CA 90077
chan@math.ucla.edu

Min Chen
209 McAllister Building
Penn State University
University Park, PA 16802
chen_m@math.psu.edu

Mathew W. Choptuik
Center for Relativity
Dept. of Physics
University of Texas
Austin, TX 78712-1081
math@ehstein.pl.utexas.edu

C. K. Chu
Dept. of Applied Physics
Columbia University
New York, NY 10027
chu@cuplvx.ap.columbia.edu

Pedro J.M. Coelho
Instituto Superior Tecnico
Av. Rovisco Pais
1096 Lisboa Codex, Portugal
D2272@BETA.IST.UTL.PT

Lawrence C. Cowsar
AT&T Bell Labs., Rm. 2C-464
600 Mountain Avenue
Murray Hill, NJ 07974-0636
cowsar@research.att.com

Eric de Sturler
Swiss Scientific Computing Ctr.
CSCS-ETHZ
Via Cantonale
Manno, CH-6928 Switzerland
sturler@cscs.ch

Mario A. Casarin
Courant Institute - NYU
251 Mercer Street
New York, NY 10012
casarin@math12.nyu.edu

Daniel C. Chan
Dept. of Aerospace Engineering
University of Southern California
Los Angeles, CA 90089
yqb532@sunshine.rdyne.rockwell.com

Julia G. Chefter
Dept. of Applied Physics
Columbia University
New York, NY 10027
jgc@appmath.columbia.edu

Hsuanjen Cheng
Mathematical Science
Courant Institute
251 Mercer Street
New York, NY 10012
chenghs@math1.cims.nyu.edu

Nikos P. Chrisochoides
NPAC, Syracuse University
111 College Place
Syracuse, NY 13244
nikos@npac.syr.edu

Patrick Philippe Ciarlet
CELV / DMA
94195 Villeneuve-St-Georges Cedex, France
ciarlet@etca.fr

William M. Coughran
AT&T Bell Laboratories
600 Mountain Avenue
Murray Hill, NJ 07974
wmc@research.att.com

Helge K. Dahle
University of Bergen
Johs. Brunsgt. 12
N-5008 Bergen, Norway

Peter Deuflhard
Konrad-Zuse-Zentrum
Heilbronner Strasse 10
Berlin, D-10711 Germany
deuflhard@sc.ZIB-Berlin.de

Zdenek Dostal
Vysoka skola banska
tr. 17. listopadu
Ostrava, CZ-70833 Czechoslovakia
HOST@nw456.vsb.cs

Maksymilian Dryja
Depart. of Math
University of Warsaw
Banacha 2
02-097 Warsaw, Poland
dryja@mimuw.edu.pl

Richard E. Ewing
Dean's Office
College of Science
Texas A&M University
College Station, TX 77843-3257
ewing@ewing.tamu.edu

Yusheng Feng
University of Texas at Austin
3500 West Balcones Center
Austin, TX 78759
feng@ticom.ae.utexas.edu

Sonia Garcia
Mathematics Dept., M/S No.9E
U.S. Naval Academy
Annapolis, MD 21402
smg@usna.navy.mil

Luc Giraud
CERFACS
42, Av. G. Coriolis
31057 Toulouse, FRANCE
giraud@cerfacs.fr

Charles I. Goldstein
Bldg. 490A
Brookhaven National Lab.
D.A.S., P.O. Box 5000
Upton, NY 11973-5696
golostel@bnl.gov

Olivier R. Gosselin
Elf Aquitaine Production Pau
Resevoir Engineering Dept. Modelling Grp
C.S.T.J.F. Avenue Larribau
64018 Pau Cedex, France
gosselin@cst092.elf-p.fr

Paul A. Gray
A209 Wells Hall
Michigan State University
East Lansing, MI 48824
pgray@math.msu.edu

Craig C. Douglas
IBM T. J. Watson Research Center
Yale University
New Haven, CT 06520
douglas-craig@cs.yale.edu

Magne S. Espedal
Inst. for Scientific Computation
326 Teague Research Center
Texas A&M University
College Station, TX 77843-3404
espedal@ise.tamu.edu

Charbel Farhat
Center for Aero Structures
University of Colorado at Boulder
Boulder, CO 80309
charbel@boulder.Colorado.EDU

Emilio Fuentes
Dept. of Nuclear Engineering
North Carolina State University
P.O. Box 7909
Raleigh, NC 27695
fuentes@nepjt.ne.ncsu.edu

Omar Ghattas
Dept. of Civil Engineering
Carnegie Mellon University
Pittsburgh, PA 15213
ghattas@cmu.edu

Roland Glowinski
Dept. of Mathematics
University of Houston
4800 Calhoun
Houston, TX 77006

Gene H. Golub
Computer Science Dept.
Stanford University
Stanford, CA 94305
golub@sccm.stanford.edu

Ivan Graham
School of Mathematical Sciences
University of Bath
Bath, Avon BA2 7AY
United Kingdom
igg@maths.bath.ac.uk

Michael Griebel
Institut fuer Informatik der
Univ. Muenchen
Postfach Muenchen
D-80290, Germany
griebel@informatik.tu-muenchen.de

Eric Grosse
AT&T Bell Labs, 2T504
600 Mountain Ave.
Murray Hill, NJ 07974
ehg@research.att.com

Jiang Heng He
Dept. of Geology and Geophysics
Mccorn Hall #405
Berkeley Campus
Berkeley, CA 94720
guoping@decollem.berkeley.edu

Michael J. Holst
Applied Mathematics 217-50
Caltech
Pasadena, CA 91125
holst@ama.caltech.edu

Valery P. Il'in
Computing Center
Russian Academy of Science
Lavrentjev str. 6
Novosibirsk 630090, Russia
ilin@comcen.nsk.su

Yimin Kang
Mathematics Dept.
Clarkson University
Potsdam, NY 13676
kangy@sun.mcs.clarkson.edu

Hideo Kawarada
Dept. of Mathematics
Faculty of Engineering
Chiba University
Chiba 263, Japan

Abdul-Qayyam M. Khaliq
Mathematics Dept.
Western Illinois University
Macomb, IL 61455
mfamk@uxa.ecn.bgu.edu

Axel Klawonn
Courant Institute
New York University
251 Mercer St.
New York, NY 10012
klawonn@cs.nyu.edu

Ralph Kornhuber
Konrad-Zuse-Zentrum
Heilbronner Strasse 10
Berlin D-10711, Germany
kornhuber@sc.zib-berlin.de

Ali Haghighat
231 Sackett Bldg.
Penn State University
University Park, PA 16802
IAW@PSUVM.PSU.EDU

Lina Hemmingsson
Dept. of Scientific Computing
Uppsala University
Box 120, S-75104 Uppsala, Sweden
lina@tdb.uu.se

Melissa A. Hunter
110 E. Foster Ave., Apt.211
State College, PA 16801
MAH118@PSUVM.PSU.EDU

Paul G. Jacobs
P.O. Box 1625
Idaho Falls, ID 83415-3730
pgj@inel.gov

Edgar Katzer
University Otto von Guericke
Postfach 4720
D-39076 Magdeburg, Germany
katzer@dmdtu11.bitnet

David E. Keyes
Dept. of Computer Science
Old Dominion University
Norfolk, VA 23529-0162
keyes@icase.edu

Seongjai Kim
Mathmematics Dept.
Purdue University
W. Lafayette, IN 47907
skim@math.purdue.edu

Andrew V. Knyazev
Mathematics Dept. & Ctr. for Comp. Math.
Univ. of Colorado at Denver
P.O. Box 173364-3364, Campus Box 170
Denver, CO 80217-3364
na.knyazev@na-net.ornl.gov

Hwar C. Ku
Applied Physics Laboratory
The Johns Hopkins University
Johns Hopkins Rd.
Laurel, MD 20723-6099
ku@aplcomm.jhuapl.edu

Yuri A. Kuznetsov
Institute of Numerical Mathematics
Russian Academy of Sciences
Leminskij Prospect, 32-a
Moscow 117334, Russia
labnumat@adonis.ias.msk.su

Yury M. Laevsky
Computing Center
Av. Lavrentiev 6
Novosibirsk 630090, Russia

Choi-Hong Lai
School of Math Statistics & Computing
University of Greenwich
Wellington St.
Woolrich, London SE18 6PF, U.K.
c.h.lai@greenwich.ac.uk

Bill Layton
Mathematics Dept.
301 Thackery Hall
University of Pittsburgh
Pittsburgh, PA 15260
wjl@vms.cis.pitt.edu

Ping Lee
Schlumberger Laboratory
8311 North 620RR
P.O. Box 200015
Austin, TX 78720
lee@austin.nam.slb.com

Hai-Xiang Lin
Dept. of Appl. Math & Informatics
Delft University of Technology
Mekelweg 4,
2628 CD, Delft, The Netherlands
lin@pa.twi.tudelft.nl

Nancy Lybeck
North Carolina State University
Ctr. for Res. in Scientific Computation
Box 8205
Raleigh, NC 27695-8205
nlybeck@crsc1.math.ncsu.edu

Jan Mandel
Center for Computational Mathematics
University of Colorado
Denver, CO 80217-3353
jmandel@tiger.denver.colorado.edu

Robert Paul Martin
10 Vario Blvd., Apt. 16-D
State College, PA 16803
martin@idcr13.psu.edu

Do Y. Kwak
Dept. of Mathematics
Korea Advanced Inst. Sci. Tech.
Yousong Ku, Kusong dong
Taejon 305-701, Korea
dykwak@math1.kaist.ac.kr

Pablo Laguna
525 Davey Laboratory
Penn State University
University Park, PA 16802

Piero Lanucara
Dip. Di Matematica-Universita
Di Roma La Sapienza
Caspur c/o CICS-P. Le Aldo Moro 2
Roma, 00185 Italy
lanucara@itcaspur.caspur.it

Daniel Lee
P.O. Box 19-136
Hsinchu Taiwan, R.O.China
cffdleff@nchc.edu.tw

Kaitai Li
Dept. of Mathematics
Xian Jiaotong University
Xian, 710049, P.R. China

Jianqun Lu
Dept. of Mathematics
218 McAllister Bldg.
University Park, PA 16802
lu_j@math.psu.edu

Yvon Maday
Analyse Numerique
Universite Pierre et Marie Curie
Tour 55 65 4, Jussieu
Paris, France
maday@ann.jussieu.fr

Donatella L. Marini
Instituto di Analisi Numerica-CNR
Corso c. Alberto 5
27100 Pavia, Italy
marini@ipvian.bitnet

William R. McKinney
Dept. of Mathematics
312 McAllister Bldg.
Penn State University
University Park, PA 16802
mckinney@math.psu.edu

Petr Mejzlik
Institute of Computer Science
Masaryk University
Buresova 20, Czechoslovakia
mejzlik@muni.cz

Ilya D. Mishev
Inst. for Scientific Computation
326 Teague Research Center
Texas A&M University
College Station, TX 77843-3404
Ilya.Mishev@math.tamu.edu

Rabi H. Mohtar
208 Agriculture Engineering Dept.
Penn State University
University Park, PA 16802
rhm8@psuvm.psu.edu

Vincent A. Mousseau
EG&G Idaho Inc.
P.O. Box 1625
Idaho Falls, ID 83415-3895
vam@inel.gov

Frederic Nataf
Centre de Mathematiques Appl.
Ecole Polytechnique
CMAP
Palaiseau Cedex 91128, France
nataf@cmapx.polytechnique.fr

Ionel M. Navon
Dept. of Mathematics and S.C.R.I.
Florida State University
Tallahassee, FL 32308-4052
navon@scri.fsu.edu

William R. O'Reilly
Martin Marietta, KAPL Inc.
P.O. Box 1072
Schenectady, NY 12301-1072

Peter Oswald
Department of Mathematics
Texas A&M University
College Station TX 77843-3368
poswald@minet.uni-jena.de

Joseph E. Pasciak
Brookhaven N.L., Bldg. 490-D
Box 5000
Upton, NY 11973-5000
pasciak@bnl.gov

Dragan Mirkovic
Dept. of Applied Mathematics & Stat.
SUNY at Stony Brook
Stony Brook, NY 11790-3600
mirkovic@ams.sunysb.edu

Hans Mittelmann
Dept. of Mathematics
Arizona State University
P.O. Box 871804
Tempe, AZ 85287-1804
mittelmann@math.la.asu.edu

Anne Morlet
Dept. of Mathematics
Ohio State University
231 W. 18th Ave.
Columbus, OH 43210-1174
morlet@math.ohio-state.edu

Mo Mu
Dept. of Mathematics
Hong Kong University of Sci. & Tech.
Clear Water Bay
Kowloon, Hong Kong
mamu@uxmail.ust.hk

Ramesh Natarajan
Thomas J. Watson Researsch Center
I.B.M.
Yorktown Heights, NY 10510
ramesh@watson.ibm.com

Sergei Nepomnyasehikh
Computing Center
No. 6 Lavrentiev
Novosibirsk, 630090 Russia
svnep@comcen.nsk.su

Maria Elizabeth Ong
Dept. of Mathematics
Univ. of California-San Diego
9500 Gilman
La Jolla, CA 92093-0112
ong@ucsd.edu

Jalil Ouazzani
Arco Fluid - IMFM
B119 Technopole Chatero
Gomert
Marseille 13451, France
jalil@mecamars.imt-mrs.fr

Abani Patra
University of Texas at Austin
3500 West Balcones Center
Austin, TX 78759
abani@ticom.ae.utexas.edu

Luca F. Pavarino
Dipartimento di Matematica
Universita' di Pavia
Via Abbiategrasso 209
27100 Pavia, Italy
pavarino@rice.edu

Jacques F. Periaux
Dassault Aviation
78 Quai Marcel Dassault
92214 Saint-Cloud, France
periaux@menusin.inria.fr

George G. Pitts
Computer Science Dept.
Virginia Tech.
549 MacBryde Hall
Blacksburg, VA 24061
pitts@igor.csvt.edu

Gerald W. Recktenwald
Mechanical Engineering Dept.
Portland State University
630 S.W. Mill St.
Portland, OR 97201-5220
gerry@me.pdx.edu

Jesus-Enrique Rodrigues
Intevep S.A.
P.O. Box 76343
Caracas, 1070 A
Venezuela
enrique@intevep.pdv.com

Ulrich J. Ruede
Inst. fuer Informatik Tech.
Arcisstr. 21
Universitaet Muenchen
Muenchen, D-80290 Germany
ruede@informatik.tu-muenchen.de

Riccardo Sacco
Dept. of Mathematics
Politecnico of Milan
Piazza Leonardo Da Vinci 32
Milan, 20133 Italy
ricsac@ipmma1.polimi.it

Marcus Sarkis
Courant Inst. of Math. Science
New York University
251 Mercer St.
New York, NY 10012
sarkis@acf4.nyu.edu

Ridgway Scott
Dept. of Mathematics
University of Houston
Houston, TX 77204
scott@uh.edu

Andrew J. Perella
Math. Dept., Science Laboratory
South Rd.
Durham University
Durham, United Kingdom
a.j.perella@uk.ac.dur

Michael Pernice
Utah Supercomputing Institute
85 55B University of Utah
Salt Lake City, UT 84102
usimap@sneffels.usi.utah.edu

Alex Pothen
Computer Science Dept.
Old Dominion University
Norfolk, VA 23529-0162
pothen@cs.odu.edu

Calvin J. Ribbens
Computer Science Dept.
Virginia Tech.
549 MacBryde Hall
Blacksburg, VA 2406
ribbens@cs.vt.edu

Francois Xavier Roux
ONERA/Div. Calcul Parallele
29 Ave. de la Division Leclerc
Chatillon 92320, France
roux@onera.fr

Torgeir Rusten
Norwegian Computing Center
P.O. Box 114 Blindern
N-0314 Oslo, Norway
Torgeir.Rusten@nr.no

Faisal Saied
Dept. of Comp. Sci. at Urbana Champaign
University of Illinois
1304 W. Springfield Ave.
Urbana, IL 61801
saied@cs.uiuc.edu

Christoph Schwab
IBM Scientific Center
Vangerowstrasse 18
Postfach 103068
D-69115 Heidelberg, Germany
schwab@dhdibm1.bitnet

Jeffrey S. Scroggs
Center for Res. in Scientific Comp.
North Carolina State University
Box 8205
Raleigh, NC 27695-8205
scroggs@wave.math.ncsu.edu

Enrico Secco
Dipartimento di Matematica
Pura ed Applicata
Universita di Padova
via Belzoni 7, 35131 Padova, Ital
maria@goedel.unipd.it

Sally Shao
Dept. of Mathematics
Cleveland State University
E. 24th. Euclid Ave.
Cleveland, OH 44115
shao@csvaxe.csuohio.edu

Ling Shen
Dept. of Mathematics
218 McAllister Bldg.
University Park, PA 16802
shen_l@math.psu.edu

Zhong-Ci Shi
Computing Center
Academia Sinica
P.O. Box 2719
Beijing 100080, China
bmaacc@ica.beijing.canet.cn

Vijay Shukla
Dept. of Mathematics
Columbia University
127 S.W. Mudd Bldg.
New York, NY 10027
vijay@rainbow.ldgo.columbia.edu

Barry F. Smith
Math and Computer Science Division
Argonne National Laboratory
Argonne, IL 60439
bsmith@mcs.anl.gov

Xue-Cheng Tai
Dept. of Mathematics
University of Jyvaskyla
P.O. Box 35
SF-40351, Jyvaskyla, Finland
tai@jylk.jyu.fi

Kian H. Tan
Dept. of Mathematics
Utrecht State University
P.O. Box 80.010
Utrecht, 3508 TA Netherlands
kian.tan@wldelft.nl

Wei-Pai Tang
Dept. of Computer Science
University of Waterloo
Waterloo, Ontario N2L 3G1, Canada
wptang@lady.waterloo.edu

Jian-Ping Shao
Center for Computational Science
325 McVey Hall, Lexington Hall
University of Kentucky
Lexington, KY 40506-0045
shao@ms.uky.edu

Jie Shen
Dept. of Mathematics
322 McAllister Bldg.
Penn State University
University Park, PA 16802
shen_j@math.psu.edu

Jonathan R. Shewchuk
Carnegie Mellon University-SCS
500 Forbes Ave.
Pittsburgh, PA 15213-3891
jrs@cs.cmu.edu

Tsi-Min Shih
Dept. of Mathematics
Hong Kong Polytechnic
Kowloon, Hong Kong
rahoshih@hkpcc.hkp.hk

Horst D. Simon
NASA Ames Research Center
Mail Stop T045-1
Moffett Field, CA 94035-1000
simon@nas.nasa.gov

Olaf Steinbach
Mathematics Institute A
University of Stuttgart
Pfaffenwaldring 57
Stuttgart, Germany
olaf@mathematik.uni-stuttgart.de

Patrick L. Tallec
University of Paris and INRIA
BP105
Le Chesnay 78153, France
letallec@menusin.inria.fr

Li-Qun Tang
Dept. of Chemical Engineering
University of Kentucky
160 Anderson Hall
Lexington, KY 40506
super204@ukcc.uky.edu

Tarek P. Mathew
Dept. of Mathematics
University of Wyoming
Laramie, WY 82071
mathew@corral.uwyo.edu

Pasi Heikki Tarvainen
Laboratory of Scientific Computing
University of Jyvaskyla
P.O. Box 35
SF-40351 Jyvaskyla, Finland
pht@math.jyu.fi

Moulay Driss Tidriri
ICASE
Mail Stop 132C
NASA Langley Research Center
Hampton, VA 23665
tidriri@icase.edu

Ezio Venturino
Mathematics Dept.
University of Iowa
Iowa City, IA 52242
venturin@math.uiowa.edu

Lev Vozovoi
Technion-IIT
Technion City
Haifa, 32000 Israel
vozovoi@cs.technion.ac.il

Hong Wang
Dept. of Mathematics
University fo South Carolina
columbia, SC 29208
hwang@math.scarolina.edu

Mary F. Wheeler
Dept. of Computational & Applied Math
Rice University
Houston, TX 77252-1898
mfw@rice.edu

Ruifeng Xie
Dept. of Mathematics
University of Wyoming
Laramie, WY 82070
xie@ledaig.uwyo.edu

David P. Young
Boeing Computer Services
P.O. Box 24346
MS 7L-21
Seattle, WA 98124-0346
dpy6629@espresso.boeing.com

Shangyou Zhang
Dept. of Mathematical Science
University of Delaware
Newark, DE 19716
szhang@math.udel.edu

Simon Tavener
Dept. of Mathematics
420 McAllister Bldg.
Penn State University
University Park, PA 16802
tavener@math.psu.edu

Tate T.H. Tsang
Dept. of Chemical Engineering
University of Kentucky
Lexington, KY 40506-0046
che159@ukcc.uky.edu

Marina Vidrascu
INRIA
BP 105
76153 Le Chesnay Cedex, France
marina.vidrascu@inria.fr

Feng Wang
Dept. of Mathematics
218 McAllister Bldg.
University Park, PA 16802
wang@math.psu.edu

Bruno D. Welfert
Dept. of Mathematics
Arizona State University
Tempe, AZ 85287-1804
bdw@venus.la.asu.edu

Olof Widlund
Courant Institute
New York University
251 Mercer St.
New York, NY 10012
widlund@widlund.cs.nyu.edu

Jinchao Xu
Dept. of Mathematics
309 McAllister Bldg.
Penn State University
University Park, PA 16802
xu@math.psu.edu

Harry Yserentant
Mathematisches Institut
Universitat Tubingen
Auf der Morgenstelle 10
D-72076, Tubingen Germany
harry@na.mathematik.uni-tuebingen.de

Xuejun Zhang
Mathematical Science Institute, Rm. 204
Cornell University
409 College Ave.
Ithaca, NY 14850
xzhang@msiadmin.cit.cornell.edu

PART I

Theory

Contemporary Mathematics
Volume **180**, 1994

Interpolation Spaces and Optimal Multilevel Preconditioners

FOLKMAR A. BORNEMANN

ABSTRACT. This article throws light on the connection between the optimal condition number estimate for the BPX method and constructive approximation theory. We provide machinery which allows us to understand the optimality as a consequence of an approximation property and an inverse inequality in $H^{1+\epsilon}$, $\epsilon > 0$. This machinery constructs so-called *approximation spaces*, which characterize a certain rate of approximation by finite elements and relates them to interpolation spaces, which characterize a certain smoothness.

1. Introduction

For simplicity we consider the following elliptic boundary problem of second order on a polygonal domain $\Omega \subset \mathbb{R}^2$:

$$-\Delta u + u = f, \qquad \partial_n u|_{\partial\Omega} = 0,$$

where $f \in L^2(\Omega)$. The weak solution $u \in H^1(\Omega)$ is given by the variational problem

$$(1.1) \qquad a(u,v) := (\nabla u, \nabla v)_{L^2} + (u,v)_{L^2} = (f,v)_{L^2} \qquad \forall v \in H^1(\Omega),$$

where we use a notation suggestive of the scalar products in $L^2(\Omega)^2$ and $L^2(\Omega)$. Let $\mathcal{T}_j$ be a sequence of nested regular quasi-uniform triangulations of Ω with mesh-size parameter

$$h_j := \max_{T \in \mathcal{T}_j} \ \mathrm{diam}(T) \approx 2^{-j}.$$

Throughout this article we use the notation $a \lesssim b$ iff there is a constant $c > 0$, such that $a \leq cb$ and $a \approx b$ iff $a \lesssim b$ and $a \gtrsim b$. The constants c will be independent of all parameters, except possibly of Ω and the shape regularity of the triangulations.

1991 *Mathematics Subject Classification.* Primary 65N55, 65N30; Secondary 41A65, 46B70.
This paper is in final form and no version of it will be submitted for publication elsewhere.

Introducing the spaces of linear finite elements

$$X_j = \{u \in C(\bar{\Omega}) : u|_T \in P_1(T) \quad \forall T \in \mathcal{T}_j\},$$

where $P_1(T)$ denotes the linear functions on the triangle T, we get

$$X_0 \subset X_1 \subset \ldots \subset X_j \subset \ldots \subset H^1(\Omega).$$

The finite element operator $A_j : X_j \to X_j$ defined by

$$(A_j u, v_j)_{L^2} = a(u, v_j) \quad \forall v_j \in X_j$$

should be *preconditioned* for efficient computation. Thus we seek for an easily invertible operator $B_j : X_j \to X_j$ such that $A_j \approx B_j$, i.e.,

$$(A_j u, u)_{L^2} \approx (B_j u, u)_{L^2} \quad \forall u \in X_j.$$

Bramble, Pasciak and Xu [4, 14] constructed the preconditioner

$$B_j = A_0 Q_0 + \sum_{k=1}^{j} 4^k (Q_k - Q_{k-1}),$$

where $Q_k : L^2(\Omega) \to X_k$ are the L^2-orthogonal projections. They were originally able to prove without any regularity assumption on the problem (1.1)

$$(1.2) \quad \frac{1}{j+1}(B_j u, u)_{L^2} \lesssim (A_j u, u)_{L^2} \lesssim (j+1)(B_j u, u)_{L^2} \quad \forall u \in X_j.$$

Their proof was based on the observation that

$$(A_k u, u)_{L^2} \approx 4^k \|u\|_{L^2}^2 \quad \forall u \in \text{range}(Q_k - Q_{k-1}),$$

which is a fairly easy consequence of the *approximation property*

$$(1.3) \qquad \|u - Q_k u\|_{L^2} \lesssim h_k \|u\|_{H^1} \quad \forall u \in H^1(\Omega)$$

and the *inverse inequality*

$$(1.4) \qquad \|u_k\|_{H^1} \lesssim h_k^{-1} \|u_k\|_{L^2} \quad \forall u_k \in X_k.$$

In the case of $H^{1+\alpha}$-regularity of the problem (1.1), $1/2 \le \alpha \le 1$, Xu [4, 14] was able to improve the factor $1/(j+1)$ of the lower bound in (1.2) to $(j+1)^{1-1/\alpha}$.

Oswald [10] was the first to observe a strong link of this method of preconditioning to approximation theory, a link which, in fact, *immediately* supplies a proof for the optimal result

$$(1.5) \qquad A_j \approx B_j,$$

without any regularity assumption on the problem (1.1). Several authors have subsequently added generalizations or constructed more or less elementary proofs [2, 3, 7, 15, 16]. The aim of this article is to clarify the link to approximation theory by making available an easily accessible framework. Moreover it will turn out that the main ingredients of the proof are inequalities like (1.3) and (1.4).

2. Approximation Spaces are Interpolation Spaces

The optimality result (1.5) would be a straightforward consequence of the following norm equivalence with *scaling exponent* $\theta = 1$:

$$(2.1) \qquad \|u\|_{H^1}^2 \approx \|u\|_{L^2}^2 + \sum_{k=0}^{\infty} \left(2^{k\theta} E_k(u)\right)^2 \qquad \forall u \in H^1(\Omega),$$

where $E_k(u)$ denotes the *error of best approximation* in $L^2(\Omega)$

$$E_k(u) = \inf_{v_k \in X_k} \|u - v_k\|_{L^2} = \|u - Q_k u\|_{L^2}.$$

We now ask the rather abstract question: Which sequences X_j of nested finite-dimensional subspaces of $L^2(\Omega)$ allow for some scaling exponent θ such that the norm equivalence (2.1) holds?

Rather than answering this question directly, we define Banach spaces $A^\theta \hookrightarrow L^2(\Omega)$ by the norms given as the right hand sides of (2.1),

$$\|u\|_{A^\theta}^2 = \|u\|_{L^2}^2 + \sum_{k=0}^{\infty} \left(2^{k\theta} E_k(u)\right)^2.$$

These *approximation spaces* A^θ, which measure by θ how well their elements can be approximated by the spaces X_j, were introduced by Peetre [6, 11] in the early sixties and have been intensively studied in approximation theory since then. It should be mentioned that the results to follow were known in a somewhat different form to the Russian school around Nikol'skiĭ as early as the fifties.

Our starting question reads now: Is there a θ such that $A^\theta = H^1(\Omega)$? This question is a key issue of approximation theory — it requires the characterization of the approximation spaces through *smoothness spaces* like the Sobolev spaces. The answer given by Peetre [6, 11] was a relation between the approximation spaces A^θ and the *interpolation spaces* $(L^2(\Omega), X)_{\sigma,2}$, where X is some "nice" space with $X_k \subset X \subset L^2(\Omega)$ for all $k \geq 0$.

THEOREM 1. *Let $\alpha > 0$. An approximation property (Jackson inequality) J_α, i.e.,*

$$(2.2) \qquad \|u - Q_k u\|_{L^2} \lesssim 2^{-k\alpha} \|u\|_X \qquad \forall k \geq 0, u \in X,$$

implies the embedding

$$\left(L^2(\Omega), X\right)_{\sigma,2} \hookrightarrow A^{\sigma\alpha} \qquad 0 < \sigma < 1.$$

Let $\beta > 0$. An inverse inequality (Bernstein inequality) B_β, i.e.,

$$(2.3) \qquad \|u_k\|_X \lesssim 2^{k\beta} \|u_k\|_{L^2} \qquad \forall k \geq 0, u_k \in X_k,$$

implies the embedding

$$A^{\sigma\beta} \hookrightarrow \left(L^2(\Omega), X\right)_{\sigma,2} \qquad 0 < \sigma < 1.$$

REMARK 1. Note that a standard interpolation argument applied to the Jackson inequality J_α for X and to the trivial estimate $\|u - Q_k u\|_{L^2} \lesssim \|u\|_{L^2}$ would only reveal an approximation property for the interpolation spaces

$$(2.4) \qquad 2^{k\alpha\sigma} \|u - Q_k u\|_{L^2} \lesssim \|u\|_{(L^2(\Omega), X)_{\sigma,2}} \qquad 0 < \sigma < 1.$$

This result would be considerably weaker than the assertion of Theorem 1, which states that the right hand sides of (2.4) are in fact square summable as a sequence of k.

If we use an appropriate method for the construction of the interpolation spaces $(L^2(\Omega), X)_{\sigma,2}$, the proof of Theorem 1 is simple. We demonstrate this for the first part concerning the Jackson inequality.

PROOF. Fix some $0 < \lambda < 1$. Using the discrete version of Peetre's K-method of interpolation [1, 6, 13] we get

$$\|u\|^2_{(L^2(\Omega), X)_{\sigma,2}} = \sum_{k \in \mathbb{Z}} \left(\lambda^{-k\sigma} K(\lambda^k, u) \right)^2.$$

The following estimates relate the K-functional with the error of best approximation by using the Jackson inequality J_α: For all $k \geq 0$

$$
\begin{aligned}
E_k(u) &\leq \inf_{v \in X} \|u - Q_k v\|_{L^2} \\
&\leq \inf_{v \in X} \left(\|u - v\|_{L^2} + \|v - Q_k v\|_{L^2} \right) \\
&\lesssim \inf_{v \in X} \left(\|u - v\|_{L^2} + 2^{-k\alpha} \|v\|_X \right) =: K(2^{-k\alpha}, u).
\end{aligned}
$$

Thus, by making the choice $\lambda = 2^{-\alpha}$, we end up with

$$\|u\|^2_{(L^2(\Omega), X)_{\sigma,2}} \gtrsim \sum_{k=0}^{\infty} \left(2^{k\sigma\alpha} E_k(u) \right)^2 + \|u\|^2_{L^2} = \|u\|^2_{A^{\sigma\alpha}}.$$

□

Let us note that the discrete version of Peetre's J-method of interpolation [1, 6, 13] turns out to be appropriate for the proof of the second part of the Theorem.

COROLLARY 1. *The approximation property is restricted by the inverse inequality, i.e., J_α and B_β imply $\alpha \leq \beta$. For $\alpha = \beta$ we get the identification*

$$A^{\sigma\alpha} = \left(L^2(\Omega), X \right)_{\sigma,2}, \qquad 0 < \sigma < 1.$$

PROOF. If J_α and B_β hold, Theorem 1 gives $A^{\sigma\beta} \hookrightarrow A^{\sigma\alpha}$ for $0 < \sigma < 1$. This embedding is equivalent to $\alpha \leq \beta$, as can easily be shown. □

3. Application to Linear Finite Elements

In order to answer the question from the beginning of our consideration, Corollary 1 leads us to the following strategy: Choose X and $0 < \sigma < 1$, such that

$$H^1(\Omega) = \left(L^2(\Omega), X\right)_{\sigma,2}.$$

In any case this requires that X is *smoother* than $H^1(\Omega)$. Interpolation theory in Sobolev spaces [**13, 1**] states for *minimally smooth* domains Ω (i.e., Ω allows a continuous extension operator $E : H^s(\Omega) \to H^s(\mathbb{R}^2)$ for all $s \geq 0$), that

$$H^1(\Omega) = \left(L^2(\Omega), H^s(\Omega)\right)_{1/s,2} \qquad \forall\, s > 1.$$

In our context it suffices to know, that a polygonal domain Ω *without slits* is minimally smooth [**12**]. Now it turns out, that the finite element spaces fulfill

$$X_k \subset H^s(\Omega) \qquad \Longleftrightarrow \qquad 0 \leq s < 3/2.$$

For the following we fix some $1 < s = 1 + \epsilon < 3/2$ and we can apply Theorem 1 — as soon as we have established an approximation property and an inverse inequality in $H^{1+\epsilon}$. We obtain the approximation property J_s, i.e.,

$$(3.1) \qquad \|u - Q_k u\|_{L^2} \lesssim h_k^s \|u\|_{H^s} \lesssim 2^{-ks} \|u\|_{H^s} \qquad \forall u \in H^s(\Omega),$$

by simple interpolation between the cases $s = 0$ and $s = 2$, as indicated in Remark 1. A little bit deeper lies the inverse inequality B_s, i.e.,

$$(3.2) \qquad \|u_k\|_{H^s} \lesssim h_k^{-s} \|u_k\|_{L^2} \lesssim 2^{ks} \|u_k\|_{L^2} \qquad \forall u_k \in X_k,$$

which can be proved using the Sobolev-Slobodeckiĭ norm

$$\|u\|_{H^{1+\epsilon}}^2 \approx \|u\|_{H^1}^2 + \int_\Omega \int_\Omega \frac{|\nabla u(x) - \nabla u(y)|^2}{|x - y|^{2+2\epsilon}} \, dx\, dy$$

of $H^s(\Omega)$, cf. [**5**]. Thus we have $\alpha = \beta = 1/\sigma$ and Corollary 1 states the equivalence, which makes the BPX preconditioner optimal:

THEOREM 2. *For linear finite elements the equivalence $A^1 = H^1(\Omega)$ holds.*

REMARK 2. This Theorem and the more general equivalence "approximation space = smoothness space", i.e, $A^s = H^s(\Omega)$ for s from some interval, holds generically for a lot of sequences X_j, like higher order finite elements, spectral methods and wavelets. Details can be found in [**8, 9**].

Our considerations show that it is reasonable to view the approximation property (3.1) and the inverse inequality (3.2) in $H^{1+\epsilon}(\Omega)$, $\epsilon > 0$ arbitrarily small, as the "chief cause" for the optimality of BPX. In the original proof of the weaker result (1.2) corresponding estimates in $H^1(\Omega)$ were the groundwork. Thus, the essential step is to use the fact that linear finite elements are a little bit smoother than one usually thinks. This essential step is hidden in one way or another in all proofs [**2, 3, 7, 10, 15, 16**] of the optimality of the BPX preconditioner.

REMARK 3. The reader should not confuse the concept of regularity $X_j \subset H^{1+\epsilon}$ of the approximating spaces X_j with the concept of $H^{1+\alpha}$-regularity of the problem (1.1). The first is a *general* property of the *method* of approximation, the second holds only for *special* cases of the underlying *problem*.

REFERENCES

1. J. Bergh and J. Löfström, *Interpolation Spaces. An Introduction*, Springer-Verlag, Berlin, Heidelberg, New York, 1976.
2. F. A. Bornemann and H. Yserentant, *A basic norm equivalence for the theory of multilevel methods*, Numer. Math. **64** (1993), 455–476.
3. J. H. Bramble and J. E. Pasciak, *New estimates for multilevel algorithms including the V-cycle*, Math. Comp. **60** (1993), 447–471.
4. J. H. Bramble, J. E. Pasciak, and J. Xu, *Parallel multilevel preconditioners*, Math. Comp. **55** (1990), 1–22.
5. _____, *The analysis of multigrid algorithms with non-imbedded spaces or non-inherited quadratic forms*, Math. Comp. **56** (1991), 1–34.
6. P. L. Butzer and K. Scherer, *Approximationsprozesse und Interpolationsmethoden*, Bibliographisches Institut, Mannheim, 1968.
7. W. Dahmen and A. Kunoth, *Multilevel preconditioning*, Numer. Math. **63** (1992), 315–344.
8. R. DeVore and G. G. Lorentz, *Constructive Approximation*. I, Springer-Verlag, Berlin, Heidelberg, New York, 1993.
9. P. Oswald, *On function spaces related to finite element approximation theory*, Z. Anal. Anwend. **9** (1990), 43–64
10. P. Oswald, *On discrete norm estimates related to multilevel preconditioners in the finite element method*, Proc. Int. Conf. Constructive Theory of Functions, Varna, 1991.
11. J. Peetre, *A Theory of Interpolation in Normed Spaces*, Notes Universidade de Brasilia, 1963.
12. E. M. Stein, *Singular Integrals and Differentiability Properties of Functions*, Princeton University Press, Princeton, 1970.
13. H. Triebel, *Interpolation Theory, Function Spaces, Differential Operators*, North-Holland Pub. Comp., Amsterdam, New York, Oxford, 1978.
14. J. Xu, *Theory of Multilevel Methods*, PhD thesis, Cornell University, May 1989.
15. J. Xu, *Iterative methods by space decomposition and subspace correction*, SIAM Review **34** (1992), 581–613.
16. X. Zhang, *Multilevel Schwarz methods*, Numer. Math. **63** (1992), 521–539.

FACHBEREICH MATHEMATIK, FREIE UNIVERSITÄT BERLIN, GERMANY
Current address: Konrad-Zuse-Zentrum Berlin, Heilbronner Str. 10, 10711 Berlin, Germany
E-mail address: bornemann@zib-berlin.de

Contemporary Mathematics
Volume **180**, 1994

Two-level Additive Schwarz Preconditioners for Nonconforming Finite Elements

SUSANNE C. BRENNER

ABSTRACT. Two-level additive Schwarz preconditioners are developed for the nonconforming finite element approximations of second order and fourth order elliptic boundary value problems. The condition numbers of the preconditioned systems are shown to be bounded independent of mesh sizes and the number of subdomains in the case of a generous overlap.

1. Introduction

We generalize the Dryja-Widlund theory of two-level additive Schwarz preconditioners (cf. [**4**], [**5**] and [**9**]) to nonconforming finite element approximations of second and fourth order elliptic boundary value problems. Compared with conforming finite elements, the nonconforming finite elements have fewer degrees of freedom. The trade-off is that the communication between grids of different sizes is more complicated. The intergrid transfer operators must be constructed carefully. We show that under certain assumptions there is a uniform bound on the condition numbers of the preconditioned systems. Details and proofs of the results can be found in [**1**]. Recent works on domain decomposition methods for nonconforming finite elements can also be found in [**3**], [**6**] and [**2**].

2. The Preconditioner

The construction of the preconditioner is based on the idea of domain decomposition. Let Ω be a bounded polygonal domain in $\mathbb{R}^2$. Write $\Omega = \cup_{j=1}^{J}\Omega_j$ as a union of overlapping subdomains. $\mathcal{T}_H$ is a triangulation of Ω and $\mathcal{T}_h$ is a subdivision of $\mathcal{T}_H$ which is aligned with each $\partial\Omega_j$.

1991 *Mathematics Subject Classification.* Primary 65F10, 65N30, 65N55.

The author was supported in part by NSF Grant #DMS-92-09332.

The final detailed version of this paper has been submitted for publication elsewhere.

We assume that there exist nonnegative C^∞ functions $\theta_1, \theta_2, \ldots, \theta_J$ in $\mathbb{R}^2$ such that (G1) $\theta_j = 0$ on $\Omega\backslash\Omega_j$, (G2) $\sum_{j=1}^J \theta_j = 1$ on $\overline{\Omega}$, (G3) $\|\nabla\theta_j\|_{L^\infty} \leq \frac{C}{\delta}$ and $\|\nabla^2\theta_j\|_{L^\infty} \leq \frac{C}{\delta^2}$, and (G4) Each point in Ω can belong to at most N_c subdomains. (Here δ measures the size of the overlap and ∇^2 is the Hessian.) These are the *geometric assumptions*. Note that from here on the generic positive constant C is independent of h, H, δ, J and N_c.

Let V_h (resp., V_H) be the finite element space associated with $\mathcal{T}_h$ (resp., $\mathcal{T}_H$) and let $V_j = \{v \in V_h : v = 0 \text{ on } \Omega\backslash\Omega_j\}$. (We only treat homogeneous Dirichlet boundary conditions.) The existence of the partition of unity implies that $V_h = \sum_{j=1}^J V_j$. Let $a_h(\cdot,\cdot)$ (resp., $a_H(\cdot,\cdot)$) be a positive-definite symmetric bilinear form on V_h (resp,. V_H).

We will describe the preconditioner in operator notation (cf. [8]). We assume that $(\cdot,\cdot)_h$ and $(\cdot,\cdot)_H$ are inner products on V_h and V_H respectively. The operators $A_h : V_h \longrightarrow V_h$, $A_j : V_j \longrightarrow V_j$, $A_H : V_H \longrightarrow V_H$, $Q_j : V_h \longrightarrow V_j$ and $P_j : V_h \longrightarrow V_j$ are defined by

$$(A_h v, w)_h = a_h(v,w), \qquad (A_j v, w)_h = a_h(v,w),$$
$$(A_H v, w)_H = a_H(v,w), \qquad (Q_j v, w)_h = (v,w)_h,$$
$$a_h(P_j v, w) = a_h(v,w).$$

These operators are related by the equation $A_j P_j = Q_j A_h$.

Let $I_H^h : V_H \longrightarrow V_h$ be the coarse-to-fine intergrid transfer operator (the construction of I_H^h for some concrete applications are given in Section 5). The fine-to-coarse intergrid transfer operator $I_h^H : V_h \longrightarrow V_H$ is defined by $(I_h^H v, w)_H = (v, I_H^h w)_h$, and the operator $P_h^H : V_h \longrightarrow V_H$ is defined by $a_H(P_h^H v, w) = a_h(v, I_H^h w)$. These operators are related by the equation $A_H P_h^H = I_h^H A_h$.

Furthermore, we assume that we have approximate solvers $R_H \sim A_H^{-1}$ on the coarse grid and $R_j \sim A_j^{-1}$ on each subregion, and that these operators are symmetric positive-definite with respect to the inner products $(\cdot,\cdot)_H$ and $(\cdot,\cdot)_h$.

The preconditioner is defined by

$$B := I_H^h R_H I_h^H + \sum_{j=1}^J R_j Q_j.$$

The discretized problem is $A_h u_h = f_h$, and the preconditioned system is $B A_h u = B f_h$. Using the relationships among the operators, we can write the operator for the preconditioned system as:

$$BA_h = I_H^h R_H I_h^H A_h + \sum_{j=1}^J R_j Q_j A_h = I_H^h R_H A_H P_h^H + \sum_{j=1}^J R_j A_j P_j.$$

It is easy to see that BA_h is symmetric positive-definite with respect to $a_h(\cdot,\cdot)$.

If we use exact solvers, i.e., if $R_H = A_H^{-1}$ and $R_j = A_j^{-1}$, then

$$BA_h = I_H^h P_h^H + \sum_{j=1}^{J} P_j = I_H^h A_H^{-1} I_h^H A_h + \sum_{j=1}^{J} P_j,$$

which shows that our preconditioner is a variant of the Dryja-Widlund preconditioner (cf. [**6**]).

3. Examples

EXAMPLE 1. Dirichlet problem for the Laplace equation:

$$-\Delta u = f \text{ in } \Omega, \qquad u|_{\partial\Omega} = 0.$$

We use the bilinear form $a_h(v_1, v_2) = \sum_{T \in \mathcal{T}_h} \int_T \nabla v_1 \cdot \nabla v_2 \, dx \quad \forall v_1, v_2 \in V_h$, where V_h is the nonconforming P1 finite element space $\{v \in L^2(\Omega) : v \in \mathcal{P}_1(T) \quad \forall T \in \mathcal{T}_h, v \text{ is continuous at the midpoints of interelement boundaries,}$ and v vanishes at the midpoints along $\partial\Omega\}$. On the coarser grid we have two choices: nonconforming P1 finite elements and conforming P1 finite elements. Our theory is applicable to both choices.

EXAMPLE 2. Dirichlet problem for the biharmonic equation:

$$\Delta^2 u = f \text{ in } \Omega, \qquad u = \frac{\partial u}{\partial n} = 0 \text{ on } \partial\Omega.$$

We use the bilinear form $a_h(v_1, v_2) = \sum_{T \in \mathcal{T}_h} \int_T \sum_{i,j=1}^{2} \frac{\partial^2 v_1}{\partial x_i \partial x_j} \frac{\partial^2 v_2}{\partial x_i \partial x_j} \, dx \quad \forall v_1, v_2 \in V_h$, where V_h is the Morley finite element space $\{v \in L^2(\Omega) : v \in \mathcal{P}_2(T) \quad \forall T \in \mathcal{T}_h, v \text{ is continuous at the vertices and vanishes at the vertices along } \partial\Omega, \frac{\partial v}{\partial n}$ is continuous at the midpoints of interelement boundaries and vanishes at the midpoints along $\partial\Omega\}$. Here we must use nonconforming finite element spaces on both grids since the Morley finite element space does not contain a conforming subspace.

4. An Abstract Theory

In what follows, k is a parameter which takes the value 1 (resp., 2) for second (resp., fourth) order problems. We use the nonconforming semi-norms

$$|v|_{H^m(\mathcal{T}_h)} := \left(\sum_{T \in \mathcal{T}_h} |v|_{H^m(T)}^2 \right)^{1/2} \quad \text{and} \quad |v|_{H^m(\mathcal{T}_{h,j})} := \left(\sum_{\substack{T \subset \Omega_j \\ T \in \mathcal{T}_h}} |v|_{H^m(T)}^2 \right)^{1/2}.$$

We make two assumptions on the variational forms:

(V1) $\sqrt{a_h(v,v)} \sim |v|_{H^k(\mathcal{T}_h)}, \quad \sqrt{a_H(v,v)} \sim |v|_{H^k(\mathcal{T}_H)}$

(V2) $a_h(v,w) \le C |v|_{H^k(\mathcal{T}_{h,j})} |w|_{H^k(\mathcal{T}_h)} \quad \forall v \in V_h, w \in V_j.$

We need two properties on the coarse-to-fine intergrid transfer operators:

(I1) $$|I_H^h v|_{H^k(\mathcal{T}_h)} \le C\, |v|_{H^k(\mathcal{T}_H)} \quad \forall\, v \in V_H \quad \text{and}$$

(I2) $$|I_H^h v - v|_{H^\ell(\mathcal{T}_h)} \le C\, H^{k-\ell}\, |v|_{H^k(\mathcal{T}_H)} \quad \forall\, v \in V_H,$$

$0 \le \ell \le k - 1.$

LEMMA. *Under assumptions* (V1)–(V2), (I1) *and* (G4),

$$\lambda_{\max}(BA_h) \le C\,\omega_1\, N_c,$$

where $\omega_1 := \max(\rho(R_H A_H), \rho(R_1 A_1), \ldots, \rho(R_J A_J)).$

To obtain a lower bound for the eigenvalues of BA_h we need a connection operator $K_h^H : V_h \longrightarrow V_H$ which satisfies the following properties.

(K1) $$|K_h^H v|_{H^k(\mathcal{T}_H)} \le C\, |v|_{H^k(\mathcal{T}_h)} \quad \forall\, v \in V_h$$

(K2) $$|K_h^H v - v|_{H^\ell(\mathcal{T}_H)} \le C\, H^{k-\ell}\, |v|_{H^k(\mathcal{T}_h)} \quad \forall\, v \in V_h,$$

$0 \le \ell \le k - 1.$

We also assume that the nodal interpolation opertor Π_h satisfies

(P1) $$|\Pi_h(\lambda v)|_{H^k(T)} \le C\, |\lambda v|_{H^k(T)} \quad \text{and}$$

(P2) $$\|\Pi_h(gv)\|_{L^2(T)} \le C\Big(\|g\|_{L^\infty(T)} + (k-1)\, h\, \|\nabla g\|_{L^\infty(T)}\Big)\|v\|_{L^2(T)}$$

for all $T \in \mathcal{T}_h, v \in \mathcal{P}(T)$ (the space of shape functions), $\lambda \in \mathcal{P}_{k-1}(T), g \in C^\infty(\overline{T}).$

LEMMA. *Under the assumptions* (P1)–(P2), (V1), (K1)–(K2), (I1)–(I2), (G1)–(G4), *given any* $v \in V_h$, *there exist* $v_0 \in V_H$, $v_j \in V_j$ $(1 \le j \le J)$ *such that*

$$v = I_H^h v_0 + \sum_{j=1}^J v_j \quad \text{and}$$

$$a_H(v_0, v_0) + \sum_{j=1}^J a_h(v_j, v_j) \le C\, N_c \left(1 + \left(\frac{H}{\delta}\right)^{2k}\right) a_h(v, v).$$

It is well-known (cf. [8]) that such a lemma implies that we have the following lower bound for the eigenvalues of BA_h.

LEMMA.
$$\lambda_{min}(BA_h) \ge C\, \frac{\omega_0}{N_c\left(1 + (\frac{H}{\delta})^{2k}\right)},$$

where $\omega_0 = \min(\lambda_{min}(R_H A_H), \lambda_{min}(R_1 A_1), \ldots, \lambda_{min}(R_J A_J))$

Putting everything together, we obtain the bound for the condition number.

THEOREM. *Under the assumptions* (G), (V), (I), (K), *and* (P), *we have*

$$\frac{\lambda_{\max}(BA_h)}{\lambda_{\min}(BA_h)} \le C\,\frac{\omega_1}{\omega_0}\, N_c^2 \left(1 + \left(\frac{H}{\delta}\right)^{2k}\right).$$

COROLLARY.

$$\frac{\lambda_{max}(BA_h)}{\lambda_{min}(BA_h)} \leq CN_c^2$$

if $\omega_1 \leq C_1$, $\quad \omega_0 \geq C_2 > 0$, $\quad \frac{H}{\delta} \leq C_3$.

In the case of a small overlap, the factor $[1 + (H/\delta)]^4$ can be reduced to $[1 + (H/\delta)]^3$ by adopting the arguments in [**7**] (cf. [**2**]).

5. The Operators I_H^h and K_h^H

We use related nested conforming spaces W_H and W_h in the constructions of the intergrid transfer and connection operators. We define I_H^h and K_h^H by the following commutative diagrams, where i is the natural injection.

$$
\begin{array}{ccc}
V_h & \xleftarrow{\ F_h\ } & W_h \\
{\scriptstyle I_H^h}\big\uparrow & & \big\uparrow{\scriptstyle i} \\
V_H & \xrightarrow[\ E_H\ ]{} & W_H
\end{array}
\qquad
\begin{array}{ccc}
V_h & \xrightarrow{\ E_h\ } & W_h \\
{\scriptstyle K_h^H}\big\downarrow & & \big\downarrow{\scriptstyle Q_h^H} \\
V_H & \xleftarrow[\ F_H\ ]{} & W_H
\end{array}
$$

Below we describe W_H, W_h, E_h, E_H, F_h, F_H and Q_h^H for the two examples in Section 3.

The P1 Nonconforming Finite Element. We take W_h to be the conforming P2 finite element space $\{w \in C(\overline{\Omega}) : w|_T \in \mathcal{P}_2(T) \quad \forall T \in \mathcal{T}_h \text{ and } w = 0 \text{ on } \partial\Omega\}$. The space W_H is defined similarly. Note that the nodal variables of the nonconforming P1 space are also nodal variables of the conforming P2 space. The operator $E_h : V_h \longrightarrow W_h$ is defined by

$$
\begin{cases}
(E_h v)(m) = v(m) \\
(E_h v)(p) = \text{average of } v_i(p)
\end{cases}
$$

where $v_i = v|_{T_i}$ and $T_i \in \mathcal{T}_h$ contains p as a vertex, and the operator $F_h : W_h \longrightarrow V_h$ is defined by

$$(F_h w)(m) = w(m).$$

The operators $E_H : V_H \longrightarrow W_H$ and $F_H : W_H \longrightarrow V_H$ are defined similarly. The operator $Q_h^H : W_h \longrightarrow W_H$ is the L^2-orthogonal projection.

The Morley Finite Element. We take $\tilde{W}_h$ to be the P5 Argyris finite element space $\{w \in C^1(\overline{\Omega}) : w|_T \in \mathcal{P}_5(T) \quad \forall T \in \mathcal{T}_h, D^\alpha w \text{ is continuous at the vertices for } |\alpha| = 2 \text{ and } w - \frac{\partial w}{\partial n} = 0 \text{ on } \partial\Omega\}$, which is contained in the larger space $W_h := \{w \in C^1(\overline{\Omega}) : w|_T \in \mathcal{P}_5(T) \quad \forall T \in \mathcal{T}_h, w = \frac{\partial w}{\partial n} = 0 \text{ on } \partial\Omega\}$. The spaces $\tilde{W}_H$ and W_H are defined similarly. The operator $E_h : V_h \longrightarrow \tilde{W}_h \subseteq W_h$ is defined by

$$
\begin{cases}
(E_h v)(p) = v(p) \\
(\partial^\alpha E_h v)(p) = \text{average of } (\partial^\alpha v_i)(p), |\alpha| = 1 \\
(\partial^\alpha E_h v)(p) = 0, |\alpha| = 2 \\
\left(\frac{\partial}{\partial n} E_h v\right)(m) = \frac{\partial v}{\partial n}(m)
\end{cases}
$$

where $v_i = v|_{T_i}$ and T_i contains p as a vertex, and the operator $F_h : W_h \longrightarrow V_h$ is defined by

$$\begin{cases} (F_h w)(p) = w(p) \\ \left(\frac{\partial}{\partial n}(F_h w)\right)(m) = \frac{\partial w}{\partial n}(m) \end{cases} .$$

Again, $Q_h^H : W_h \longrightarrow W_H$ is the L^2-orthogonal projection.

The estimates (I1)–(I2) and (K1)–(K2) are obtained from the corresponding estimates of E_h, E_H, F_h, F_H and Q_h^H.

6. Stationary Stokes Equations

Our theory can also be applied to the stationary Stokes equations

$$-\Delta \underset{\sim}{u} + \operatorname{grad} p = \underset{\sim}{f} \quad \text{in } \Omega$$

$$\operatorname{div} \underset{\sim}{u} = 0 \quad \text{in } \Omega$$

$$\underset{\sim}{u} = \underset{\sim}{0} \quad \text{on } \partial\Omega$$

using the divergence-free P1 nonconforming finite element. The main difficulty is the divergence-free constraint. It can be circumvented by the connection between the divergence-free P1 nonconforming finite element space and the Morley finite element space (cf. [1]).

References

1. S.C. Brenner, *Two-level additive Schwarz preconditioners for nonconforming finite element methods*, preprint.
2. ———, *A two-level additive Schwarz preconditioner for nonconforming plate elements*, preprint.
3. L.C. Cowsar, *Domain decomposition methods for nonconforming finite elements spaces of Lagrange-type*, Technical Report TR93-11, Rice University (1993).
4. M. Dryja and O.B. Widlund, *An additive variant of the Schwarz alternating method in the case of many subregions*, Technical Report 339, Department of Computer Science, Courant Institute (1987).
5. ———, *Some domain decomposition algorithms for elliptic problems*, Technical Report 438, Department of Computer Science, Courant Institute (1989).
6. M. Sarkis, *Two-level Schwarz methods for nonconforming finite elements and discontinuous coefficients*, Technical Report 629, Department of Computer Science, Courant Institute (1993).
7. O.B. Widlund, *Some Schwarz methods for symmetric and nonsymmetric elliptic problems*, Fifth International Symposium on Domain Decomposition Methods for Partial Differential Equations (D.E. Keyes, et.al., eds.), SIAM, Philadelphia, 1991, pp. 19–36.
8. J. Xu, *Iterative methods by space decomposition and subspace correction*, SIAM Review **34** (1992), 581–613.
9. X. Zhang, *Studies in Domain Decomposition: Multi-level Methods and the Biharmonic Dirichlet Problem*, Dissertation, (Technical Report 584, Department of Computer Science) Courant Institute (1991).

Department of Mathematics and Computer Science, Clarkson University, Potsdam, NY 13699

Current address: Department of Mathematics, University of South Carolina, Columbia, SC 29208

E-mail address: brenner@math.scarolina.edu

Contemporary Mathematics
Volume **180**, 1994

Two Proofs of Convergence for the Combination Technique for the Efficient Solution of Sparse Grid Problems

H.-J. BUNGARTZ, M. GRIEBEL, D. RÖSCHKE, AND C. ZENGER

ABSTRACT. For a simple model problem — the Laplace equation on the unit square with a Dirichlet boundary function vanishing for $x = 0$, $x = 1$, and $y = 1$, and equaling some suitable $g(x)$ for $y = 0$ — we present a proof of convergence for the combination technique, a modern, efficient, and easily parallelizable sparse grid solver for elliptic partial differential equations that has recently gained importance in fields of applications like computational fluid dynamics. For full square grids with meshwidth h and $O(h^{-2})$ grid points, the order $O(h^2)$ of the discretization error using finite differences was shown in [**5**], if $g(x) \in C^2[0,1]$. In this paper, we show that the finite difference discretization error of the solution produced by the combination technique on a sparse grid with only $O\left((h^{-1}\log_2(h^{-1}))\right)$ grid points is of the order $O\left(h^2\log_2(h^{-1})\right)$, if the Fourier coefficients b_k of $\tilde{g}$, the 2-periodic and 0-symmetric extension of g, fulfill $|b_k| \le c_g \cdot k^{-3-\varepsilon}$ for some arbitrary small positive ε. If $0 < \varepsilon \le 1$, this is valid for $g \in C^4[0,1]$ and $g(0) = g(1) = g''(0) = g''(1) = 0$, for example. A simple transformation even shows that $g \in C^4[0,1]$ is sufficient. We present results of numerical experiments with functions g of varying smoothness.

1. Introduction

Since their presentation in 1990 [**7**], sparse grids have turned out to be a very interesting tool for the efficient solution of elliptic boundary value problems. Besides the implementation of a hierarchical finite element algorithm on sparse grids [**1**], it has been first of all the combination technique [**4**] that attained attention in the sparse grid context. One of the main advantages of the combination technique stems from the properties of sparse grids [**1**]: In comparison to the standard full grid approach, the number of grid points can be reduced significantly from $O(N^d)$ to $O(N(\log_2(N))^{d-1})$ in the d-dimensional case, whereas the accuracy of the calculated approximation to the solution is only slightly deteriorated from $O(N^{-2})$ [**5**] to

---

1991 *Mathematics Subject Classification.* 35J05, 65N06, 65N15, 65N22, 65N55.

This work is supported by the Bayerische Forschungsstiftung via FORTWIHR — The Bavarian Consortium for High Performance Scientific Computing.

This paper is in final form, and no version of it will be submitted for publication elsewhere.

$O(N^{-2}(\log_2(N))^{d-1})$ [1, 4]. Additional advantages have to be seen in the simplicity of the combination concept, in its inherent parallel structure, and in the fact that it is a framework which allows the integration of existing PDE solvers. Up to now, the combination technique has been applied to several types of elliptic PDEs including problems originating from computational fluid dynamics. Its excellent parallelization properties have been verified on different parallel architectures and on workstation networks (see the references in [3]).

For a proof of the approximation properties mentioned above, in the 2D case the existence of an error splitting of the type

$$(1) \quad u_{h_x,h_y}(x,y) - u(x,y) = C_1(x,y,h_x)h_x^2 + C_2(x,y,h_y)h_y^2 + C(x,y,h_x,h_y)h_x^2h_y^2$$

with $|C_1(x,y,h_x)|$, $|C_2(x,y,h_y)|$, and $|C(x,y,h_x,h_y)|$ bounded by some positive $B(x,y)$ for all meshwidths h_x and h_y has to be shown [4]. Here, $u(x,y)$ is the exact solution of the given boundary value problem, and $u_{h_x,h_y}(x,y)$ denotes the solution on the rectangular full grid with meshwidths h_x and h_y resulting from a finite difference discretization.

We present two proofs for the existence of such an error splitting in the case of Laplace's equation on the unit square, if the Dirichlet boundary function satisfies certain smoothness requirements. Both times, we first split the discretization error $u_{h_x,h_y}(x,y) - u(x,y)$ into three error terms also involving the solutions $u_{h_x,0}$ and u_{0,h_y} of the Laplacian discretized in only one direction. Then, the first approach represents these error terms (differences) as integrals over differential quotients and, thus, reduces statements on $u_{h_x,h_y}(x,y) - u(x,y)$ to statements on some partial derivatives of $u(x,y)$ and $u_{h_x,h_y}(x,y)$. The second proof works in a more straight-forward way using the mean value theorem to get derivatives instead of differences. Here, we just outline the essential steps. For details, see [2] and [3].

2. The Combination Technique

Let L be an elliptic operator of second order. Consider the partial differential equation $Lu = f$ on the unit square $\Omega =]0,1[^2$ with appropriate boundary conditions. Furthermore, let $G_{i,j}$ be the rectangular grid on $\bar{\Omega}$ with meshwidths $h_x := 2^{-i}$ in x- and $h_y := 2^{-j}$ in y-direction $(i,j \in \mathbb{N})$. In [4], a technique has been introduced which combines the solutions u_{h_x,h_y} of the discrete problems $L_{h_x,h_y}u_{h_x,h_y} = f_{h_x,h_y}$ associated to $Lu = f$ on different rectangular grids $G_{i,j}$:

$$(2) \quad u_n^c := \sum_{i+j=n+1} u_{2^{-i},2^{-j}} - \sum_{i+j=n} u_{2^{-i},2^{-j}}.$$

The resulting solution u_n^c is given on the sparse grid $\tilde{G}_{n,n}$ with $O(2^n n)$ grid points instead of the usual full grid $G_{n,n}$ with $O(4^n)$ points. Fig. 1 shows the sparse grids $\tilde{G}_{3,3}$ and $\tilde{G}_{4,4}$. For a detailed introduction to sparse grids, see [7] and [1]. Note that we have to solve n different problems with $O(2^{n+1})$ grid points $(i+j = n+1)$ and $n-1$ different problems with $O(2^n)$ grid points $(i+j = n)$. These $2n-1$

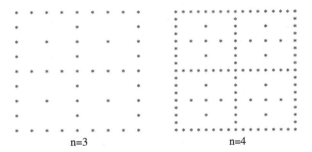

FIGURE 1. Sparse grids $\tilde{G}_{3,3}$ and $\tilde{G}_{4,4}$.

problems are independent from one another and can be solved in parallel. Finally, their bilinearly interpolated solutions are combined according to (2).

Assuming that, for each inner point $(x, y) \in \Omega$, the pointwise error of the solution $u_{h_x,h_y} = u_{2^{-i},2^{-j}}$ obtained on the rectangular grid $G_{i,j}$ is of the form (1), it has been shown in [4] that the error $e_n^c(x, y)$ of the combined solution $u_n^c(x, y)$ fulfills

$$(3) \qquad |e_n^c(x, y)| := |u_n^c(x, y) - u(x, y)| \leq B(x, y) \cdot h_n^2 \cdot \left(1 + \frac{5}{4}\log_2(h_n^{-1})\right),$$

where $h_n = 2^{-n}$. In this paper, we deal with the question of which smoothness requirements have to be fulfilled by the Dirichlet boundary function in order to get a behaviour of the error $u_{h_x,h_y}(x, y) - u(x, y)$ on $G_{i,j}$ as indicated in (1).

For the remainder, we concentrate on the Laplace equation

$$(4) \qquad \Delta u(x, y) = 0, \ (x, y) \in \Omega,$$

with the Dirichlet boundary condition

$$(5) \qquad u(x, y) = \begin{cases} g(x) : (x, y) \in \delta\bar{\Omega}, y = 0 \\ 0 \ : (x, y) \in \delta\bar{\Omega}, y > 0. \end{cases}$$

Throughout the paper, we assume that the Fourier coefficients b_k of $\tilde{g}$, the 2-periodic and 0-symmetric extension of g, fulfill

$$(6) \qquad |b_k| := \left|\int_{-1}^{1} \tilde{g}(x)\sin(k\pi x)dx\right| = \left|2\int_{0}^{1} g(x)\sin(k\pi x)dx\right| \leq \frac{c_g}{k^{3+\varepsilon}}$$

for some arbitrary small positive ε and a constant c_g depending only on g. For $0 < \varepsilon \leq 1$, e.g., it is a well-known fact from Fourier theory that (6) is valid if $\tilde{g} \in C^4(\mathbb{R})$ or if $g \in C^4[0, 1]$ and $g(0) = g(1) = g''(0) = g''(1) = 0$.

3. Explicit Solutions and Error Splitting

In this section, we are looking for explicit representations of the solutions u, u_{h_x,h_y}, $u_{h_x,0}$, and u_{0,h_y} of the Laplace equation (4) and the following three associated discretized problems with meshwidths $h_x = 2^{-i}$ and $h_y = 2^{-j}$:

$$(7) \quad 0 = \frac{u_{h_x,h_y}(x-h_x,y) - 2u_{h_x,h_y}(x,y) + u_{h_x,h_y}(x+h_x,y)}{h_x^2}$$
$$+ \frac{u_{h_x,h_y}(x,y-h_y) - 2u_{h_x,h_y}(x,y) + u_{h_x,h_y}(x,y+h_y)}{h_y^2},$$

$$(8) \quad 0 = \frac{u_{h_x,0}(x-h_x,y) - 2u_{h_x,0}(x,y) + u_{h_x,0}(x+h_x,y)}{h_x^2} + \frac{\partial^2 u_{h_x,0}(x,y)}{\partial y^2},$$

$$(9) \quad 0 = \frac{\partial^2 u_{0,h_y}(x,y)}{\partial x^2} + \frac{u_{0,h_y}(x,y-h_y) - 2u_{0,h_y}(x,y) + u_{0,h_y}(x,y+h_y)}{h_y^2}.$$

Thus, u_{h_x,h_y} solves the discrete problem resulting from the use of central finite differences in x- and y-direction. Accordingly, $u_{h_x,0}$ and u_{0,h_y} denote the solutions of the problems discretized only in x- or in y-direction, respectively. Defining

$$(10) \quad T(t,y) := \frac{\sinh(t\pi(1-y))}{\sinh(t\pi)},$$

we know from [6] that the solution u of (4) fulfilling (5) can be written as

$$(11) \quad u(x,y) := \sum_{k=1}^{\infty} b_k \sin(k\pi x) T(k,y).$$

As shown in [3], the three discrete problems (7)–(9) are solved by

$$(12) \quad u_{h_x,h_y}(x,y) := \sum_{k=1}^{\infty} b_k \sin(k\pi x) T(\mu_k,y),$$

$$(13) \quad u_{h_x,0}(x,y) := \sum_{k=1}^{\infty} b_k \sin(k\pi x) T(\nu_k,y),$$

$$(14) \quad u_{0,h_y}(x,y) := \sum_{k=1}^{\infty} b_k \sin(k\pi x) T(\lambda_k,y),$$

where

$$(15) \quad \mu_k(h_x,h_y) := \frac{2}{\pi h_y} \cdot \operatorname{arcsinh}\left(\frac{h_y}{h_x} \cdot \sin\left(\frac{k\pi h_x}{2}\right)\right),$$
$$\nu_k(h_x) := \lim_{h_y \to 0} \mu_k(h_x,h_y), \qquad \lambda_k(h_y) := \lim_{h_x \to 0} \mu_k(h_x,h_y).$$

In each sum (12), (13), and (14), $\sin(k\pi x)T(.,y)$ solves the corresponding problem (7), (8), and (9), respectively. The series converge, because $T(t,y) \leq 1$ and because of our smoothness assumptions (6). Furthermore, u_{h_x,h_y}, $u_{h_x,0}$, and u_{0,h_y} fulfill the same boundary condition (5) that u does. This results from the fact that we always use the Fourier coefficients b_k of the continuous problem.

Now, we use the explicit representations of u, u_{h_x,h_y}, $u_{h_x,0}$, and u_{0,h_y} to split the discretization error $u_{h_x,h_y} - u$ into three terms:

$$(16) \quad u_{h_x,h_y} - u = \Gamma_{h_x}^{(1)} + \Gamma_{h_y}^{(2)} + \Gamma_{h_x,h_y},$$

where

(17)
$$\Gamma^{(1)}_{h_x} := u_{h_x,0} - u, \qquad \Gamma^{(2)}_{h_y} := u_{0,h_y} - u,$$
$$\Gamma_{h_x,h_y} := u_{h_x,h_y} - u_{h_x,0} - u_{0,h_y} + u.$$

In the following, we show that the smoothness requirement (6) is sufficient for

(18)
$$\Gamma^{(1)}_{h_x} = h_x^2 \cdot C_1(x,y,h_x), \qquad \Gamma^{(2)}_{h_y} = h_y^2 \cdot C_2(x,y,h_y),$$
$$\Gamma_{h_x,h_y} = h_x^2 h_y^2 \cdot C(x,y,h_x,h_y)$$

to hold, where $|C_1|$, $|C_2|$, and $|C|$ are bounded from above by some $B(x,y)$.

4. Order of the Error Terms

We restrict ourselves to Γ_{h_x,h_y}. Looking at the explicit representations (11)–(14) of u, $u_{h_x,0}$, u_{0,h_y}, and u_{h_x,h_y}, we see that Γ_{h_x,h_y} can be written as a series, too:

(19) $$\Gamma_{h_x,h_y} = \sum_{k=1}^{\infty} b_k \sin(k\pi x) \cdot \Big(T(\mu_k,y) - T(\nu_k,y) - T(\lambda_k,y) + T(k,y) \Big).$$

Here, the crucial task is to transform the differences in (19) to derivatives, since the latter ones are easier to deal with. We study two possible ways of doing so.

For the first approach, we note that $\Gamma_{h_x,h_y} = \sum_{l=0}^{\infty} \sum_{m=0}^{\infty} \gamma_{h_x \cdot 2^{-l}, h_y \cdot 2^{-m}}$, where $\gamma_{h_x,h_y} = u_{h_x,h_y} - u_{h_x,h_y/2} - u_{h_x/2,h_y} + u_{h_x/2,h_y/2}$. Then, introducing $v(x,y,s,t) := u_{2^{-s},2^{-t}}(x,y)$ for arbitrary $s,t \geq 1$, we get

(20) $$\gamma_{h_x,h_y} = \gamma_{2^{-i},2^{-j}} = \int_i^{i+1} \int_j^{j+1} \frac{\partial^2 v(x,y,\sigma,\tau)}{\partial\sigma\partial\tau} d\sigma d\tau.$$

In a last step, we have to find estimates for the partial derivative $\frac{\partial^2 v}{\partial\sigma\partial\tau}$. This can be done profiting from the tools of symbolic computation (see [2] for details).

The second proof starts with studying properties of T, ν_k, λ_k, and μ_k introduced in (10) and (15). As a result, applying the mean value theorem twice, we get

(21)
$$T(\mu_k,y) - T(\nu_k,y) - T(\lambda_k,y) + T(k,y)$$
$$= \frac{\partial}{\partial h_y} \Big(T_t\big(F(\xi_k,h_y),y\big) \cdot F_t(\xi_k,h_y) \Big)\Big|_{\alpha h_y} \cdot h_y \cdot (\nu_k - k),$$

where $F(t,h) := \operatorname{arcsinh}(th\frac{\pi}{2})/(h\frac{\pi}{2})$, $\zeta_k \in]\nu_k, k[$, and $\alpha \in]0,1[$. Finally, we have to find bounds for the occurring partial derivatives of T and F (see [3] for details).

5. Numerical Experiments

We present some numerical results for the Laplacian on $\Omega =]0,1[^2$ with Dirichlet boundary conditions and the solutions $u^{(\alpha)}(x,y) := Im\big((x - \frac{1}{2} + iy)^{\alpha}\big)$ and $u(x,y) := \sin(\pi y) \cdot \sinh(\pi(1-x))/\sinh(\pi)$. Here, α indicates the smoothness of $u^{(\alpha)}$ at the critical point $(\frac{1}{2},0)$. If, e.g., $\alpha \geq 4$, then $u^{(\alpha)}(x,0) \in C^4([0,1])$. Note that these examples can be reduced to the situation of (4), (5). For the discretization on

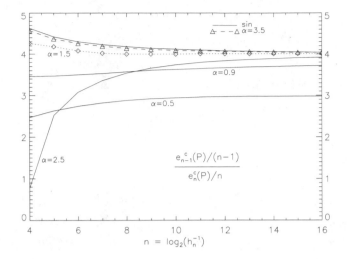

FIGURE 2. Decrease in $e_n^c(P)/n$ proceeding from level $n-1$ to level n.

the full grids $G_{i,j}$, we used finite differences as indicated in (7). The resulting systems were solved with the help of the NAG library routine D03EDF written by P. Wesseling (see [3] for details). Fig. 2 shows the decrease in $e_n^c(P)/n$ proceeding from level $n-1$ to level n for our examples. Here, $e_n^c(x,y) := u(x,y) - u_n^c(x,y)$ denotes the error of the combined solution, and $P := (\frac{1}{4}, \frac{1}{4})$. We see the $O(h_n^2 \log_2(h_n^{-1}))$-behaviour for $\alpha \in \{1.5, 2.5, 3.5\}$ and for u and a worse convergence if $\alpha < 1$.

REFERENCES

1. H.-J. Bungartz, *Dünne Gitter und deren Anwendung bei der adaptiven Lösung der dreidimensionalen Poisson-Gleichung*, Dissertation, Technische Universität München, 1992.
2. H.-J. Bungartz, M. Griebel, D. Röschke, and C. Zenger, *A proof of convergence for the combination technique for the Laplace equation using tools of symbolic computation*, Proc. Internat. IMACS Symp. on Symbolic Computation (G. Jacob, N. Oussous, and S. Steinberg, eds.), IMACS / Université des Sciences et Technologies de Lille, Lille, 1993, pp. 56–61 (also: SFB-Report 342/4/93 A, Institut für Informatik der Technischen Universität München, 1993).
3. H.-J. Bungartz, M. Griebel, D. Röschke, and C. Zenger, *Pointwise convergence of the combination technique for Laplace's equation*, SFB-Report 342/16/93 A, Institut für Informatik der Technischen Universität München, 1993.
4. M. Griebel, M. Schneider, and C. Zenger, *A combination technique for the solution of sparse grid problems*, Proc. IMACS Internat. Symp. on Iterative Methods in Linear Algebra (P. de Groen and R. Beauwens, eds.), Elsevier, Amsterdam, 1992, pp. 263–281.
5. P. Hofmann, *Asymptotic expansions of the discretization error of boundary value problems of the Laplace equation in rectangular domains*, Numerische Mathematik **9** (1967), pp. 302–322.
6. J. Walsh and D. Young, *On the degree of convergence of solutions of difference equations to the solution of the Dirichlet problem*, J. Math. & Phys. **33** (1954), pp. 80–83.
7. C. Zenger, *Sparse grids*, Parallel algorithms for partial differential equations: Proc. Sixth GAMM-Seminar (Kiel, 1990), Notes on Numerical Fluid Mechanics, vol. 31 (W. Hackbusch, ed.), Vieweg, Braunschweig, 1991.

INSTITUT FÜR INFORMATIK DER TU MÜNCHEN, D-80290 MÜNCHEN, GERMANY
E-mail address: bungartz, griebel, roeschke, zenger@informatik.tu-muenchen.de

Contemporary Mathematics
Volume **180**, 1994

Domain Decomposition Methods for Monotone Nonlinear Elliptic Problems

XIAO-CHUAN CAI AND MAKSYMILIAN DRYJA

ABSTRACT. In this paper, we study several overlapping domain decomposition based iterative algorithms for the numerical solution of some nonlinear strongly elliptic equations discretized by the finite element methods. In particular, we consider additive Schwarz algorithms used together with the classical inexact Newton methods. We show that the algorithms converge and the convergence rates are independent of the finite element mesh parameter, as well as the number of subdomains used in the domain decomposition.

1. Introduction

Schwarz type overlapping domain decomposition methods have been studied extensively in the past few years for linear elliptic finite element problems, see e. g., [**2, 4, 5, 11, 9**]. In this paper, we extend some of the theory and methods to the class of nonlinear strongly elliptic finite element problems. The first study of the classical Schwarz alternating method for nonlinear elliptic equations appeared in the paper of P. L. Lions [**14**], in which the class of continuous monotonic elliptic problems was investigated. There are basically two approaches that a domain decomposition method can be used to solve a nonlinear problem. The first approach is to locally linearize the nonlinear equation via a Newton-like algorithm and then to solve the resulting linearized problems at each nonlinear iteration by a domain decomposition method. The second approach is to use domain decomposition, such as the Schwarz alternating method, directly on the

1991 *Mathematics Subject Classification.* 65F10, 65N30, 65N55.
Key words and phrases. nonlinear elliptic problems, preconditioner, finite elements, overlapping Schwarz algorithms, Newton's method.
The work of Cai was supported in part by the National Science Foundation and the Kentucky EPSCoR Program under grant STI-9108764. This work of Dryja was supported in part by the Center for Computational Sciences, University of Kentucky, in part by the Polish Scientific Grant 211669101.
This paper is in final form and no version of it will be submitted elsewhere.

nonlinear problems. In this case, a number of smaller nonlinear problems need to be solved per domain decomposition iteration. In this paper, we focus on the first approach. We show under certain assumptions that the mesh parameters independent convergence can be obtained. Certain related multilevel approaches can be found in [1, 16].

Let $\Omega \subset R^d (d = 2, 3)$ be a polygonal domain with boundary $\partial\Omega$ and $a(u, v) = (\nabla u, \nabla v)_{L^2(\Omega)}$. Here $u, v \in V_h$ and V_h is the usual triangular finite element subspace of $H_0^1(\Omega)$(inner product $a(\cdot, \cdot)$ and norm $\| \cdot \|_a = a(\cdot, \cdot)^{1/2}$) consisting of continuous piecewise linear functions. Following the Dryja-Widlund construction of the overlapping decomposition of V_h (cf. [11]), the triangulation of Ω is introduced as follows. The region is first divided into nonoverlapping substructures Ω_i, $i = 1, \cdots, N$, whose union forms a coarse subdivision of Ω. Then all the substructures Ω_i, which have diameter of order H, are divided into elements of size h. The assumption, common in finite element theory, that all elements are shape regular is adopted. To obtain an overlapping decomposition of the domain, we extend each subregion Ω_i to a larger region Ω_i', i.e., $\Omega_i \subset \Omega_i'$. We assume that the overlap is uniform and $V_i \subset V_h$ is the usual finite element space over Ω_i'. Let $V_0 \subset V_h$ be a triangular finite element subspace defined on the coarse grid. It is clear that $\Omega = \bigcup_i \Omega_i'$ and $V_h = V_0 + \cdots + V_N$.

Base on the decomposition of V_h discussed above, we introduce and analyze some algorithms for the finite element solution of the following quasilinear elliptic problem with Dirichlet boundary condition:

$$\mathcal{L}u = \sum_{i=1}^{d} \frac{\partial}{\partial x_i} a_i(x, u, \nabla u) + a_0(x, u, \nabla u) = f(x).$$

The corresponding variational problem reads as following: Find $u^* \in V_h$, such that

(1) $b(u^*, v) = (f, v) \quad \forall v \in V_h,$

where

$$b(u, v) = \int_\Omega \left(\sum_{i=1}^{d} a_i(x, u, \nabla u) \frac{\partial v}{\partial x_i} + a_0(x, u, \nabla u)v \right) dx.$$

The existence and uniqueness of the continuous problem are understood under certain assumptions, see e.g., Ladyzhenskaya and Ural'Tseva [13]. Let $u_i(x, p_0, p_1, p_2) = a_i(x, u, u_x, u_y)$, $p = (p_0, p_1, p_2)$ and $p' = (p_1, p_2)$. The basic assumptions are, for some positive constants c and C,

(A1) $a_i \in C^1(\Omega \times R^3)$;

(A2) $\max \left\{ |a_i|, \left| \frac{\partial a_i}{\partial x_j} \right|, \left| \frac{\partial a_i}{\partial p_k} \right| \right\} \leq C$, for $i, k = 0, \cdots, d$, and $j = 1, \cdots, d$;

(A3) the operator is strongly elliptic; i.e.,

$$\sum_{i,j=0}^{d} \frac{\partial a_i(x, p)}{\partial p_j} \xi_i \xi_j \geq c \sum_{i=0}^{d} \xi_i^2.$$

As a direct consequence of assumptions (A1-3), we can prove the following lemmas, which will be used extensively in the convergence analysis in the subsequent sections of this paper.

LEMMA 1. *The functional* $b(\cdot, \cdot) : H_0^1(\Omega) \times H_0^1(\Omega) \to R$ *satisfies the strong monotonicity condition, i.e., there exists a constant* $c > 0$, *such that for any* $u, v \in H_0^1(\Omega)$,

$$(2) \qquad b(u, u - v) - b(v, u - v) \geq c\|u - v\|_a^2$$

or, equivalently, for any $v, z \in H_0^1(\Omega)$,

$$b(v + z, z) - b(v, z) \geq c\|z\|_a^2.$$

LEMMA 2. *The functional* $b(\cdot, \cdot)$ *is uniformly bounded in the sense that there exists a constant* $C > 0$, *such that*

$$(3) \qquad |b(u, w) - b(v, w)| \leq C\|u - v\|_a\|w\|_a,$$

for any $u, v, w \in H_0^1(\Omega)$.

Let $V_h = span\{\phi_1, \cdots, \phi_n\}$ and the finite element solution $u^* = \sum_{i=1}^n u_i\phi_i$. Define

$$b_i(u_1, \cdots, u_n) = b\left(\sum_{j=1}^n u_j\phi_j, \phi_i\right), \qquad f_i = (f, \phi_i)$$

$B = (b_1, \cdots, b_n)^T$ and $\hat{f} = (f_1, \cdots, f_n)^T$. The rest of the paper is devoted to the solution of the following nonlinear algebraic equation

$$(4) \qquad G(u) = B(u) - \hat{f} = 0.$$

Here and in the remainder of the paper, we use u (or v, w, z) to denote either a function in V_h or its corresponding vector representation in terms of the basis functions, i.e., $u = (u_1, ..., u_n)^T \in R^n$ and $u = \sum_{i=1}^n u_i\phi_i \in V_h$. We consider the well-known Newton-like method [7].

2. A Simple Poisson-Schwarz-Newton Method

In this section, we discuss a simple algorithm that combines the Schwarz preconditioning technique with a Newton's method. The preconditioner is defined by using the Poisson operator(i.e., using $u(\cdot, \cdot)$), which generally has nothing to do with the nonlinear problem to be solved. We show that with a properly chosen relaxation parameter λ the algorithm converges at an optimal rate, which is independent of the mesh parameters. The involvement of the parameter λ makes the algorithm not very practical, but nevertheless, it provides some theoretical insight to the preconditioning process.

For each subspace V_i, let us define an operator $Q_i : V_h \to V_i$, by

$$a(Q_i(u), v) = b(u, v), \quad \forall u \in V_h, \ v \in V_i.$$

$Q_i(u)$ can also be understood in the matrix form $Q_i(u) = R_i^T A_i^{-1} R_i B(u)$, where A_i is the subdomain discretization of $a(\cdot, \cdot)$ and $R_i : V_h \to V_i$ is a restriction operator, [3]. To define the additive Schwarz method, let us define

$$Q = Q_0 + Q_1 + \cdots + Q_N.$$

We note that the operators Q_i and Q are not linear in general. We shall show that the following nonlinear equation

$$(5) \qquad \tilde{G}(u) \equiv Q(u) - \tilde{g} = 0$$

has a unique solution, and is equivalent to equation (4), i.e., they have the same solution. Here the right-hand vector $\tilde{g} = \sum_{i=0}^N \tilde{g}_i$, and $\tilde{g}_i = Q_i u^*$. These $\tilde{g}_i$ can be pre-computed without knowing the exact solution u^*, as illustrated in [11]. Let us define

$$M^{-1} = \sum_{i=0}^N R_i^T A_i^{-1} R_i \text{ and } M = \left(\sum_{i=0}^N R_i^T A_i^{-1} R_i \right)^{-1}.$$

From the additive Schwarz theory of Dryja and Widlund [11], we understand that M is symmetric and positive definite and the norm generated by M ($\| \cdot \|_M$) is equivalent to the norm $\| \cdot \|_a$.

ALGORITHM 1 (ADDITIVE-SCHWARZ-RICHARDSON). *For a properly chosen parameter* λ, *iterate for* $k = 0, 1, \cdots$ *until convergence*

$$u^{k+1} = u^k + \lambda s^k,$$

where $s^k = -\tilde{G}(u^k) = -M^{-1} G(u^k)$.

We note that the algorithm can also be written as $u^{k+1} = u^k - \lambda \left(Q(u^k) - \tilde{g} \right)$. The following technical lemma plays a key role in our optimal convergence theory.

LEMMA 3. *There exists two constants* δ_0 *and* δ_1, *such that*

$$(6) \qquad (Q(u+z) - Q(u), z)_M \geq \delta_0 \|z\|_M^2, \quad \forall u, z \in V_h,$$

and

$$(7) \qquad \|Q(u+z) - Q(u)\|_M^2 \leq \delta_1 \|z\|_M^2, \quad \forall u, z \in V_h.$$

The optimal convergence of the Algorithm 1 is stated in the main theorem of this section.

THEOREM 1. *If we choose* $0 < \lambda < 2\delta_0/\delta_1$, *where* δ_0 *and* δ_1 *are both defined in Lemma 3, then Algorithm 1 converges optimally in the sense that*

$$(8) \qquad \|u^k - u^*\|_a \leq C \rho^k \|u^0 - u^*\|_a.$$

Here $\rho^2 = 1 - \lambda \delta_1 (2\delta_0/\delta_1 - \lambda) < 1$ *and* C *are independent of the mesh paramenters* h *and* H. *The optimal* $\lambda_{opt} = \delta_0/\delta_1$ *and* $\rho_{opt}^2 = 1 - \delta_0^2/\delta_1$.

3. A Newton-Krylov-Schwarz Method (NKS)

In this section, we study an outer-inner iterative method for solving (1). Classical Newton is used as the outer iterative method, and a Schwarz preconditioned Krylov subspace method is used as the inner iterative method. We prove that under certain conditions that if the number of inner iterations is sufficiently large, then the outer iteration converges at a rate independent of the finite element mesh parameters, and the number of subdomains.

At each point $u \in V_h$, let us define

$$M_{AS}^{-1}(u) = \sum_{i=0}^{N} R_i^T L_i^{-1}(u) R_i,$$

as the additive Schwarz preconditioner corresponding to the Jacobi operator $L(u)$ of $B(u)$. Here $L_i^{-1}(u)$ is the inverse of $L(u)$ in the subspace V_i and $R_i : V_h \to V_i$ is the restriction operator. To solve for the kth Newton correction, we use n_k steps of a Schwarz-preconditioned Krylov subspace iterative method with initial guess $v^0 = 0$. Let F_k be the iteration operator, i.e., at the lth Krylov iteration, the error is given by

$$(9) \qquad\qquad v^l - v = F_k(v^0 - v).$$

Or, equivalently, we have $v = (I - F_k)^{-1} v^l$. For the simplicity of presentation, we replace the Krylov iterative method by a simpler Richardson's method. The operator F_k has the form

$$F_k(u_k) = \left(I - \tau_k M_{AS}^{-1}(u_k) L(u_k) \right)^l,$$

where the τ_k are relaxation parameters. We assume that the operator F_k is bounded, i.e., there exists a constant $0 < \rho_k < 1$, such that

$$(10) \qquad\qquad \|F_k\|_a \leq \rho_k.$$

The estimate (10) is satisfied for a number of Krylov space methods, such as GMRES [15]. In the rest this section, we study the convergence of the following NKS algorithm.

ALGORITHM 2 (NEWTON-KRYLOV-ADDITIVE-SCHWARZ ALGORITHM). *For any given $u_0 \in V_h$, iterate with $k = 0, 1, \ldots$ until convergence*

$$(11) \qquad\qquad L(u_k)(I - F_k)^{-1}(u_{k+1} - u_k) = -B(u_k) + \hat{f}.$$

In practice, a damping parameter can usually be used in each outer iteration to accelerate the convergence of the Newton method. The parameters can be selected by using either a line search or a trust region approach, see e.g. [7]. Since we are interested mostly in theoretical aspects of the algorithm, the selection of parameters is omitted from its description.

Before giving the main result, we present a few auxiliary lemmas. Let $A = \{a(\phi_i, \phi_j)\}$, $i, j = 1, \cdots n$. We assume that $L(u)$ satisfies the Lipschitz condition, i.e.,

$$\|L(u) - L(v)\|_{A^{-1}} \leq \gamma \|u - v\|_A.$$

LEMMA 4. *There exist two constants γ_0 and γ_1, such that for any $v, w \in V_h$,*

$$(12) \qquad a(L^{-1}(v)w, L^{-1}(v)w) \leq \gamma_0 a(A^{-1}w, A^{-1}w)$$

and

$$(13) \qquad a(A^{-1}L(v)w, A^{-1}L(v)w) \leq \gamma_1 a(w, w).$$

LEMMA 5.

$$(14) \qquad \|B(v + z) - B(v) - L(v)z\|_{A^{-1}} \leq C\|z\|_A^2$$

LEMMA 6. *Let $e_k = u^* - u_k$, we then have*

$$(15) \qquad \|e_{k+1}\|_a \leq C_0 \left(\|e_k\|_a^2 + \delta_k \right),$$

where $\delta_k \leq \rho_k (1 - \rho_k)^{-1} \|u_{k+1} - u_k\|_a$.

Based on this lemma, we prove that

THEOREM 2. *There exist constants c_1 and c_2, both sufficiently small, such that if $\|u^* - u_0\|_a \leq c_1$ and $\rho_k \leq c_2$, for all k, then*

$$\|u^* - u_k\|_a \leq \rho^k \|u^* - u_0\|_a.$$

Here $0 \leq \rho < 1$ is a constant independent of the mesh parameters. In addition, if $\rho_k \to 0$ in such a way that $\rho_k \leq \{C\|e_k\|_a, 1/(2C_0)\}$, then the convergence is quadratic, i.e.,

$$\|u^* - u_{k+1}\|_a \leq C\|u^* - u_k\|_a^2,$$

where C is independent of the mesh parameters.

REFERENCES

1. R. E. Bank and D. J. Rose, *Analysis of a multilevel iterative method for nonlinear finite element equations*, Math. Comp., 39 (1982), pp. 453–465.
2. J. H. Bramble, J. E. Pasciak, J. Wang and J. Xu, *Convergence estimates for product iterative methods with applications to domain decomposition and multigrid*, Math. Comp., 57 (1991), pp. 1–22.
3. X.- C. Cai, W. D. Gropp and D. E. Keyes, *A comparison of some domain decomposition algorithms for nonsymmetric elliptic problems*, Numer. Lin. Alg. Appl., June, 1994. (to appear)
4. X.- C. Cai, W. D. Gropp, D. E. Keyes and M. D. Tidriri, *Parallel implicit methods for aerodynamics*, these proceedings, 1994.
5. X. -C. Cai and O. B. Widlund, *Multiplicative Schwarz algorithms for nonsymmetric and indefinite elliptic problems*, SIAM J. Numer. Anal., 30, (1993), pp. 936–952.
6. P. G. Ciarlet, *The Finite Element Method for Elliptic Problems*, North-Holland, New York, 1978.
7. J. E. Dennis, Jr. and R. B. Schnabel, *Numerical Methods for Unconstrained Optimization and Nonlinear Equations*, Prentice-Hall, Englewood Cliffs, NJ, 1983.

8. J. Douglas and T. Dupont, *A Galerkin method for a nonlinear Dirichlet problem*, Math. Comp., 29 (1975), pp. 689–696.
9. M. Dryja, B. F. Smith and O. B. Widlund, *Schwarz analysis of iterative substructuring algorithms for elliptic problems in three dimensions*, TR-638, Courant Institute, New York Univ., 1993. (SIAM J. Numer. Anal., to appear)
10. M. Dryja and O. B. Widlund, *An additive variant of the Schwarz alternating method for the case of many subregions*, Tech. Rep. 339, Courant Inst., New York Univ., 1987.
11. M. Dryja and O. B. Widlund, *Towards a unified theory of domain decomposition algorithms for elliptic problems*, in Third International Symposium on Domain Decomposition Methods for Partial Differential Equations, T. F. Chan, R. Glowinski, J. Périaux, and O. B. Widlund, eds., SIAM, Philadelphia (1990).
12. M. Dryja and O. B. Widlund, *Domain decomposition algorithms with small overlap*, SIAM J. Sci. Comp., 15 (1994).
13. O. A. Ladyzhenskaya and N. N. Ural'Tseva, *Linear and Quasilinear Elliptic Equations*, Academic Press, 1968
14. P. L. Lions, *On the Schwarz alternating method I*, First International Symposium on Domain Decomposition Methods for Partial Differential Equations, R. Glowinski, G. H. Golub, G. A. Meurant, and J. Périaux, eds., SIAM, Philadelphia, 1988.
15. Y. Saad and M. H. Schultz, *GMRES: A generalized minimal residual algorithm for solving nonsymmetric linear systems*, SIAM J. Sci. Stat. Comp., 7 (1986), pp. 865–869.
16. J. Xu, *Two-grid finite element discretization for nonlinear elliptic equations*, SIAM J. Numer. Anal., 1993. (to appear)

DEPARTMENT OF COMPUTER SCIENCE, UNIVERSITY OF COLORADO AT BOULDER
E-mail address: cai@cs.colorado.edu

DEPARTMENT OF MATHEMATICS, WARSAW UNIVERSITY, POLAND
E-mail address: dryja@mimuw.edu.pl

Contemporary Mathematics
Volume **180**, 1994

Cascadic Conjugate Gradient Methods for Elliptic Partial Differential Equations: Algorithm and Numerical Results

PETER DEUFLHARD

ABSTRACT. Cascadic conjugate gradient methods for the numerical solution of elliptic partial differential equations consist of Galerkin finite element methods as an outer iteration and (possibly preconditioned) conjugate gradient methods as an inner iteration. Both iterations are known to minimize the energy norm of the arising iteration errors. The present paper derives a unified framework in which to study the relative merits of different preconditioners versus the case of no preconditioning. Surprisingly, in the numerical experiments the cascadic conjugate gradient method *without any preconditioning* (to be called CCG method) turns out to be not only simplest but also fastest. It appears that the cascade principle in itself already realizes some kind of preconditioning. A theoretical explanation of the observed iteration pattern will be given elsewhere.

Introduction

This paper deals with *cascadic preconditioned conjugate gradient methods* — hereinafter called CPCG methods — for the solution of general elliptic boundary value problems for partial differential equations. Any such method is based on the so–called *cascade principle* which involves the cascade-like numerical solution of a sequence of linear systems of equations associated with a sequence of finite element spaces on successively finer grids. In this setting, the *coarse* grid linear system (up to moderate size) is assumed to be solved *directly* — say, by a (sparse) elimination technique. *Finer* grid systems are solved iteratively by *preconditioned conjugate methods* — hereinafter called PCG methods. Starting values for the PCG iteration on a given discretization level are just the (approximate) finite element solutions of the previous level. The successive finite element spaces are constructed *adaptively* based on local energy error estimators. Within each discretization level the PCG termination criterion aims at keeping the iteration error below the expected discretization error.

1991 *Mathematics Subject Classification.* Primary 65F10, 65N30.

This paper is in final form and no version of it will be submitted for publication elsewhere.

In former realizations of this concept [9, 5, 10, 12, 13], the whole iteration control mechanism was based on some cheap but not very satisfactory *energy error norm approximation*, which led to a close coupling between the local error estimators and the PCG termination criterion. That approximation has been recently replaced by another more satisfactory one, which is also cheap – compare [7] or the more thorough discussion in [8]. In Section 1, the new energy error control is applied to the (nested) CPCG iteration as a whole. The sequence of Galerkin approximations on successively finer grids is interpreted as an *outer iteration*, which minimizes the energy error norms over a sequence of finite element spaces. Arbitrary PCG iterations, which minimize the energy error norms over a sequence of Krylov spaces, are interpreted as *inner iterations*. An efficient strategy for the matching of inner versus outer energy error norms is developed. In Section 2, the proposed CPCG method is illustrated by numerical experiments on the Laplace equation. Comparison runs with the hierarchical basis preconditioner due to YSERENTANT [15], the multilevel preconditioner due to XU [14], BRAMBLE, PASCIAK and XU [6], and no preconditioning are presented. Two different CPCG modes are exemplified, an *adaptive* mode — including an adaptive mesh refinement strategy in the spirit of [1] based on the edge oriented error estimator of [9] — and a *uniform* mode. A convergence analysis explaining the surprising numerical findings of Section 2 will be given in a forthcoming paper.

1. Energy Error Control in Cascadic Preconditioned CG Methods

Consider an elliptic PDE problem given in the weak formulation

$$(1) \qquad a(u,v) = \langle f, v \rangle, \qquad v \in H,$$

wherein H is the appropriate Hilbert space, $u \in H$ is the solution to be computed, $a(\cdot, \cdot)$ is a symmetric H-elliptic bilinear form with

$$(2) \qquad \| \cdot \|_A = a(\cdot, \cdot)^{\frac{1}{2}}$$

the associated energy norm, and $\langle \cdot, \cdot \rangle$ the L_2 inner product. Consider further a Galerkin method for a sequence of nested finite element spaces $S_0 \subset S_1 \subset \ldots S_i \subset H$. This generates a sequence of linear systems

$$(3) \qquad A_j u_j = b_j, \qquad j = 0, 1, \ldots, i.$$

All matrices A_j are symmetric positive definite so that PCG methods can be applied. Let $n_j = \dim S_j$. For ease of writing we will not distinguish between the solutions u_j of the Galerkin equations within subspace S_j and the exact solutions u_j of the corresponding systems of linear equations (3). The meaning will be clear from the context.

1.1. Galerkin method as outer iteration. At discretization level j, let

$$\epsilon_j = \|u - u_j\|_A^2, \qquad \delta_j = \|u_{j+1} - u_j\|_A^2$$

denote the various discretization error norms. Then, for *nested* spaces S_j, the *orthogonality relation*

$$(4) \qquad \epsilon_{j+1} = \epsilon_j - \delta_j,$$

is well–known to hold. For an appropriate sequence of subspaces of the Hilbert space H we have convergence $u_j \to u$ and therefore $\epsilon_\infty = 0$, so that the recursion (4) can be solved to yield

$$(5) \qquad \epsilon_j = \sum_{l=j}^{\infty} \delta_l, \qquad j = 0, 1, \dots .$$

As for the convergence of the FE method, i.e. the outer iteration, we will naturally require the theoretical termination criterion

$$(6) \qquad \epsilon_i \leq TOL \, \epsilon_0,$$

with some error tolerance parameter TOL to be prescribed by the user and some index i to count for the finest actually computed discretization level. Upon proceeding as in [7, 8], we will try to replace the not implementable termination criterion (6) by a *sufficient and implementable* termination criterion. As for the above right-hand side, we will replace ϵ_0 by the lower bound

$$(7) \qquad \epsilon_0^{(i)} = \|u_{i+1} - u_0\|_A^2 = \epsilon_0 - \epsilon_{i+1} = \sum_{j=0}^{i} \delta_j \leq \epsilon_0.$$

As for the left-hand side of (6), let Θ denote a contraction factor understood to satisfy

$$\epsilon_{j+1} \leq \Theta \epsilon_j, \qquad \Theta < 1 \qquad j > j_0$$

for some threshold index j_0. Then equation (4) implies that

$$(8) \qquad \epsilon_j \leq \bar{\epsilon}_j = \frac{\delta_j}{1 - \Theta}.$$

For illustration purposes, consider the Laplace equation on some polygonal domain Ω in d–dimensional space with $d = 2$ or $d = 3$. This means that H is now some space H_0^1 wherein the subscript 0 indicates the fact that we assume Dirichlet boundary conditions on a sufficient part of the boundary $\partial\Omega$. In this case we know that *uniform* mesh refinement and *linear* finite elements in the regular case lead to

$$(9) \qquad \Theta \doteq \frac{1}{4}, \qquad \bar{\epsilon}_j \doteq \frac{4}{3}\delta_j.$$

In the case of *adaptive* meshes — assuming *energy error equidistribution* and once more *linear* finite elements — we expect a comparable estimated contraction factor

$$(10) \qquad \Theta \doteq \bar{\Theta}_j = \left(\frac{n_{j-1}}{n_j}\right)^{2/d}, \qquad j > j_0,$$

on the basis of theoretical results of [2]. The latter choice is used in the numerical experiments below. Further improvements of this factor should be possible in close combination with the adaptive mesh refinement strategy based on local extrapolation [1]. Such a device is presently under investigation.

Summarizing, we end up with the *sufficient* condition

$$(11) \qquad\qquad \bar{\epsilon}_i \leq TOL\, \epsilon_0^{(i)}$$

for termination at finite element level i. The implementation of this criterion requires the iterative quantities δ_j, which — as we will see below — can be obtained cheaply from the PCG iteration.

1.2. PCG method as inner iteration. In previous versions of cascade type algorithms — such as [9, 5, 10] — the criterion (11) has been used in connection with the edge–oriented discretization error estimator due to [9]. In what follows, an alternative technique based on considerations from [7, 8] will be worked out.

With the outer iterates u_j from the Galerkin method, we now need two indices $u_{j,k}$ for the PCG iterates. On levels $j = 1, 2, \dots, i$, the iteration index k formally runs within $k = 0, \dots, n_j$. On the coarse grid level $j = 0$, direct linear equation solving supplies some u_0 assumed to be exact. On finer levels $j > 0$, the cascade principle realizes

$$u_{j,0} = u_{j-1}, \qquad u_j = u_{j,n} \quad \text{for } n = n_j.$$

It is known from [7, 8] that the iterative energy error contributions

$$\delta_{j,k} = \|u_{j,k+1} - u_{j,k}\|_A^2$$

are cheaply available on levels $j > 0$. Moreover, since

$$\|u_j - u_{j-1}\|_A^2 = \|u_{j,n} - u_{j,0}\|_A^2 = \sum_{k=0}^{n_j - 1} \|u_{j,k+1} - u_{j,k}\|_A^2,$$

we can express the iterative discretization errors as

$$(12) \qquad\qquad \delta_{j-1} = \|u_j - u_{j-1}\|_A^2 = \sum_{k=0}^{n_j - 1} \delta_{j,k}.$$

In words: *The (exact) PCG iteration on discretization level j supplies the energy norm of the iterative discretization error of the preceding level $j - 1$.*

In actual computation, things are slightly more complicated, since instead of the above exact PCG iterates $u_{j,k}$ we have *perturbed* iterates $\tilde{u}_{j,k}$ obtained from truncated PCG iterations, which yield perturbed Galerkin approximations $\tilde{u}_j$. As in the exact case, we start from the direct solution $\tilde{u}_0 = u_0$. On finer levels $j > 0$, however, we continue according to

$$(13) \quad \tilde{u}_{j,0} = \tilde{u}_{j-1}, \qquad \tilde{u}_j = \tilde{u}_{j,m+1} \quad \text{for some truncation index } m = m_j.$$

With the cheaply available quantities

$$\tilde{\delta}_{j,k} = \|\tilde{u}_{j,k+1} - \tilde{u}_{j,k}\|_A^2$$

instead of the $\delta_{j,k}$ we obtain the analog of result (12) now in the form

(14)
$$\tilde{\delta}_{j-1} = \|\tilde{u}_j - \tilde{u}_{j-1}\|_A^2 = \sum_{k=0}^{m_j} \tilde{\delta}_{j,k}.$$

The associated discretization errors

(15)
$$\tilde{\epsilon}_j = \|u - \tilde{u}_j\|_A^2$$

can be seen to satisfy

(16)
$$\tilde{\epsilon}_j = \|u - \tilde{u}_j\|_A^2 \geq \min_{v \in S_j} \|u - v\|_A^2 = \|u - u_j\|_A^2 = \epsilon_j.$$

Throughout this paper the FE spaces are (technically) assumed not to depend on the sequence of truncation indices of the PCG iteration — an assumption which is only realistic if utmost care is taken in the realization of the whole scheme including the adaptive mesh refinement strategy. As in (4) the orthogonality relation

(17)
$$\tilde{\epsilon}_{j+1} = \tilde{\epsilon}_j - \tilde{\delta}_j$$

holds — this time not as a consequence of the Galerkin minimization property, but due to the cascade property (13) and the orthogonality of the PCG iterative corrections within the Krylov spaces also in the perturbed case. As in the exact case, we want to replace the unavailable term $\tilde{\epsilon}_0$ by computationally available terms of the form

(18)
$$\tilde{\epsilon}_0^{(i)} = \sum_{j=0}^{i} \tilde{\delta}_j.$$

With $\tilde{\epsilon}_0 = \epsilon_0$ — due to the direct solution on the coarse grid — and (16) we obtain

$$\tilde{\epsilon}_0^{(i)} = \tilde{\epsilon}_0 - \tilde{\epsilon}_{i+1} = \epsilon_0 - \tilde{\epsilon}_{i+1} \leq \epsilon_0 - \epsilon_{i+1} = \epsilon_0^{(i)}.$$

This means that — in view of sufficiency — the right–hand side term $\epsilon_0^{(i)}$ in the termination criterion (11) can be replaced by $\tilde{\epsilon}_0^{(i)}$.

We now turn to the approximation of the left–hand side of (11), which means that we have to consider an approximation of the quantity ϵ_j at the recursive levels j. Unfortunately, this quantity cannot be bounded on either side by its associated estimate $\tilde{\epsilon}_j$, since the starting values $\tilde{u}_{j,0}$ for the perturbed PCG iteration differ from the exact starting values $u_{j,0}$ on the finer grids $j > 0$. Moreover, at level j only the discretization error $\tilde{\delta}_{j-1}$ for the previous level is actually available. In this situation, we recur to (10) and (15) to obtain

(19)
$$\tilde{\epsilon}_j \doteq \bar{\Theta}_j \tilde{\epsilon}_{j-1}, \qquad \tilde{\epsilon}_{j-1} \leq \frac{\tilde{\delta}_{j-1}}{1 - \bar{\Theta}_j}.$$

Accordingly, we approximate the termination criterion (11) by the criterion

(20)
$$\hat{\epsilon}_i = \frac{\bar{\Theta}_i \tilde{\delta}_{i-1}}{1 - \bar{\Theta}_i} \leq TOL \, \tilde{\epsilon}_0^{(i)}$$

to be satisfied at final level i. This is now the implementable *termination criterion for the outer iteration* to be used for the determination of the actually needed discretization level i according to the user prescribed relative accuracy TOL. (The terms $\hat{\epsilon}_j$ will be returned to the user as *recursive discretization error estimate* on level j.)

We are still left with the decision of how to control the inner PCG iterations in such a way that the iterative discretization error estimates $\hat{\epsilon}_j$ are sufficiently reliable. In view of [7, 8] we will require a condition of the kind

$$(21) \qquad \frac{\tilde{\delta}_{j,m}}{1 - \bar{\Theta}_{j,m}} \leq \hat{\rho}_j \bar{\Theta}_j \tilde{\delta}_{j-1} \qquad \text{for some truncation index } m = m_j.$$

Herein the estimate of the contraction factors $\bar{\Theta}_{j,m}$ may cause difficulties in actual computation and therefore require some additional heuristics. The safety factors $\hat{\rho}_j$ need to be chosen as internal default parameters. In former realizations [9, 5], the choice $\hat{\rho}_j = \text{const}$ had been made, which led to a condition most stringent on the *final* level i. Recall, however, the global error relation

$$(22) \qquad \tilde{\epsilon}_i = \|u - \tilde{u}_i\|_A^2 = \epsilon_0 - \sum_{j=1}^{i} \sum_{k=0}^{m_j} \tilde{\delta}_{j,k}.$$

This relation seems to indicate that the final level restriction from (21) should already be observed on coarser levels to avoid unnecessary more costly iterations on the finer levels. Therefore we suggest to replace (21) by

$$(23) \qquad \frac{\tilde{\delta}_{j,m}}{1 - \bar{\Theta}_{j,m}} \leq \hat{\rho} \, TOL \, \tilde{\epsilon}_0^{(j)} \qquad \text{for some truncation index } m = m_j.$$

This criterion is more stringent on *coarser* levels. Numerical experiments (with $\hat{\rho} = \frac{1}{16}$ throughout) strongly confirmed the expectation that criterion (23) saves costly iterations on finer levels compared to the criterion (21). This is now the desired implementable *termination criterion for the inner iteration.*

With the two termination criteria (20) for the Galerkin outer iteration and (23) for the inner PCG iteration, the whole CPCG iteration error control is now complete. Note that it applies independent of any special choice of preconditioner or even without any preconditioner.

For the convenience of the reader, we now summarize the whole CCG iteration (without preconditioning) in the form of a *pseudocode* — see Table 1. The notation for the CG iteration follows the one given in [7].

TABLE 1. **Pseudo-code: cascadic conjugate gradient method**

PROCEDURE CCG (OUTER ITERATION):

input TOL   // prescribed tolerance
$j = 0$
while $(j \leq maxLevel)$
{

      assemble linear system A,b ;
      CG(A,u,b);   // conjugate gradient iteration
      calculate $\hat{\epsilon}_j, \tilde{\epsilon}_0^{(j)}$
      if $\left(\hat{\epsilon}_j \leq TOL\, \tilde{\epsilon}_0^{(j)}\right)$ **end;**   // termination criterion (20)
      refine mesh;
      $j = j + 1;$

}

PROCEDURE CG(A,x,b) (INNER ITERATION):

input $j, TOL, \hat{\rho}, \tilde{\epsilon}_0^{(j)}$
$p_0 = r_0 = b - Ax_0,$
$\sigma_0 = (r_0, r_0),$
$k = 0$
while $(k \leq maxIter)$
{

$$\alpha_k = \frac{\langle Ap_k, p_k \rangle}{\sigma_k},$$

$$x_{k+1} = x_k + \frac{1}{\alpha_k} p_k,$$

$$\tilde{\delta}_{j,k} = \sigma_k / \alpha_k;$$
      calculate $\bar{\Theta}_{j,k}, \tilde{\delta}_{j-1}, ;$

      if $\left(\dfrac{\tilde{\delta}_{j,k}}{1 - \bar{\Theta}_{j,k}} \leq \hat{\rho} TOL\, \tilde{\epsilon}_0^{(j)}\right)$ **return;**   // termination criterion (23)

$$r_{k+1} = r_k - \frac{1}{\alpha_k} Ap_k,$$

$$\sigma_{k+1} = \langle r_{k+1}, r_{k+1} \rangle, \quad \beta_{k+1} = \frac{\sigma_{k+1}}{\sigma_k},$$

$$p_{k+1} = r_{k+1} + \beta_{k+1} p_k;$$
      $k = k + 1;$

}

2. Numerical Experiments

Up to now, numerical tests were only made for the *Laplace equation with linear finite elements*, both in 2–D and in 3–D. The picture in 3–D appeared to be essentially the same as in 2–D (though slightly less reliable in the adaptive case). As computing times and storage requirements for the 3–D test runs blew up considerably, the subsequent illustration is restricted to the 2–D case. Out of the three test examples given in [7] only one will be considered here for reasons of restricted space.

Peak Problem. Given the PDE

(24) $$-\Delta u = f \ ,$$

Dirichlet boundary conditions are imposed such that

$$u = (x+1)(x-1)(y+1)(y-1)e^{-100(x^2+y^2)}$$

is the solution.

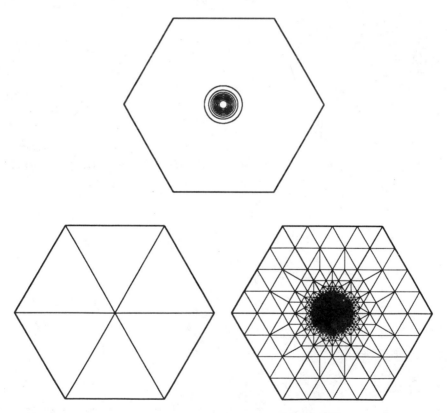

FIGURE 1. Solution of peak problem on level $j = 7$, $n = 1975$ nodes (adaptive mode) and grids on levels $j = 0$ and $j = 7$.

The comparative performance of the following three variants of the CPCG algorithm are presented:

— CPCG–HB: algorithm with *hierarchical basis* preconditioner [15]

— CPCG–BPX: algorithm with *multilevel* BPX preconditioner [14] , [6]

— CCG: algorithm *without any* preconditioner

In the simple case of the Laplace equation, the HB preconditioner is known to give rise to an $O(j^2)$-bound on the condition number in 2–D — see [15], whereas the BPX preconditioner is known to lead to $O(1)$ independent of the space dimension — see [11]. In 2–D, the expected numerical efficiency will be nearly the same for both preconditioners. Based on the subtle condition number estimates of XU [14], the expectation for the algorithm without any preconditioning would be that it might be asymptotically disastrous in rather uniform grids (which exhibit a *geometric* increase of the number of nodes) and not too bad in highly non–uniform grids (with an *arithmetic* increase of the number of nodes).

In actual computation, the explicit formulation of the FE problem (numerical quadrature for evaluation of inner products), will dominate the whole computing time. In order to make the differences between the three algorithmic variants visible, the subsequent comparison runs will mostly quote the pure *iteration times* and the *number of required iterations*. Since the solutions of all examples above are explicitly known, the directly computed iterative errors and the errors estimated from the CPCG iterations could be compared: the discrepancies were marginal on lower levels and tolerable (in most cases) on the finest levels. For this reason, only the estimated accuracies are documented here — which is the realistic case. The accuracy is measured in terms of the improvement factor from initial to final energy error norms; since, in the best case, we can expect an iterative improvement of one bit of accuracy for the Laplace equation with linear finite elements, all Figures below will use binary digits. Numerical experiments were run on a SUN Sparc Workstation 10/41 using the g++ C compiler.

2.1. Adaptive Mode. In this section we will arrange comparative results for the three algorithmic variants running in the adaptive mode, which means that a *refinement strategy* is applied to generate a sequence of *possibly highly non uniform meshes*. Any such mesh refinement strategy will naturally aim at equidistributing the energy error. In the earlier version [9] of the cascade principle a mean value strategy due to BANK [3] has been used. This strategy, however, sometimes produced unsatisfactory meshes in critical examples. Therefore, the more advanced versions [10] and [5] realized a mesh refinement technique in the spirit of BABUŠKA and RHEINBOLDT [1]. It is based on *local extrapolation* of energy error contributions from the edges as obtained by the *edge oriented discretization error indicator* due to [9]. The subsequent numerical experiments are run with a slightly modified heuristic – for details see [7].

In Fig. 2 the comparative results for the three codes are represented graphically in terms of *iteration computing times*. The surprise is that the code CCG without any preconditioning is fastest. The comparison in terms of *number of iterations* is given in Fig. 3. Obviously, the asymptotic behavior of both CCG and CPCG–BPX is the same. Between start and end all three variants show some iteration number bump, coming from the global accuracy requirement (23), which is more stringent on coarser levels. The bump is largest for CCG and smallest for BPX. The HB variant ranges in between. The different picture in terms of computing times is explained by the fact that each CG iteration with BPX preconditioning (even in a rather efficient implementation — see e.g. [4]) costs a rough factor of 3 more than each pure CG iteration.

Remark 1. It should be mentioned that the above Fig. 3 does *not* contradict Fig. 10, p. 3198 in [5], wherein the effect of BPX preconditioning versus no pre-conditioning has been exemplified as well. There, however, the iteration has been continued *far below the discretization error.* In this setting, the number of itera-tions without preconditioning drifted off far above the number of iterations with BPX preconditioning.

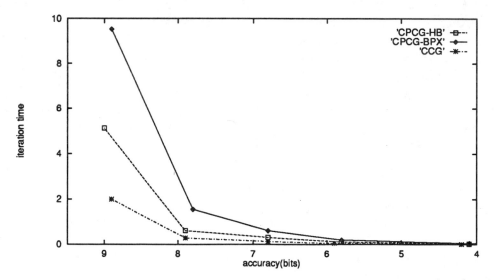

FIGURE 2.   Comparative iteration times, adaptive mode.

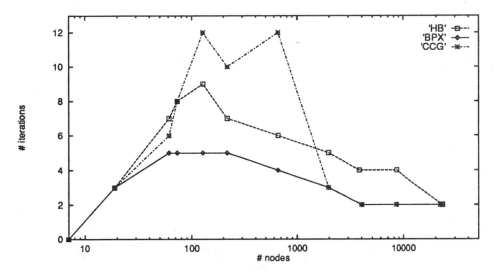

FIGURE 3.   Comparative iteration number patterns, adaptive mode.

Remark 2. For the sake of completeness, we should mention that part of the above orthogonality relations require the successive FE spaces to be *nested* — a condition, which is not satisfied whenever so-called *green edges* (cf. [3]) are dissolved from one level to the next. We have corrected the above formulas so affected in terms of the energy error differences thus introduced. However, the effect was so minor that this modification was ultimately omitted.

2.2. Uniform Mode. We now illustrate the three algorithmic variants of the CPCG method in the non-adaptive or uniform mode. In this mode, *uniform mesh refinement* is performed without making actual use of any discretization error estimator or indicator. The associated considerable amount of overall computing time is therefore saved. We exemplify this mode only for the peak problem above, which would certainly require a highly non-uniform mesh (compare Fig. 1).

PETER DEUFLHARD

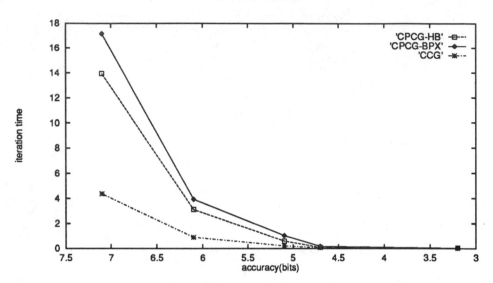

FIGURE 4. Comparative iteration times, uniform mode.

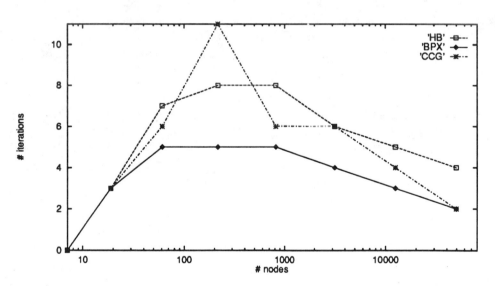

FIGURE 5. Comparative iteration number patterns, uniform mode.

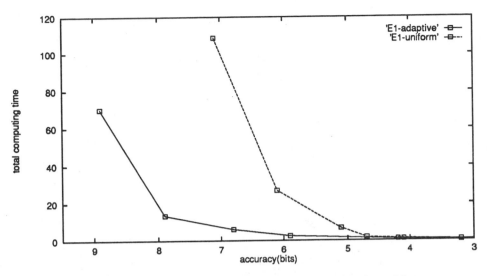

FIGURE 6.   Uniform versus adaptive mode, peak problem.

Fig. 4 shows the comparative iteration times and Fig. 5 the comparative number of iterations. The effects in the uniform mode are obviously the same as in the adaptive mode before. Once more, the CCG variant without preconditioning is clearly superior to the two preconditioners HB and BPX. Note that the CCG variant is much simpler to implement and does not need any analytic pre–investigations, which typically involve a high technical level of sophistication.

The above experiments should not leave the impression that adaptivity does not pay off. For this reason, Fig. 6 compares the total amount of computing time (full Galerkin approximation) for the adaptive and the uniform CCG mode as a function of the achieved discretization error accuracy. As can be seen, storage and computing time limitations lead to rather stringent accuracy limitations for the uniform mode. This factor is even more limiting in 3–D!

Conclusion

The present paper derives a simple but efficient strategy to control the discretization errors of the Galerkin FEM in combination with the iteration errors of the PCG method in terms of the energy norm. The relative merits of different preconditioners versus the case of no preconditioning have been compared. It appeared that the cascadic conjugate gradient method without any preconditioning (called CCG herein) was not only simplest but also fastest compared to the HB and BPX preconditioned case. In the 2–D comparison runs, HB was second and BPX was third, whereas in the 3–D runs (not documented herein) the two preconditioners interchanged their role – as expected from theory. The asymptotic behavior of the CCG method turned out to be the same as the one of the CPCG method with BPX preconditioning. Moreover, the effects were the same both for the adaptive and the non–adaptive mode — and therefore independent of any possible energy error equidistribution.

Summarizing, the numerical results seem to indicate that the cascade principle in itself already realizes some kind of preconditioning. A theoretical study of this feature is in progress.

Acknowledgements. The author gratefully acknowledges invaluable computational assistance from R. Roitzsch and B. Erdmann and helpful comments and careful reading of the manuscript by R. Beck. He also wishes to thank E. C. Körnig for her quick and patient help with the revision of the manuscript.

References

1. I. Babuška, W.C. Rheinboldt: *Estimates for Adaptive Finite Element Computations.* SIAM J. Numer. Anal. **15** (1978), pp. 736–754.
2. I. Babuška, R.B. Kellogg, J. Pitkaränta: *Direct and Inverse Error Estimates for Finite Elements with Mesh Refinements.* Numer. Math. **33** (1979), pp. 447–471.
3. R.E. Bank: *A-posteriori Error Estimates, Adaptive Local Mesh Refinement and Multigrid Iteration.* In: W. Hackbusch, U. Trottenberg (eds.): Multigrid Methods II. Springer Verlag, Berlin, Heidelberg, New York: Lect. Notes Math. Vol. **1228** (1986), pp. 7–23.
4. F. Bornemann: *An Adaptive Multilevel Approach to Parabolic Equations in Two Space Dimensions.* Dissertation, FU Berlin (1991).
5. F. Bornemann, B. Erdmann, R. Kornhuber: *Adaptive Multilevel–Methods in Three Space Dimensions.* Int. J. Numer. Methods Eng. **36** (1993), pp. 3187–3203.
6. J.H. Bramble, J.E. Pasciak, J. Xu: *Parallel Multilevel Preconditioners.* Math. Comp.**55** (1990), pp. 1–22.
7. P. Deuflhard: *Cascadic Conjugate Gradient Methods for Elliptic Differential Equations I. Algorithm and Numerical Results.* Konrad–Zuse–Zentrum Berlin (ZIB), Preprint SC 93–23 (1993).
8. P. Deuflhard: *On Error Control in Preconditioned CG Iterations.* In preparation.
9. P. Deuflhard, P. Leinen, H. Yserentant: *Concepts of an Adaptive Hierarchical Finite Element Code.* IMPACT **1** (1989), pp. 3–35.
10. B. Erdmann, R. Roitzsch, F. Bornemann: *KASKADE numerical experiments.* Technical Report TR 91–1, Konrad–Zuse–Zentrum Berlin (ZIB), (1991).
11. P. Oswald: *On Discrete Norm Estimates Related to Multilevel Preconditioners in the Finite Element Method.* In: Proceedings of the International Conference on the Constructive Theory of Functions, Varna 1991 (to appear).
12. R. Roitzsch: *KASKADE User's Manual.* Konrad–Zuse–Zentrum Berlin (ZIB), Technical Report TR 89–4 (1989)
13. R. Roitzsch: *KASKADE programmer's manual.* Konrad–Zuse–Zentrum Berlin (ZIB), Technical Report TR89–5 (1989).
14. J. Xu: *Theory of Multilevel Methods.* Report No. AM 48, Department of Mathematics, Pennsylvania State University (1989).
15. H. Yserentant: *On the Multilevel Splitting of Finite Element Spaces.* Numer. Math. **49** (1986), pp. 379–412 .

Konrad-Zuse-Zentrum Berlin, Heilbronner Str.   10, D-10711 Berlin-Wilmersdorf, Germany

E-mail address: deuflhard@sc.zib-berlin.de

Contemporary Mathematics
Volume **180**, 1994

Multilevel Methods for Elliptic Problems with Discontinuous Coefficients in Three Dimensions

MAKSYMILIAN DRYJA

ABSTRACT. Multilevel Schwarz methods are developed for a conforming approximation of second order elliptic problems. We focus on problems in three dimensions and with possibly large jumps in the coefficients across the interface separating the subregions. We establish a condition number estimate for the iterative operator which is independent of the coefficients and grows at most as the square of the number of levels. We also characterize a class of distributions of the coefficients, called quasi-monotone, for which the weighted L^2-projection is stable and for which we can use the standard piecewise linear function to construct a coarse space. In this case, we obtain optimal methods.

1. Introduction

In this paper, we discuss methods known as BPX algorithms (cf. Bramble, Pasciak and Xu [1] and Xu [9]) or multilevel Schwarz methods with one dimensional subspaces; see Zhang [10], and Dryja and Widlund [5]. It is well known that these methods are optimal when the coefficients are regular. A challenging problem is to extend these methods to problems which have very highly discontinuous coefficients. In [5], the BPX method was modified and applied to a Schur complement systems. In that case the condition number of the preconditioned system is bounded by $C_1 (1 + \log (H/h))^2$, where H and h are the parameters of the coarse and fine triangulations, respectively. In this paper, we obtain the same estimate for multilevel additive methods with several exotic coarse spaces; see Widlund [8]. For multiplicative versions such as V-cycle multigrid, we obtain rates of convergence bounded from above by $1 - C_2 (1 + \log H/h)^{-2}$; see further Sarkis [6], and Dryja, Sarkis, and

1991 *Mathematics Subject Classification.* Primary 65F10, 65N30, 65N55.
This work has been supported in part by the National Science Foundation under Grant NSF-CCR-9204255, in part by Polish Scientific Grant 211669101, and in part by the Center for Computational Sciences of the University of Kentucky at Lexington.
This paper is in final form and no version of it will be submitted for publication elsewhere.

Widlund [2]. In this paper, all constants C_i are independent of the variables appearing in the inequalities and the parameters related to meshes, spaces and, especially, the weight ρ.

This brief paper represents joint work with Marcus Sarkis and Olof Widlund and all proofs and details can be found in [2].

2. Differential and Finite Element Model Problems

We consider the following selfadjoint second order problem:
Find $u \in H_0^1(\Omega)$, such that

$$(1) \qquad\qquad a(u, v) = f(v) \ \forall \ v \in \ H_0^1(\Omega),$$

where

$$a(u, v) = \int_\Omega \rho(x) \, \nabla u \cdot \nabla v \, dx \text{ and } f(v) = \int_\Omega fv \, dx \text{ for } f \in \ L^2(\Omega).$$

Let Ω be a bounded Lipschitz region in $\Re^3$ with a diameter of order 1. A triangulation of Ω is introduced by dividing the region into nonoverlapping simplices $\{\Omega_i\}_{i=1}^N$, with diameters of order H, which are called substructures or subdomains. This partitioning induces a coarse triangulation associated with the parameter H.

We assume that $\rho(x) > 0$ is constant in each substructure with possibly large jumps occurring only across substructure boundaries. Therefore, $\rho(x) = \rho_i = \text{const}$ in each substructure Ω_i. The analysis of the methods introduced here can easily be extended to the case when $\rho(x)$ varies moderately in each subregion.

We define a sequence of nested triangulations $\{T^k\}_{k=0}^\ell$ as follows. We start with the coarse triangulation $T^0 = \{\Omega_i\}_{i=1}^N$ and let $h_0 = H$. A triangulation T^k on level k is obtained by subdividing each individual element in T^{k-1} into several elements. The assumptions on the regularity of the refinements are standard; see Zhang [10]. For each level of triangulation, we define a finite element space $V_0^k(\Omega)$ as the space of continuous piecewise linear functions associated with the triangulation T^k and which vanish on $\partial\Omega$, the boundary of Ω. We denote $V_0^h(\Omega) = V_0^\ell(\Omega)$. The discrete problem associated with (1) is given by:
Find $u \in V_0^h(\Omega)$, such that

$$(2) \qquad\qquad a(u, v) = f(v) \ \forall \ v \in \ V_0^h(\Omega).$$

The bilinear form $a(u, v)$ is directly related to a weighted Sobolev space $H_\rho^1(\Omega)$ defined by the seminorm

$$|u|_{H_\rho^1(\Omega)}^2 = a(u, u).$$

We also define a weighted L^2 norm by:

$$(3) \qquad\qquad \|u\|_{L_\rho^2(\Omega)}^2 := \int_\Omega \rho(x) \, |u(x)|^2 \, dx \text{ for } u \in L^2(\Omega).$$

3. Multilevel Additive Schwarz Methods

The multilevel methods that we consider are based on the MDS-multilevel diagonal scaling introduced by Zhang [10], enriched with a coarse space V_{-1} as in Dryja and Widlund [4], and Dryja, Smith, and Widlund [3].

Let $\mathcal{N}^k$ be the set of nodal points associated with the space V_0^k. Let ϕ_j^k be a standard nodal basis function of V_0^k, and let $V_j^k = \text{span}\{\phi_j^k\}$. We decompose V_0^h as

$$V_0^h = V_{-1}^X + \sum_{k=0}^{\ell} V_0^k = V_{-1}^X + \sum_{k=0}^{\ell} \sum_{j \in \mathcal{N}^k} V_j^k.$$

We note that this decomposition is not a direct sum and that $\dim(V_j^k) = 1$. Four different coarse spaces V_{-1}^X and associated bilinear forms $b_{-1}^X(u,v) : V_{-1}^X \times V_{-1}^X \to \Re$, $X = F, E, NN$, and W are considered; see next section.

We introduce operators $P_j^k : V_0^h \to V_j^k$, by

$$a(P_j^k u, v) = a(u, v) \ \forall \, v \in V_j^k,$$

and an operator $T_{-1}^X : V^h \to V_{-1}^X$, by

$$b_{-1}^X(T_{-1}^X u, v) = a(u, v) \ \forall \, v \in V_{-1}^X.$$

Let

$$(4) \qquad T^X = T_{-1}^X + \sum_{k=0}^{\ell} \sum_{j \in \mathcal{N}^k} P_j^k.$$

We now replace (2) by

$$(5) \qquad T^X u = g, \ g = T_{-1}^X u + \sum_{k=0}^{\ell} \sum_{j \in \mathcal{N}^k} P_j^k u.$$

The equation (5) is typically solved by a conjugate gradient method. In order to estimate its rate of convergence, we need to obtain upper and lower bounds for the spectrum of T^X.

THEOREM 1. For $u \in V_0^h(\Omega)$, we have

$$C_3 \left(1 + \log(H/h)\right)^{-2} a(u, u) \le a(T^X u, u) \le C_4 \, a(u, u).$$

4. Coarse Spaces and Bilinear Forms

Let $\mathcal{F}_{ij}$ represent the open face which is shared by two substructures Ω_i and Ω_j. Let $\mathcal{W}_i$ denote the wire basket of the subdomain Ω_i, i.e. the union of the closures of the edges of $\partial\Omega_i$. We define the *wire basket* by $\mathcal{W} = \cup \mathcal{W}_i \backslash \partial\Omega$. The sets of nodes on $\mathcal{F}_{ij}$, $\mathcal{W}$, and $\mathcal{W}_i$ are denoted by $\mathcal{F}_{ij,h}$, $\mathcal{W}_h$, and $\mathcal{W}_{i,h}$.

• **A face and wire basket based coarse space.** The first coarse space is denoted by V_{-1}^F, and is based on the wire basket $\mathcal{W}_h$ and the average over each face $\mathcal{F}_{ij,h}$. This space can conveniently be defined as the range of an interpolation operator $I_h^F : V_0^h \to V_{-1}^F$, defined by

$$I_h^F u(x)_{|\bar{\Omega}_i} = \sum_{x_p \in \mathcal{W}_{i,h}} u(x_p)\varphi_p(x) + \sum_{\mathcal{F}_{ij} \subset \partial\Omega_i} \bar{u}_{\mathcal{F}_{ij}} \theta_{\mathcal{F}_{ij}}(x).$$

Here, $\varphi_p(x)$ is the discrete harmonic function into Ω_i which equals 1 at x_p and vanishes elsewhere on $\partial\Omega_{i,h}$. $\bar{u}_{\mathcal{F}_{ij}}$ is the average value of u on $\mathcal{F}_{ij,h}$, and $\theta_{\mathcal{F}_{ij}}(x)$ the discrete harmonic function in Ω_i which equals 1 on $\mathcal{F}_{ij,h}$ and is zero on $\partial\Omega_{i,h}\backslash\mathcal{F}_{ij,h}$.

We define the bilinear form by

$$\begin{aligned} b_{-1}^F(u,u) \ = \ & \sum_i \rho_i \Big\{ \sum_{x_p \in \mathcal{W}_{i,h}} h\,(u(x_p) - \bar{u}_i)^2 \\ & + H(1 + \log H/h) \sum_{\mathcal{F}_{ij} \subset \partial\Omega_i} (\bar{u}_{\mathcal{F}_{ij}} - \bar{u}_i)^2 \Big\}, \end{aligned}$$

where $\bar{u}_i$ is the average of the discrete values of u over $\partial\Omega_{i,h}$.

• **A face, edge, and vertex based coarse space.** We can decrease the dimension of the coarse space given above and define another coarse space denoted by V_{-1}^E. Rather than using the values of all the nodes on the edges as degrees of freedom, only one degree of freedom per edge, an average value is used; see [**3**].

• **A Neumann-Neumann coarse space.** We can also consider the coarse space V_{-1}^{NN}; see [**4**]. This space is of minimal dimension with only one degree of freedom per substructure.

• **A wire basket based coarse space.** Finally, we consider a coarse space V_{-1}^W, due to Barry Smith see [**7**], or [**3**]. It is based only on the wire basket $\mathcal{W}_h$.

REMARK 1. *We can decrease the complexity of our algorithm by considering approximate discrete harmonic extension given by simple explicit formulas in* [**2**].

5. Quasi-Monotone Coefficients and an Optimal Algorithm

In this section, we indicate that if the coefficients ρ_i satisfy certain assumptions, the L_ρ^2-projection is stable and we can use the space of piecewise linear functions $V^H(\Omega)$ as a coarse space to obtain an *optimal* multilevel preconditioner.

Let $\{\mathcal{V}_m\}_{m=1}^M$ be the set of substructure vertices. We also include the vertices that are on $\partial\Omega$. Let Ω_{m_i}, $i = 1, \cdots, s(m)$, denote the substructures that have the vertex $\mathcal{V}_m$ in common, and let the ρ_{m_i} denote their coefficients. Let Ω_m' be the interior of the closure of the union of these substructures Ω_{m_i}, i.e. the interior of $\cup_{i=1}^{s(m)}\bar{\Omega}_{m_i}$. By using the fact that all substructures are simplices, we see that each Ω_{m_i} has a whole face in common with $\partial\Omega_m'$. Thus, the vertex $\mathcal{V}_m$ is the only internal cross point in $\bar{\Omega}_m'$.

DEFINITION 1. *For each Ω_m', we order its substructures such that*

$$\rho_{m_1} = \max_{i:1,\cdots,s(m)} \rho_{m_i}.$$

We say that a distribution of ρ_{m_i} is quasi-monotone on Ω'_m if for each $i = 1, \cdots, s(m)$, there exists a sequence i_j, $j = 1, \cdots, R$, with

$$(6) \qquad \rho_{m_i} = \rho_{m_{i_R}} \le \cdots \le \rho_{m_{i_{j+1}}} \le \rho_{m_{i_j}} \le \cdots \le \rho_{m_{i_1}} = \rho_{m_1},$$

where the substructures $\Omega_{m_{i_j}}$ and $\Omega_{m_{i_{j+1}}}$ have a face in common. If the vertex $\mathcal{V}_m \in \partial\Omega$, then we additionally assume that $\partial\Omega_{m_1} \cap \partial\Omega$ contains a face for which $\mathcal{V}_m$ is a vertex.

A distribution ρ_i on Ω is quasi-monotone if it is quasi-monotone on each Ω'_m.

THEOREM 2. *For a quasi-monotone distribution of the coefficients on Ω, we have*

$$(7) \qquad \|(I - Q_\rho^H)u\|_{L^2_\rho(\Omega)} \preceq H \, |u|_{H^1_\rho(\Omega)} \; \forall u \in V_0^h(\Omega).$$

Here, Q_ρ^H is the weighted L^2-projection from $V_0^h(\Omega)$ to $V_0^H(\Omega)$.

THEOREM 3. *Let $T^H = T^X$ be defined by (4) with $V_{-1}^X = V^H(\Omega)$ and $b_{-1}(\cdot, \cdot) = a(\cdot, \cdot)$. For a quasi-monotone distribution of the coefficients on Ω, we have*

$$C_5 \, a(u, u) \le a(T^H u, u) \le C_6 \, a(u, u) \; \forall u \in V_0^h(\Omega).$$

REMARK 2. *The analysis can be extended to problems with Neumann or mixed boundary conditions, and quasi-monotone coefficients. In this case, we also obtain an optimal method.*

REFERENCES

1. James H. Bramble and Joseph E. Pasciak and Jinchao Xu, *Parallel Multilevel Preconditioners*, Math. Comp. 55(1990), 1–22.
2. Maksymilian Dryja, Marcus Sarkis, and Olof B. Widlund, *Multilevel Schwarz Methods for Elliptic Problems with Discontinuous Coefficients in Three Dimensions*, TR #662, Courant Institute, NYU, March, 1994.
3. Maksymilian Dryja, Barry F. Smith, and Olof B. Widlund, *Schwarz Analysis of Iterative Substructuring Algorithms for Elliptic Problems in Three Dimensions*, TR #638, CS Department, Courant Institute, May 1993. To appear in SIAM J. Numer. Anal.
4. Maksymilian Dryja and Olof B. Widlund, *Schwarz Methods of Neumann-Neumann Type for Three-Dimensional Elliptic Finite Element Problems*, TR #626, CS Department, Courant Institute, March 1993. To appear in Comm. Pure Appl. Math.
5. Maksymilian Dryja and Olof B. Widlund, *Additive Schwarz Methods for Elliptic Finite Element Problems in Three Dimensions*, in Fifth International Symposium on Domain Decomposition Methods for Partial Differential Equations, Tony F. Chan, David E. Keyes, Gérard A. Meurant, Jeffrey S. Scroggs, and Robert G. Voigt, editors, SIAM, Philadelphia, PA, 1992,
6. Marcus Sarkis *Multilevel Methods for P_1 Nonconforming Finite Elements and Discontinuous Coefficients in Three Dimensions*, these proceedings.
7. Barry F. Smith, *A Domain Decomposition Algorithm for Elliptic Problems in Three Dimensions*, Numer. Math. 60(1991), 219–234.
8. Olof B. Widlund *Exotic Coarse Spaces for Schwarz Methods for Lower Order and Spectral Finite Elements*, these proceedings.
9. Jinchao Xu, *Iterative Methods by Space Decomposition and Subspace Correction*, SIAM Review 34(1992), 581–613.
10. Xuejun Zhang, *Multilevel Schwarz Methods*, Numer. Math. 63(1992), 521–539.

DEPARTMENT OF MATHEMATICS, WARSAW UNIVERSITY, BANACHA 2, 02-097 WARSAW, POLAND.
E-mail address: dryja@mimuw.edu.pl

Contemporary Mathematics
Volume 180, 1994

Multilevel Methods for Elliptic Problems on Domains not Resolved by the Coarse Grid

RALF KORNHUBER AND HARRY YSERENTANT

ABSTRACT. Elliptic boundary value problems are frequently posed on complicated domains, which cannot be covered by a simple coarse initial grid as is needed for multigrid–like iterative methods. In the present article, this problem is resolved for selfadjoint second order problems and Dirichlet boundary conditions. The idea is to construct appropriate subspace decompositions of the corresponding finite element spaces by way of an embedding of the domain under consideration into a simpler domain like a square or a cube. Then the general theory of subspace correction methods can be applied.

1. Introduction

By definition, a multilevel method for finite element equations is based on a sequence of refined triangulations. One starts with a coarse initial mesh crudely reflecting the properties of the boundary value problem under consideration. For the usual mathematical test problems like the Laplace equation on the unit square or on an L–shaped domain, this initial triangulation consists of only very few elements. Real–life problems, on the other hand, are often posed on very complicated regions which can only be described by hundreds or thousands of finite elements.

A possibility for construction of fast solvers for the resulting linear systems is to disregard any refinement history of the underlying grids and to decompose these grids a posteriori. This leads to some kind of algebraic multigrid method. A recent approach is described in the paper of Bank and Xu in these proceedings.

In the present article, we follow the opposite direction of approximating complicated geometries in the course of the refinement process. We restrict our attention to second–order selfadjoint elliptic boundary value problems and Dirichlet boundary conditions. For this special class of problem we are able to construct

1991 *Mathematics Subject Classification.* Primary 65N55; Secondary 65F10, 65N30 .
This paper is in final form, and no version of it will be submitted for publication elsewhere.

nearly optimal iterative methods that do not depend on the regularity of the boundary. For plane domains, even unphysical boundary conditions at a single point (which have no continuous counterpart) are allowed.

The idea is to construct appropriate subspace decompositions of the corresponding finite element spaces by way of an embedding of the domain under consideration into a simpler domain. Then the general theory of additive and multiplicative subspace correction methods can be applied directly. For a survey of this machinery, see the review articles of Xu [10] and of Yserentant [14], for example.

Our construction has originally been motivated by the numerical solution of obstacle problems; see [6], [5], [7]. In this application, the domain on which linear elliptic problems have to be solved is the subdomain where the current approximate solution does not touch the obstacle. This domain is unknown in advance of the computation and for this reason usually has no exact representation on coarser grids.

2. The discrete elliptic problem

Let $\Omega \subseteq \mathbf{R}^2$ be a simple polygonal domain, e.g., a square. Let $\mathcal{T}_0$ be a coarse initial triangulation of Ω, which is refined several times, giving a sequence of triangulations $\mathcal{T}_0, \mathcal{T}_1, \ldots, \mathcal{T}_j$. Despite the fact that all techniques and estimates in this paper can easily be generalized to the case of nonuniformly refined grids, we assume for ease of presentation that the triangles are uniformly refined. The triangles in $\mathcal{T}_{k+1}$ are generated from the triangles in $\mathcal{T}_k$ by subdividing these triangles in the usual fashion into four congruent subtriangles.

With the triangulations $\mathcal{T}_k$ we associate finite element spaces $\mathcal{S}_k$, consisting of the continuous functions, which are linear on the triangles in $\mathcal{T}_k$. The functions $u \in \mathcal{S}_k$ are given by their values at the nodes $x_i^{(k)}$, which are the vertices of the triangles in $\mathcal{T}_k$. With every node $x_i^{(k)}$ we associate a basis function $\psi_i^{(k)}$ of $\mathcal{S}_k$, taking the value 1 at this node and the value 0 at all other nodes.

Let Ω' be an arbitrarily complicated, nasty subdomain of Ω, possibly without any regularity property. Consider the nested subspaces

$$(2.1) \qquad \mathcal{S}_k' = \operatorname{span}\{\, \psi_i^{(k)} \mid \psi_i^{(k)}(x) = 0 \text{ for } x \notin \Omega'\,\}$$

of the spaces $\mathcal{S}_k$ and in particular the space $\mathcal{S}' = \mathcal{S}_j'$. Our aim is to construct and to analyze fast iterative solution procedures for the discrete boundary value problem to find the function $u \in \mathcal{S}'$ satisfying

$$(2.2) \qquad a(u, v) = f^*(v), \quad v \in \mathcal{S}'.$$

Here f^* is a given linear functional on $\mathcal{S}'$. $a(u, v)$ denotes a symmetric coercive bilinear form on $\mathcal{S}'$ with the property that

$$(2.3) \qquad \delta |u|_1^2 \leq a(u, u) \leq M |u|_1^2$$

holds for all $u \in \mathcal{S}'$. δ and M are positive constants and $|u|_1 = |u|_{1;\Omega'}$ denotes the usual seminorm on $H^1(\Omega')$ given by

$$(2.4) \qquad |u|_1^2 = \sum_{i=1}^{2} \int_{\Omega'} |D_i u(x)|^2 \mathrm{d}x.$$

We assume that $|\cdot|_1$ is a norm on $\mathcal{S}'$.

3. Subspace correction methods

Let $\langle u, v \rangle$ be an arbitrary inner product on $\mathcal{S}'$. Define the operator $A : \mathcal{S}' \to \mathcal{S}'$ by the relation

$$(3.1) \qquad \langle Au, v \rangle = a(u, v), \quad v \in \mathcal{S}',$$

and determine the right–hand side $f \in \mathcal{S}'$ by

$$(3.2) \qquad \langle f, v \rangle = f^*(v), \quad v \in \mathcal{S}'.$$

Then the discrete boundary value problem (2.2) is equivalent to the operator equation

$$(3.3) \qquad Au = f.$$

To solve (3.3) iteratively, one specifies subspaces

$$(3.4) \qquad \mathcal{W}_k \subseteq \mathcal{S}'_k, \quad k = 0, 1, \dots, j,$$

where the case $\mathcal{W}_k = \mathcal{S}'_k$ is explicitly included. Introducing the a–orthogonal projections $P_k : \mathcal{S}' \to \mathcal{W}_k$ by

$$(3.5) \qquad a(P_k u, w_k) = a(u, w_k), \quad w_k \in \mathcal{W}_k,$$

the basic building blocks of the iterative methods considered here, are the *subspace corrections*

$$(3.6) \qquad \widetilde{u} \leftarrow \widetilde{u} + P_k(u - \widetilde{u})$$

with respect to the spaces $\mathcal{W}_k$. After the subspace correction (3.6), the error $u - \widetilde{u}$ between the exact solution u and the new approximation $\widetilde{u}$ is a–orthogonal to the space $\mathcal{W}_k$. Utilizing the projections $Q_k : \mathcal{S}' \to \mathcal{W}_k$, given by

$$(3.7) \qquad \langle Q_k u, w_k \rangle = \langle u, w_k \rangle, \quad w_k \in \mathcal{W}_k,$$

and the operators $A_k : \mathcal{W}_k \to \mathcal{W}_k$, defined analogously to $A : \mathcal{S}' \to \mathcal{S}'$, the correction step (3.6) can be written as

$$(3.8) \qquad \widetilde{u} \leftarrow \widetilde{u} + A_k^{-1} Q_k(f - A\widetilde{u}).$$

For large subspaces $\mathcal{W}_k$, as considered here, the correction steps (3.8) are too expensive. Therefore, one replaces these correction steps by the approximate correction steps

$$(3.9) \qquad \widetilde{u} \leftarrow \widetilde{u} + B_k^{-1} Q_k (f - A\widetilde{u})$$

with symmetric and positive definite operators $B_k : \mathcal{W}_k \rightarrow \mathcal{W}_k$. These operators should have the property that the correction term

$$(3.10) \qquad d_k = B_k^{-1} Q_k (f - A\widetilde{u})$$

can easily be computed as the solution of the linear system

$$(3.11) \qquad \langle B_k d_k, w_k \rangle = \langle f - A\widetilde{u}, w_k \rangle, \quad w_k \in \mathcal{W}_k.$$

It should be noted that the evaluation of

$$(3.12) \qquad \langle f - A\widetilde{u}, w_k \rangle = f^*(w_k) - a(\widetilde{u}, w_k)$$

requires neither an explicit knowledge of the abstract operator A nor of the right hand side f.

If the single subspace correction steps (3.9) are repeated in a cyclic order, one gets a *multiplicative subspace correction method*. These methods generalize the classical Gauss–Seidel iteration, where the subspaces are one–dimensional and are spanned by basis functions. The corresponding *additive subspace correction method*

$$(3.13) \qquad \widetilde{u} \leftarrow \widetilde{u} + \sum_{k=0}^{j} B_k^{-1} Q_k (f - A\widetilde{u})$$

is a Jacobi–type iteration. It is usually applied in form of a preconditioner for the conjugate gradient method.

A classical multigrid method for the solution of (2.2) and (3.3), respectively, would correspond to the choice $\mathcal{W}_k = \mathcal{S}_k'$ and to a simple Jacobi– or symmetric Gauss–Seidel iteration B_k^{-1}.

The general framework outlined here arose from the abstract formulation of domain decomposition methods. Breakthroughs in the analysis of these methods were the papers [2] and [3] of Bramble, Pasciak, Wang and Xu, in which the first satisfying convergence proof for the multiplicative case was given. For detailed references and a thorough discussion, we refer to [10], [14], or to Oswald's paper in these proceedings.

4. Subspace decompositions

The basic step in the convergence analysis of the subspace correction methods is to find a subspace decomposition

$$(4.1) \qquad \mathcal{S}' = \mathcal{V}_0 \oplus \mathcal{V}_1 \oplus \ldots \oplus \mathcal{V}_j$$

of the space $\mathcal{S}' = \mathcal{S}'_j$ into subspaces

$$(4.2) \qquad \qquad \mathcal{V}_k \subseteq \mathcal{W}_k, \quad k = 0, 1, \ldots, j,$$

such that

$$(4.3) \qquad \qquad \sum_{k=0}^{j} \langle B_k v_k, v_k \rangle \leq K_1 \, \| \sum_{k=0}^{j} v_k \|^2$$

holds for all $v_k \in \mathcal{V}_k$. The norm on the right hand side of this equation is the energy norm

$$(4.4) \qquad \qquad \|u\| = a(u, u)^{1/2}$$

associated with the boundary value problem under consideration. The constant K_1 (possibly still depending on the number j of refinement levels) describes the *stability of the decomposition* (4.1).

If the B_k are taken as usual (Jacobi method, symmetrized Gauss–Seidel iterations, etc.) and if the level 0 correction is exactly computed, the estimate (4.3) is equivalent to the estimate

$$(4.5) \qquad \qquad |v_0|_1^2 + \sum_{k=1}^{j} 4^k \|v_k\|_0^2 \leq \tilde{K}_1 \, | \sum_{k=0}^{j} v_k |_1^2$$

for the functions $v_k \in \mathcal{V}_k$, or follows at least from this estimate. The (semi–) norm on the right–hand side of (4.5) is given by (2.4), and the inner product

$$(4.6) \qquad \qquad (u, v) = \sum_{T \in \mathcal{T}_0} \frac{1}{\operatorname{area}(T)} \int_T uv \, dx$$

induces the norm $\|v\|_0 = (v, v)^{1/2}$. The purpose of the weights $1/\operatorname{area}(T)$ is to make the estimates independent of the size of the triangles in the initial triangulation. In the three–dimensional case, these factors have to be replaced by other factors behaving like $1/\operatorname{diam}(T)^2$. The factors $4^k = (2^k)^2$ arise from the fact that the diameters of the triangles shrink by the factor 2 from one refinement level to the next. For a detailed exposition of the relation between (4.3) and (4.5), see [14].

For the analysis of the additive version one needs the second essential condition that

$$(4.7) \qquad \qquad \| \sum_{k=0}^{j} w_k \|^2 \leq K_2 \sum_{k=0}^{j} \langle B_k w_k, w_k \rangle$$

holds for all $w_k \in \mathcal{W}_k$, or equivalently, as above, the estimate

$$(4.8) \qquad \qquad | \sum_{k=0}^{j} w_k |_1^2 \leq \tilde{K}_2 \Big\{ |w_0|_1^2 + \sum_{k=1}^{j} 4^k \|w_k\|_0^2 \Big\}.$$

As (4.8) is known to hold for all functions $w_k \in \mathcal{S}_k$ (the spaces associated with the basic domain Ω) with a constant $\tilde{K}_2$ not depending on j, nothing has to be shown here. For a proof of (4.8), see [1] or [14]. Similarly, the Cauchy–Schwarz inequality, needed for the analysis of the multiplicative procedure (see [10] or [14], for example), is a direct consequence of the corresponding property for the full spaces $\mathcal{S}_k$.

The speed of convergence of the optimally scaled additive method (3.13), or of its conjugate gradient–accelerated version, can be estimated in terms of the constants K_1 and K_2 in (4.3) and (4.7). Similar results hold for the multiplicative version. For a detailed exposition, we refer again to the survey articles [10] and [14].

5. A subspace decomposition by interpolation operators

The remaining task is to construct a decomposition (4.1) of the discrete solution space $\mathcal{S}'$ with the property (4.5). In this section, we consider decompositions generated by interpolation–like operators $I'_k : \mathcal{S}' \to \mathcal{S}'_k$ given by

$$(5.1) \qquad (I'_k u)(x_i^{(k)}) = \begin{cases} u(x_i^{(k)}) & \psi_i^{(k)} \in \mathcal{S}'_k \\ 0 & \text{otherwise}. \end{cases}$$

Recall that such splittings are related to the hierarchical basis method [11], [13]. Because of the decomposition

$$(5.2) \qquad u = I'_0 u + \sum_{k=1}^{j} (I'_k u - I'_{k-1} u)$$

of the functions $u \in \mathcal{S}'$, the space $\mathcal{S}'$ is the direct sum of $\mathcal{S}'_0$ and of the subspaces

$$(5.3) \qquad \mathcal{V}_k = \{ I'_k u - I'_{k-1} u \mid u \in \mathcal{S}' \}$$

of the spaces $\mathcal{S}'_k$.

The analytic foundation for the proof of the stability of this decomposition is the following

LEMMA 5.1. There exists a constant c depending only on the shape regularity of the triangles $T \in \mathcal{T}_k$ such that

$$(5.4) \qquad |u(x) - u(y)| \leq c \sqrt{j-k+1} \, |u|_{1;T}$$

holds for all functions $u \in \mathcal{S}_j$ and all points $x, y \in T$.

This estimate can be proved along the lines given in [11]. The fact that we are dealing with two space dimensions enters the proof of this lemma. For three space dimensions, the estimate (5.4) does not hold.

With Lemma 5.1, we can estimate the norm of the modified interpolation operators (5.1). This is, in a certain sense, the key result of this section.

LEMMA 5.2. There exists a constant c depending only on the shape regularity of the triangles in $\mathcal{T}_0$ such that

$$(5.5) \qquad\qquad |I_k' u|_1 \le c \sqrt{j-k+1}\, |u|_1$$

holds for all functions u in the subspace $\mathcal{S}' = \mathcal{S}_j'$ of $\mathcal{S}_j$.

PROOF. We estimate $|I_k' u|_{1;T}^2$ for the triangles $T \in \mathcal{T}_k$. Two cases have to be distinguished. The first case is that T is an "interior" triangle of Ω', i.e., that the basis functions $\psi_i^{(k)}$ associated with all three vertices of T belong to $\mathcal{S}_k'$. In this case, the restriction of $I_k' u$ to T is simply the linear interpolant of u at the vertices of T. Therefore, the estimate

$$(5.6) \qquad\qquad |I_k' u|_{1;T} \le c_1 \sqrt{j-k+1}\, |u|_{1;T}$$

follows from Lemma 5.1 by a simple scaling argument.

If there is a basis function $\psi_i^{(k)}$ associated with a vertex $x_i^{(k)}$ of T that does not belong to $\mathcal{S}_k'$, the situation is slightly more complicated. In this case, there exists at least one point $\bar{x} \notin \Omega'$ in $T_1 = T$ or in another triangle $T_1 \in \mathcal{T}_k$ with vertex $x_i^{(k)}$. The functions in $\mathcal{S}'$ vanish at $\bar{x}$. Therefore, for every $x \in T$, one gets

$$
\begin{aligned}
|u(x)| &\le\; |u(x) - u(x_i^{(k)})| + |u(x_i^{(k)}) - u(\bar{x})| \\
&\le\; c\sqrt{j-k+1}\,\{\,|u|_{1;T} + |u|_{1;T_1}\,\}.
\end{aligned}
$$

As above, this yields the estimate

$$(5.7) \qquad\qquad |I_k' u|_{1;T} \le c_2 \sqrt{j-k+1}\, |u|_{1;T \cup T_1}.$$

As each triangle in $\mathcal{T}_k$ intersects only a limited number of other triangles in $\mathcal{T}_k$, the proposition follows from (5.6) and (5.7). □

The functions v_k in the space $\mathcal{V}_k$ satisfy the estimate

$$(5.8) \qquad\qquad 4^k \|v_k\|_0^2 \le c\, |v_k|_1^2$$

with a constant c depending again only on the shape regularity of the triangles under consideration. This estimate relies on the observation that every node $x_i^{(k)}$ has a neighbor $x_l^{(k)}$ of first or second degree at which the functions in $\mathcal{V}_k$ vanish. The scaling factor 4^k depends only on the number k of refinement levels, because the L_2–like norm induced by the inner product (4.6) is scaled by the areas of the triangles in the initial triangulation.

As an immediate consequence of Lemma 5.2 and of (5.8) (compare [11], [12]), we can state

THEOREM 5.3. There exists a constant C depending only on the shape regularity of the triangles in $\mathcal{T}_0$ (and not on the domain Ω'!) such that

$$(5.9) \qquad |I_0'u|_1^2 + \sum_{k=1}^{j} 4^k \|I_k'u - I_{k-1}'u\|_0^2 \le C j^2 |u|_1^2$$

holds for all functions $u \in \mathcal{S}_j'$.

Hence, the decomposition of $\mathcal{S}'$ into $\mathcal{S}_0'$ and the subspaces (5.3) is stable in the sense of (4.3) with a constant

$$(5.10) \qquad K_1 \sim j^2$$

growing only logarithmically with $1/h \sim 2^j$. Therefore, for any choice

$$(5.11) \qquad \mathcal{V}_k \subseteq \mathcal{W}_k \subseteq \mathcal{S}_k',$$

one gets a nearly optimal multilevel method. The number of iteration steps, needed to reduce the error by a given factor, increases at most logarithmically, when the gridsize tends to zero. Classical multigrid methods correspond to the choice $\mathcal{W}_k = \mathcal{S}_k'$, whereas the other extreme $\mathcal{W}_k = \mathcal{V}_k$ leads hierarchical basis type iterative methods. This considerably generalizes related results in [6].

Note that absolutely no regularity assumption concerning the boundary of the domain $\Omega' \subseteq \Omega$ enters. Even unphysical boundary conditions at a single point, which have no continuous counterpart, are allowed. On the other hand, the construction works only for two space dimensions.

6. The decomposition of the solution space by L_2–like projections

Assuming a certain regularity of Ω', one can also utilize the L_2–like decomposition

$$(6.1) \qquad u = Q_0'u + \sum_{k=1}^{j}(Q_k'u - Q_{k-1}'u)$$

of the functions $u \in \mathcal{S}_j'$, where the $Q_k' : \mathcal{S}' \to \mathcal{S}_k'$ are the orthogonal projections with respect to the inner product (4.6). Recall that a decomposition of this type played a crucial role in [4]. It turns out that the decomposition (6.1) of $\mathcal{S}_j'$ into $\mathcal{S}_0'$ and the subspaces

$$(6.2) \qquad \mathcal{V}_k = \{Q_k'u - Q_{k-1}'u) \mid u \in \mathcal{S}'\}$$

is stable, if $\mathbf{R}^2 \backslash \Omega'$ is "rich enough". Unlike the construction in the last section, this approach works also for three space dimensions.

For simplicity, let $\mathcal{S}_k'$ be a subspace of $H_0^1(\Omega) \subseteq H^1(\mathbf{R}^2)$. We call $T \in \mathcal{T}_k$ a boundary triangle of Ω', if at least one basis function $\psi_i^{(k)}$ associated with a vertex $x_i^{(k)}$ of T is not contained in $\mathcal{S}_k'$. We make the following regularity assumption on Ω': For every boundary triangle, there exists a circle B such that every triangle in $\mathcal{T}_k$ intersecting T is completely overlapped by B and such that the area of B can be estimated as

$$(6.3) \qquad \text{meas } B \leq c_1 \text{ meas } B \backslash \Omega',$$

and the diameter of B as

$$(6.4) \qquad \text{diam } B \leq c_2 \text{ diam } T.$$

The property (6.3) excludes that the complement of Ω' consists of single points or lines. This was allowed in the last section. Nevertheless the condition is extremely weak. It covers domains which some people would call "fractal". Oswald [9] discusses a related condition for the solution of our problem.

For a given boundary triangle $T \in \mathcal{T}_k$ and for the associated circle B, one can define the operator $\Pi : L_2(B) \to L_2(B)$ by

$$(6.5) \qquad (\Pi u)(x) = \frac{1}{\text{meas } B \backslash \Omega'} \int_{B \backslash \Omega'} u(y) \, dy .$$

It maps the functions in $L_2(B)$ to constants. The square of the L_2–norm of this operator is

$$(6.6) \qquad \frac{\text{meas } B}{\text{meas } B \backslash \Omega'} \leq c_1 .$$

As Π reproduces constant functions, (6.6) yields

$$(6.7) \qquad \|u - \Pi u\|_{L_2(B)} \leq (1 + \sqrt{c_1}) \inf_{\alpha \in \mathbf{R}} \|u - \alpha\|_{L_2(B)}.$$

With help of the Poincaré–inequality for the circle B and with (6.4), one obtains the error estimate

$$(6.8) \qquad \|u - \Pi u\|_{0;B} \leq c \, 2^{-k} |u|_{1;B}, \quad u \in H^1(B),$$

with respect to the weighted L_2–norm induced by the inner product (4.6) on the left–hand side. Therefore, the functions $u \in H^1(\mathbf{R}^2)$ vanishing on $B \backslash \Omega'$ and in particular the functions in $u \in S'$ satisfy

$$(6.9) \qquad \|u\|_{0;B} \leq c \, 2^{-k} |u|_{1;B} .$$

The constant c in this estimate depends only on the constants c_1 and c_2 in (6.3) and (6.4), respectively, and on the shape regularity of the triangles.

Next, we introduce the L_2–bounded quasi–interpolants $M_k' : S' \to S_k'$ by

$$(6.10) \qquad M_k' u = \sum_{\psi_i^{(k)} \in S_k'} \frac{(u, \psi_i^{(k)})}{(1, \psi_i^{(k)})} \, \psi_i^{(k)} .$$

Then, utilizing (6.9) and the Poincaré–inequality, one can show the error estimate

$$(6.11) \qquad \|u - M_k' u\|_0 \leq \hat{c} \, 2^{-k} |u|_1 .$$

The proof relies essentially on the fact that the operators M'_k reproduce functions on a triangle $T \in \mathcal{T}_k$, which are constant in a neighborhood of T. Details on this technique can be found in [12].

The estimate (6.11) implies the error estimate

$$(6.12) \qquad \|u - Q'_k u\|_0 \leq \hat{c}\, 2^{-k} |u|_1$$

for the orthogonal projections $Q'_k : \mathcal{S}' \to \mathcal{S}'_k$. Using in addition that

$$(6.13) \qquad |u - Q'_0 u|_1^2 \leq \tilde{K}_1 \sum_{k=1}^{j} 4^k \|Q'_k u - Q'_{k-1} u\|_0^2$$

holds for $u \in \mathcal{S}'$ (this is a consequence of (4.8)), one finally obtains

THEOREM 6.1. Provided that the subregion Ω' has the properties described at the beginning of this section, there exists a constant C such that

$$(6.14) \qquad |Q'_0 u|_1^2 + \sum_{k=1}^{j} 4^k \|Q'_k u - Q'_{k-1} u\|_0^2 \leq C j \, |u|_1^2$$

holds for all functions $u \in \mathcal{S}'$.

We remark that the optimality of the decomposition (6.1) can be shown in a similar way; see [9]. Trying to keep the conditions on the boundary of Ω' as weak (and simple) as possible, we did not attempt to prove such a result here.

Based on Theorem 6.1, one obtains nearly optimal subspace correction methods for $\mathcal{W}_k = \mathcal{S}'_k$; again the number of iteration steps needed to reduce the error by a given factor grows only at most logarithmically, when the gridsize decreases.

7. Numerical experiences and final remarks

As an illustrating example, we consider the unit square $\Omega = (0,1) \times (0,1)$ with an initial triangulation consisting of 16 congruent rectangular triangles and with final triangulations $\mathcal{T}_j$ obtained by a successive uniform refinement of $\mathcal{T}_0$ as described in Section 2. The bilinear form $a(u,v)$ of Section 2 is the Dirichlet–integral, leading to a boundary value problem for the Laplace equation. The right–hand side $f^*(v)$ is given by the integral of v over Ω.

We compared the convergence rates of a classical multigrid method (the multiplicative subspace correction method corresponding to $\mathcal{W}_k = \mathcal{S}'_k$ and the symmetric Gauss–Seidel–method as approximate solver B_k^{-1}) for $\Omega'_1 = \Omega$ with the rates obtained for $\Omega'_2 = (0, 1-2^{-(j+2)}) \times (0,1)$ as the other extreme. The observed convergence rates are shown in Table 1.

The convergence rates for the two domains differ considerably, a fact which can easily be explained. Of course, the unit square Ω'_1 gives an optimal performance, whereas, for Ω'_2, the supports of the functions in $\mathcal{S}'_k$ overlap only the rectangle $(0, 1-2^{-(k+2)}) \times (0,1)$ so that $\Omega'_2 = (0, 1-2^{-(j+2)}) \times (0,1)$ is not exhausted very well. The convergence rates for Ω_2' are typical for the convergence rates that we observed for many other domains with critical boundaries.

TABLE 1. The convergence rates for the first example

	j=2	j=3	j=4	j=5	j=6	j=7
Ω_1'	0.28	0.30	0.31	0.32	0.33	0.33
Ω_2'	0.49	0.58	0.65	0.71	0.75	0.78

One possibility to improve the convergence rate is to use enlarged correction spaces $\mathcal{W}_k = \mathcal{S}_k^\vee$, as obtained from the extension of $\mathcal{S}_k'$ by truncated basis functions of $\mathcal{S}_k$ (c.f. [7], [8]). In this way, one reaches nearly the same convergence rates as for regular problems. For the domain Ω_2' above, the convergence rates are asymptotically equal to the rates for $\Omega_1' = \Omega$.

As a second example, we consider the Laplacian on a subdomain Ω' of the unit square Ω with a "fractal" boundary. Starting with the same initial triangulation $\mathcal{T}_0$ as in the first example, the domain Ω' is approximated by a sequence of domains Ω_j', which are triangulated by subsets $\mathcal{T}_j' \subset \mathcal{T}_j$. The boundaries of Ω_4' and Ω_5' are shown in Figure 1. We solved the boundary value problem on these domains.

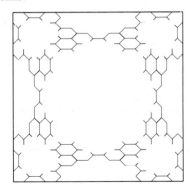 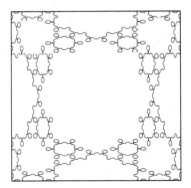

FIGURE 1. The approximate boundaries of the fractal domain

The convergence rates of the multiplicative methods with $\mathcal{W}_k = \mathcal{S}_k'$ and with suitably extended spaces $\mathcal{W}_k = \mathcal{S}_k^\vee$ are given in Table 2.

TABLE 2. The convergence rates for the fractal domain

	j=2	j=3	j=4	j=5	j=6	j=7
$\mathcal{S}_k'$	0.30	0.61	0.58	0.68	0.65	0.71
$\mathcal{S}_k^\vee$	0.28	0.30	0.32	0.32	0.33	0.34

Difficulties in the analysis of the modified version do not arise from the stability estimate (4.3), because one can use the same subspaces $\mathcal{V}_k \subseteq \mathcal{S}_k' \subseteq \mathcal{S}_k^\vee$ as before, but from the estimates (4.7), (4.8) and the generalized Cauchy–Schwarz–inequality mentioned above, respectively. The optimal estimate (4.8) and this generalized Cauchy–Schwarz–inequality do not transfer from the spaces $\mathcal{S}_k$ to the spaces $\mathcal{W}_k = \mathcal{S}_k^\vee$ as in the case of correction spaces $\mathcal{W}_k \subseteq \mathcal{S}_k$.

However, utilizing a result like Theorem 5.4 in [14] (which is essentially taken from [2], [3]), these problems can easily be remedied at the cost of an additional power of j for the case of multiplicative methods. In addition, a very crude argument, using only the triangle inequality, shows that (4.7) still holds with a constant $K_2 \sim j$. Hence, a similar result can also be proven for the additive version.

References

1. Bornemann, F.A., Yserentant, H., *A basic norm equivalence for the theory of multilevel methods*, Numer. Math. **64** (1993), 445–476.
2. Bramble, J.H., Pasciak, J.E., Wang, J., Xu, J., *Convergence estimates for product iterative methods with application to domain decomposition*, Math. Comp. **57** (1991), 1–21.
3. Bramble, J.H., Pasciak, J.E., Wang, J., Xu, J., *Convergence estimates for multigrid algorithms without regularity assumptions*, Math. Comp. **57** (1991), 23–45.
4. Bramble, J.H., Pasciak, J.E., Xu, J., *Parallel multilevel preconditioners*, Math. Comp. **55** (1990), 1–22.
5. Erdmann, B., Hoppe, R.H.W., Kornhuber, R., *Adaptive multilevel–methods for obstacle problems in three space dimensions*, to appear in: Proceedings of the Ninth GAMM–Seminar (W. Hackbusch and G. Wittum, eds.), Notes on Numerical Fluid Mechanics Vol. 46, Vieweg, Braunschweig and Wiesbaden, 1994.
6. Hoppe, R.H.W., Kornhuber, R., *Adaptive multilevel–methods for obstacle problems*, SIAM J. Numer. Anal. **31** (1994), 301–323.
7. Kornhuber, R., *Monotone multigrid methods for elliptic variational inequalities I*, to appear in Numer. Math.
8. Kornhuber, R., *Monotone multigrid methods for elliptic variational inequalities II*, Preprint SC 93–19, Konrad–Zuse–Zentrum für Informationstechnik Berlin, Berlin, 1993.
9. Oswald, P., *Stable subspace splittings for Sobolev spaces and their applications*, Forschungsergebnisse der Friedrich–Schiller–Universität Jena Math/93/7, Jena, 1993.
10. Xu, J., *Iterative methods by space decomposition and subspace correction*, SIAM Review **34** (1992), 581–613.
11. Yserentant, H., *On the multi–level splitting of finite element spaces*, Numer. Math. **49** (1986), 379–412.
12. Yserentant, H., *Two preconditioners based on the multi–level splitting of finite element spaces*, Numer. Math. **58** (1990), 163–184.
13. Yserentant, H., *Hierarchical bases*, In: ICIAM 91 (R.E. O'Malley, ed.), SIAM, Philadelphia, 1992, 256–276.
14. Yserentant, H., *Old and new convergence proofs for multigrid methods*, In: Acta Numerica 1993 (A. Iserles, ed.), Cambridge University Press, Cambridge, 1993, 285–326.

Konrad–Zuse–Zentrum für Informationstechnik Berlin, 10711 Berlin, Germany

Mathematisches Institut, Universität Tübingen, 72076 Tübingen, Germany

Contemporary Mathematics
Volume **180**, 1994

Three-dimensional Domain Decomposition Methods with Nonmatching Grids and Unstructured Coarse Solvers

P. LE TALLEC, T. SASSI, M. VIDRASCU

December 1993

ABSTRACT. This paper deals with finite element approximations which are defined independently on each subdomain and which do not match at the interfaces. Weak matching conditions then impose that the solution on two neighboring subdomains share the same L^2 projection on the mortar space which is defined on their common interface. For elliptic problems, we will prove that such discretization strategies lead to optimal approximation errors.

On the numerical side, the resulting discrete problem can be reduced to a problem set on the interface and associated to a generalized Schur complement matrix. This interface problem can then be solved by a preconditioned Conjugate Gradient method, using a Neumann-Neumann preconditioner with coarse grid correction. This technique, illustrated on several numerical examples, is proved to be optimal in such cases.

1. Introduction

Variational methods for decomposing and solving elliptic problems by domain decomposition techniques are well established. Most applications use discretization grids which are defined globally over the whole domain and then split into subdomains. In mechanics, this results into an overall conforming approximation of the velocity field. However, it might be more convenient and efficient to use approximations which are defined independently on each subdomain and which do not match at the interfaces. This allows the user to make local and adaptive change of designs, models, approximation strategies, or grids on one domain without modifying the other ones, provided that the user has found an adequate

---

1991 *Mathematics Subject Classification.* Primary 65F30; Secondary 65F10.
The paper is in its final version and will not be submitted for publication elsewhere.

way of imposing the weak continuity of both the fluxes and the velocities across such nonconforming interfaces.

In this paper, we will solve this problem by introducing a three-dimensional variant of the so-called mortar spaces. On the mathematical side, this technique imposes that the solution on two neighboring subdomains share the same L^2 projection on the mortar space which is defined on their common interface. For elliptic problems, we will prove that such discretization strategies lead to optimal approximation errors.

On the practical side, the resulting discrete problem can be reduced to a problem set on the interface and associated to a generalized Schur complement matrix. This interface problem can then be solved by a preconditioned Conjugate Gradient method, using for example a Neumann-Neumann preconditioner. This algorithm is very flexible and can be used both in a conforming and in a nonconforming framework. In both cases, we will add an unstructured coarse grid solver when using decompositions with a large number of subdomains, and prove the optimality of the resulting preconditioner.

2. Construction of the discrete problem

Nonoverlapping domain decomposition algorithms compute interface values which usually are the values of the unknowns on interface nodes shared by neighboring subdomains. But, it might be more efficient to consider interface nodes which do not match across the interfaces (Figure 1) [13], [8], [7], [9].

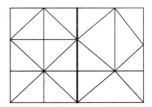

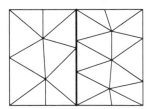

FIGURE 1. *Matching and Nonmatching Grids*

In such situations, the problem is then to impose global continuity of the unknowns. For this purpose, let us define the different subdomains Ω_i of Ω

either by direct juxtaposition of existing local meshes or by automatic partitioning of an existing global mesh. Such an automatic partitioning can be obtained for example by using spectral dissection techniques [18], [11] or K-means techniques (dynamic clusters) as classically used in data analysis. On all resulting subdomains, let us then introduce independent finite element spaces $_h(\Omega_i)$. Functions of these local finite element spaces are supposed to belong to $^1(\Omega_i)$ and to vanish on the local Dirichlet boundary $\partial\Omega_D \cap \partial\Omega_i$. To match all these spaces together, we finally introduce finite dimensional *mortar* spaces $_{ijh}$ defined on each interface F_{ij}, $i < j$, and weak traces Tr_{ijh} and Tr_{jih} defined by L^2 projection :

$$(2.1) \qquad \int_{F_{ij}} (Tr_{ijh}v_i - v_i)\mu_h = 0, \forall \mu_h \in \ _{ijh}, \forall v_i \in \ _h(\Omega_i), Tr_{ijh}v_i \in \ _{ijh},$$

$$(2.2) \qquad \int_{F_{ij}} (Tr_{jih}v_j - v_j)\mu_h = 0, \forall \mu_h \in \ _{ijh}, \forall v_j \in \ _h(\Omega_j), Tr_{jih}v_j \in \ _{ijh}.$$

Then, the space of finite element approximations of u over Ω can be defined by

$$(2.3) \qquad _h(\Omega) = \{(v_{ih})_i \in \prod_i \ _h(\Omega_i), Tr_{ijh}v_{ih} = Tr_{jih}v_{jh} \text{ on } F_{ij}, \forall i < j\}.$$

On this space, we wish to solve the second order elliptic variational problem

Find $u_h \in \ _h(\Omega)$ solution of

$$(2.4) \qquad a(u_h, v_h) := \sum_i \int_{\Omega_i} \sigma(\nabla u_h) : \nabla v_h = L(v_h), \forall v_h \in \ _h(\Omega).$$

Above, $\sigma(\nabla u_h)$ denotes a given symmetric elliptic linear function of ∇u_h. For example, the classical case of isotropic linear elasticity corresponds to the choice

$$\sigma(\nabla u_h) = \lambda Tr(\nabla u_h)Id + \mu(\nabla u_h + \nabla^T u_h)$$

which relates the deformation tensor $(\nabla u_h + \nabla^T u_h)/2$ to the stress tensor σ. Moreover, the right-hand side $L(v_h)$ is usually of the form

$$L(v_h) = \int_\Omega f \cdot v_h.$$

By introducing the Lagrange multipliers of the interface continuity constraints $Tr_{ijh}v_{ih} = Tr_{jih}v_{jh}$, this global problem takes the mixed form :

Find $(u_{ih}) \in \prod_i \ _h(\Omega_i)$, and $(\lambda_{ij}) \in \prod_{i<j} \ _{ijh}$ solution of

$$\int_{\Omega_i} \sigma(\nabla u_{ih}) : \nabla v_{ih} + \sum_{i<j} \int_{F_{ij}} \lambda_{ij}v_{ih} - \sum_{i>j} \int_{F_{ji}} \lambda_{ji}v_{ih} = L(v_{ih}),$$

$$\forall v_{ih} \in \ _h(\Omega_i), \forall i,$$

$$\int_{F_{ij}} (u_{ih} - u_{jh})\mu_h = 0, \forall \mu_h \in \ _{ijh}, \forall i < j.$$

In algebraic form, we thus recover the classical subdomain by subdomain writing

$$A_i U_i + \sum_{i<j} Tr_{ijh}^T \Lambda_{ij} - \sum_{i>j} Tr_{ijh}^T \Lambda_{ji} = F_i, \forall i,$$

$$Tr_{ijh} U_i = \bar{U}_{ij}, \forall j \neq i.$$

This sequence of local problems can now be solved by any classical substructuring algorithms. Indeed, after elimination of U_i and Λ_{ij}, and with respect to the interface unknowns $\bar{U} = (\bar{U}_{ij})_{i<j} = (Tr_{ijh}U)_{i<j}$, the above problem takes the standard form:

$$(2.5) \qquad \sum_i \tilde{R}_i^T \begin{pmatrix} A_i & Tr_{ijh}^T \\ Tr_{ijh} & 0 \end{pmatrix}^{-1} \tilde{R}_i \bar{U} = F.$$

Compared to the case of matching grids, the nonmatching case finally leads to the same algorithms with three major changes :

- the pointwise traces are replaced on each face F_{ij} by local L^2 averaged traces Tr_{ijh};
- the pointwise interface restriction $\bar{R}_i$ is replaced by a global restriction operator $\tilde{R}_i$ which maps the global trace $\bar{U}$ into the local right-hand side

$$\begin{pmatrix} 0 \\ \bar{U}_{ij} \end{pmatrix}, \forall j \neq i;$$

- the space of global traces $\bar{U}$ is now defined face by face in the (smaller) product space $\prod_{i<j} \;_{ijh}$. Edges and vertices do not play any role in this definition. In fact, the definition of the trace on any geometric vertex or edge will no longer be unique and will depend of the particular face $\bar{F}_{ij}$ on which it is taken.

REMARK 2.1. *At the limit where the local spaces* $_h(\Omega_i)$ *become dense in* $H^1(\Omega_i)$, *a straightforward integration by parts shows that the solution of the proposed discrete problem satisfies the interface flux continuity requirement:*

For each interface F_{ij}, *there exists a traction force* $\lambda_{ij} \in \;_{ijh}$ *such that*

$$\int_{F_{ij}} v_i(\sigma(u_i).n - \lambda_{ij}) = 0, \forall v_i \in \;_h(\Omega_i),$$

$$\int_{F_{ij}} v_j(\sigma(u_j).n - \lambda_{ij}) = 0, \forall v_j \in \;_h(\Omega_j).$$

We thus observe that the multiplier unknown λ_{ij} *plays the role of a common traction force or generalized normal derivative.*

3. Error analysis

The above framework reduces and simplifies the interface algebraic problem, improves the flexibility of the numerical method, but changes the discrete problem. In particular, the proposed discrete solution is not pointwise continuous across the different interfaces. Nevertheless, we prove in this section that this new approximate solution still converges optimally in the sense that

$$\|u - u_h\|_H \le Ch^k \|u\|_{H^{k+1}(\Omega)}.$$

Here k is the order of the finite elements which are used, and u denotes the continuous solution of the original elliptic problem set on the space

$$(\Omega) = \{v \in H_0^1(\Omega), v = 0 \text{ on } \partial\Omega_D\}.$$

Moreover, the norm $\|u - u_h\|_H$ is a sum of local H^1 subdomain norms :

$$\|u - u_h\|_H = \left(\sum_i \|u - u_h\|_{H^1(\Omega_i)}^2 \right)^{1/2}.$$

To prove such an optimal convergence result, we need two assumptions on the mortar spaces $_{ijh}$.

ASSUMPTION 3.1. *The space $_{ijh}$ is a consistent approximation of the dual of $H^{\frac{1}{2}}(F_{ij})$ in the sense that*

$$\inf_{\lambda_{ij} \in \ _{ijh}} \|\mu - \lambda_{ij}\|_{(H^{1/2}(F_{ij}))'} \le Ch^k \|\mu\|_{H^{k-1/2}(F_{ij})}.$$

In practice, it is sufficient that $_{ijh}$ be a good approximation of $H^{k+1}(F_{ij})$ in L^2. This means that functions of $_{ijh}$ may be discontinuous but must be allowed to take non zero values next to the boundary ∂F_{ij}.

ASSUMPTION 3.2. *On each interface F_{ij}, one neighboring subdomain k ($k = i$ or $k = j$) has a regular triangulation and satisfies*

$$\inf_{\lambda_{ij} \in \ _{ijh}} \sup_{v_h \in Tr \ _h(\Omega_k) \cap H_0^1(F_{ij})} \frac{\int_{F_{ij}} \lambda_{ij} v_h}{\|\lambda_{ij}\|_{L^2(F_{ij})} \|v_h\|_{L^2(F_{ij})}} \ge \beta.$$

This assumption means that $_{ijh}$ must be small compared to the space of traces of $_h(\Omega_k)$. We refer to Maday [14] for examples of finite element spaces satisfying this assumption. From the technical point of view, the above assumption is written in the L^2 norm which makes its verification much easier ([13]). With these assumptions, we can now prove :

LEMMA 3.1. *Under Assumption 3.1, we have*

$$\sup_{w_h \in \mathcal{V}_h(\Omega)} \frac{|a(u, w_h) - L(w_h)|}{\|w_h\|_H} \leq Ch^k \left(\sum_{i<j} \|\sigma.n\|_{H^{k-1/2}(F_{ij})} \right),$$

and under Assumption 3.2 , we have

$$\inf_{v_h \in \mathcal{V}_h(\Omega)} \|u - v_h\|_H \leq Ch^k \|u\|_{H^{k+1}(\Omega)}.$$

Proof. By integration by parts, and since u is by construction the solution of the partial differential equation

$$\mathrm{div}\,(\sigma) + f = 0 \,\mathrm{on}\, \Omega,$$

we first get

$$
\begin{aligned}
R &= a(u, w_h) - L(w_h) \\
&= \sum_i - \int_{\Omega_i} \{ div\sigma + f \}.w_{ih} + \sum_{i<j} \int_{F_{ij}} (\sigma.n).(w_{ih} - w_{jh}) \\
&= \sum_{i<j} \int_{F_{ij}} (\sigma.n).(w_{ih} - w_{jh}), \forall w_h \in \mathcal{V}_h(\Omega).
\end{aligned}
$$

Since w_h belongs to $\mathcal{V}_h(\Omega)$, we have $Tr_{ijh}w_{ih} = Tr_{jih}w_{jh}$, and thus we can rewrite the above inequality as

$$
\begin{aligned}
R &= \sum_{i<j} \int_{F_{ij}} (\sigma.n).((w_{ih} - w_{jh}) - (Tr_{ijh}w_{ih} - Tr_{jih}w_{jh})) \\
&= \sum_{i<j} \int_{F_{ij}} (\sigma.n - \mu_h) \cdot ((w_{ih} - w_{jh}) - (Tr_{ijh}w_{ih} - Tr_{jih}w_{jh})),
\end{aligned}
$$

for any function μ_h in $\mathcal{M}_{ijh}$. Hence, we deduce

$$
\begin{aligned}
|R| &\leq \sum_{i<j} \inf_{\mu_h \in \mathcal{M}_{ijh}} \|\sigma.n - \mu_h\|_{H^{-1/2}(F_{ij})} \|(w_{ih} - w_{jh})\|_{H^{1/2}(F_{ij})}, \\
&\leq Ch^k \left(\sum_{i<j} \|\sigma.n\|_{H^{k-1/2}(F_{ij})} \right) \|w_h\|_H.
\end{aligned}
$$

For the second estimate, we construct the function v_h by

$$(v_h)_{|\Omega_i} = \mathcal{I}_i u - \sum_{ij \in J(i)} Tr^{-i} \circ Tr_{ijh}^{-i} \circ (Tr_{ijh}\mathcal{I}_i u - Tr_{jih}\mathcal{I}_j u).$$

Above, $\mathcal{I}_i$ is the usual interpolation operator,

$$Tr_{ijh}^{-i}: \mathcal{M}_{ijh} \to Tr\,\mathcal{V}_h(\Omega_i) \cap H_0^1(F_{ij})$$

is a continuous inverse of Tr_{ijh}, which is well defined in L^2 from Assumption 3.2, and Tr^{-i} is a continuous inverse of the usual trace operator. Moreover, the set $J(i)$ on which the summation is carried is defined as the set of all interfaces F_{ij} for which $_h(\Omega_i)$ satisfies Assumption 3.2.

On each face $ij \in J(i)$, we first have by construction

$$
\begin{aligned}
Tr_{ijh}(v_h) &= Tr_{ijh}\mathcal{I}_i u - (Tr_{ijh}\mathcal{I}_i u - Tr_{jih}\mathcal{I}_j u) \\
&= Tr_{jih}\mathcal{I}_j u = Tr_{jih} v_h
\end{aligned}
$$

which guarantees that the above function v_h does belong to $_h(\Omega)$.

On the other hand, from the inverse Sobolev inequality, the contracting properties of the L^2 projection Tr_{ijh} and from standard results on interpolation, we get :

$$
\begin{aligned}
\|u - v_h\|^2_{H^1(\Omega_i)} &\leq C\|u - \mathcal{I}_i u\|^2_{H^1(\Omega_i)} \\
&\quad + \sum_{ij \in J(i)} \|Tr^{-i}\|^2 \|Tr^{-i}_{ijh} \circ (Tr_{ijh}\mathcal{I}_i u - Tr_{jih}\mathcal{I}_j u)\|^2_{H^{1/2}(F_{ij})} \\
&\leq C\|u - \mathcal{I}_i u\|^2_{H^1(\Omega_i)} + \sum_{ij \in J(i)} \frac{C}{h_i} \|Tr^{-i}_{ijh} \circ (Tr_{ijh}\mathcal{I}_i u - Tr_{jih}\mathcal{I}_j u)\|^2_{L^2(F_{ij})} \\
&\leq C\|u - \mathcal{I}_i u\|^2_{H^1(\Omega_i)} + \sum_{ij \in J(i)} \frac{C}{h_i} \|(\mathcal{I}_i u - \mathcal{I}_j u)\|^2_{L^2(F_{ij})} \\
&\leq C h_i^{2k} \|u\|^2_{H^{k+1}(\Omega_i)} + \sum_{ij \in J(i)} \frac{C}{h_i} (h_i^{2k+1} + h_j^{2k+1}) \|u\|^2_{H^{k+1/2}(F_{ij})} \\
&\leq C h_i^{2k} \|u\|^2_{H^{k+1}(\Omega_i)} + \sum_{ij \in J(i)} \left(h_i^{2k} \|u\|^2_{H^{k+1}(\Omega_i)} + \frac{h_j^{2k+1}}{h_i} \|u\|^2_{H^{k+1}(\Omega_j)} \right)
\end{aligned}
$$

which yields the desired estimate.

We are now ready to prove our final convergence result.

THEOREM 3.1. *For any continuous uniformly elliptic second order operator $a(.,.)$, and under the assumptions 3.1 and 3.2, the error between the nonconforming discrete solution u_h and the exact solution u is bounded by*

$$
\|u - u_h\|_H \leq C h^k \left(\|u\|^2_{H^{k+1}(\Omega)} + (\sum_{i<j} \|\sigma.n\|^2_{H^{k-1/2}(F_{ij})})^{1/2} \right).
$$

Proof. From a classical lemma of Strang and Fix [**17**], any nonconforming finite element approximation of the solution of a continuous uniformly elliptic second order problem satisfies

$$
\sum_i \|u - u_h\|^2_{H^1(\Omega_i)} \leq \sum_i \|u - v_h\|^2_{H^1(\Omega_i)} + \sup_{w_h \in V_{0h}} \frac{|a(u, w_h) - L(w_h)|}{\|w_h\|_H}.
$$

The conclusion follows then by a direct application of the above lemma. Observe finally that for smooth coefficients, we have

$$
\begin{aligned}
\|\sigma.n\|_{H^{k-1/2}(F_{ij})} &= \left\| \sum_l a_l \frac{\partial u}{\partial n_L} \right\|_{H^{k-1/2}(F_{ij})} \\
&\leq C\|u\|^2_{H^{k+1/2}(F_{ij})} \leq C\|u\|_{H(\Omega_i)}.
\end{aligned}
$$

4. Neumann-Neumann algorithm

4.1. Basic algorithm. Using either matching or nonmatching grids yields the same discrete interface problem :

$$
(4.1) \qquad \sum_i \tilde{R}_i^T \begin{pmatrix} A_i & Tr_{ijh}^T \\ Tr_{ijh} & 0 \end{pmatrix}^{-1} \tilde{R}_i \bar{U} = F.
$$

The presence of nonmatching grids simply replaces the pointwise operator Tr_{ij} by the global projection Tr_{ijh}.

In any case, this problem can be solved by the usual Neumann-Neumann preconditioned conjugate gradient algorithm which acts on any given weak trace $\bar{U} = (\bar{U}_{ij})_{i<j} = (Tr_{ijh}U)_{i<j}$ as follows :

- On each subdomain, solve in parallel the local mixed problem

$$
(4.2) \qquad \begin{pmatrix} A_i & Tr_{ijh}^T \\ Tr_{ijh} & 0 \end{pmatrix} \begin{pmatrix} U_i \\ \Lambda_{ij}^i \end{pmatrix} = \begin{pmatrix} 0 \\ \bar{U}_{ij} \end{pmatrix}.
$$

- On each interface $F_{ij}, i < j$, compute the residual jump of the generalized local normal derivatives

$$
\Lambda_{ij} = \Lambda_{ij}^i + \Lambda_{ij}^j = \left((S_i + S_j)\bar{U} \right)_{|F_{ij}} = \left(S\bar{U} \right)_{|F_{ij}}.
$$

- Project this residual onto the original space by solving in parallel the local Neumann problems

$$
A_i \Psi_i = -\frac{\rho_i}{\rho_i + \rho_j} Tr_{ijh}^T \Lambda_{ij}.
$$

- Update $\bar{U}$ by the preconditioned residual

$$
\mathcal{F}(\bar{U})_{|F_{ij}} = \frac{\rho_i}{\rho_i + \rho_j} Tr_{ijh} \Psi_i + \frac{\rho_j}{\rho_i + \rho_j} Tr_{jih} \Psi_j.
$$

The above algorithm is very flexible and has good localization properties. The coefficient ρ_i is a local average value of the coefficients of the elliptic operator $\sigma(\nabla u)$ and cancels the effects of large jumps of coefficients across the different interfaces. The stiffness matrix A_i is the usual finite element matrix of the local space $_h(\Omega_i)$. For matching grids, the interface matrix Tr_{ijh} is a restriction matrix with a unique nonzero element per row. For nonmatching grids, the element kl of this matrix is given by the L^2 interface scalar product of the

k finite element shape function ϕ_k of (Ω_i) and of the l finite element shape function ψ_l of the mortar space $_{ij}$:

$$(Tr_{ijh})_{kl} = \int_{F_{ij}} \phi_k \psi_l.$$

4.2. Coarse Grid Solver. In the original algorithm, the Neumann subproblems are defined to within a rigid body motion and the condition number of the associated preconditioned operator grows with the inverse of the diameters H of the subdomains $(cond(M^{-1}S) = C/H^2)$. J. Mandel [15] proposed computing these arbitrary rigid body motions in order to optimize the quality of the preconditioner. For this purpose, we first build a small space Z_i on each subdomain. This space must contain the kernel of A_i (the rigid body motions) and any other local function whose energy scales badly with the size of the subdomain. The extended (balanced) Neumann-Neumann preconditioner then adds to the original local preconditioner Ψ_i the elements z_i of Z_i minimizing

$$\|\frac{\rho_i}{\rho_i + \rho_j} Tr_{ijh}(\Psi_i + z_i) \frac{\rho_j}{\rho_i + \rho_j} Tr_{jih}(\Psi_j + z_j) - S^{-1}\Lambda\|_S$$

This optimisation problem in z_i is a coarse problem with very few unknowns per subdomain (6 for three-dimensional linear elasticity). It can be written for all type of partitions and operators. The resulting algorithm amounts to projecting the interface problem onto the orthogonal of the coarse space $\prod Z_i$, which cancels the bad influence of the elements of $\prod Z_i$ on the preconditioner. Its convergence now becomes independent of the number of subdomains or of the coefficient discontinuities :

THEOREM 4.1. *For either matching or nonmatching grids, the condition number of the above balanced Neumann-Neumann algorithm is bounded by*

$$cond(M^{-1}S) \leq C\left(1 + \log(H/h)\right)^2.$$

The constant C above is independent of the number of subdomains, independent of the averaged coefficients ρ_i, but does depend on the aspect ratio of the different subdomains. In case of nonmatching grids, we have to assume that the weak trace has a continuous inverse in the following sense :

$$\inf_{(\lambda_{ij}) \in \prod_j} \sup_{ijh \ v_h \in Tr \ _h(\Omega_i)} \frac{\sum_j \int_{F_{ij}} \lambda_{ij} v_h}{(\sum_j \|\lambda_{ij}\|_{L^2(F_{ij})}) \|v_h\|_{L^2(\partial\Omega_i)}} \geq \beta.$$

Proof. For two subdomains, we have formally

$$M^{-1}S = (S_1^{-1} + S_2^{-1})(S_1 + S_2) = 2I + S_1^{-1}S_2 + S_2^{-1}S_1.$$

Similarly in the general case, we can prove (Mandel [15], Le Tallec [12])

$$cond(M^{-1}S) = C \sup_{U_j \in V_j/Z_j} \frac{< S_i \frac{\rho_j}{\rho_i + \rho_j} U_j, \frac{\rho_j}{\rho_i + \rho_j} U_j >}{< S_j U_j, U_j >}.$$

The term $\frac{\rho_j}{\rho_i+\rho_j}S_iU_j$ is bounded in two steps:

i) interface mirror. Viewed from Ω_i, the function $\frac{\rho_j}{\rho_i+\rho_j}U_j$ has a singularity at the corner and thus, it is only bounded by [**10**] :

$$\rho_i^{1/2}\|\frac{\rho_j}{\rho_i+\rho_j}U_j\|_{H^{1/2}(\partial\Omega_i)} \leq C(1+\log(H/h))\rho_j^{1/2}\|U_j\|_{H^{1/2}(\partial\Omega_j)};$$

ii) weak harmonic extension. Introducing as in Lemma 2 the bounded extension Tr_h^{-1} defined on Ω_j by

$$Tr_h^{-1} = Tr^{-i} \circ \left(P_i - Tr_h^{-i}(\prod_k (P_i - I)_{|F_{ik}})\right),$$

we can easily derive the upper bound

$$< S_iU_i, U_i > \leq \rho_i\|Tr_h^{-1}\|^2\|U_i\|^2_{H^{1/2}(\Gamma_j)}.$$

Here, P_i denotes the L^2 projection on the trace of $_h(\Omega_i)$ on $\partial\Omega_i$. The norm of Tr_h^{-1} is independent of the subdomain diameter H because we work on a quotient space V_j/Z_j [**14**], [**12**]. On the other hand, it depends strongly on the aspect ratio of the subdomain Ω_i.

5. Numerical Results

5.1. Matching grids. The following tests aim at comparing the original Neumann-Neumann algorithm and the new *global version* with coarse grid solver. The first example considers a three-dimensional cantilever beam. The domain was partitioned successively in 4, 8, 32, and 128 identical subdomains. Both slices and boxes were treated in the 4 domain case. The table below displays the characteristics of the partition (local number of elements NE and of degrees of freedom NTDL, number of words used for matrix storage LMUA, size of the coarse grid problem LRIGI, aspect ratio ASP) and the number of iterations which were required to obtain a residual below 10^{-6}. Two numbers are given, one without coarse solver and one with coarse grid solver (given in parenthesis).

Nber of subdomains	NE	NTDL	LMUA	LRIGI	ASP	NITER
4 = 1*1*4 (slices)	512	8019	6 712 830	18	0.8	23(5)
4 = 2*2*1 (boxes)	512	8331	3 970 029	12	0.1	34(8)
8 = 2*2*2	256	4347	1 914 113	36	0.2	62(10)
32 = 2*2*8	64	1275	295 571	168	0.8	157(13)
128 = 4*4*8	16	423	39 665	672	0.4	791(30)

FIGURE 2. *Description of the different partitions*

The second example describes a three-dimensional complex elastic structure, made of aluminium, fixed on three lateral bolts, and twisted through an imposed rotation of its internal axis. The finite element mesh and final deformed shape is depicted on Figure 3. It contains 46, 133 first order $P1$ tetrahedral finite elements, 31, 143 degrees of freedom, among which 4, 248 lie on subdomain interfaces. This mesh was automatically partitioned into 24 subdomains, and the calculation was performed using 1 or 24 processors of a KSR-1 parallel computer.

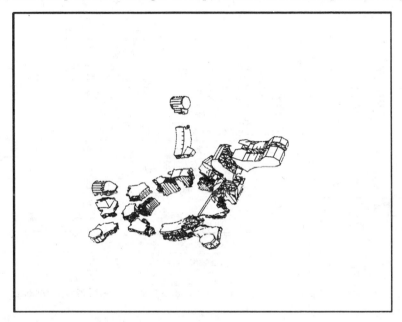

FIGURE 3.   *Finite Element mesh of the structure*

The final solution was obtained after 116 iterations of the Neumann-Neumann algorithm without coarse grid solver or after 37 iterations of the Neumann-Neumann algorithm with coarse grid solver. On one processor, the calculation and assembly of the local stiffness matrices took 224 s, their factorisation 224 s, the construction of the interface data structure 3 s, the conjugate gradient initialisation 12.62 s, the local subdomain solves 1, 322 s, and the interface scalar products 47.15 s. After parallelisation of the subdomain solves on 24 processors, the timings for initialisation, local solves and interface scalar products were of 3.75 s, 58.24 s and 54.97 s, respectively. All these figures show the nice parallel properties of the Neumann-Neumann algorithm even for complex three-dimensional structures.

5.2. Nonmatching Grids without coarse grid solver. The domain considered is a beam of section 0.5m × 0.2m and length 1m or 2m. The beam is made of a quasi-incompressible material with $E = 10^{11}$MPa (Young modulus) and $\nu = 0.49$ (Poisson coefficient). As our main interest lies in the numerical

solver, and not too much in the accuracy of the discretised problem, the beam is
simply partitioned into first order tetrahedral finite elements. The beam has been
sliced either along its leading dimension (nocross) or following a two-dimensional
pattern, with edges and cross-points (cross)(see Figure 5). We show in the next
table the effect of the number of subdomains p and of the mesh step h on the
convergence rate of the Neumann preconditioned conjugate gradient algorithm
(NPGC). The number of iterations does not appear to be very sensitive to the
nonmatching character of the grid.

Number of subdomains and step size	iter	d.o.f in Ω	d.o.f in Γ
$p = 2\ h = h_1$ (matching, nocross)	20	9180	270
$p = 4\ h = h_1$ (matching, nocross)	39	9720	810
$p = 4\ h = h_1$ (nonmatching, nocross)	50	8430	765
$p = 8\ h = h_1$ (matching, nocross)	127	10800	2160
$p = 8\ h = h_1$ (nonmatching, nocross)	107	9480	1785
$p = 4\ h = h_2$ (matching, nocross)	67	2400	360
$p = 4\ h = h_2$ (nonmatching, nocross)	76	1800	225
$p = 4\ h = h_2/2$ (matching, nocross)	66	14580	1215
$p = 4\ h = h_2/2$ (nonmatching, nocross)	88	10629	1080
$p = 4\ h = h_2$ (nonmatching, cross)	41	2145	330
$p = 4\ h = h_2/2$ (nonmatching, cross)	54	12852	972

FIGURE 4. *Test over the number of subdomains and the step size*

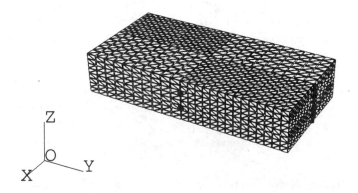

FIGURE 5. *Nonmatching finite element decomposition*

6. Conclusion

We have introduced and tested a theoretical and algorithmic framework which can handle nonmatching grids in three-dimensional situations. This approach leads to smaller interface problem because they are set on the product space $\prod_{ij}$, which has a better structure (the notion of corners and edges have disappeared), and an optimal order of approximation error.

We have also presented and tested a parallel implementation of the Neumann-Neumann preconditioner with coarse grid solver. This implementation handles three-dimensional elasticity and plates operators, matching or nonmatching grids, and any kind of unstructured partition of the mesh. We obtain such partitions by using automatic mesh partitioning strategies such as K-means techniques.

We would like now to extend these techniques to complex three-dimensional CFD problems. But two problems remain open in this direction :

- Which implicit solver to pick for a Navier-Stokes implementation? This choice does not affect the approximation strategy but has a direct consequence on the choice of the substructuring algorithm.

- What is a consistent nonmatching grid approximation of stabilized advection problems, especially in the limit of vanishing viscosity? Is it consistent with a nice multidomain approximation of the pure hyperbolic limit?

REFERENCES

1. R. Glowinski, G. Golub, G. Meurant and J. Periaux (eds.), *Proceedings of the First International Symposium on Domain Decomposition Methods for Partial Differential Equations, Paris, France, January 7-9, 1987*, (SIAM, Philadelphia, 1988).
2. T. Chan, R. Glowinski, J. Periaux and O. Widlund (eds.), *Proceedings of the Second International Symposium on Domain Decomposition Methods for Partial Differential Equations, Los Angeles, California, January 14-16, 1988*, (SIAM, Philadelphia, 1989).
3. T. Chan, R. Glowinski (eds.), *Proceedings of the Third International Symposium on Domain Decomposition Methods for Partial Differential Equations, Houston, Texas, March 20-22, 1989*, (SIAM, Philadelphia, 1990).
4. R. Glowinski, Y. Kuznetsov, G. Meurant, J. Periaux and O. Widlund (eds.), *Proceedings of the fourth international symposium on Domain Decomposition Methods for Partial Differential Equations, Moscow, June 1990*, (SIAM, Philadelphia, 1991).
5. T. Chan, D. Keyes, G. Meurant, J. Scroggs and R. Voigt (eds.), *Proceedings of the fifth international symposium on Domain Decomposition Methods for Partial Differential Equations, Norfolk, May 1991*, (SIAM, Philadelphia, 1992).
6. A. Quarteroni (ed), *Proceedings of the sixth international symposium on Domain Decomposition Methods for Partial Differential Equations, Como, June 1992*, (AMS, Providence, 1993).
7. Y. Achdou and O. Pironneau, *A fast solver for Navier-Stokes Equations in the laminar regime using mortar finite element and boundary element methods*, Technical Report 93-277 (Centre de Mathématiques Appliquées, Ecole Polytechnique, Paris, 1993)
8. C. Bernardi, Y. Maday and A. Patera, *A new nonconforming approach to domain decomposition: the mortar element method*, in H. Brezis and J.L. Lions (eds.) Nonlinear Partial Differential Equations and their Applications (Pitman, 1989).
9. J. Bramble, R. Ewing, R. Parashkevov and J. Pasciak, *Domain decomposition methods for problems with partial refinement*, SIAM J. Sci. Stat. Comp. **13** (1992) 397–410.

10. Y. H. De Roeck and P. Le Tallec, *Analysis and test of a local domain decomposition preconditioner*, in [4].
11. C. Farhat and M. Lesoinne, *Automatic Partitioning of Unstructured Meshes for the Parallel Solution of Problems in Computational Mechanics*, Int. J. Num. Meth. in Eng. **36** (1993) 745–764.
12. P. Le Tallec, *Domain Decomposition Methods in Computational Mechanics*, Computational Mechanics Advances (1994).
13. P. Le Tallec and T. Sassi, *Domain Decomposition with Nonmatching Grids: Schur Complement Approach*, Rapport CEREMADE 9323, Université de Paris Dauphine (1993).
14. Y. Maday, Proceedings of this Conference.
15. Jan Mandel, *Balancing Domain Decomposition*, Communications in Numerical Methods in Engineering, **9** (1993), 233–241.
16. Jan Mandel and M. Brezina, *Balancing Domain Decomposition: theory and performances in two and three dimensions*, Technical report (Computational Mathematics Group, University of Colorado at Denver, March 1993).
17. G. Strang and G. Fix, *An Analysis of the Finite Element Method*, Prenctice Hall (1973).
18. H. Simon, *Partitioning of unstructured problems for parallel processing*, Comp. Systems in Eng. **2** (1991) 135–148.

UNIVERSITÉ PARIS DAUPHINE AND INRIA - ROCQUENCOURT, BP105, 78 153 LE CHESNAY, FRANCE
    E-mail address: letallec@menusin.inria.fr

Contemporary Mathematics
Volume **180**, 1994

Domain Decomposition for Elliptic Problems with Large Condition Numbers

S. V. NEPOMNYASCHIKH

ABSTRACT. This paper suggests a technique for the construction of preconditioning operators for the iterative solution of systems of grid equations approximating elliptic boundary value problems with strong singularities in the coefficients. The technique suggested is based on the decomposition of the original domain into subdomains in which the singularity of coefficients is characterized by some parameter. The convergence rate of the preconditoned iterative process is independent of both the mesh size and the coefficients.

1. Introduction

In this paper, we design preconditioning operators for the system of grid equations approximating the following boundary value problem:

$$(1.1) \quad \begin{cases} -\sum_{i,j=1}^{2} \frac{\partial}{\partial x_i} a_{ij}(x) \frac{\partial u}{\partial x_j} + a_0(x)u = f(x), & x \in \Omega, \\ u(x) = 0, & x \in \Gamma \end{cases}$$

We assume that Ω is a bounded, polygonal region and Γ is its boundary. Let Ω be a union of n nonoverlapping subdomains Ω_i,

$$\overline{\Omega} = \bigcup_{i=1}^{n} \overline{\Omega}_i, \quad \Omega_i \bigcap \Omega_j = \varnothing, \quad i \neq j,$$

where Ω_i are polygons and Γ_i are their boundaries. Let Ω^h

$$\Omega^h = \bigcup_{i=1}^{n} \Omega_i^h$$

be a regular triangulation of Ω which is characterized by a parameter h.

1991 *Mathematics Subject Classification.* Primary 65N55, 65N30.

This paper is in final form and no version of it will be submitted for publication elsewhere.

Let us introduce the weighted Sobolev spaces $H_\alpha^1(\Omega)$ with the norms [11]

(1.2)
$$\|u\|^2_{H_\alpha^1(\Omega)} = \|u\|^2_{L_2(\Omega)} + |u|^2_{H_\alpha^1(\Omega)},$$
$$\|u\|^2_{L_2(\Omega)} = \int_\Omega u^2(x)dx,$$
$$|u|^2_{H_\alpha^1(\Omega)} = \int_\Omega \left(\frac{|\nabla u(x)|}{(\varrho(x))^{\alpha(x)}}\right)^2 dx.$$

Here

$$\alpha(x) \equiv \alpha_i = \text{ const}, \ x \in \Omega_i$$

and $\varrho(x)$ is the distance between the point $x \in \Omega_i$ and the boundary Γ_i of the subdomain Ω_i. We assume that

(1.3)
$$|\alpha| < \frac{1}{2}.$$

Denote by $\overset{o}{H}_\alpha^1(\Omega)$ the subspace of $H_\alpha^1(\Omega)$ with zero trace on Γ and introduce the bilinear form

$$a(u,v) = \int_\Omega \left(\sum_{i,j=1}^2 a_{ij}(x)\frac{\partial u}{\partial x_j}\frac{\partial v}{\partial x_i} + a_0(x)uv\right) dx$$

and the linear functional

$$l(v) = \int_\Omega f(x)vdx.$$

We assume that the coefficients of the problem (1.1) are such that $a(u,v)$ is a symmetric, coercive and continuous form on $\overset{o}{H}_\alpha^1(\Omega) \times \overset{o}{H}_\alpha^1(\Omega)$, and that the linear functional $l(v)$ is continuous in $\overset{o}{H}_\alpha^1(\Omega)$.

Denote by W a space of real–valued continuous functions linear on triangles of the triangulation Ω^h. A weak formulation of (1.1) is: Find $u \in \overset{o}{H}_\alpha^1(\Omega)$ such that

(1.4)
$$a(u,v) = l(v), \quad \forall v \in \overset{o}{H}_\alpha^1(\Omega).$$

Using the finite element method, we can pass from (1.4) to the linear algebraic system

(1.5)
$$Au = f.$$

The condition number of the matrix A depends on h, α and can be large. Our purpose is the design of a preconditioner B for the problem (1.5) such that the following inequalities are valid:

(1.6)
$$c_1(Bu,u) \le (Au,u) \le c_2(Bu,u), \quad \forall u \in R^N.$$

Here N is the dimension of W, the positive constants c_1, c_2 are independent of h and α, and the action of B^{-1} on a vector can be realized at low cost.

2. Additive Schwarz method for singular problems

The construction of the preconditioner for the system (1.5) will be realized on the basis of the additive Schwarz method [1, 3, 4]. To design the preconditioning operator B, we follow [7, 10] and decompose the space W into a sum of subspaces

$$W = W_0 + W_1.$$

To this end, divide the nodes of the triangulation Ω^h into two groups: those which lie inside of Ω_i^h and those which lie on boundaries of Ω_i^h. The subspace W_0 corresponds to the first set. Let

$$S = \bigcup_{i=1}^{n} \partial\Omega_i^h,$$

$$W_0 = \left\{ u^h \in W \;\middle|\; u^h(x) = 0, x \in S \right\},$$

$$W_{0,i} = \{ u^h \in W_0 \;\middle|\; u^h(x) = 0, \quad x \overline{\in} \Omega_i^h \}, \quad i = 1, 2, \ldots, n.$$

It is clear that W_0 is the direct sum of the orthogonal subspaces $W_{0,i}$ with respect to the scalar product in $\overset{\circ}{H}_\alpha^1(\Omega)$:

$$W_0 = W_{0,1} \oplus \cdots \oplus W_{0,n}.$$

The subspace W_1 corresponds to the second group of nodes Ω^h and can be defined in the following way. First, define V which is the space of traces of functions from W on S:

$$V = \{ \varphi^h \mid \varphi^h(x) = u^h(x), \quad x \in S, \quad u^h \in W \}.$$

To define the subspace W_1, we need a norm preserving extension operator of functions given at S into Ω^h. The basis of the further construction is the following trace theorem for the weighted Sobolev spaces $H_\alpha^1(\Omega)$ [11]:

THEOREM 2.1. *Let Ω be a bounded domain with piecewise–smooth boundary Γ from the class C^2 satisfying the Lipschitz condition and α is a constant such that $|\alpha| < \frac{1}{2}$. Then there exists a positive constant c_1 independent of α, such that*

$$\|\varphi\|_{H^{\frac{1}{2}+\alpha}(\Gamma)} \le c_1 \|u\|_{H_\alpha^1(\Omega)}$$

for any function $u \in H_\alpha^1(\Omega)$, where $\varphi \in H^{\frac{1}{2}+\alpha}(\Gamma)$ is the trace of u at the boundary Γ. Conversely, there exists a positive constant c_2, independent of α, such that for any function $\varphi \in H^{\frac{1}{2}+\alpha}(\Gamma)$ there exist $u \in H_\alpha^1(\Omega)$ such that

$$u(x) = \varphi(x), \quad x \in \Gamma,$$
$$\|u\|_{H_\alpha^1(\Omega)}^2 \le c_2 \|\varphi\|_{H^{\frac{1}{2}+\alpha}(\Gamma)}^2.$$

Here $\|\varphi\|_{H^{\frac{1}{2}+\alpha}(\Gamma)}$ is the norm in the Sobolev space $H^{\frac{1}{2}+\alpha}(\Gamma)$:

$$
\begin{aligned}
(2.1) \qquad \|\varphi\|^2_{H^{\frac{1}{2}+\alpha}(\Gamma)} &= \|\varphi\|^2_{L^2(\Gamma)} + |\varphi|^2_{H^{\frac{1}{2}+\alpha}(\Gamma)}, \\
\|\varphi\|^2_{L^2(\Gamma)} &= \int_\Gamma \varphi^2(x)\,dx, \\
|\varphi|^2_{H^{\frac{1}{2}+\alpha}(\Gamma)} &= \int_\Gamma \int_\Gamma \frac{(\varphi(x)-\varphi(y))^2}{|x-y|^{2+2\alpha}}\,dxdy.
\end{aligned}
$$

Denote

$$
H(S) = \left\{ \varphi \mid \; \varphi\Big|_{\Gamma_i} = \varphi_i, \quad \varphi_i \in H^{\frac{1}{2}+\alpha_i}(\Gamma_i) \right\},
$$

$$
\|\varphi\|^2_{H(S)} = \sum_{i=1}^n \|\varphi\|^2_{H^{\frac{1}{2}+\alpha_i}(\Gamma_i)}.
$$

To define the bounded extension operator for the finite element case from V into W, we need mesh counterparts of the norms (1.2) and (2.1). To this end, let us split the triangles $\mathcal{T}_j$ of the triangulation Ω^h into three groups. Denote by M_1 a set of such $\mathcal{T}_j$ that $\mathcal{T}_j$ do not have vertices on S, denote by M_2 a set of such $\mathcal{T}_j$ that $\mathcal{T}_j$ have only one vertex on S, and denote by M_3 a set of such $\mathcal{T}_j$ that $\mathcal{T}_j$ have more than one vertex on S. Set

$$
\begin{aligned}
\|u^h\|^2_{H^1_{\alpha,h}(\Omega)} &= \sum_{i=1}^n \|u^h\|^2_{H^1_{\alpha_i,h}(\Omega_i)}, \\
\|u^h\|^2_{H^1_{\alpha_i,h}(\Omega_i)} &= \|u^h\|^2_{L_{2,h}(\Omega_i)} + |u^h|^2_{H^1_{\alpha_i,h}(\Omega_i)}, \\
\|u^h\|^2_{L_{2,h}(\Omega_i)} &= \sum_{z_j \in \Omega_i^h} (u^h(z_j))^2 h^2, \\
|u^h|^2_{H^1_{\alpha_i,h}(\Omega_i)} &= \sum_{\mathcal{T}_j \in M_1 \cap \Omega_i} \frac{(u_{j1}-u_{j2})^2 + (u_{j2}-u_{j3})^2 + (u_{j3}-u_{j1})^2}{(\varrho(\mathcal{T}_j,\Gamma_i))^{2\alpha_i}} + \\
&\quad | \sum_{\mathcal{T}_j \in M_2 \cap \Omega_i} \frac{(u_{j1}-u_{j2})^2 + (u_{j2}-u_{j3})^2 + (u_{j3}-u_{j1})^2}{h^{2\alpha_i}} + \\
&\quad + \sum_{\mathcal{T}_j \in M_3 \cap \Omega_i} \frac{(u_{j1}-u_{j2})^2 + (u_{j2}-u_{j3})^2 + (u_{j3}-u_{j1})^2}{(1-2\alpha_i)h^{2\alpha_i}}, \forall u^h \in W.
\end{aligned}
$$

Here z_i are vertices of Ω^h, u_{j1}, u_{j2}, u_{j3} are values of u^h at vertices of $\mathcal{T}_j$, and $\varrho(\mathcal{T}_j, \Gamma_i)$ is the distance between $\mathcal{T}_j$ and Γ_i.

Using the natural order of nodes on Γ_i, let us put for each node $z_j \in \Gamma_i$ into correspondence the node z_{j+1}, which is a node neighboring upon z_i, and set

$$\|\varphi^h\|^2_{H_h(S)} \;=\; \sum_{i=1}^{n}\|\varphi^h\|^2_{H_h^{\frac{1}{2}+\alpha_i}(\Gamma_i)},$$

$$\|\varphi^h\|^2_{H_h^{\frac{1}{2}+\alpha_i}(\Gamma_i)} \;=\; \|\varphi^h\|^2_{L_{2,h}(\Gamma_i)} + |\varphi^h|^2_{H_h^{\frac{1}{2}+\alpha_i}(\Gamma_i)},$$

$$\|\varphi^h\|^2_{L_{2,h}(\Gamma_i)} \;=\; \sum_{z_j\in\Gamma_i}(\varphi^h(z_j))^2 h,$$

$$|\varphi^h|_{H_h^{\frac{1}{2}+\alpha_i}(\Gamma_i)} \;=\; \sum_{z_j\in\Gamma_i}\sum_{\substack{z_k\in\Gamma_i\\ j\neq k}} \frac{(\varphi^h(z_j)-\varphi^h(z_k))^2}{|z_j-z_k|^{2+2\alpha_i}}h^2 +$$

$$+ \sum_{z_j\in\Gamma_i^h}\frac{(\varphi^h(z_j)-\varphi^h(z_{j+1}))^2}{(1-2\alpha_i)h^{2\alpha_i}}, \quad \forall\varphi^h\in V.$$

The following lemmas are valid.

LEMMA 2.1. *There exist positive constants c_3 and c_4, independent of α and h, such that*

$$c_3\|u^h\|_{H^1_{\alpha_i}(\Omega_i)} \le \|u^h\|_{H^1_{\alpha_i,h}(\Omega_i)} \le c_4\|u^h\|_{H^1_{\alpha_i}(\Omega_i)}, \quad \forall u^h\in W, \quad i=1,2,\ldots,n.$$

LEMMA 2.2. *There exist positive constants c_5 and c_6, independent of α and h, such that*

$$c_5\|\varphi^h\|_{H^{\frac{1}{2}+\alpha_i}(\Gamma_i)} \le \|\varphi^h\|_{H_h^{\frac{1}{2}+\alpha_i}(\Gamma_i)} \le c_6\|\varphi^h\|_{H^{\frac{1}{2}+\alpha_i}(\Gamma_i)}, \quad \forall\varphi^h\in V, \quad i=1,2,\ldots,n.$$

To define W_1, let us use the explicit extension operator

(2.2) $$t^h : V \to W,$$

which was suggested for regular elliptic second order problems. The definition and the realization algorithm were done in [5, 6, 8] and briefly can be described in the following way. Let us introduce the near–boundary coordinate system (s,n) which is defined in a δ–neighborhood of Γ_i. Here s defines a point P at Γ_i and n is the distance between the given point and Γ_i along the internal pseudonormal, whose direction at the angular points coincides with the bisectrix of the angle and along the smooth part the vector changes, for example, linearly. Set

(2.3)
$$t : H(S) \to H^1_\alpha(\Omega),$$
$$t\varphi = u,$$
$$u(s,n) = \left(1-\tfrac{n}{\delta}\right)\int\limits_{s}^{s+n}\frac{\varphi(t)}{n}\,dt,$$

where the function u is extended by zero in the rest of Ω. Using the auxiliary mesh, which is topologically equivalent to a uniform rectangular mesh, we can define the finite–element analogue t^h (2.2) of the operator t from (2.3). The following theorem is valid.

THEOREM 2.2. *There exists a positive constant c_7, independent of α and h, such that*

$$\|u^h\|_{H^1_{\alpha,h}(\Omega)} = \|t^h\varphi^h\|_{H^1_{\alpha,h}(\Omega)} \leq c_7\|\varphi^h\|_{H^{\frac{1}{2}+\alpha}_h(S)}, \quad \forall \varphi^h \in V.$$

Remark 2.1. The cost of the actions of t^h and $(t^h)^*$ on vectors is $O(h^{-2})$ arithmetic operations (see [5] for details).

At last, we can define subspace W_1

$$W_1 = \left\{u^h \mid u^h = t^h\varphi^h, \quad \varphi^h \in V\right\}.$$

It is obvious that

$$W = W_0 + W_1$$

and this decomposition of the space W is regular in the following sense.

THEOREM 2.3. *There exists a positive constant c_8, independent of α and h, such that for any function $u^h \in W$ there exist $u_i \in W_i, i = 0,1$, such that*

$$u_0 + u_1 = u,$$

$$\|u_0\|_{H^1_\alpha(\Omega)} + \|u_1\|_{H^1_\alpha(\Omega)} \leq c_8\|u\|_{H^1_\alpha(\Omega)}.$$

According to [8], we can construct a preconditioner for the subspace W_1 of the following form

$$B_1^+ = t^h\Sigma^{-1}(t^h)^*,$$

where Σ has to satisfy

$$c_9\|\varphi^h\|^2_{H(S)} \leq (\Sigma\varphi, \varphi) \leq c_{10}\|\varphi^h\|^2_{H(S)}, \quad \forall\varphi^h \in V.$$

Here the components of the vector φ are equal to the values of the function φ^h in corresponding nodes. The constants c_9, c_{10} should be independent of α and h. The construction of the easily invertible operator (matrix) Σ can be done, using [4, 6, 7]. The cost of the action B_1^+ on vectors is $O(h^{-2})$ arithmetic operations.

3. Preconditioning operator for W_0.

The goal of this section is the design of the preconditioning operator for the space W which was defined in Section 2. Since W_0 is the direct sum of the orthogonal subspaces $W_{0,i}, i = 1, 2, \ldots, n$ which correspond to the subdomains Ω_i, we can design preconditioners independently for each subdomain Ω_i with the boundary Γ_i. For the sake of simplicity, we omit the subscript i. To construct the preconditioner, we use the additive Schwarz method. Let us decompose the domain Ω into two overlapping parts

$$\Omega = \Omega_{in}\bigcup\Omega_b,$$

(3.1) $$\Omega_b = \{x = (s,n) \in \Omega \mid 0 \leq s \leq L, \quad 0 < n < 2\delta\},$$
$$\text{dist}(\Gamma, \partial\Omega_{in}) = \delta.$$

Here (s,n) is the near-boundary coordinate system; $\delta = 0(1)$ is independent of h; for the sake of simplicity, we assume that Ω is the simply connected domain and L is the length of Γ. Then the triangulation Ω^h can be decomposed into two overlapping parts

$$\Omega^h = \Omega^h_{in} \bigcup \Omega^h_b,$$

$$\Omega^h_{in} = \bigcup_{T_j \subset \Omega_{in}} T_j,$$

$$\Omega^h_b = \bigcup_{T_j \subset \Omega_b} T_j$$

and the finite element space W_0 can be decompose into two overlapping subspaces

$$W_0 = W_{in} + W_b,$$
$$W_{in} = \left\{ u^h \in W_0 \mid u^h(x) = 0, \quad x \overline{\in} \Omega^h_{in} \right\},$$
$$W_b = \left\{ u^h \in W_0 \mid u^h(x) = 0, \quad x \overline{\in} \Omega^h_b \right\}.$$

Using (3.1), it is easy to see that there exists a positive constant c_1, independent of α and h, such that for any $u^h \in W_0$ there exists $u^h_{in} \in W_{in}$ and $u^h_b \in W_b$ such that

$$u^h_{in} + u^h_b = u^h,$$

$$\|u^h_{in}\|_{H^1_\alpha(\Omega)} + \|u^h_b\|_{H^1_\alpha(\Omega)} \le c_1 \|u\|_{H^1_\alpha(\Omega)}.$$

Then, according to [3], we can define the preconditioner in the following form

$$B^{-1} = B^+_{in} + B^+_b,$$

where B_{in} and B_b are such that

$$B_{in} : W \to W_{in},$$

$$c_2 \|u^h\|^2_{H^1_\alpha(\Omega)} \le (B_{in} u, u) \le c_3 \|u^h\|^2_{H^1_\alpha(\Omega)}, \quad \forall u^n \in W_{in},$$

$$B_b : W \to W_b,$$

$$c_2 \|u^h\|^2_{H^1_\alpha(\Omega)} \le (B_b u, u) \le c_3 \|u^h\|^2_{H^1_\alpha(\Omega)}, \quad \forall u^n \in W_b.$$

Here c_2 and c_3 are independent of α and h. From (3.1) we have that there exist positive constants c_4 and c_5, independent of α and h, such that

$$c_4 \|u^h\|_{H^1(\Omega)} \le \|u^h\|_{H^1_\alpha(\Omega)} \le c_5 \|u^h\|_{H^1(\Omega)}, \quad \forall u^n \in W_{in}.$$

This implies that the construction of B_{in} is equivalent to the construction of preconditioners for regular elliptic problems. For instance, using combinations of the domain decomposition and fictitious domain methods, the construction of effective preconditioners was studied in [4, 5, 10]. A new element of the construction of the preconditioner B is the construction of B_b. To this end, let us decompose Ω_b into overlapping parts

$$\Omega_b = \bigcup_{i=L}^{l} D_i,$$

$$D_i = \left\{ x = x(s,n) \in \Omega_b \mid (i-1)\frac{L}{l} < s < (i+1)\frac{L}{l}, 0 < n < \delta \right\},$$

where $l = O(1)$, i.e. the number of subdomains is fixed, and $x(L + s, n) = x(s, n)$. Then the triangulation Ω_b^h can be decomposed into overlapping parts

$$\Omega_b^h = \bigcup_{i=1}^{l} D_i^h,$$

$$D_i^h = \bigcup_{\mathcal{T}_j \subset D_i} \mathcal{T}_j$$

and the space W_b can be decomposed into overlapping subspaces

$$W_{in} = \sum_{i=1}^{l} U_i,$$

$$U_i = \left\{ u^h \in W_{in} \mid \quad u^h(x) = 0, x \bar{\in} D_i^h \right\}.$$

The following lemma is valid.

LEMMA 3.1. *Let Ω be a rectangular domain*

$$\Omega = \{ (x_1, x_2) \mid \quad -1 < x_1 < 1, 0 < x_2 < 1 \}$$

and Ω^h be a regular triangulation of Ω. Denote by W a space of real-valued continuous functions linear on triangles of the triangulation Ω^h. Then, there exists a positive constant c_6, independent of α and h, such that $\forall u^h \in W$

$$\int\limits_{-1}^{1} \int\limits_{0}^{1} \left(\frac{1}{x_2^{2\alpha}} |\nabla \tilde{u}^h|^2 + (\tilde{u}^h)^2 \right) dx_2 dx_1 \leq c_6 \int\limits_{0}^{1} \int\limits_{0}^{1} \left(\frac{1}{x_2^{2\alpha}} |\nabla u^h|^2 + (u^h)^2 \right) dx_2 dx_1,$$

for any constant $\alpha : |\alpha| < 1/2$. Here the function $\tilde{u}^h \in W$ is defined in the following way:

$$\tilde{u}^h(z_i) = \tilde{u}^h(x_{1i}, x_{2i}) = \begin{cases} u^h(z_i), z_i \in [0, 1] \times [0, 1], \\ (1 + x_{1i}) u^h(\tilde{z}_i), z_i \in [-1, 0] \times [0, 1]. \end{cases}$$

Here $\tilde{z}_i$ is a node of Ω^n which is the nearest for the point with the coordinates $(-x_{1i}, x_{2i})$.

Using Lemma 3.1, it is easy to see that there exists a positive constant c_7, independent of α and h, such that for any $u^h \in W_{in}$ there exists $u_i^h \in U_i$:

(3.2)
$$u_1^h + \cdots + u_l^h = u^h,$$
$$c_7 \left(\|u_1^h\|_{H_\alpha^1(\Omega)} + \cdots + \|u_l^h\|_{H_\alpha^1(\Omega)} \right) \leq \|u^h\|_{H_\alpha^1(\Omega)}.$$

According to [3] and (3.2), to define the operator B_{in}, we can define the easily invertible norms for subspaces $V_i, i = 1, 2, \ldots, l$. To this end, we use the fictitious space lemma [5, 9].

To design the easily invertible norm in U_i, we consider an auxiliary topologically uniform mesh. For the sake of simplicity, we omit the subscript i.

Let us assume that the domain D in the near-boundary coordinate system (s, n) has the following representation

$$D = \{ x = (s, n) \mid \quad 0 < s < \xi, \quad 0 < n < \delta \}.$$

Introduce in the domain D the auxiliary mesh Q^h with the mesh size h_0 and the nodes z_{ij}

$$z_{ij} = (s_i, n_j), \quad s_i = i \cdot h_0, \quad n_j = j \cdot h_0$$

$$i = 0, 1, \ldots, n, \quad j = 0, 1, \ldots, m,$$

$$n \cdot h_0 = \xi, \quad m \cdot h_0 = \delta.$$

Assume that $h_0 \leq h_{min}/2$, where h_{min} is the length of the minimal side of triangles of the triangulation D^h. Denote the cells of the mesh Q^h by Q_{ij}

$$\begin{aligned} Q_{ij} &= \{x = (s, n) \mid \ s_i \leq s < s_{i+1}, \ n_j \leq n < n_{j+1}\}, \\ &i = 0, 1, \ldots, n-1 \\ &j = 0, 1, \ldots, m-1. \end{aligned}$$

On the mesh Q^h we will consider the mesh function $V(z_{ij})$ vanishing at the boundary nodes of the mesh Q^h. We will identify the mesh Q^h and the triangulation of D with the nodes z_{ij}. Denote by F a space of real-valued continuous functions V^h linear on triangles of the triangulation Q^h.

Using the tensor product of matrices, introduce

$$B = A \otimes J + I_{n-1} \otimes T,$$

where the tridiagonal matrix A of order $n-1$ approximates the second derivative and J is a diagonal matrix of order $m-1$

$$J = \operatorname{diag}\left(\frac{1}{h_0^{2\alpha}}, \frac{1}{(2h_0)^{2\alpha}}, \ldots, \frac{1}{((m-1)h_0)^{2\alpha}}\right);$$

The matrix I_{n-1} is the identity matrix of the order $n-1$; the tridiagonal matrix T of the order $m-1$:

$$\begin{bmatrix} \frac{1}{(1-2\alpha)h_0^{2\alpha}} + \frac{1}{(2h_0)^{2\alpha}} & -\frac{1}{(2h_0)^{2\alpha}} & & \\ -\frac{1}{(2h_0)^{2\alpha}} & \frac{1}{(2h_0)^{2\alpha}} + \frac{1}{(3h_0)^{2\alpha}} & & \\ & \ddots & \ddots & \ddots \\ & & -\frac{1}{((m-1)h_0)^{2\alpha}} \\ & -\frac{1}{((m-1)h_0)^{2\alpha}} & \frac{1}{((m-1)h_0)^{2\alpha}} + \frac{1}{(mh_0)^{2\alpha}} \end{bmatrix}$$

The following Lemma is valid.

LEMMA 3.2. *There exist positive constants* c_8, c_9, *independent of* α *and* h, *such that*

$$c_8 \|V^h\|^2_{H^1_\alpha(D)} \leq (BV, V) \leq c_9 \|V^h\|^2_{H^1_\alpha(D)}, \quad \forall V^h \in F$$

where the components of the vector V *are equal to the values of the function* V^h *in the corresponding nodes.*

Remark 3.1. Using the spectral decomposition of the matrix A

$$A = Q \Lambda Q^T,$$

we can invert the matrix B :

$$B^{-1} = Q \otimes I_{m-1} (\Lambda^{-1} \otimes J^{-1} + I_{n-1} \otimes T^{-1}) Q^T \otimes I_{m-1}$$

Then the multiplication of vectors by B^{-1} can be performed in $O(h^{-2} \log h^{-1})$ arithmetic operations using the Fast Fourier Transform. To define the easily invertible norm in the space V, using [**9**], we need to define operators R and T:

$$R : F \to V, \quad T : V \to F.$$

Let us define the operator R which puts into the correspondence to each function $V^h(z_{ij}) \in F$ a function $u^h \in V$ in the following way. Let z_L be a node of the triangulation D^h and let $z_e \in Q_{ij}$. Set

$$u^h(z_l) = V^h(z_{ij}).$$

Note that by the assumption on h_0 only one node z_l of the triangulation D^h belonging to the cell Q_{ij} can exist. Then the operator T is defined as follows. If the cell Q_{ij} contains a node z_l of the triangulation D^h, we set

$$V^h(z_{ij}) = u^h(z_l).$$

At other nodes of the mesh Q^h the function $V^h(z_{ij})$ can be defined in a sufficiently arbitrary way, for instance, as follows. Let the node z_{ij} belong to the triangle T_l of the triangulation D^h with the vertices z_{l_1}, z_{l_2} and z_{l_3}. Set

$$V^h(z_{ij}) = \frac{1}{3}(u^h(z_{l_1}) + u^h(z_{l_2}) + u^h(z_{l_3})),$$

It is easy to see that the above-defined operators R and T satisfy the hypothesis of Lemma 4.3 while the constants c_{10} and c_{11} are independent of α and h:

$$\|RV^h\|^2_{H^1_\alpha(D)} \le c_{10}(BV, V), \quad \forall V^h \in F,$$
$$(BTu^h, Tu^h) \le c_{11}\|u^h\|_{H^1_\alpha(D)}, \quad \forall u^h \in V.$$

Then, the following theorem is valid:

THEOREM 3.1. *There exist positive constants c_{12} and c_{13}, independent of α and h , such that*

$$c_{12}\|u^h\|^2_{H^1_\alpha(D)} \le (C^{-1}u, u) \le c_{13}\|u^h\|^2_{H^1_\alpha(D)}, \quad \forall u^h \in V,$$

$$C = RB^{-1}R^T,$$

where the components of the vector u are equal to the values of the functions u^h in the corresponding nodes.

References

1. P.L. Lions, *On the Schwarz alternating method,* I. First International Symposium on Domain Decomposition Methods for Partial Differential Equations, (R.Glowinski, G.H. Golub, G. Meurant and J. Périaux, eds.), SIAM, Philadelphia,1988

2. G.I. Marchuk,*Methods of Numerical Mathematics,* Springer, New York, 1982

3. A.M. Matsokin and S.V. Nepomnyaschikh, *Schwarz alternating method in subspaces,* Soviet Mathematics, 29 (1985), pp. 78–84.

4. A.M. Matsokin and S.V. Nepomnyaschikh, *Norms in the space of traces of mesh functions,* Sov. J. Numer. Anal. Math. Modeling, 3 (1988), 199–216.

5. A.M. Matsokin and S.V. Nepomnyaschikh, *Method of fictitious space and explicit extension operators,* Zh. Vychisl. Mat. Mat. Fiz , 33(1993), 52–68.

6. S.V. Nepomnyaschikh, *Domain decomposition and Schwarz methods in a subspace for the approximate solution of elliptic boundary value problems,* Thesis, Computing Center of the Siberian Branch of the USSR Academy of Sciences, Novosibirsk, USSR, 1986.

7. S.V. Nepomnyaschikh, *Domain decomposition method for elliptic problems with discontinuous coefficients,* Proc. 4th Conference on Domain Decomposition methods for Partial Differential Equations, Philadelphia, PA, SIAM, 1991, 242–251.

8. S.V. Nepomnyaschikh, *Method of splitting into subspaces for solving elliptic boundary value problems in complex-form domains,* Sov. J. Numer. Anal. Math. Modelling, 6(1991), 151–168.

9. S.V. Nepomnyaschikh, *Mesh theorems on traces, normalization of function traces and their inversion,* Sov. J. Numer. Anal. Math. Modelling, 6(1991), 223–242.

10. S.V. Nepomnyaschikh, *Decomposition and fictitious domain methods for elliptic boundary value problems,* 5th Conference on Domain Decomposition Methods for Partial Differential Equations, Philadelphia, PA , SIAM, 1992.

11. S.M. Nikol'ski, *Approximation of functions of many variables and embedding theorems,* Nauka, Moscow, 1977 (Russian).

12. G.N. Yakovlev, *On the traces of functions from spaces W_p^l on piecewise-smooth surfaces,* Mat. Sbornik 74(1967), 526–543, (Russian).

RUSSIAN ACADEMY OF SCIENCES, COMPUTING CENTER, LAVRENTIEV AV. 6, NOVOSIBIRSK, 630090, RUSSIA.

E-mail address: svnep@comcen.nsk.su

Contemporary Mathematics
Volume **180**, 1994

Stable Subspace Splittings for Sobolev Spaces and Domain Decomposition Algorithms

P. OSWALD

ABSTRACT. The notion of a stable subspace splitting is basic for the theoretical understanding of some modern iterative methods for solving variational problems. For Sobolev spaces over polyhedral domains, examples of such splittings into finite element subspaces are given along with typical applications to multilevel and domain decomposition algorithms.

1. Introduction

Many modern iterative algorithms for solving elliptic p.d.e. discretizations can be interpreted as additive (Jacobi-like) or multiplicative (Gauss-Seidel-like) subspace correction methods, see [**27, 30, 10**]. The key to their analysis is the study of some metric properties of the underlying splitting of the discretization space V into a sum of subspaces V_j and of the variational problem on V into auxiliary problems on these subspaces. In Section 2, we start with a brief overview of the abstract theory for the symmetric positive definite case based on our joint paper with M. Griebel [**12**].

Investigation of such splittings for the solution of variational problems on Sobolev spaces benefits from already existing experience with decomposing elements of function spaces into simple building blocks. Approximation theory, Fourier analysis, and the theory of function spaces are helpful in this respect. Some examples of useful splittings of $H^s(\Omega)$ with respect to finite element subspaces over polyhedral domains in R^d are given in Section 3. We restrict ourselves to applications to second order elliptic boundary value problems (and some problems that are closely related), an analogous theory holds for fourth order problems, see [**17, 18, 32, 6**], for similar developments involving wavelets we refer to [**5, 7**] and the papers cited therein.

1991 *Mathematics Subject Classification.* Primary 65F10, 65F35; Secondary 65N20, 65N30.
This paper is in final form and no version of it will be submitted for publication elsewhere.

In the final Section 4 we show how some domain decomposition algorithms can be derived along the lines of our approach.

2. Abstract Schwarz methods

Let V be some fixed Hilbert space, with the scalar product given by a continuous symmetric positive definite (s.p.d.) form $a(\cdot, \cdot) : V \times V \to R$. Note that at this stage V may be finite- or infinite-dimensional. Consider an arbitrary additive representation of V by the sum of a finite or infinite number of subspaces $V_j \subset V$:

$$(2.1) \qquad V = \sum_j V_j \,,$$

this means that any $u \in V$ possesses at least one V-converging representation $u = \sum_j u_j$ where $u_j \in V_j$ for all j. Suppose that the V_j are equipped with auxiliary continuous s.p.d. forms $b_j(\cdot, \cdot) : V_j \times V_j \to R$. We call

$$(2.2) \qquad \{V; a\} = \sum_j \{V_j; b_j\} \,,$$

stable splitting of $\{V; a\}$ if the quantity

$$(2.3) \qquad \||u|\| = \inf_{u_j \in V_j \,:\, u = \sum_j u_j} \sqrt{\sum_j b_j(u_j, u_j)}$$

defines an equivalent norm on V, i.e. if the bounds

$$(2.4) \qquad \lambda_{\min} = \inf_{u \neq 0} \frac{a(u, u)}{\||u|\|^2} \,, \qquad \lambda_{\max} = \sup_{u \neq 0} \frac{a(u, u)}{\||u|\|^2} \,,$$

are positive and finite. The quantity

$$(2.5) \qquad \kappa \equiv \kappa(\{V; a\} = \sum_j \{V_j; b_j\}) = \frac{\lambda_{\max}}{\lambda_{\min}}$$

will be called *stability constant* or simply *condition number* of the splitting (2.2). Note that if the splitting (2.2) is into a finite number of subspaces then it is automatically stable and the difference is only in the size of κ. Also, stability and condition do not change if we change the ordering of the subspaces.

Introduce the operators $T_j : V \to V_j$ by the auxiliary variational problems

$$(2.6) \qquad b_j(T_j u, v_j) = a(u, v_j) \qquad \forall \, v_j \in V_j \,.$$

If the splitting (2.2) is stable then it is easy to show that the associated additive Schwarz operator

$$(2.7) \qquad P = \sum_j T_j : V \to V$$

is well-defined and s.p.d. on V, with exact lower and upper bounds for its spectrum given by the constants $\lambda_{\min}$ and $\lambda_{\max}$ from (2.4). Moreover, if for a given linear continuous functional Φ on V we define elements $\phi_j \in V_j$ and $\phi = \sum_j \phi_j \in V$ via

(2.8) $b_j(\phi_j, v_j) = \Phi(v_j) \qquad \forall\, v_j \in V_j$,

then the given variational problem

(2.9) $find\ u \in V\ such\ that \qquad a(u, v) = \Phi(v) \qquad \forall\, v \in V$,

is equivalent to the operator equation

(2.10) $find\ u \in V\ such\ that \qquad Pu = \phi$.

This is the so-called additive Schwarz formulation of (2.9) associated with the splitting (2.2), see [27, 30, 10, 12] for historical references.

In [12] we gave another reformulation of (2.9) as operator equation in the product Hilbert space $\tilde{V} = \times_j V_j$ which is useful in connection with the treatment of additive and multiplicative subspace correction methods. In order not to talk about computationally irrelevant situations, let from now on (2.2) be a finite splitting, i.e. let $j = 0, \dots, J$, and suppose in addition that V is finite-dimensional. The additive algorithm associated with the splitting (2.2) is typically defined as the Richardson method applied to (2.10):

(2.11) $u^{(l+1)} = u^{(l)} - \omega(Pu^{(l)} - \phi) = u^{(l)} - \omega \sum_{j=0}^{J} (T_j u^{(l)} - \phi_j)$,

$l = 0, 1, \dots$, with $u^{(0)} \in V$ any given initial approximation to the solution u of (2.9) resp. (2.10), and ω a relaxation parameter. Alternatively, one may apply the conjugate gradient method to the equation (2.10), relying on the same theoretical analysis.

In contrast to the parallel incorporation of the subspace corrections $r_j^{(l)} = T_j u^{(l)} - \phi_j$ into the iteration (2.11), the multiplicative algorithm uses them in a sequential way:

(2.12) $v^{(l+(j+1)/(J+1))} = v^{(l+j/(J+1))} - \omega(T_j v^{(l+j/(J+1))} - \phi_j)$,

where $j = 0, \dots, J$, $l = 0, 1, \dots$.

The simple observation which was made in [12] is that the analysis of the abstract methods (2.11), (2.12) can be carried out in almost the same spirit as in the traditional block-matrix situation if one switches from the operator P to the matrix-operator

(2.13) $\tilde{P} = \{T_i|_{V_j}\}$

acting in the auxiliary Hilbert space $\tilde{V}$. Let $\tilde{P} = \tilde{L} + \tilde{D} + \tilde{U}$ be the decomposition of $\tilde{P}$ into lower triangular, diagonal, and upper triangular parts (note that $\tilde{U} =$

$\tilde{L}^T$ since $\tilde{P}$ is symmetric positive semi-definite in $\tilde{V}$). The following result which explains also the central role of the above stability concept is contained in [**12**], see [**14**] for statements of this type in the matrix case.

THEOREM 2.1. *Suppose that V is finite-dimensional, and that the splitting (2.2) is finite. Let the characteristic numbers $\lambda_{\max}, \lambda_{\min}$, and κ of the splitting be given by (2.4), (2.5).*
(a) *The additive method (2.11) converges for $0 < \omega < 2/\lambda_{\max}$, with the optimal asymptotic convergence rate*

$$(2.14) \qquad \rho_{as}^* = 1 - \frac{2}{1+\kappa} \qquad (\omega = \frac{2}{\lambda_{\max} + \lambda_{\min}}) .$$

(b) *The multiplicative method (2.12) converges for $0 < \omega < 2/\|\tilde{D}\|$, with a bound for the optimal asymptotic convergence rate given by*

$$(2.15) \quad (\rho_{ms}^*)^2 \leq 1 - \frac{2}{(\sqrt{q^2+1}+1)\kappa} , \qquad (\omega = \frac{2(\sqrt{q^2+1}-1)}{\lambda_{\max}}) ,$$

*where the quantity $q = 2\|\tilde{L}\|/\lambda_{\max}$ can be further estimated under additional assumptions, e.g. assuming the validity of strengthened Cauchy-Schwarz inequalities for the splitting (see [**27**, **30**, **10**]). Without additonal assumptions, we have a guaranteed worst case estimate $q \leq \lceil \log_2(4J) \rceil$ which leads to*

$$(2.16) \qquad \rho_{ms}^* \leq 1 - \frac{1}{\alpha_J \kappa} , \qquad \alpha_J \approx \log_2 J , \ J \to \infty .$$

For full proofs and more details, see [**12**]. Note that several modifications of the above basic additive and multiplicative schemes, e.g. the analog of symmetric SOR, may be studied along the same lines. Though the estimate (2.16) is asymptotically (for $J \to \infty$) best possible in the general case [**21**], it is too rough to explain the better convergence rates of the multiplicative scheme observed in practical applications to special problem classes.

Concluding this section, we want to emphasize the crucial role played by the condition number of the splitting for both the additive and multiplicative algorithms. In our opinion, the derivation of a computationally suitable algorithm should include a thorough analysis of the behavior of the stability constants in order to make sure that the method is close to an optimal one. In the remaining sections we implement this strategy in a particular situation: we briefly present basic splittings for typical variational problems in Sobolev spaces, and apply them to derive some known domain decomposition algorithms.

3. Splittings for C^0 finite elements

Let $\Omega \subset R^d$ be a bounded polyhedral domain equipped with a nested sequence of partitions

$$(3.1) \qquad \mathcal{T}_0 \prec \mathcal{T}_1 \prec \ldots \prec \mathcal{T}_j \prec \ldots$$

into d-dimensional simplices (or, if the domain is rectangular-like, into d-dimensional rectangles etc.) which are regular and quasi-uniform, with constants that are independent of j. Suppose that

$$(3.2) \qquad h_j \equiv \max_{\Delta \in \mathcal{T}_j} \operatorname{diam}(\Delta) \approx 2^{-j} , \quad j = 0, 1, \ldots .$$

In practice, (3.1) is often produced by regular dyadic refinement from an initial partition, and the constants characterizing regularity and quasi-uniformity in (3.1), (3.2) depend only on $\mathcal{T}_0$.

Consider the sequence of linear C^0 Lagrange finite element subspaces

$$(3.3) \qquad S_0 \subset S_1 \subset \ldots \subset S_j \subset \ldots$$

corresponding to (3.1). The usual nodal basis of S_j will be denoted by $\mathcal{N}_j = \{N_{j,P} : P \in \mathcal{V}_j\}$ where $\mathcal{V}_j$ is the set of all vertices (or nodal points) of $\mathcal{T}_j$. Let $S_{j,P}$ be the one-dimensional subspace of S_j spanned by the nodal basis function $N_{j,P}$ ($P \in \mathcal{V}_j$, $j = 0, 1, \ldots$).

For the definition of the Sobolev spaces $H^s(\Omega)$ we refer to [25, 26, 13]). The following theorem is essentially contained in [15], see also [22] or our survey [19], other proofs have recently been given by Bramble, Pasciak [2], Xu [27], Zhang [31], Dahmen, Kunoth [5], Bornemann, Yserentant [1].

THEOREM 3.1. *Let* $0 < s < 3/2$. *Suppose that* $a(\cdot, \cdot)$ *is a symmetric* H^s-*elliptic bilinear form on* $H^s(\Omega)$. *Then, under the above assumptions on* (3.3), *the following splittings are stable:*

$$(3.4) \qquad \{H^s(\Omega); a(\cdot, \cdot)\} = \sum_{j=0}^{\infty} \{S_j; 2^{2sj}(\cdot, \cdot)_{L_2(\Omega)}\}$$

$$(3.5) \qquad \{H^s(\Omega); a(\cdot, \cdot)\} = \sum_{j=0}^{\infty} \sum_{P \subset \mathcal{V}_j} \{S_{j,P}; 2^{2sj}(\cdot, \cdot)_{L_2(\Omega)}\}$$

$$(3.6) \qquad \{H^s(\Omega); a(\cdot, \cdot)\} = \sum_{j=0}^{\infty} \sum_{P \in \mathcal{V}_j} \{S_{j,P}; a(\cdot, \cdot)\}$$

The characteristic constants $\lambda_{\min}$, $\lambda_{\max}$, *and* κ *for these splittings depend only on the constants characterizing the regularity and quasi-uniformity of the partitions, on* s, *and on the ellipticity constants of the bilinear form.*

We will call these splittings *basic* since many other results about computationally relevant splittings can be deduced from Theorem 3.1. Note that (3.5) and (3.6) are consequences of (3.4). Indeed, the L_2-stability of the nodal basis and a comparison of the L_2 and H^s norms for nodal basis functions gives the stability of the splittings

$$(3.7) \quad \{S_j; 2^{2sj}(\cdot,\cdot)_{L_2(\Omega)}\} = \sum_{P \in \mathcal{V}_j} \{S_{j,P}; 2^{2sj}(\cdot,\cdot)_{L_2(\Omega)}\} = \sum_{P \in \mathcal{V}_j} \{S_{j,P}; a(\cdot,\cdot)\}\,,$$

with uniformly bounded condition numbers κ_j in both cases. Thus, it remains to substitute (3.7) into (3.4) to get the remaining splittings of Theorem 3.1.

We do not state the immediate corollaries of Theorem 3.1 concerning splittings for the trace spaces

$$H^s(\Omega)|_\Gamma = \{h \in L_2(\Gamma) \,:\, \exists\, f \in H^s(\Omega) : h = f|_\Gamma,\ \|h\|_{H^s|_\Gamma} = \inf_{h=f|_\Gamma} \|f\|_{H^s(\Omega)}\}$$

(this class essentially coincides with $H^{s-1/2}(\Gamma)$), and for the spaces

$$H^s_\Gamma(\Omega) = \{f \in H^s(\Omega) \,:\, f|_\Gamma = 0\}$$

which are necessary to handle Dirichlet boundary conditions for second order elliptic problems (in both cases it is assumed that $1/2 < s < 3/2$, and that Γ is the union of some $(d-1)$-dimensional faces of simplices from $\mathcal{T}_0$). Instead, we quote some computationally relevant finite splittings which fit the assumptions of Theorem 2.1 and lead to fast subspace correction methods for solving elliptic finite element discretizations.

THEOREM 3.2. *Let the assumptions of Theorem 3.1 be fulfilled. Then, for all $0 \le j_0 < J < \infty$, the following finite splittings possess uniform condition number estimates depending only on s and on the κ in Theorem 3.1:*

$$(3.8) \qquad \{S_J; a(\cdot,\cdot)\} = \{S_{j_0}; a(\cdot,\cdot)\} + \sum_{j=j_0+1}^{J} \sum_{P \in \mathcal{V}_j} \{S_{j,P}; 2^{2sj}(\cdot,\cdot)_{L_2(\Omega)}\}$$

$$(3.9) \qquad \{S_J; a(\cdot,\cdot)\} = \{S_{j_0}; a(\cdot,\cdot)\} + \sum_{j=j_0+1}^{J} \sum_{P \in \mathcal{V}_j} \{S_{j,P}; a(\cdot,\cdot)\}$$

$$(3.10) \qquad \{S_J; a(\cdot,\cdot)\} = \{S_{j_0}; a(\cdot,\cdot)\} + \sum_{P \in \mathcal{V}_J} \{S_{P,j_0,J}; a(\cdot,\cdot)\}$$

where $S_{P,j_0,J} = span\,\{N_{j,P} \,:\, \forall\, j_0 < j \le J : P \in \mathcal{V}_j\}$. In all cases, $a(\cdot,\cdot)$ is the natural restriction of the bilinear form defined on $H^s(\Omega)$ onto the respective subspaces. The statement holds for $j_0 = -1$ if the first term in the splittings is dropped ($S_{-1} = \{0\}$). The extension of the results to the subspaces $S_{J,\Gamma} = \{g \in S_J \,:\, g|_\Gamma = 0\}$ of $H^s_\Gamma(\Omega)$, $1/2 < s < 3/2$, is immediate.

The algorithm behind (3.8) was introduced by Bramble, Pasciak, Xu [4], see also [29, 1], (3.9) goes back to X.Zhang [31]. The third splitting (3.10) was introduced and analyzed (for $j_0 = -1$) by Griebel, see [11]. For details of the proof of Theorem 3.2, see [19, 22].

All splittings presented in Theorem 3.2 are overlapping. One may ask for uniformly stable nonoverlapping splittings for $\{S_J; a(\cdot, \cdot)\}$. Stable nonoverlapping splittings of a Hilbert space into one-dimensional subspaces are actually equivalent to unconditional Schauder bases (or Riesz bases) which leads to a classical problem that is interesting on its own. Orthonormal bases are particular cases but hard to construct, especially for Sobolev spaces on domains. It is still an open question to define and implement a prewavelet-like system of locally supported finite element functions corresponding to (3.3) such that it forms an unconditional basis in $L_2(\Omega)$ for arbitrary regular and quasi-uniform sequences of partitions (3.1) (see [19] for the consequences of the existence of such a system in the spirit of Theorem 3.2). For nice domains and sequences of shift-invariant partitions, such constructions are essentially known.

Another early attempt to use splittings into a direct sum of one-dimensional subspaces is the hierarchical basis method introduced by Yserentant [28, 29]. The condition number estimates for the hierarchical basis splitting can also be obtained as consequences of Theorem 3.1, see [16].

Let us briefly mention the case of higher degree Lagrange elements. There are two ways to deal with them: on the one hand, we can develop the whole machinery of infinite splittings of Sobolev spaces for these elements, see [15]. On the other hand, if we are mostly interested in algorithms, we can simply reduce the construction of iterative methods for the new element types to the case of linear elements on the same sequence of partitions by a procedure which corresponds to condensation of inner variables. This latter approach seems to be preferable for several reasons, applications to nonconforming schemes have been described in [20]. However, we do not know about a serious performance comparison of these two possibilities. For Hermite or serendipity elements where the monotonicity of the family of finite element subspaces (3.3) is violated, one is recommended to use the second strategy.

It is well-known that iterative methods based on subspace splittings of the above type can be carried over to an adaptive environment. What is not so easy (and, therefore, a drawback of our approach) is to overcome the restrictions on the geometry of the domain and on the construction of the partitions implicitly contained in (3.1), (3.2). Domains that do not allow for a simple initial partition into a few simplices of diameter ≈ 1 or cannot be reduced to this situation after a dilation tend to produce theoretically larger condition number estimates. Since the underlying triangulation of a finite-element discretization space may be produced by some grid-generation or -optimization method which does not care about having a sequence of partitions (3.1) but rather provides us with some $\mathcal{T}_J$ we may have some trouble. Also, non-polyhedral domains and the treatment of isoparametric elements which occur in engineering problems require additional ideas and arguments to get a smooth theory (this does not necessarily mean that the algorithms, if properly adopted, will not work for these new situations).

4. Applications to domain decomposition

The use of Theorem 3.1 and 3.2 for domain decomposition methods is quite obvious, similar ideas are contained in [23, 24]. Throughout this section, we assume that the conditions on the domain, on the sequence of partitions (3.1), (3.2), and on the sequence of linear finite element subspaces (3.3) are the same as in Section 2. Moreover, to simplify the notation, we will consider a symmetric H^1-elliptic variational problem $a(\cdot,\cdot)$ on $H^1(\Omega)$, the case of essential boundary conditions is completely analogous.

4.1. Nonoverlapping domain decomposition schemes. Suppose that Ω is decomposed into nonoverlapping subdomains. In order to keep the considerations simple and to use the basic splittings in a straightforward way, we assume that the subdomains are identical with the simplices of some of the partitions $\mathcal{T}_{j_0}$ (we may actually allow groups of less than a fixed number of such simplices to form the subdomains). We denote the subdomains by Ω_l, and introduce the notations $S_{j,l}$ for the set of finite element functions from S_j with support in Ω_l, $\mathcal{V}_{j,l}$ for the set of nodal points interior to Ω_l, and set $\mathcal{V}_{j,\gamma} = \mathcal{V}_j \setminus \cup_l \mathcal{V}_{j,l}$ for the part of $\mathcal{V}_j$ located on $\gamma = \cup_l(\partial\Omega_l \cap \Omega)$. Here, $j \geq j_0$, by our assumptions the subspaces $S_{j_0,l}$ are trivial, accordingly, $\mathcal{V}_{j_0,l}$ is empty.

For any $J > j_0$, consider the splitting (3.8) of Theorem 3 and group the one-dimensional subspaces as follows:

$$(4.1) \quad \{S_J; a(\cdot,\cdot)\} = \sum_l \left(\sum_{j=j_0+1}^{J} \sum_{P\in\mathcal{V}_{j,l}} \{S_{j,P}; 2^{2j}(\cdot,\cdot)_{L_2(\Omega_l)}\} \right)$$

$$+ \left(\{S_{j_0}; a(\cdot,\cdot)\} + \sum_{j=j_0+1}^{J} \sum_{P\in\mathcal{V}_{j,\gamma}} \{S_{j,P}; 2^{2j}(\cdot,\cdot)_{L_2(\Omega)}\} \right).$$

The first sum contains groups of subspaces that form splittings of the $S_{J,l}$. Using a simple scaling argument, we can apply Theorem 3, (3.8), on the subdomains Ω_l:

$$\{S_{J,l}; a(\cdot,\cdot)\} = \sum_{j=j_0+1}^{J} \sum_{P\in\mathcal{V}_{j,l}} \{S_{j,P}; 2^{2j}(\cdot,\cdot)_{L_2(\Omega_l)}\}$$

are stable splittings, with a common bound for their condition numbers which is independent of l. Thus, we arrive at another stable splitting

$$(4.2) \qquad \{S_J; a(\cdot,\cdot)\} = \sum_l \{S_{J,l}; a(\cdot,\cdot)\}$$

$$+ \left(\{S_{j_0}; a(\cdot,\cdot)\} + \sum_{j=j_0+1}^{J} \sum_{P \in \mathcal{V}_{j,\gamma}} \{S_{j,P}; 2^{2j}(\cdot,\cdot)_{L_2(\Omega)}\} \right).$$

Before going on with rewriting our basic splittings, let us introduce the Schur complement form $s_J(\cdot,\cdot)$ on $S_J|_\gamma$ induced by the form $a(\cdot,\cdot)$ and corresponding to the given subdomain structure, by setting

$$(4.3) \qquad s_J(u_\gamma, u_\gamma) = \inf_{u \in S_J \,:\, u_\gamma = u|_\gamma} a(u,u) \quad \forall u_\gamma \in S_J|_\gamma.$$

This bilinear form corresponds to the Schur complement problem on γ which arises if in (2.9) the unknowns corresponding to interior nodal points (i.e. $P \in \mathcal{V}_{j,l}$ for some l) are eliminated.

THEOREM 4.1. *Under the above assumptions on γ and the bilinear form, the following splitting for the Schur complement problem on γ is stable, with a condition number that is independent of j_0 (coarse level) and J:*

$$(4.4)\; \{S_J|_\gamma; s_J(\cdot,\cdot)\} = \{S_{j_0}|_\gamma; s_{j_0}(\cdot,\cdot)\} + \sum_{j=j_0+1}^{J} \sum_{P \in \mathcal{V}_{j,\gamma}} \{S_{j,P}|_\gamma; 2^j(\cdot,\cdot)_{L_2(\Omega)}\}.$$

The proof of the stability of (4.4) runs as follows. By (4.3) and (4.2)

$$s_J(u_\gamma, u_\gamma) = \inf_{u \in S_J \,:\, u_\gamma = u|_\gamma} a(u,u)$$

$$\approx \inf_{u \in S_J \,:\, u_\gamma = u|_\gamma} \; \inf_{u = u_{j_0} + \sum_{j=j_0+1}^{J} \sum_{P \in \mathcal{V}_{j,\gamma}} u_{j,P}} \left(a(u_{j_0}, u_{j_0}) \right.$$

$$\left. + \sum_{j=j_0+1}^{J} \sum_{P \in \mathcal{V}_{j,\gamma}} 2^{2j}(u_{j,P}, u_{j,P})_{L_2(\Omega)} \right)$$

$$\approx \inf_{u_\gamma = u_{j_0,\gamma} + \sum_{j=j_0+1}^{J} \sum_{P \in \mathcal{V}_{j,\gamma}} u_{j,P,\gamma}} \left(a(u_{j_0}, u_{j_0}) \right.$$

$$\left. + \sum_{j=j_0+1}^{J} \sum_{P \in \mathcal{V}_{j,\gamma}} 2^{2j}(u_{j,P}, u_{j,P})_{L_2(\Omega)} \right).$$

Here we have already used that $u_{j,l}|_\gamma = 0$ for all subdomains. u_{j_0} resp. $u_{j,P}$ denote the unique extensions of $u_{j_0,\gamma} \in S_{j_0,\gamma}$ resp. $u_{j,P,\gamma} \in S_{j,P}|_\gamma$ to functions in S_{j_0} resp. $S_{j,P}$ (note that the latter are one-dimensional). Now it remains to express the bilinear forms in terms of $u_{j_0,\gamma}$ resp. $u_{j,P,\gamma}$ which leads to (4.4). To this end, use the definition (4.3) (with J replaced by j_0), and

$$\|N_{j,P}|_\gamma\|^2_{L_2(\gamma)} \approx 2^j \|N_{j,P}\|^2_{L_2(\Omega)}, \quad P \in \mathcal{V}_{j,\gamma}$$

where, once again, the regularity and quasi-uniformity of the partitions comes in.

One can prove analogs of (3.9) or (3.10) as well. Another possibility is to group the subspaces of (4.4) according to the geometrical structure of γ. E.g., in 2D applications one could associate groups with each edge of γ and with each vertex of $\mathcal{T}_{j_0}$. Then the computations of the subspace corrections for edges can be carried out independently, e.g. on different processors, the same is true for the vertex components if one allows for a certain redundancy in the computations. There were different proposals to neglect the vertex components to simplify the computations, but it is clear that this results in an increase of the condition number of the reduced splitting. In 3D, the corresponding substructures are faces and the wirebasket (composed of edges and vertices). One can introduce a lot of modifications (especially on the wirebasket), and also change the auxiliary problems on the substructures. For some of the many algorithms of this type which have been derived and analyzed by other researchers using different techniques, we refer to work by Dryja, Widlund et.al. (see [8, 10] and the references cited therein). Note that these authors deal also with more complicated situations which are not covered by our reasoning.

4.2. Domain decomposition methods with overlap. In the domain decomposition methods with overlap, the subregions Ω_l are enlarged to a certain extent giving domains on which local Dirichlet problems are solved (these are the subspace problems involved in such type of algorithms). More precisely, under the same assumptions on $\{\Omega_l\}$ as in the previous subsection we will compose the enlarged regions $\hat{\Omega}_{l,j_1}$ of all simplices (rectangles etc.) in $\mathcal{T}_{j_1}$ that intersect with the closure of Ω_l. The number j_1 is chosen between j_0 and J and characterizes the amount of overlap (if $j_1 = j_0$ the overlap is called generous, if $j_1 = J$ we have minimal overlap). Let $\hat{S}_{j;l,j_1}$ denote the subspace consisting of all finite element functions from S_j that vanish outside of $\hat{\Omega}_{l,j_1}$.

Consider first $j_1 = j_0$ or $j_1 = j_0 + 1$. In this case, any of the one-dimensional subspaces $S_{j,P}$ ($P \in \mathcal{V}_j$, $j = j_0 + 1, \ldots, J$) belongs to at least one and at most $d + 1$ (if simplicial triangulations are used) of the $\hat{S}_l \equiv \hat{S}_{J;l,j_1}$. Since, generally,

$$\{V; a(\cdot, \cdot)\} = \sum_{j=1}^{m} \{V; a(\cdot, \cdot)\}$$

is a stable splitting with $\lambda_{\min} = \lambda_{\max} = m$ and condition number exactly 1, we can refine the splitting (3.8) from Theorem 3.2 by adding the necessary number of copies of one-dimensional subspaces without destroying the condition number too much, i.e. under the assumptions of Theorem 3 we get the uniform stability of the splittings

$$\{S_J; a(\cdot, \cdot)\} = \{S_{j_0}; a(\cdot, \cdot)\} + \sum_{l} (\sum_{j=j_0+1}^{J} \sum_{supp\, N_{j,P} \in \hat{\Omega}_{l,j_1}} \{S_{j,P}; 2^{2j}(\cdot, \cdot)_{L_2(\Omega)}\}).$$

Now it remains to apply once again a scaled version of Theorem 3.2, (3.8), to $\hat{\Omega}_{l,j_1}$. This leads to the particular cases $j_1 = j_0, j_0 + 1$ of the following

THEOREM 4.2. *Under the above assumptions, the condition numbers of the splitting*

$$(4.5) \qquad \{S_J; a(\cdot, \cdot)\} = \{S_{j_0}; a(\cdot, \cdot)\} + \sum_l \{\hat{S}_l; a(\cdot, \cdot)\},$$

behave like $\approx 2^{j_1 - j_0}$, *with constants that are independent of* $j_0 \le j_1 \le J$. *The result extends to* $S_{J;\Gamma}$.

This result is known, see the recent papers [8, 9] where the effect of smaller up to minimal overlap is studied in a different way. Our approach to the case of general j_1 is first to switch from (3.8) to the subsplitting

$$\{S_J; a(\cdot, \cdot)\} = \{S_{j_0}; a(\cdot, \cdot)\} + \sum_{j=j_0+1}^{J} \sum_{P \in \mathcal{V}_j \, : \, \exists \, l \; supp \, N_{j,P} \in \hat{\Omega}_{l,j_1}} \{S_{j,P}; 2^{2j}(\cdot, \cdot)_{L_2(\Omega)}\},$$

where the condition number degenerates by the factor $O(2^{j_1 - j_0})$, and then to apply the above arguments to the groups of subspaces corresponding to the subdomains $\hat{\Omega}_{l,j_1}$. Due to the lack of space, we omit the technical details.

REFERENCES

1. F. A. Bornemann and H. Yserentant, *A basic norm equivalence for the theory of multilevel methods*, Numer. Math. **64** (1993), 455-476.
2. J. H. Bramble and J. E. Pasciak, *New estimates for multilevel methods including the V-cycle*, Math. Comp. **60** (1993), 447-471.
3. J. H. Bramble, J. E. Pasciak, J. Wang, and J. Xu, *Convergence estimates for product iterative methods with applications to domain decomposition*, Math. Comp. **57** (1991), 1-21.
4. J. H. Bramble, J. E. Pasciak, and J. Xu, *The analysis of multigrid algorithms with nonested spaces or noninherited quadratic forms*, Math. Comp. **56** (1991), 1-34.
5. W. Dahmen and A. Kunoth, *Multilevel preconditioning*, Numer. Math. **63** (1992), 315-344.
6. W. Dahmen, P. Oswald, and X.-Q. Shi, C^1 *hierarchical bases*, J. Comp. Appl. Math. (to appear).
7. W. Dahmen, S. Prößdorf, and R.Schneider, *Wavelet approximation methods for pseudodifferential operators II: Matrix compression and fast solution*, Advances in Computational Mathematics **1** (1993), 259-335.
8. M. Dryja and O. Widlund, *Some recent results on Schwarz type domain decomposition algorithms*, Proc. Sixth International Conference on Domain Decomposition in Science and Engineering (Como 1992), Tech.Rep. 615, Courant Inst., New York University, September 1992.
9. _____ , *Domain decomposition algorithms with small overlap*, SIAM J. Sci. Stat. Comput. (to appear).
10. M. Dryja, B. F. Smith, and O. Widlund, *Schwarz analysis of iterative substructuring algorithms for elliptic problems in three dimensions* Preprint, Courant Inst., New York Univ., May 1993.
11. M. Griebel, *Punktblock-Multilevelmethoden zur Lösung elliptischer Differentialgleichungen*, Habilitation, Inst. für Informatik, TU Munich 1993, Teubner-Skripten zur Numerik, Teubner, Stuttgart, 1994. Teubner, Stuttgart 1994.
12. M. Griebel and P. Oswald, *Remarks on the abstract theory of additive and multiplicative Schwarz methods*, Numer. Math. (to appear), preliminary version: Report TUM-I9314, TU Munich, July 1993.

13. P. Grisvard, *Elliptic problems in non-smooth domains*, Pitman Monogr. v. 24, Longman Sci.& Techn., Harlow 1985.

14. W. Hackbusch, *Iterative Lösung großer schwachbesetzter Gleichungssysteme*, Teubner, Stuttgart 1991.

15. P. Oswald, *On function spaces related to finite element approximation theory*, Z. Anal. Anwendungen **9** (1990), 43-64.

16. _____, *On discrete norm estimates related to multilevel preconditioners in the finite element method*, Constructive Theory of Functions, Proc. Int. Conf. Varna 1991 (K.G.Ivanov, P.Petrushev, B.Sendov, eds.), Bulg. Acad. Sci., Sofia, 1992, 203-214.

17. _____, *Hierarchical conforming finite element methods for the biharmonic equation*, SIAM J. Numer. Anal. **29** (1992), 1610-1625.

18. _____, *Multilevel preconditioners for discretizations of the biharmonic equation by rectangular finite elements*, J. Numer. Lin. Alg. Appl. (submitted), preliminary version: Preprint Math/91/3, FSU Jena, November 1991.

19. _____, *Multilevel finite element approximation: theory & applications*, Lecture notes, TU Munich, March/April 1993, Teubner-Skripten zur Numerik, Teubner, Stuttgart, 1994 (to appear).

20. _____, *Preconditioners for nonconforming elements*, Math. Comp. (submitted), preliminary version: Preprint Math/93/3, FSU Jena, June 1993.

21. _____, *On the convergence rate of SOR: a worst case estimate*, Computing **52** (1994), 245-255.

22. _____, *Stable subspace splittings for Sobolev spaces and their applications*, Preprint Math/93/7, FSU Jena, September 1993.

23. B. Smith and O. Widlund, *A domain decomposition algorithm using a hierarchical basis*, SIAM J. Sci. Stat. Comput. **11** (1990), 1212-1226.

24. C. H. Tong, T. F. Chan, and C. C. J. Kuo, *A domain decomposition preconditioner based on a change to a multilevel nodal basis*, SIAM J. Sci. Stat. Comput. **12** (1991), 1486-1495.

25. H. Triebel, *Interpolation theory, Function spaces, Differential operators*, Dt. Verl. Wiss., Berlin, 1978 - North-Holland, Amsterdam ,1978.

26. _____, *Theory of function spaces* II. Birkhäuser, Basel ,1992.

27. J. Xu, *Iterative methods by space decomposition and subspace correction*, SIAM Review **34** (1992), 581-613.

28. H Yserentant, *On the multi-level splittings of finite element spaces*, Numer. Math. **49** (1 ?6), 379-412.

29. _____, *Two preconditioners based on the multi-level splitting of finite element spaces*, Nun ?. Math. **58** (1990), 163-184.

30. _____, *Old and new convergence proofs for multigrid methods*, Acta Numerica 1993, Cambriuge Univ. Press, New York, 1993, 285-326.

31. X. Zhang, *Multilevel Schwarz methods*, Numer. Math. **63** (1992), 521-539.

32. _____, *Multilevel Schwarz methods for the biharmonic Dirichlet problem*, Tech. Report UMIACS-TR-92-52, Univ. Maryland, College Park, May 1992.

INSTITUT FÜR ANGEWANDTE MATHEMATIK, FSU JENA, D-07740 JENA, GERMANY
E-mail address: poswald@minet.uni-jena.de

Contemporary Mathematics
Volume **180**, 1994

A Wire Basket Based Method for
Spectral Elements in Three Dimensions

LUCA F. PAVARINO

ABSTRACT. A two-level iterative substructuring algorithm with a wire basket based coarse space is proposed and analyzed. The three dimensional
model problem is scalar, elliptic, with discontinuous coefficients, and is
discretized by conforming spectral elements. The condition number of the
resulting iteration operator is bounded by $C(1 + \log p)^2$, where the constant
is independent of the number of elements, their diameters, their degree p,
and the size of the jumps across element boundaries of the coefficients of
the elliptic operator. The results of this paper have been obtained jointly
with Olof B. Widlund.

1. Introduction

Iterative substructuring methods are two-level domain decomposition methods based on nonoverlapping subregions. For the h-version finite element method,
extensive research has been conducted in the last decade and many algorithms
have been proposed for three dimensional problems; see e.g. Bramble, Pasciak,
and Schatz [**2**], Dryja [**4**], Dryja and Widlund [**6**], Smith [**15**], and Le Tallec, De
Roeck and Vidrascu [**7**]. A recent paper by Dryja, Smith, and Widlund [**5**] sum-
marizes the current knowledge of the h-version case. See also Chan and Mathew
[**3**] for an overview of domain decomposition algorithms.

For p-version finite elements and spectral methods, the construction of iterative substructuring methods is more challenging, since the stiffness matrices
can be much more ill-conditioned and different mathematical tools are needed.
See Babuška, Craig, Mandel, and Pitkäranta [**1**] for two dimensional problems,
Mandel [**10**] for three dimensional problems, Pavarino [**12, 11**] for overlapping

1991 *Mathematics Subject Classification.* 41A10, 65N30, 65N35, 65N55.
This work was supported by the U.S. Department of Energy under contract DE-FG-05-
92ER25142 and by the State of Texas under contract 1059.
The final detailed version of this paper has been submitted for publication elsewhere.

methods in two and three dimensions, Rønquist [14] and Maday and Patera [8] for spectral element methods.

In this paper, we propose a wire basket based algorithm (in the terminology of Dryja, Smith, and Widlund [5]) with condition number bounded by $C(1+\log p)^2$, where p is the degree of the spectral elements. This method is directly inspired by a method developed for the h-version by Smith [15, 16]. Complete proofs of the results can be found in Pavarino and Widlund [13].

2. The elliptic problem and its discretization

We consider a three dimensional domain $\Omega = \bigcup_{i=1}^{N} \Omega_i$, a union of elements which are cubes or smooth images of a reference cube. We consider a model, elliptic problem on Ω with Dirichlet boundary conditions on $\Gamma_D \subset \partial\Omega$ and Neumann boundary conditions on $\partial\Omega - \Gamma_D$: find $u \in V = H^1_{\Gamma_D}(\Omega)$ such that

$$(1) \qquad a(u,v) = \sum_{i=1}^{N} \int_{\Omega_i} \rho_i \nabla u \cdot \nabla v \, dx = \int_{\Omega} fv \, dx \qquad \forall v \in V.$$

The values $\rho_i > 0$ can be very different in different subregions. (1) is discretized by a continuous, piecewise Q_p Galerkin method, using conforming spectral elements. The discrete space $V^p \subset V$ is given by:

$$V^p = \{v_p \in C^0(\Omega) : v_p|_{\Omega_i} \in Q_p, \ i = 1, \cdots, N\} \cap H^1_{\Gamma_D}(\Omega).$$

The finite element problem obtained is turned into a linear system of algebraic equations, $K\underline{u} = \underline{b}$, by choosing a basis in V^p. A basis particularly useful in the convergence analysis of the method will be given in the next sections, but more practical hierarchical bases can also be used. As usual in the literature for spectral and p-version finite elements, we distinguish between interior basis functions, with support in the interior of an element and interface basis functions, with support intersecting the interface $\Gamma = \bigcup_{i=1}^{N} \partial\Omega_i$ (these can be further divided into face, edge and vertex basis functions). The coefficients of the unknown functions are partitioned accordingly, $\underline{u} = (\underline{u}_I, \underline{u}_B)$. As in most iterative substructuring algorithms, the unknowns $\underline{u}_I$ associated to the interior basis functions are eliminated first. The reduced Schur complement obtained in this way, $S\underline{u}_B = \tilde{\underline{b}}$, is solved with a preconditioned conjugate gradient method (or a more general Krylov method for nonsymmetric or indefinite problems).

3. The new method

The Schur complement system corresponds to a discrete variational problem posed in the discrete harmonic subspace $\tilde{V}^p$ of V^p, with the inner product $s(u,v) = \underline{u}_B^T S \underline{v}_B$. The functions in $\tilde{V}^p$ are $a(\cdot, \cdot)$-orthogonal to the interior basis functions and they are completely specified by their interface values. The space $\tilde{V}^p$ is decomposed into the direct sum of the following subspaces:

$$\tilde{V}^p = V_0 + \sum_{ij} V_{ij}.$$

Here V_{ij} is the discrete harmonic subspace of $V^p \cap H_0^1(\Omega_{ij})$, where $\Omega_{ij} = \Omega_i \cup F_{ij} \cup \Omega_j$, (one space for each face), and V_0 is a coarse space consisting of piecewise discrete harmonic functions defined solely by their values on the wire baskets. The values on the faces are given by an interpolation operator I^W defined by eq. (3), Section 5. On V_0, we use the bilinear form $\tilde{s}_0(\cdot, \cdot) = (1 + \log p) \sum_i \rho_i \inf_{c_i} \|u - c_i\|_{L_2(\mathcal{W}_i)}^2$. We introduce $s(\cdot, \cdot)$-orthogonal projections P_{ij} onto V_{ij} by $s(P_{ij}u, v) = s(u, v)$, $\forall v \in V_{ij}$ and an approximate projection T_0 onto V_0 by $\tilde{s}_0(T_0 u, v) = s(u, v)$, $\forall v \in V_0$. Our iterative substructuring method consists in solving

$$(2) \qquad Tu = (T_0 + \sum_{ij} P_{ij})u = g.$$

by a conjugate gradient method. This is equivalent to an additive Schwarz preconditioner for the original Schur complement system.

The following is the main result of the paper.

THEOREM 1. *For the iterative substructuring method just introduced,*

$$\kappa(T) \leq const.(1 + \log p)^2.$$

Here the constant is independent of the number of elements, their diameters, the degree p, and the size of the jumps of the coefficient ρ_i across element boundaries.

4. Special sets of polynomials and a basis for $Q_p([-1,1]^3)$

We denote by P_0^p the space of degree p polynomials on $[-1, 1]$ that vanish at the endpoints.

DEFINITION 1. *Let φ_0^+ be the degree p polynomial satisfying*

$$\min_{\varphi} \|\varphi\|_{L_2(-1,1)}, \quad \varphi(-1) = 0, \quad \varphi(1) = 1.$$

DEFINITION 2. *Let $\Phi_i \in P_0^p$ and $\lambda_i, i = 1, \cdots, p - 1$, be the eigenfunctions (normalized with unit H^1-norm) and eigenvalues defined by*

$$\int_{-1}^{1} \frac{d\Phi_i(x)}{dx} \frac{dv(x)}{dx}\, dx = \lambda_i \int_{-1}^{1} \Phi_i(x)v(x)\, dx \qquad \forall v \in P_0^p.$$

DEFINITION 3. *Given the eigenvalues $\{\lambda_i\}_{i=1}^{p-1}$ of Definition 2, define a set $\{\varphi_i^+\}_{i=1}^{p-1}$ of degree p polynomials by $\varphi_i^+(-1) = 0$, $\varphi_i^+(1) = 1$ and*

$$\int_{-1}^{1} \frac{d\varphi_i^+(x)}{dx} \frac{dv(x)}{dx}\, dx + \frac{\lambda_i}{2} \int_{-1}^{1} \varphi_i^+(x)v(x)\, dx = 0 \qquad \forall v \in P_0^p.$$

DEFINITION 4. *Given the eigenvalues* $\{\lambda_i\}_{i=1}^{p-1}$ *of Definition 2, define a set* $\{\varphi_{ij}^+\}_{i,j=1}^{p-1}$ *of degree p polynomials by* $\varphi_{ij}^+(-1) = 0$, $\varphi_{ij}^+(1) = 1$ *and*

$$\int_{-1}^{1} \frac{d\varphi_{ij}^+(x)}{dx} \frac{dv(x)}{dx} \, dx + (\lambda_i + \lambda_j) \int_{-1}^{1} \varphi_{ij}^+(x) v(x) \, dx = 0 \qquad \forall v \in P_0^p.$$

Analogous families of polynomials satisfying opposite boundary conditions, obtained by changing x into $-x$, will be denoted by φ_0^-, $\{\varphi_i^-\}$ and $\{\varphi_{ij}^-\}$. By computing the Legendre expansion of φ_0, it is possible to prove that $\|\varphi_0^+\|_{L^2(-1,1)}^2 = \frac{2}{p(p+2)}$ and $\|\varphi_0^+\|_{L^\infty(-1,1)}^2 \le 1$. We can now define a basis for $Q_p([-1,1]^3)$.

- *Interior functions:* $\Phi_i(x)\Phi_j(y)\Phi_k(z)$, $i,j,k = 1, \cdots, p-1$.
  They are $a(\cdot,\cdot)-$ and L_2-orthogonal.
- *Face functions:* $\varphi_{i,j}^+(x)\Phi_i(y)\Phi_j(z)$, $i,j = 1, \cdots, p-1$,
  for the face defined by $x = 1$.
- *Edge functions:* $e_i^{(1)}(x,y,z) = \varphi_i^+(x)\varphi_i^+(y)\Phi_i(z)$, $i = 1, \cdots, p-1$,
  for the edge E_1 defined by $x = 1, y = 1$.
- *Vertex functions:* $v_i(x,y,z) = \varphi_0^\pm(x)\varphi_0^\pm(y)\varphi_0^\pm(z)$,
  for the vertices $V_i = (\pm 1, \pm 1, \pm 1)$.

The following result follows from a direct computation.

LEMMA 1. *Face, edge and vertex basis functions are discrete harmonic.*

5. Extension from the wire basket

In order to carry out a local analysis of the algorithm (i.e. global bounds from local bounds on individual elements), it is crucial to include the constants in the coarse space V_0, see Lemma 2.2 in Dryja, Smith and Widlund [5] or Theorem 5.1 in Mandel [9]. V_0 can be seen as the range of an interpolation operator. We first introduce locally a preliminary interpolation operator $\tilde{I}^W : V^p \to V_0$ by $\tilde{I}^W u = u_V + u_E$. Here $u_V = \sum_{i=1}^{8} u(V_i)\varphi_0^\pm\varphi_0^\pm\varphi_0^\pm$ is the sum of the vertex components of u and $u_E = \sum_{i=1}^{12}\sum_{j=1}^{p-1} \alpha_j^{(i)} \underbrace{\Phi_j \varphi_j^\pm \varphi_j^\pm}_{permutations}$ is the sum of the edge components of u, with coefficients $\alpha_j^{(i)} = \lambda_j \int_{-1}^{1}(u - u_V)\Phi_j dx$. The range of this interpolation operator does not contain the constants. We therefore consider $\mathcal{F} = \tilde{I}^W 1$, the image of the function identically equal to 1 on the wire basket. $\mathcal{F}$ is not equal to 1 on the faces. In order to recover the constants, consider the function $\kappa = 1 - \mathcal{F}$, which vanishes on the wire basket. It can be split into six discrete harmonic components, each with nonzero values only on one face: $\kappa = \sum_{i=1}^{6} \kappa_i$. The new interpolation operator is then defined by

$$(3) \qquad I^W u = \tilde{I}^W u + \sum_{i=1}^{6} \overline{u}_{\partial F_i} \kappa_i,$$

where $\overline{u}_{\partial F_i} = \frac{1}{8}\int_{\partial F_i} u$. If $u \equiv 1$ on W, then $I^W u \equiv 1$ on $\partial\Omega$. This interpolation operator defines a change of basis in the wire basket space, by mapping edge and vertex basis functions into:

(4)
$$\begin{aligned}
\tilde{e}_j^{(k)} &= I^W e_j^{(k)} = e_j^{(k)} + \sum_i \bar{e}_{j,\partial F_i}^{(k)} \kappa_i, \\
\tilde{v}_j &= I^W v_j = v_j + \sum_i \bar{v}_{j,\partial F_i} \kappa_i.
\end{aligned}$$

6. Matrix form of the preconditioner

We are solving the Schur complement system $S\underline{u}_B = \underline{\tilde{b}}$ obtained by eliminating the interior unknowns. Let us order the face basis functions first and then those of the wire basket. The contribution to the Schur complement S attributable to the element Ω_i can be written as:

$$S^{(i)} = \begin{pmatrix} S_{FF}^{(i)} & S_{FW}^{(i)} \\ S_{FW}^{(i)^T} & S_{WW}^{(i)} \end{pmatrix}.$$

The preconditioner $\hat{S}$ is similarly obtained by subassembly of local contributions $\hat{S}^{(i)}$, constructed for individual substructures. We first change basis for the wire basket space by using the new edge and vertex basis functions defined by (4). The face basis functions remain the same. The transformation from the old to the new basis can be represented by $\begin{pmatrix} I & 0 \\ R^{(i)} & I \end{pmatrix}$. In the new basis $S^{(i)}$ is transformed into

$$\begin{pmatrix} I & 0 \\ R^{(i)} & I \end{pmatrix} \begin{pmatrix} S_{FF}^{(i)} & S_{FW}^{(i)} \\ S_{FW}^{(i)^T} & S_{WW}^{(i)} \end{pmatrix} \begin{pmatrix} I & R^{(i)^T} \\ 0 & I \end{pmatrix} = \begin{pmatrix} S_{FF}^{(i)} & \text{nonzero} \\ \text{nonzero} & \tilde{S}_{WW}^{(i)} \end{pmatrix}.$$

We construct the local preconditioner by replacing $S_{FF}^{(i)}$ with its block diagonal part $\hat{S}_{FF}^{(i)}$ with one block for each face and by dropping the coupling between face and wire basket spaces. $\tilde{S}_{WW}^{(i)}$ is replaced by a rank-one perturbation of a multiple of the identity. It turns out that, in our basis, $\hat{S}_{FF}^{(i)}$ is the diagonal of $S_{FF}^{(i)}$ because each diagonal block is diagonal. We then return to the old basis:

$$\hat{S}^{(i)} = \begin{pmatrix} I & 0 \\ -R^{(i)} & I \end{pmatrix} \begin{pmatrix} \hat{S}_{FF}^{(i)} & 0 \\ 0 & \hat{S}_{WW}^{(i)} \end{pmatrix} \begin{pmatrix} I & -R^{(i)^T} \\ 0 & I \end{pmatrix}.$$

The preconditioner is obtained by subassembly:

$$\hat{S} = \begin{pmatrix} I & 0 \\ -R & I \end{pmatrix} \begin{pmatrix} \hat{S}_{FF} & 0 \\ 0 & \hat{S}_{WW} \end{pmatrix} \begin{pmatrix} I & -R^T \\ 0 & I \end{pmatrix}.$$

Therefore

(5)
$$\hat{S}^{-1} S = R_0 \hat{S}_{WW}^{-1} R_0^T S + \sum_i R_{F_i} \hat{S}_{F_i F_i}^{-1} R_{F_i}^T S,$$

where $R_0 = (R, I)$ (see Dryja, Smith, and Widlund [5]). Clearly this is an additive preconditioner with independent parts associated with the wire basket and each face. (5) is the matrix form of the operator T of (2).

REFERENCES

1. I. Babuška, A. Craig, J. Mandel, and J. Pitkäranta, *Efficient preconditioning for the p-version finite element method in two dimensions*, SIAM J. Numer. Anal. **28** (1991), 624–661.
2. J. H. Bramble, J. E. Pasciak, and A. H. Schatz, *The construction of preconditioners for elliptic problems by substructuring, IV*, Math. Comp. **53** (1989), 1–24.
3. T. F. Chan and T. P. Mathew, *Domain decomposition algorithms*, Acta Numerica (1994), 61–143.
4. M. Dryja, *A method of domain decomposition for 3-D finite element problems*, First International Symposium on Domain Decomposition Methods for Partial Differential Equations (Philadelphia, PA) (R. Glowinski, G. H. Golub, G. A. Meurant, and J. Périaux, eds.), SIAM, 1988.
5. M. Dryja, B. F. Smith, and O. B. Widlund, *Schwarz analysis of iterative substructuring algorithms for elliptic problems in three dimensions*, Technical Report 638, Courant Institute of Mathematical Sciences, May 1993. To appear in SIAM J. Numer. Anal.
6. M. Dryja and O. B. Widlund, *Towards a unified theory of domain decomposition algorithms for elliptic problems*, Third International Symposium on Domain Decomposition Methods for Partial Differential Equations, held in Houston, Texas, March 20-22, 1989 (T. Chan, R. Glowinski, J. Périaux, and O. Widlund, eds.), SIAM, Philadelphia, PA, 1990.
7. P. Le Tallec, Y.-H. De Roeck, and M. Vidrascu, *Domain-decomposition methods for large linearly elliptic three dimensional problems*, J. of Comput. and Appl. Math. **34** (1991), 93-117.
8. Y. Maday and A. T. Patera, *Spectral element methods for the Navier-Stokes equations*, State of the Art Surveys in Computational Mechanics (New York) (A.K. Noor and J.T. Oden, eds.), ASME, 1989.
9. J. Mandel, *Iterative solvers by substructuring for the p-version finite element method*, Comput. Meth. Appl. Mech. Engrg. **80** (1990), 117–128.
10. J. Mandel, *Iterative solvers for p-version finite element method in three dimensions*, Second Int. Conf. on Spectral and High Order Methods for PDE's, 1993, Proceedings of ICOSAHOM 92, a conference held in Montpellier, France, June 1992. To appear in Comput. Meth. Appl. Mech. Engrg.
11. L. F. Pavarino, *Domain decomposition algorithms for the p-version finite element method for elliptic problems*, Ph.D. thesis. Technical Report 616, Courant Institute of Mathematical Sciences, September 1992.
12. L. F. Pavarino, *Additive Schwarz methods for the p-version finite element method*, Numer. Math. **66** (1994), 493–515.
13. L. F. Pavarino and O. B. Widlund, *A polylogarithmic bound for an iterative substructuring method for spectral elements in three dimensions*, Technical Report 661, Courant Institute of Mathematical Sciences, March 1994.
14. E. M. Rønquist, *A domain decomposition method for elliptic boundary value problems: Application to unsteady incompressible fluid flow*, Fifth International Symposium on Domain Decomposition Methods for Partial Differential Equations (Philadelphia, PA) (T. F. Chan, D. E. Keyes, G. A. Meurant, J. S. Scroggs, and R. G. Voigt, eds.), SIAM, 1992.
15. B. F. Smith, *A domain decomposition algorithm for elliptic problems in three dimensions*, Numer. Math. **60** (1991), 219–234.
16. B. F. Smith, *A parallel implementation of an iterative substructuring algorithm for problems in three dimensions*, SIAM J. Sci. Comput. **14** (1993), 406–423.

DEPARTMENT OF COMPUTATIONAL AND APPLIED MATHEMATICS, RICE UNIVERSITY, HOUSTON, TX 77251

E-mail address: pavarino@rice.edu

Contemporary Mathematics
Volume **180**, 1994

AN ANALYSIS OF SPECTRAL GRAPH PARTITIONING VIA QUADRATIC ASSIGNMENT PROBLEMS

ALEX POTHEN

ABSTRACT. Recently a spectral algorithm for partitioning graphs has been widely used in many applications including domain decomposition. Following some work of Rendl and Wolkowicz, we describe a mathematical programming formulation of the graph partitioning problem, and obtain lower bounds on the number of edges cut by a partition. We also show that finding a nearest feasible solution to the partitioning problem from an infeasible solution that attains the lower bound leads to a justification of the spectral algorithm.

1. INTRODUCTION

A fundamental problem in the solution of systems of equations by domain decomposition is the problem of partitioning the domain into a given number of subdomains, such that the subdomains have approximately equal amounts of work and few edges cross the subdomains. (Other criteria to measure the "communication" requirements may be used, but for simplicity we consider here only the number of edges.) This problem can be formulated as the problem of partitioning a graph into subsets of specified sizes such that few edges join different subsets. The graph partitioning problem has applications in other contexts such as the data- and task-mapping problem in parallel computation, the ordering problem in direct methods for sparse matrix factorization, etc.

A large number of methods have been proposed in recent years for the graph partitioning problem. Here we consider a widely-used spectral graph partitioning algorithm that was motivated by earlier work of Alan Hoffman, Miroslav Fiedler, Earl Barnes, Bojan Mohar, *inter alios*. The work in [6] showed that the spectral partitioning algorithm computes high-quality partitions for large finite element meshes; this paper also contains a brief survey of earlier work, and additional theoretical results. Since then the spectral method has been carefully implemented, extended, and employed to partition problems in several application contexts; for instance, [5, 7, 10, 11]. However, despite its good computational behavior, a sound theoretical justification of the method has been lacking.

In this paper we employ a mathematical programming formulation to obtain lower bounds on the number of edges cut by a partition, and to justify the spectral

1991 *Mathematics Subject Classification*. Primary 68R10, 65K05, 65Y05.

This work was supported by the National Science Foundation grant CCR-9024954, by the U. S. Department of Energy grant DE-FG02-91ER25095, and by the Canadian Natural Sciences and Engineering Research Council under grant OGP0008111 at the University of Waterloo.

This paper is in final form and no version of it will be submitted for publication elsewhere.

partitioning algorithm. The lower bounds are obtained by a projection technique for quadratic assignment problems (QAPs) developed by Hadley, Rendl, and Wolkowicz [4], and applied to the graph partitioning problem by Rendl and Wolkowicz [8], who formulated the problem in terms of the adjacency matrix (perturbed by a diagonal matrix). Computational results from their formulation are reported in [2]. The formulation of the bipartition problem is given here in terms of the Laplacian matrix, and although it is equivalent to the formulation with respect to the adjacency matrix (i.e., there is a choice of the diagonal perturbation that leads to the Laplacian eigenvalues), we believe this treatment is more direct and easier to understand. Due to space limitations, we do not provide extensive computational results here. None of these results are new; they can all be derived from the earlier results of Rendl and Wolkowicz [8].

The QAP formulation has also been applied to envelope-size minimization and the related 2-sum minimization problem [3]. Another justification for the spectral partitioning algorithm may be found in [1].

2. BIPARTITION AND THE QUADRATIC ASSIGNMENT PROBLEM

Consider the problem of partitioning a graph $G = (V, E)$ into two subgraphs of m_1 and m_2 vertices such that the number of edges "cut", i.e., the number of edges joining one subgraph to the other, is minimized. We denote the two vertex subsets by V_1 and V_2, with $|V_j| = m_j$ for $j = 1, 2$, and $m_1 + m_2 = n$. This problem can be formulated as a quadratic assignment problem (QAP) involving the Laplacian matrix Q of the graph G. Recall that the Laplacian $Q = D - A$, where D is a diagonal matrix of vertex-degrees, and A is the adjacency matrix of G.

2.1. Formulation of bipartition as a QAP. Let $X = \left(\begin{array}{cc} \underline{x}_1 & \underline{x}_2 \end{array} \right)$ be an $n \times 2$ *partition matrix* consisting of the two indicator vectors $\underline{x}_j$ (for $j = 1, 2$), where x_{ij} is equal to one if vertex i belongs to the set V_j, and is zero otherwise. Then

$$\underline{x}_j{}^T Q \underline{x}_j = \sum_{i=1}^{n} \sum_{k=1}^{n} x_{ij} q_{ik} x_{kj} = \sum_{v \in V_j} d(v) - 2|E(V_j, V_j)|,$$

where $d(v)$ is the number of vertices adjacent to v, and $E(V_j, V_j)$ is the set of edges in E with both endpoints in V_j. We denote the edges joining V_1 and V_2 by the set $\delta(V_1, V_2)$, and recall that the *trace* of a square matrix Q, denoted by $tr\, Q$, is the sum of its diagonal elements. Then

$$
\begin{aligned}
tr\, X^T Q X &= \sum_{v \in V} d(v) - 2|E(V_1, V_1)| - 2|E(V_2, V_2)| \\
&= 2|E| - 2|E(V_1, V_1)| - 2|E(V_2, V_2)| = 2|\delta(V_1, V_2)|.
\end{aligned}
$$

(1)

Thus the problem of minimizing the number of edges cut by a bipartition with part sizes equal to m_1 and m_2 can be written as

(2)

$$|\delta_{min}(V_1, V_2)| \equiv \min\{|\delta(V_1, V_2)| : |V_1| = m_1, |V_2| = m_2\} = (1/2) \min_X tr\, X^T Q X,$$

where X varies over partition matrices with exactly m_j ones in the jth column.

Let $\underline{u}_n = (1/\sqrt{n})\begin{pmatrix} 1 & 1 & \cdots & 1 \end{pmatrix}^T$ denote the n-vector of all ones, scaled to have 2-norm one. (We will write $\underline{u}$ instead of $\underline{u}_n$ when the dimension is clear from the context.) A partition matrix X is characterized by the following three conditions:

(3) $$X\underline{u}_2 = \sqrt{(n/2)}\,\underline{u}_n; \quad X^T\underline{u}_n = (1/\sqrt{n})\begin{pmatrix} m_1 \\ m_2 \end{pmatrix};$$

(4) $$X^TX = \begin{pmatrix} m_1 & 0 \\ 0 & m_2 \end{pmatrix} \equiv M;$$

(5) $$x_{ij} \geq 0 \quad \text{for } i = 1,\dots,n; j = 1,2.$$

The first part of the first condition states that each row sum of a partition matrix is one, signifying that each vertex belongs to exactly one of the parts V_1 or V_2. The second part shows that there are m_j vertices in the jth part V_j. The second condition indicates that the columns of a partition matrix are orthogonal, and the third that the elements of a partition matrix are nonnegative.

Scaling $X = YM^{1/2}$ simplifies the following exposition and exposes the structure of the problem. With this scaling, the conditions on X are transformed to the following conditions on Y:

(6) $$Y\underline{m} = \underline{u}_n; \quad Y^T\underline{u}_n = \underline{m}, \quad \text{where } \underline{m} = (1/\sqrt{n})\begin{pmatrix} \sqrt{m_1} \\ \sqrt{m_2} \end{pmatrix};$$

(7) $$Y^TY = I_2;$$

(8) $$\left(YM^{1/2}\right)_{ij} \geq 0 \quad \text{for } i = 1,\dots,n; j = 1,2.$$

The objective function $tr\, X^TQX$ becomes $tr\, M^{1/2}Y^TQYM^{1/2} = tr\, MY^TQY$. In the last transformation we have used the identity $tr\, MN = tr\, NM$, where M is $n \times k$, and N is $k \times n$.

Minimizing this objective function subject to these constraints is NP-complete. Hence we obtain lower bounds on the number of edges cut by relaxing the third of these conditions.

2.2. Projected lower bounds. It is convenient to impose (6) on Y by projecting the problem to the subspace orthogonal to the manifold defined by this condition. Note that the two parts of this condition yield $YY^T\underline{u}_n = \underline{u}_n$, and $Y^TY\underline{m} = \underline{m}$. Thus we find that $\underline{m}$ is a right singular vector and that $\underline{u}_n$ is a left singular vector of Y corresponding to the singular value one. Choose an $n \times n$ orthogonal matrix $P_1 = \begin{pmatrix} \underline{u} & V \end{pmatrix}$, and a 2×2 orthogonal matrix $P_2 = \begin{pmatrix} \underline{m} & \underline{v} \end{pmatrix}$. The first step of the singular value decomposition of Y is

$$P_1^TYP_2 = \begin{pmatrix} \underline{u}^TY\underline{m} & \underline{u}^TY\underline{v} \\ V^TY\underline{m} & V^TY\underline{v} \end{pmatrix} = \begin{pmatrix} \underline{u}^T\underline{u} & \underline{m}^T\underline{v} \\ V^T\underline{u} & V^TY\underline{v} \end{pmatrix}$$

(9) $$= \begin{pmatrix} 1 & 0 \\ \underline{0} & \underline{z} \end{pmatrix}, \quad \text{where } \underline{z} \equiv V^TY\underline{v}.$$

Thus if we choose Y to be

(10) $$Y = P_1\begin{pmatrix} 1 & 0 \\ \underline{0} & \underline{z} \end{pmatrix}P_2^T = \underline{u}\,\underline{m}^T + V\underline{z}\,\underline{v}^T,$$

then (6) is satisfied.

Substitution of this representation of Y in the orthogonality condition (7), followed by pre-multiplication with P_2^T and post-multiplication with P_2, shows that

$$\begin{pmatrix} 1 & 0 \\ 0 & \underline{z}^T \underline{z} \end{pmatrix} = I_2,$$

and hence we obtain the condition $\underline{z}^T \underline{z} = 1$.

Substituting for Y from (10) in the objective function $tr\ MY^T QY$, we find that since $Q\underline{u} = \underline{0}$, only one of the four terms survives, and it becomes

$$
\begin{aligned}
tr\ MY^T QY &= tr\ M\underline{v}\,\underline{z}^T V^T QV\underline{z}\,\underline{v}^T &= tr\ \underline{v}^T M\underline{v}\,\underline{z}^T \widehat{Q}\underline{z} \\
(11) \qquad &= (\underline{v}^T M\underline{v})\,(\underline{z}^T \widehat{Q}\underline{z}),
\end{aligned}
$$

where $\widehat{Q} \equiv V^T QV$ is the projected Laplacian.

The first term on the right-hand-side, $\underline{v}^T M\underline{v}$, is easily computed to be $2m_1 m_2/n$, since $\underline{m} = (1/\sqrt{n}) \begin{pmatrix} \sqrt{m_1} \\ \sqrt{m_2} \end{pmatrix}$ implies that $\underline{v} = \pm (1/\sqrt{n}) \begin{pmatrix} \sqrt{m_2} \\ -\sqrt{m_1} \end{pmatrix}$. Thus we obtain the result $|\delta(V_1, V_2)| = (1/2)\,(2m_1 m_2/n)\,\underline{z}^T \widehat{Q}\underline{z}$.

The bipartition problem is hence

$$
\begin{aligned}
|\delta_{min}(V_1, V_2)| &= (m_1 m_2/n) \min_{\underline{z}} \underline{z}^T \widehat{Q}\underline{z} \\[6pt]
\text{subject to} \quad & \underline{z}^T \underline{z} = 1, \\[4pt]
(12) \qquad & \left(\underline{u}\,\underline{m}^T M^{1/2} + V\underline{z}\,\underline{v}^T M^{1/2} \right)_{ij} \geq 0.
\end{aligned}
$$

Though this problem is intractable, a lower bound may be obtained by relaxing the second constraint, and thus we find

$$(13) \qquad |\delta_{min}(V_1, V_2)| \geq (m_1 m_2/n)\lambda_1(\widehat{Q}) = (m_1 m_2/n)\lambda_2(Q),$$

since the eigenvalues of $\widehat{Q}$ are the $n-1$ nonzero eigenvalues of Q. The lower bound is attained by the corresponding eigenvector $\underline{z}_0 = V^T \underline{x}_2$, where $\underline{x}_2$ is the second Laplacian eigenvector. Hence the orthogonal matrix attaining the lower bound is

$$(14) \qquad Y_0 = \underline{u}\,\underline{m}^T + VV^T \underline{x}_2\,\underline{v}^T.$$

2.3. Diagonal perturbations. The lower bound on the number of cut edges can be improved further by considering diagonal perturbations of the Laplacian. Let $Q(\underline{d}) = Q + Diag(\underline{d})$, where $\underline{d}$ is an n-vector whose components sum to zero. It can be verified that $tr\ X^T Q(\underline{d})X = tr\ X^T QX$, so that this perturbation has no effect on the number of cut edges. Proceeding as in the unperturbed case, we can show that

$$
\begin{aligned}
|\delta_{min}(V_1, V_2)| \geq \max_{\underline{d}} \min_{\underline{z}} \Big\{ & (m_1 m_2/n)\underline{z}^T \widehat{Q}(d)\underline{z} \\
+ \ & (1/2n\sqrt{n})(m_1 - m_2)\sqrt{m_1 m_2}\,\underline{d}^T \underline{z} \Big\},
\end{aligned}
$$

$$(15) \qquad\qquad \text{subject to} \qquad \underline{z}^T \underline{z} = 1.$$

The lower bound can be computed by nondifferentiable optimization techniques [9].

The lower bounds in terms of the unperturbed Laplacian are still weak for "well-shaped" finite-element meshes that possess partitions with few edges being cut. It would be interesting to see how much the bounds improve when diagonal perturbations are included.

2.4. Closest partition matrix. Which partition matrix Z is "closest" to the orthogonal matrix $X_0 = Y_0 M^{1/2}$ attaining the lower bound (see (14)) in the bipartition problem? We can answer this question by considering the objective function of the bipartition problem, which we may write as

$$(16) \qquad \min_Z tr\, Z^T(Q + \alpha I)Z = \min_Z \|(Q + \alpha I)^{1/2}Z\|_F^2.$$

Here we have shifted the Laplacian by a small positive multiple of the identity to make the matrix $Q + \alpha I$ positive definite, so that its square root is nonsingular. This is necessary to obtain a weighted norm. It can be verified that this shifts the objective function by the constant αn, and hence has no effect on the minimizer. We now expand Z about X_0 to obtain

$$
\begin{aligned}
(17) \qquad & \min_Z \|(Q + \alpha I)^{1/2}\left(X_0 + (Z - X_0)\right)\|_F^2 \\
& - \min_Z \|(Q + \alpha I)^{1/2}X_0\|_F^2 + 2\,tr\, X_0^T(Q + \alpha I)(Z - X_0) \\
& + \|(Q + \alpha I)^{1/2}(Z - X_0)\|_F^2.
\end{aligned}
$$

Here we have used the identity

$$\|A + B\|_F^2 = \|A\|_F^2 + \|B\|_F^2 + 2\,tr\, A^T B,$$

for real matrices A and B.

On the right-hand-side of (17), the first term is a constant since X_0 is a fixed orthogonal matrix; we ignore the third term, which is a quadratic in the difference $(Z - X_0)$, to obtain a linear approximation. Hence we consider the problem

$$(18) \qquad \min_Z tr\, X_0^T(Q + \alpha I)Z = \min_Z tr\, M^{1/2}Y_0^T(Q + \alpha I)Z.$$

Substituting for Y_0 from (14), and noting that $\underline{u}^T Q = \underline{0}^T$, the problem becomes

$$(19) \qquad \min_Z tr\, M^{1/2}\left(\underline{v}\,\underline{x}_2^T V V^T QZ + \alpha\,\underline{m}\,\underline{u}^T Z + \alpha\,\underline{v}\,\underline{x}_2^T V V^T Z\right).$$

The second term in the right-hand-side is constant since $\underline{u}^T Z = (1/\sqrt{n})\,\begin{pmatrix} m_1 & m_2 \end{pmatrix}$ (by (3)), and hence the problem reduces to

$$(20) \qquad \min_Z tr\, M^{1/2}\underline{v}\,\underline{x}_2^T V V^T(Q + \alpha I)Z.$$

Replacing $V V^T = P_1 P_1^T - \underline{u}\,\underline{u}^T = (I_n - \underline{u}\,\underline{u}^T)$, noting again that $\underline{u}$ is an eigenvector of Q corresponding to the zero eigenvalue (or that $\underline{x}_2^T\underline{u} = 0$), the objective function simplifies to

$$\min_Z tr\, M^{1/2}\underline{v}\,\underline{x}_2^T(Q + \alpha I)Z = \min_Z (\lambda_2(Q) + \alpha)\, tr\, M^{1/2}\underline{v}\,\underline{x}_2^T Z.$$

Further simplifications are possible. An important observation is that in the bipartition problem $Z = \begin{pmatrix} \underline{z}_1 & \sqrt{n}\underline{u} - \underline{z}_1 \end{pmatrix}$, where $\underline{z}_1$ is an indicator vector with m_1 ones and remaining elements equal to zeros, and the second column of Z is the complement of $\underline{z}_1$ with respect to the vector of all ones. Also note that $M^{1/2}\underline{v} = \pm\sqrt{m_1 m_2 / n}\begin{pmatrix} 1 \\ -1 \end{pmatrix}$. Putting these observations together, the problem becomes

$$\left((\lambda_2(Q) + \alpha)\,\sqrt{m_1 m_2 / n}\,\right) \min_{\underline{z}_1} tr \pm \left(\begin{matrix} \underline{x}_2{}^T \underline{z}_1 & \underline{x}_2{}^T(\sqrt{n}\underline{u} - \underline{z}_1) \\ -\underline{x}_2{}^T \underline{z}_1 & -\underline{x}_2{}^T(\sqrt{n}\underline{u} - \underline{z}_1) \end{matrix} \right)$$

$$(21) \quad = \quad \left(2(\lambda_2(Q) + \alpha)\,\sqrt{m_1 m_2 / n}\,\right) \min_{\underline{z}_1} \pm \underline{x}_2{}^T \underline{z}_1.$$

In going from the first to the second line we have used $\underline{x}_2{}^T\,\underline{u} = 0$.

Thus the algebraic manipulations in this subsection come to a glorious conclusion! One solution to (21), and hence to a linear approximation to the nearest partition matrix problem, is obtained by choosing $\underline{z}_1$ to have ones in rows corresponding to the smallest (most negative) m_1 eigenvector components of $\underline{x}_2$. A second solution is obtained by choosing the rows corresponding to the largest (most positive) m_1 eigenvector components. (These two solutions correspond to the choice of the sign in (21).) Hence we obtain a justification of the spectral algorithm for the bipartition problem which partitions the graph with respect to the m_1th smallest or largest second eigenvector component.

Acknowledgments. I thank Professor Stan Eisenstat of Yale University for his comments and questions, and for pointing out errors in a draft of this paper.

References

1. T. F. Chan and W. K. Szeto. *On the near optimality of the recursive spectral bisection method for graph partitioning.* Manuscript, Feb. 1993.
2. J. Falkner, F. Rendl, and H. Wolkowicz. *A computational study of graph partitioning.* Math. Programming, 1994. To appear.
3. A. George and A. Pothen. *An analysis of the spectral approach to envelope reduction via a quadratic assignment formulation.* In preparation, 1994.
4. S.W. Hadley, F. Rendl, and H. Wolkowicz. *A new lower bound via projection for the quadratic assignment problem.* Mathematics of Operations Research, 17(1992), 727–739.
5. B. Hendrickson and R. Leland. *An improved spectral graph partitioning algorithm for mapping parallel computations.* Technical Report SAND92-1460, UC-405, Sandia Natl. Lab., Albuquerque, N.M., Sep. 1992.
6. A. Pothen, H. D. Simon, and K. P. Liou. *Partitioning sparse matrices with eigenvectors of graphs.* SIAM J. Matrix Anal. Appl., 11(1990), 430–452.
7. A. Pothen and L. Wang. *An improved spectral nested dissection algorithm.* In preparation, 1994.
8. F. Rendl and H. Wolkowicz. *A projection technique for partitioning the nodes of a graph.* Annals of Operations Research, 1994. This paper was written in 1990, and will appear in a special issue devoted to APMOD93.
9. H. Schramm and J. Zowe. *A version of the bundle idea for minimizing a nonsmooth function: Conceptual idea, convergence analysis, numerical results.* SIAM J. Opt., 2(1992), 121–152.
10. H. D. Simon. *Partitioning of unstructured problems for parallel processing.* Computing Systems in Engineering, 2(1991), 135–148.
11. Lie Wang. *Spectral Nested Dissection.* Ph.D. thesis, The Pennsylvania State University, University Park, PA, 1994.

Computer Science Department, Old Dominion University, Norfolk, VA 23529-0162, and ICASE, MS 132C, NASA-LaRC, Hampton, VA 23681-0001.

E-mail address: pothen@icase.edu

Contemporary Mathematics
Volume **180**, 1994

Error estimators based on stable splittings

U. RÜDE

ABSTRACT. Stable splittings have been used successfully to describe and analyze the performance of iterative solvers based on subspace corrections. The same theoretical foundation can be used to construct abstract upper and lower error estimates. This approach leads to a unified treatment of discretization and iteration errors and can therefore be used to guide iteration and mesh refinement strategies. The estimate does not require the exact solution of a fixed finite element problem, but can be used for any approximation that may be available. The estimate is based only on quantities that occur naturally in the solution process, so no extra work is required.

1. Introduction

Multilevel- and domain decomposition methods are generally based on a *splitting* of the solution space. Such subspaces, together with their Hilbert space structure, define elementary operations, that can in turn be used to construct iterative methods and preconditioners, the so-called *subspace correction methods*. This framework includes many iterative methods like classical relaxation schemes, domain decomposition algorithms, and multilevel preconditioners. The discussion of algorithms in this setup turns out to be useful, because the performance of solvers and preconditioners depends on a single abstract feature of the subspace system, the so-called *stability* of the splitting.

Additionally, the splitting of the space can be used to derive error estimates. If the subspace splitting is stable, we obtain uniformly bounded lower and upper error estimates. Depending on the interpretation of the spaces, the bounds apply to the (algebraic) iteration error or the (continuous) discretization error such that the error estimate combines in a natural way *discretization errors* and *algebraic errors*. This dual viewpoint distinguishes our approach from conventional error estimators. In contrast to more conventional estimates, this approach does not

---

1991 *Mathematics Subject Classification.* 65N22, 65N50, 65N55.
Key words and phrases. Fully adaptive multigrid, multilevel iteration, error estimate.
This paper is in final form and no version of it will be submitted for publication elsewhere.

require the exact solution of a finite element system, but can be applied to an approximate solution during the iterative solution process. It therefore provides useful criteria for the development of a strategy for switching from iteration to refinement, and vice versa.

Since iteration and discretization errors are estimated by a uniform approach, it becomes natural to consider algorithms that combine iteration, error estimation, and mesh refinement. An implementation of these ideas has been introduced as the the *virtual global grid* refinement technique and the *multilevel adaptive iteration* in Rüde [7, 8]. Our error estimate is based on subspace corrections, as they are computed in each iteration step such that the estimate involves no extra cost but only the correct interpretation of quantities occuring in the iterative solution process.

The integration of error estimates in a multilevel solution process has been discussed in several papers, see e.g. Bank, Smith, and Weiser [2, 1], Deuflhard, Leinen, and Yserentant [3], and Verfürth [9]. In this paper we will derive new error estimates based on the theory of stable splittings.

2. Stable splittings

We use an abstract setting, where V is a Hilbert space equipped with a scalar product $\langle \cdot, \cdot \rangle_V$ and a norm

$$\|u\|_V = \langle u, u \rangle_V^{1/2}.$$

Given a V-elliptic, symmetric, continuous bilinear form $a : V \times V \longrightarrow \mathbb{R}$ with constants $0 < c_1 \leq c_2 < \infty$, such that

$$(2.1) \qquad\qquad c_1 \langle v, v \rangle_V \leq a(v, v) \leq c_2 \langle v, v \rangle_V$$

for all $v \in V$, we study the elliptic problem: Find $u \in V$ such that

$$(2.2) \qquad\qquad a(u, v) = \Phi(v)$$

for all $v \in V$, where the functional $\Phi \in V^*$ is a continuous linear form.

To introduce a multilevel structure, we consider a finite or infinite collection $\{V_j\}_{j \in J}$ of subspaces of V, each with its own scalar product $(\cdot, \cdot)_{V_j}$ and the associated norm

$$\|u\|_{V_j} = (u, u)_{V_j}^{1/2}.$$

We further assume that the full space V can be represented as the sum of the subspaces V_j, $j \in J$,

$$(2.3) \qquad\qquad V = \sum_{j \in J} V_j$$

and assume that the spaces are nested, that is $J \subset \mathbb{N}_0$, $V_i \subset V_j$, if $i \leq j$.

In typical applications, $\| \cdot \|_V$ is the H^1-Sobolev norm and the subspace norms $\| \cdot \|_{V_j}$ are properly scaled L_2-norms. Typically, $\| \cdot \|_{V_j} = 2^j \| \cdot \|_{L_2}$.

Any element of V can be represented as a sum of elements in V_j, $j \in J$. Generally, this representation is non-unique. The *additive Schwarz norm* $||| \cdot |||$ in V with respect to the collection of subspaces $\{V_j\}_{j \in J}$ is defined by

$$(2.4) \qquad |||v||| \overset{\text{def}}{=} \inf \left\{ \left(\sum_{j \in J} \|v_j\|_{V_j}^2 \right)^{\frac{1}{2}} \,\middle|\, v_j \in V_j, \sum_{j \in J} v_j = v \right\}.$$

A collection of spaces $\{V_j\}_{j \in J}$ satisfying (2.3) is called a *stable splitting* of V, if $\| \cdot \|_V$ is equivalent to the additive Schwarz norm of V, that is, if there exist constants $0 < c_3 \leq c_4 < \infty$ such that

$$(2.5) \qquad c_3 \|v\|_V^2 \leq |||v|||^2 \leq c_4 \|v\|_V^2$$

for all $v \in V$. The number

$$(2.6) \qquad \kappa(V, \{V_j\}_{j \in J}) \overset{\text{def}}{=} \inf(c_4/c_3),$$

that is the infimum over all possible constants in (2.5), is called the *stability constant* of the splitting $\{V_j\}_{j \in J}$.

Next, we introduce auxiliary V_j-elliptic, symmetric, bilinear forms

$$b_j : V_j \times V_j \longrightarrow \mathbb{R}$$

in the spaces V_j, respectively. In classical multigrid terminology, the choice of b_j determines which kind of smoother we use. For the theory, we require that the b_j are uniformly equivalent to the respective inner product of the subspace, that is that there exist constants $0 < c_5 \leq c_6 < \infty$ such that

$$(2.7) \qquad c_5(v_j, v_j)_{V_j} \leq b_j(v_j, v_j) \leq c_6(v_j, v_j)_{V_j},$$

for all $v_j \in V_j$, $j \in J$.

Multilevel algorithms are now described in terms of *subspace corrections* $P_{V_j} : V \longrightarrow V_j$, mapping the full space V into each of the subspaces V_j. P_{V_j} is defined by

$$(2.8) \qquad b_j(P_{V_j} u, v_j) = a(u, v_j),$$

for all $v_j \in V_j$, $j \in J$. Analogously, we define $\phi_j \subset V_j$ by

$$(2.9) \qquad b_j(\phi_j, v_j) = \Phi(v_j),$$

for all $v_j \in V_j$, $j \in J$.

The *additive Schwarz operator* (also called BPX-operator) $P_V : V \longrightarrow V$ with respect to the multilevel structure on V is defined by

$$(2.10) \qquad P_V = \sum_{j \in J} P_{V_j}$$

and

$$\phi = \sum_{j \in J} \phi_j.$$

With suitable bilinear forms b_j it is possible to evaluate P_V efficiently based on its definition as a sum. The explicit construction of P_V is not required.

The hierarchical structure in the subspace system seems to be essential for obtaining a stable splitting. Otherwise, the complexity of the original problem must be captured in the bilinear forms $b_j(\cdot, \cdot)$, and then the evaluation of the P_{V_j} is as expensive as the solution of the original problem.

The importance of the abstract multilevel structure for practical applications is indicated by the following theorem.

THEOREM 2.1. *Assume that the subspaces* V_j, $j \in J$ *of a Hilbert space* V *are a stable splitting. Assume further that* P_{V_j} *and* P_V *are defined as above with bilinear forms* b_j *satisfying (2.7). The variational problem (2.2) is equivalent to the operator equation*

$$(2.11) \qquad\qquad P_V u = \phi,$$

and the spectrum of P_V *can be estimated by*

$$(2.12) \qquad\qquad \frac{c_1}{c_4 c_6} \le \lambda_{\min}(P_V) \le \lambda_{\max}(P_V) \le \frac{c_2}{c_3 c_5}.$$

For a proof see Oswald [**4, 5**] or Rüde [**8**].

Remark. Results similar to Theorem 2.1 have been developed within the domain decomposition and multigrid literature. The interested reader is also referred to the survey articles of Xu [**10**], Yserentant [**11**], and the references given therein.

3. Multilevel Error Estimators

Besides providing the theoretical basis for the fast iterative solution of discretized PDEs, the multilevel splittings can also be used to provide error estimates.

The **scaled residuals** $\bar{r}_j \in V_j$ of u are defined by

$$(3.1) \qquad\qquad b_j(\bar{r}_j, v_j) = a(u - u^*, v_j) = a(u, v_j) - \Phi(v_j)$$

for all $v_j \in V_j$, $j \in J$, where u^* is the solution of (2.2) in V. The following theorem is an abstract and more general version of a result given in [**6**], see also [**8**].

THEOREM 3.1. *Assume that the collection of spaces* $\{V_j\}_{j \in J}$ *is a stable splitting of* V, *and that* u^* *is the solution of (2.2) in* V. *Then there exist constants* $0 < c_0 \le c_1 < \infty$ *such that*

$$(3.2) \qquad\qquad c_0 \sum_{j \in J} \|\bar{r}_j\|_{V_j}^2 \le \|u - u^*\|_V^2 \le c_1 \sum_{j \in J} \|\bar{r}_j\|_{V_j}^2.$$

Proof. With inequalities (2.1) and equation (2.7) it suffices to show that there exist constants $0 < \bar{c}_0 \leq \bar{c}_1 < \infty$ such that

$$(3.3) \qquad \bar{c}_0 \sum_{j \in J} b_j(\bar{r}_j, \bar{r}_j) \leq a(u - u^*, u - u^*) \leq \bar{c}_1 \sum_{j \in J} b_j(\bar{r}_j, \bar{r}_j).$$

From Theorem 2.1, we know that there exist such constants with

$$\bar{c}_0 a(P_V(u - u^*), u - u^*) \leq a(u - u^*, u - u^*) \leq \bar{c}_1 a(P_V(u - u^*), u - u^*).$$

Additionally,

$$
\begin{aligned}
a(P_V(u - u^*), u - u^*) &= \sum_{j \in J} a(u - u^*, P_{V_j}(u - u^*)) \\
&= \sum_{j \in J} b_j(\bar{r}_j, P_{V_j}(u - u^*)) \\
&= \sum_{j \in J} a(\bar{r}_j, (u - u^*)) \\
&= \sum_{j \in J} b_j(\bar{r}_j, \bar{r}_j),
\end{aligned}
$$

which concludes the proof. □

The value of Theorem 3.1 is that it estimates an unknown quantity, the error $u - u^*$, by known quantities, the *residuals* $P_{V_j}u - \phi_j$.

In contrast to the usual error estimators used for finite elements, (3.2) uses a sum of residuals from all levels. In general, this sum is infinite, so that, at a first glance, the practical usefulness seems to be limited. Estimate (3.2), however, gives valuable insight into the nature of adaptive processes because it relates the residuals from all levels of a hierarchical representation of the solution. Thus, iteration errors, which result in residuals on coarser levels, are included in the estimate. This information can now be used to guide the switching from iteration to refinement and vice versa. Suitable algorithms have been proposed in Rüde [8], where the idea is to treat refinement and iteration as essentially the same process.

Finite element nodes must be relaxed whenever the associated residuals are large relative to the overall error estimate. Using the virtual global grid data structure (see [8]), this strategy can be extended to unknowns that are not yet included in the finite element system. Of course, such unknowns must be generated by some refinement algorithm, before they can be relaxed.

4. Estimates based on residuals on one level

Based on (3.2) we will now develop an error estimator that uses the residuals of one level only. In view of Theorem 3.1 we need an additional assumption that can be used to bound the error contribution of an infinite sequence of levels by

one of them alone. This can be accomplished by introducing the well known
saturation condition (see, e.g., Bank and Smith [1])

(4.1) $$a(u_{j+1}^* - u^*, u_{j+1}^* - u^*) \leq \gamma_j a(u_j^* - u^*, u_j^* - u^*),$$

for a constant $\gamma_j \leq \gamma < 1$, where u_j^* denotes the *exact* solution of the level j
equations defined by

(4.2) $$a(u_j^*, v_j) = \Phi(v_j),$$

for all $v_j \in V_j$.

The saturation condition (4.1) makes assumptions about the speed of convergence of u_j^* to u^* with respect to the multilevel system. Note that if V_{j+1} is a
space with higher approximation order than V_j, then typically, $\gamma_j \to 0$, when
$j \to \infty$ and if u^* is sufficiently smooth. In our paper we do not assume higher
order approximation spaces, so that γ_j will tend to a constant like $1/4$ for linear
finite elements and smooth solutions. Consequently, our error estimate cannot
be asymptotically exact.

THEOREM 4.1. *Let $\bar{r}_{j+1}^*$ denote the residual of u_j on level $j + 1$, defined by*

(4.3) $$b_{j+1}(\bar{r}_{j+1}^*, v_{j+1}) = a(u_j^* - u^*, v_{j+1})$$

*for all $v_{j+1} \in V_{j+1}$. If the saturation condition (4.1) holds, then there exist
constants $0 < c_1 \leq c_2 < \infty$ such that*

(4.4) $$c_1 b_{j+1}(\bar{r}_{j+1}^*, \bar{r}_{j+1}^*) \leq a(u_j^* - u^*, u_j^* - u^*) \leq c_2 b_{j+1}(\bar{r}_{j+1}^*, \bar{r}_{j+1}^*).$$

Proof. Using the theorem of Pythagoras

$$a(u_j^* - u^*, u_j^* - u^*) = a(u_{j+1}^* - u_j^*, u_{j+1}^* - u_j^*) + a(u_{j+1}^* - u^*, u_{j+1}^* - u^*)$$

and the saturation condition (4.1), we obtain

$$a(u_{j+1}^* - u_j^*, u_{j+1}^* - u_j^*) \leq a(u_j^* - u^*, u_j^* - u^*) \leq \frac{1}{1 - \gamma_j} a(u_{j+1}^* - u_j^*, u_{j+1}^* - u_j^*).$$

Next, we must estimate the difference $u_{j+1}^* - u_j^*$ in terms of $\bar{r}_{j+1}^*$. First, note that
in V_{j+1} equipped with the $\|\cdot\|_V$-norm, the finite system of spaces $V_0, V_1, \ldots, V_{j+1}$,
each with its own norm $\|\cdot\|_{V_0}, \|\cdot\|_{V_1}, \ldots, \|\cdot\|_{V_{j+1}}$ is a stable splitting (see Rüde
[8]). The finite additive Schwarz operator

$$\sum_{k=0}^{j+1} P_{V_k}$$

is spectrally equivalent to the identity in V_{j+1}, so that

$$a(u_{j+1}^* - u_j^*, u_{j+1}^* - u_j^*) \leq c_1 a \left(\sum_{k=0}^{j+1} P_{V_k}(u_{j+1}^* - u_j^*), u_{j+1}^* - u_j^* \right)$$

$$= c_1 \sum_{k=0}^{j+1} b_{j+1}(P_{V_k}(u_{j+1}^* - u_j^*), P_{V_k}(u_{j+1}^* - u_j^*)).$$

Noting that

$$P_{V_k}(u_j^* - u^*) = 0$$

for $k \leq j$, and

$$P_{V_{j+1}}(u_{j+1}^* - u^*) = \bar{r}_{j+1}^*,$$

we find the upper bound in (4.4). The proof of the lower bound is analogous. $\square$

5. Conclusions

In this paper we have briefly outlined how multilevel error estimates can be derived from the theory of stable splittings, and how they are therefore linked to fast iterative solvers. We believe that this is of not only theoretical interest, because it can be used directly to construct and analyze efficient, adaptive multilevel solvers.

Future research must extend these ideas to more general problems, including non-selfadjoint and nonlinear equations. Additionally, the estimate must be tested for realistic problems and must be compared experimentally to alternative techniques.

REFERENCES

1. R. Bank and K. Smith, *A posteriori estimates based on hierarchical basis*, SIAM J. Numer. Anal. **30** (1993), no. 4, 921–935.
2. R.E. Bank and A. Weiser, *Some a-posteriori estimators for elliptic partial differential equations*, Math. Comp **44** (1985), 283–301.
3. P. Deuflhard, P. Leinen, and H. Yserentant, *Concepts of an adaptive hierarchical finite element code*, IMPACT of Computing in Science and Engineering **1** (1989), 3–35.
4. P. Oswald, *Norm equivalencies and multilevel Schwarz preconditioning for variational problems*, Forschungsergebnisse Math/92/01, Friedrich Schiller Universität, Jena, January 1992.
5. _____, *Stable splittings of Sobolev spaces and fast solution of variational problems*, Forschungsergebnisse Math/92/05, Friedrich Schiller Universität, Jena, May 1992.
6. U. Rüde, *On the robustness and efficiency of the fully adaptive multigrid method*, Proceedings of the Sixth International Conference on Domain Decomposition in Science and Engineering, Como, Italy, June 15-19, 1992 (Providence, RI, USA) (A. Quarteroni, ed.), Amer. Math. Soc., 1992.
7. _____, *Fully adaptive multigrid methods*, SIAM Journal on Numerical Analysis **30** (1993), no. 1, 230–248.
8. _____, *Mathematical and computational techniques for multilevel adaptive methods*, Frontiers in Applied Mathematics, vol. 13, SIAM, Philadelphia, PA, USA, 1993.
9. R. Verfürth, *A review of a posteriori error estimation and adaptive mesh refinement techniques*, Lecture Notes of a Compact Seminar held at the TU Magdeburg, June 2 – 4, 1993, Report Universität Zürich, 1993.

10. J. Xu, *Iterative methods by space decomposition and subspace correction*, SIAM Review **34** (1992), no. 4, 581 – 613.
11. H. Yserentant, *Old and new convergence proofs for multigrid methods*, Acta Numerica (1993), 285–326.

(U. Rüde) INSTITUT FÜR INFORMATIK, TECHNISCHE UNIVERSITÄT MÜNCHEN, D-80290 MÜN-CHEN, GERMANY

Current address: Fachbereich Mathematik, Technische Universität Chemnitz-Zwickau, D-09009 Chemnitz, Germany

E-mail address: ruede@informatik.tu-muenchen.de

Contemporary Mathematics
Volume **180**, 1994

Multilevel Methods for P_1 Nonconforming Finite Elements and Discontinuous Coefficients in Three Dimensions

MARCUS SARKIS

ABSTRACT. We introduce multilevel Schwarz preconditioners for solving the discrete algebraic equations that arise from a nonconforming P_1 finite element method approximation of second order elliptic problems. For the additive multilevel version, we obtain a condition number bounded by $C_1 (1 + \log H/h)^2$ from above, and for the multiplicative versions, such as the V-cycle multigrid methods using Gauss Seidel and damped Jacobi smoothers, we obtain a rate of convergence bounded from above by $1 - C_2 (1 + \log H/h)^{-2}$.

1. Introduction

There are many engineering applications in which the main goal is to find a good approximation for $q = \rho \nabla u$. Here, u is the solution of an elliptic problem with coefficient ρ. We can find an approximation for q by finding an approximation for u and then applying the operator $\rho \nabla$. This procedure may generate serious errors since when ρ becomes more discontinuous, the solution u becomes more singular and the operator $\rho \nabla$ more numerically unstable. Furthermore, we note that in the interior of Ω we have, formally, $\operatorname{div} q = f$. Therefore, we expect $q(x)$ to be less sensitive than $u(x)$ to variations of $\rho(x)$. For instance, if we consider the one-dimensional case with $f = 0$ and inhomogeneous Dirichlet data, we obtain $q = $ constant. This is why mixed methods are introduced in order to approximate $\rho \nabla u$ and u, independently. Our motivation for considering the P_1-nonconforming space comes from the fact that there is an equivalence

1991 *Mathematics Subject Classification.* 65F10, 65N30, 65N55.

This work has been supported in part by a graduate student fellowship from Conselho Nacional de Desenvolvimento Cientifico e Tecnologico - CNPq, in part by Dean's Dissertation New York University Fellowship, and in part by the National Science Foundation under Grant NSF-CCR-9204255 and the U. S. Department of Energy under contract DE-FG02-92ER25127.

The final detailed version of this paper will be submitted for publication elsewhere.

between mixed methods and nonconforming methods; see Arnold and Brezzi [1].

The first multigrid methods for nonconforming finite elements were introduced by Braess and Verfüth [2], and Brenner [3]. The existing convergence results are based on the H^2-regularity assumption for the continuous problem. Later, Oswald [12], Vassilevski and Wang [15] proposed optimal multilevel BPX-preconditioners for nonconforming P_1 elements in three-dimensional case by using a sequence of nested conforming subspaces. No additional regularity assumption beyond the H^1 is used. We note, however, that we cannot guarantee that the rate of convergence of these methods are insensitive to large variations in the coefficients of the differential equation; see also [11], [16], and [10]. In this paper, we modify the Oswald preconditioner by introducing our nonstandard coarse spaces Sarkis [13], and establish that its condition number grows at most as the square of the number of levels, and does not depend on the number of substructures and the jumps of the coefficients. To analyze our methods, we introduce nonstandard local interpolators [13, 14] in order to convert results from the conforming to the nonconforming case. We note that an operator similar to ours has independently been introduced in Cowsar, Mandel, and Wheeler [4].

This paper is very closely related to those of Dryja [6], and Dryja, Sarkis, and Widlund [8] and we refer to [6] for some of our notation. The proofs of our results can be found in [14].

2. Notation

A coarse triangulation of Ω is introduced by dividing the region into nonoverlapping simplicial substructures Ω_i, $i = 1, \cdots, N$, with diameters of order H. The barycenters of faces of tetrahedra $\tau_j^h \in \mathcal{T}^h$ are called CR nodal points. The sets of CR nodal points belonging to Ω, $\partial\Omega$, $\mathcal{F}_{ij}$, $\partial\Omega_i$, and Γ are denoted by Ω_h^{CR}, $\partial\Omega_h^{CR}$, $\mathcal{F}_{ij,h}^{CR}$, $\partial\Omega_{i,h}^{CR}$, and Γ_h^{CR}, respectively.

DEFINITION 1. *The nonconforming P_1 element spaces on the h-mesh (cf. Crouzeix and Raviart [5]) are given by*

$$\tilde{V}^h(\Omega) := \{v \,|\, v \text{ linear in each tetrahedron } \tau_j^h \in \mathcal{T}^h,$$
$$\text{and } v \text{ continuous at the nodes of } \Omega_h^{CR}\}, \text{ and}$$
$$\tilde{V}_0^h(\Omega) := \{v \,|\, v \in \tilde{V}^h(\Omega) \text{ and } v = 0 \text{ at the nodes of } \partial\Omega_h^{CR}\}.$$

Note that $\tilde{V}_0^h(\Omega)$ is nonconforming since $\tilde{V}_0^h(\Omega) \not\subset H_0^1(\Omega)$.

For $u \in \tilde{V}_0^h(\Omega)$, we define a nonconforming discrete weighted energy norm with $\rho = \rho_i$ on Ω_i by

$$(1) \qquad\qquad |u|^2_{H^1_{\rho,h}(\Omega)} := a^h(u, u),$$

where

$$(2) \qquad a^h(u,v) = \sum_{i=1}^{N} \sum_{\tau_j^h \in \Omega_i} \int_{\tau_j^h} \rho_i \, \nabla u \cdot \nabla v \, dx = \sum_{i=1}^{N} a_{\Omega_i}^h(u,w).$$

Our discrete problem is given by:
Find $u \in \tilde{V}_0^h(\Omega)$, such that

$$(3) \qquad a^h(u,v) = f(v) \ \forall \ v \in \ \tilde{V}_0^h(\Omega).$$

3. Nonconforming Coarse Spaces

In this section, we introduce two different types of coarse spaces which make it possible to design efficient domain decomposition methods for problems with discontinuous coefficients in three dimensions.

3.1. A face based coarse space. The first coarse space to be considered, $\tilde{V}_{-1}^F \subset \tilde{V}_0^h(\Omega)$, is based on the average over each face $\mathcal{F}_{ij,h}^{CR}$. Let $\bar{u}_{\mathcal{F}_{ij}^{CR}}$ and $\bar{u}_{\partial\Omega_i^{CR}}$ be the average values of u over $\mathcal{F}_{ij,h}^{CR}$ and $\partial\Omega_{i,h}^{CR}$, respectively, and let $\theta_{\mathcal{F}_{ij}}^{CR}$ be the discrete nonconforming harmonic function in Ω_i, in the sense of $a_{\Omega_i}^h(\cdot,\cdot)$, which equals 1 on $\mathcal{F}_{ij,h}^{CR}$ and is zero on $\partial\Omega_{i,h}^{CR}\backslash\mathcal{F}_{ij,h}^{CR}$.

The space $\tilde{V}_{-1}^F$ can conveniently be defined as the range of am interpolation operator $\tilde{I}_h^F : \tilde{V}_0^h(\Omega) \to \tilde{V}_{-1}^F$, defined by

$$\tilde{I}_h^F u(x)_{|\bar{\Omega}_i} = \sum_{\mathcal{F}_{ij} \subset \partial\Omega_i} \bar{u}_{\mathcal{F}_{ij}^{CR}} \, \theta_{\mathcal{F}_{ij}}^{CR}(x).$$

The associated bilinear form is defined by:

$$b_{-1,F}^{CR}(u,u) = \sum_{i} \rho_i \{ H(1 + \log H/h) \sum_{\mathcal{F}_{ij} \subset \partial\Omega_i} (\bar{u}_{\mathcal{F}_{ij}^{CR}} - \bar{u}_{\partial\Omega_i^{CR}})^2 \}.$$

3.2. Neumann-Neumann coarse spaces. We consider a family of coarse spaces with only one degree of freedom per substructure; see [13]].

For each $\beta \geq 1/2$, we define the pseudo inverses $\mu_{i,\beta}^+$, $i = 1, \cdots, N$, by

$$\mu_{i,\beta}^+(x) = \frac{1}{(\rho_i)^\beta + (\rho_j)^\beta}, \ x \in \mathcal{F}_{ij,h}^{CR} \ \forall \mathcal{F}_{ij} \subset (\partial\Omega_i\backslash\partial\Omega_!)$$

and

$$\mu_{i,\beta}^+(x) - 0, \ x \in (\Gamma_h^{CR}\backslash\partial\Omega_{i,h}^{CR}) \cup \partial\Omega_h^{CR}.$$

We extend $\mu_{i,\beta}^+$ elsewhere in Ω as a nonconforming discrete harmonic function with data on $\Gamma_h^{CR}\cup\partial\Omega_h^{CR}$. The resulting functions belong to $\tilde{V}_0^h(\Omega)$ and are also denoted by $\mu_{i,\beta}^+$.

We can now define the coarse space $\tilde{V}_{-1}^{NN} \subset \tilde{V}_0^h(\Omega)$ by

$$\tilde{V}_{-1}^{NN} = \text{span}\{\rho_i^\beta \, \mu_{i,\beta}^+\},$$

where the span is taken over all the substructures Ω_i.

We note that $\tilde{V}_{-1}^{NN}$ is also the range of the interpolation operator $\tilde{I}_h^{NN}$ given by

(4)
$$u_{-1} = \tilde{I}_h^{NN} u(x) = \sum_i u_{-1}^{(i)} = \sum_i \bar{u}_{\partial\Omega_i^{CR}} (\rho_i)^\beta \mu_{i,\beta}^+.$$

We can even define a Neumann-Neumann coarse space for $\beta = \infty$ [8] by considering the limit of the space $\tilde{V}_{-1}^{NN}$ when β approaches ∞, i.e.

$$\tilde{V}_{-1}^{NN} = \text{span}\{ \lim_{\beta\to\infty} \rho_i^\beta \, \mu_{i,\beta}^+ \}.$$

The associated bilinear form is defined by:

$$b_{-1,NN}^{CR}(u,u) = a^h(u,u).$$

LEMMA 1. *Let $X = F$ or NN. Then, for any $u \in \tilde{V}_0^h(\Omega)$*

$$a^h(\tilde{I}_h^X u, \tilde{I}_h^X u) \le C_3 \left(1 + \log H/h\right) a^h(u,u).$$

All constants C_i in this paper are independent of the mesh parameters, β, and the jumps of the coefficients across the interfaces separating the substructures.

4. Multilevel Additive Schwarz Method

Any Schwarz method can be defined by the underlying splitting of the discretization space $\tilde{V}_0^h(\Omega)$ into a sum of subspaces, and by bilinear forms associated with these subspaces. Let $X = F$ or NN. The splitting of $\tilde{V}_0^h(\Omega)$ that we consider is given by

$$\tilde{V}_0^h = \tilde{V}_{-1}^X + \sum_{k=0}^{\ell} \sum_{j\in\mathcal{N}^k} V_j^k + \sum_{j\in\mathcal{N}_h^{CR}} \tilde{V}_j^h.$$

Here, $\mathcal{N}_h^{CR}$ is the set of CR nodal points associated with the space $\tilde{V}_0^h(\Omega)$. The space $\tilde{V}_j^h \subset \tilde{V}_0^h(\Omega)$ is the one-dimensional space spanned by $\tilde{\phi}_j^h$, the standard P_1-nonconforming basis functions associated with the nodes $j \in \mathcal{N}_h^{CR}$. For the definitions of V_j^k and $\mathcal{N}^k$ see [6].

We introduce the following operators:

 i) $\tilde{T}_{-1}^X : \tilde{V}_0^h \to \tilde{V}_{-1}^X$, is given by

$$b_{-1,X}^{CR}(\tilde{T}_{-1}^X u, v) = a^h(u,v), \ \forall v \in \tilde{V}_{-1}^X.$$

 ii) $\tilde{P}_j^k : \tilde{V}_0^h \to V_j^k$, $k = 0, \cdots, \ell$, $j \in \mathcal{N}^k$ is given by

$$a^h(\tilde{P}_j^k u, v) = a^h(u,v), \ \forall v \in V_j^k.$$

 iii) $\tilde{P}_j^h : \tilde{V}_0^h \to \tilde{V}_j^h$, $j \in \mathcal{N}_h^{CR}$, is given by

$$a^h(\tilde{P}_j^h u, v) = a^h(u,v), \ \forall v \in \tilde{V}_j^h.$$

Let

$$(5) \qquad \tilde{T}^X = \tilde{T}^X_{-1} + \sum_{k=0}^{\ell} \sum_{j \in \mathcal{N}^k} \tilde{P}^k_j + \sum_{j \in \mathcal{N}^{CR}_h} \tilde{P}^h_j.$$

THEOREM 1. *For any* $u \in \tilde{V}^h_0(\Omega)$

$$C_4 \left(1 + \log H/h \right)^{-2} a^h(u, u) \le a^h(\tilde{T}^X u, u) \le C_5 \, a^h(u, u).$$

5. A Multiplicative Version

We consider two versions:

$$(6) \qquad E_G = (\prod_{j \in \mathcal{N}^{CR}_h} (I - \tilde{P}^h_j))(\prod_{k=0}^{\ell} \prod_{j \in \mathcal{N}^k} (I - \tilde{P}^k_j))(I - \tilde{T}^X_{-1}),$$

and

$$(7) \qquad E_J = (I - \eta \sum_{j \in \mathcal{N}^{CR}_h} \tilde{P}^h_j)(\prod_{k=0}^{\ell} (I - \eta \sum_{j \in \mathcal{N}^k} \tilde{P}^k_j))(I - \eta \tilde{T}^X_{-1}),$$

where η is a damping factor such that

$$\|\eta \tilde{T}^X_{-1}\|_{H^1_{\rho,h}}, \ \|\eta \sum_{j \in \mathcal{N}^{CR}_h} \tilde{P}^h_j\|_{H^1_{\rho,h}}, \ \|\eta \sum_{j \in \mathcal{N}^k} \tilde{P}^k_j\|_{H^1_{\rho,h}} \le w < 2.$$

When the product is arranged in an appropriate order, the operators E_G and E_J correspond to the error propagation operator of V-cycle multigrid methods using Gauss Seidel and damped Jacobi smoothers, respectively; see Zhang [17]. The norm of the error propagation operators $\|E_G\|_{H^1_{\rho,h}}$ and $\|E_F\|_{H^1_{\rho,h}}$ can be estimated from above by $1 - C_2 \left(\log \left(H/h \right) \right)^{-2}$.

REMARK 1. *In* [13], *we analyzed a two-level additive Schwarz method for discontinuous coefficients. There, we cover* Ω *by overlapping subregions by extending each substructure* Ω_i *to a larger region. We can modify that method by considering inexact local solvers and by covering* Ω *in a different way. The analysis of our methods suggests two attractive ways of covering* Ω *: by face regions* $\Omega_{ij} = \Omega_i \cup \mathcal{F}_{ij} \cup \Omega_j$ *, or by cross point regions* Ω'_m *; see* [6]. *We again obtain condition number estimates which are polylogarithmically on the number of degree of freedom of individual local subproblems.*

REMARK 2. *We can decrease the complexity of our algorithm by considering approximate discrete nonconforming harmonic extension given by simple explicit formulas in* [13].

REMARK 3. *In a case in which we have quasi-monotone coefficients* [6] *and use* V^H *, the piecewise linear function, as a coarse space, all algorithms in this paper are optimal.*

Acknowledgments. I would like to thank my advisor Olof Widlund and Maksymilian Dryja for many suggestions concerning this work.

REFERENCES

1. D. N. Arnold and F. Brezzi, *Mixed and nonconforming finite element methods: implementation, postprocessing and error estimates*, Math. Model. and Numer. Anal. **19** (1985), 7–32.
2. D. Braess and R. Verfürth, *Multigrid methods for nonconforming finite element methods*, SIAM J. Numer. Anal. **27** (1990), 979–986.
3. Suzanne C. Brenner, *An optimal-order multigrid method for P1 nonconforming finite elements*, Math. Comp. **53** (1989), 1–15.
4. Lawrence C. Cowsar and Jan Mandel and Mary F. Wheeler, *Balancing domain decomposition for mixed finite elements*, Technical Report TR93-08, Department of Mathematical Sciences, Rice University, March, 1993.
5. Crouseix, M. and Raviart, P.A. *Conforming and nonconforming finite element methods for solving the stationary Stokes equations.*, RAIRO Anal. Numer. **7** (1973), 33–76.
6. Makysmilian Dryja, *Multilevel methods for elliptic problems with discontinuous coefficients in three dimensions*, These Proceedings.
7. Maksymilian Dryja and Olof B. Widlund, *Schwarz Methods of Neumann-Neumann Type for Three-Dimensional Elliptic Finite Element Problems*, TR #626, CS Department, Courant Institute, March 1993. To appear in Comm. Pure Appl. Math.
8. Maksymilian Dryja, Marcus Sarkis, and Olof B. Widlund, *Multilevel Schwarz methods for elliptic problems with discontinuous coefficients in three dimensions*, TR 662, CS Department, Courant Institute, March, 1994.
9. M. Dryja, B. F. Smith, and O. B. Widlund, *Schwarz analysis of iterative substructuring algorithms for elliptic problems in three dimensions*, Technical Report 638, Courant Institute of Mathematical Sciences, May 1993. To appear in SIAM J. Numer. Anal.
10. C. O. Lee, *A nonconforming multigrid method using conforming subspaces*, In proceedings of the 1993 Copper Mountain Conference on Multigrid Methods.
11. Peter Oswald, *On a hierarchical basis on the multilevel method with nonconforming P1 elements*, Numer. Math. **62** (1992), 189–212.
12. Peter Oswald, *On a BPX-preconditioner for P1 elements*, Math./91/2, Friedrich Schiller Universität, Jena, Germany, 1991.
13. Marcus Sarkis, *Two-Level Schwarz methods for nonconforming finite elements and discontinuous coefficients*, TR 629, CS Department, Courant Institute, March, 1993. Also in the proceedings of the 1993 Copper Mountain Conference on Multigrid Methods.
14. Marcus Sarkis, *Multilevel methods for a hybrid-mixed formulation of elliptic problems with discontinuous coefficients in three dimensions.* TR, CS Department, Courant Institute, to appear.
15. P. S. Vassilevski and J. Wang, *Multilevel methods for nonconforming finite element methods for elliptic problems*, Preprint, 1993.
16. B. Wohlmuth and R. H. W. Hoppe, *Multilevel approaches to nonconforming finite element discretizations of linear second order elliptic boundary value problems*, TUM-M9320, Inst. f. Informatik, TU München, October, 1993.
17. Xuejun Zhang, *Multilevel Schwarz methods*, Numer. Math. **63** (1992), 521–539.

COURANT INSTITUTE OF MATHEMATICAL SCIENCES, NEW YORK UNIVERSITY, 251 MERCER STREET, NEW YORK, N.Y. 10012
E-mail address: sarkis@acf4.nyu.edu

Contemporary Mathematics
Volume **180**, 1994

On Generalized Schwarz Coupling Applied to Advection-Dominated Problems

K.H. TAN AND M.J.A. BORSBOOM

ABSTRACT. Schwarz methods can be interpreted as domain decomposition methods in which a local mechanism at interfaces is used to restore iteratively the connection between artificially decoupled subproblems. The speed of convergence of these iterative processes can be improved significantly by the choice of a proper coupling mechanism. If known, properties of the discretized problem can be taken into account in such an optimization of the coupling.

In this paper a flexible local coupling mechanism is proposed that can easily be tailored to properties of the discrete problem. It consists of parameterized interface conditions, which are formulated with the use of virtual unknowns. Besides a Dirichlet and a Neumann part, tangential and mixed derivatives are involved in these interface conditions. The coupling technique is combined with the block-Jacobi iteration, resulting in a generalized additive Schwarz method. Its convergence properties are analyzed and optimized for a number of simple time-dependent advection-dominated problems, discretized by means of central finite differences.

1. Introduction

Schwarz methods can be introduced and investigated in both a continuous and discrete setting. Consider the continuous problem $Lu = g$ in a domain Ω that has been subdivided into two non-overlapping subdomains $\Omega = \Omega_1 \cup \Omega_2$ with common interface Γ, $\Omega_1 \cap \Omega_2 = \Gamma$. Then in a continuous setting this problem is typically reformulated as:

$$
\begin{align}
Lu_1 &= g_1 & &\text{in } \Omega_1 \,, \tag{1} \\
\Phi_i(u_1) &= \Phi_i(u_2) & &\text{on } \Gamma_i \subset \Gamma, \ i = 1, ..., I \,, \tag{2} \\
\Psi_j(u_2) &= \Psi_j(u_1) & &\text{on } \Gamma_j \subset \Gamma, \ j = 1, ..., J \,, \tag{3} \\
Lu_2 &= g_2 & &\text{in } \Omega_2 \,, \tag{4}
\end{align}
$$

1991 *Mathematics Subject Classification.* 65M55, 65F10

This work was sponsored by the Stichting Nationale Computerfaciliteiten (National Computing Facilities Foundation, NCF) for the use of supercomputer facilities, with financial support from the Nederlandse Organisatie voor Wetenschappelijk Onderzoek (Netherlands Organization for Scientific Research, NWO).

The first author was supported by a grant from Delft Hydraulics, the Netherlands.

This paper is in final form and no version of it will be submitted for publication elsewhere.

complemented with appropriate boundary conditions on $\partial\Omega_1\backslash\Gamma$ and $\partial\Omega_2\backslash\Gamma$. If the coupling equations are chosen such that unique solutions u_1 and u_2 exist for which holds that $u_1 = u|_{\Omega_1}$ and $u_2 = u|_{\Omega_2}$, the Schwarz formulation is said to be equivalent with the original problem formulation. In general, many Φ_i and Ψ_j lead to equivalence. This freedom is exploited in optimizing the convergence speed of domain decomposition methods derived from (1)–(4). A suitable choice, resulting into fast convergence, should be made from the many coupling operators that guarantee equivalence. For some recent examples of operators proposed in the literature we refer to [1] and [3].

In the construction of coupling operators that maximize the convergence speed of domain decomposition methods for discretized problems, the discrete equivalent of (1)–(4), rather than the continuous problem itself, is to be considered. For this reason we describe in section 2 a discrete equivalent of (1)–(4). It is formulated with the use of so-called enhanced systems of equations, as introduced in [3]. In section 3, a simplified model problem is formulated. For this problem, a specific form of a parameterized enhancement is proposed. In section 4, the convergence behavior as a function of the parameters of the enhanced system will be examined and optimized. The resulting convergence rates are considerably better than the ones that can be obtained with the usual combinations of Neumann and Dirichlet conditions, as will be illustrated with some numerical results in section 5.

2. A discrete Schwarz formulation

Consider a system of linear equations $Au = f$ for which a unique solution exists. To introduce a discrete counterpart of (1)–(4), let this system be partitioned as:

$$(5) \qquad \begin{pmatrix} A_{11} & A_{12}^l & A_{12}^r & A_{13} \\ \hat{A}_{21} & \hat{A}_{22}^l & \hat{A}_{22}^r & \hat{A}_{23} \\ \check{A}_{21} & \check{A}_{22}^l & \check{A}_{22}^r & \check{A}_{23} \\ A_{31} & A_{32}^l & A_{32}^r & A_{33} \end{pmatrix} \begin{pmatrix} u_1 \\ u_I^l \\ u_I^r \\ u_2 \end{pmatrix} = \begin{pmatrix} f_1 \\ f_I^l \\ f_I^r \\ f_2 \end{pmatrix} .$$

The dimensions of the subvectors are n_1, n_I^l, n_I^r, and n_2, respectively. Usually, the partitioning will be such that n_I^l, $n_I^r \ll n_1$, n_2.

Consider the following enhanced system of equations:

$$(6) \qquad \begin{pmatrix} A_{11} & A_{12}^l & A_{12}^r & 0 & 0 & A_{13} \\ \hat{A}_{21} & \hat{A}_{22}^l & \hat{A}_{22}^r & 0 & 0 & \hat{A}_{23} \\ 0 & R^l & S^l & -R^l & -S^l & 0 \\ 0 & -R^r & -S^r & R^r & S^r & 0 \\ \check{A}_{21} & 0 & 0 & \check{A}_{22}^l & \check{A}_{22}^r & \check{A}_{23} \\ A_{31} & 0 & 0 & A_{32}^l & A_{32}^r & A_{33} \end{pmatrix} \begin{pmatrix} c_1 \\ c_I^l \\ \tilde{c}_I^r \\ \tilde{c}_I^l \\ c_I^r \\ c_2 \end{pmatrix} = \begin{pmatrix} f_1 \\ f_I^l \\ 0 \\ 0 \\ f_I^r \\ f_2 \end{pmatrix} ,$$

where $\tilde{c}_I^l$ and $\tilde{c}_I^r$ are vectors of dimension n_I^l and n_I^r, respectively.

Let $\bar{c}_1 \doteq (c_1, c_I^l, c_I^r, c_2)^T$, $\bar{c}_2 = (\tilde{c}_I^l, \tilde{c}_I^r)^T$, $g_1 = (f_1, f_I^l, f_I^r, f_2)^T$ and $g_2 = (0, 0)^T$.

Equation (6) can be written as:

$$
(7) \qquad \begin{pmatrix} B_{11} & B_{12} \\ B_{21} & B_{22} \end{pmatrix} \begin{pmatrix} \bar{c}_1 \\ \bar{c}_2 \end{pmatrix} = \begin{pmatrix} g_1 \\ g_2 \end{pmatrix} ,
$$

where $B_{11} = \begin{pmatrix} A_{11} & A_{12}^l & 0 & A_{13} \\ \hat{A}_{21} & \hat{A}_{22}^l & 0 & \hat{A}_{23} \\ \check{A}_{21} & 0 & \check{A}_{22}^r & \check{A}_{23} \\ A_{31} & 0 & A_{32}^r & A_{33} \end{pmatrix}$, $B_{12} = \begin{pmatrix} 0 & A_{12}^r \\ 0 & \hat{A}_{22}^r \\ \check{A}_{22}^l & 0 \\ A_{32}^l & 0 \end{pmatrix}$,

$B_{21} = \begin{pmatrix} 0 & R^l & -S^l & 0 \\ 0 & -R^r & S^r & 0 \end{pmatrix}$, and $B_{22} = \begin{pmatrix} -R^l & S^l \\ R^r & -S^r \end{pmatrix}$. We have the following theorem from [2].

THEOREM 1. *Enhanced system of equations (6) has an unique solution c for which holds* $c_k = u_k$ *for* $k = 1, 2$, $c_I^l = \tilde{c}_I^l = u_I^l$ *and* $c_I^r = \tilde{c}_I^r = u_I^r$, *if and only if* B_{22}^{-1} *exists.*

Proof: If B_{22}^{-1} exists, the Schur complement system $(B_{11} - B_{12}B_{22}^{-1}B_{21})\bar{c}_1 = g_1$ is precisely the original system of equations $A\bar{c}_1 = f$. As A was assumed non-singular, it follows that $\bar{c}_1 = u$, and hence $\tilde{c}_I^l = u_I^l$ and $\tilde{c}_I^r = u_I^r$. Conversely, let c be the (unique) solution of (6) for which the equalities of theorem 1 hold. Then $\begin{pmatrix} -R^l & S^l \\ R^r & -S^r \end{pmatrix} \begin{pmatrix} c_I^l - \tilde{c}_I^l \\ c_I^r - \tilde{c}_I^r \end{pmatrix} = \begin{pmatrix} 0 \\ 0 \end{pmatrix}$ has only the trivial solution. Hence B_{22} is non-singular. $\qquad \square$

Enhanced system (6) is the discrete counterpart we were looking for. The two middle rows represent the (additional) coupling equations, the first two and last two rows represent the original discrete problem, while the existence of B_{22}^{-1} guarantees equivalence with the original problem $Au = f$.

To solve $Au = f$, we propose the domain decomposition method that consists of block-Jacobi iteration, applied to an enhanced system of the form (6). In this block-Jacobi iteration the blocks correspond with the substructures, indicated in (6).

For discretizations with local support, c_I^l and c_I^r can be chosen such that A_{13}, $\hat{A}_{23}$, $\check{A}_{21}$ and A_{31} vanish. The optimization of the convergence speed boils down to a proper choice of R^l, S^l, R^r and S^r. The only constraint on this choice is given by the condition of existence of B_{22}^{-1}, which is a condition independent of the discretization. For a cell-vertex scheme using three-times-three-point computational molecules on a two-dimensional structured grid, these submatrices already disappear if c_I^l and c_I^r are chosen as the vectors consisting of the unknowns defined on two adjacent grid lines, with the interface in between.

3. A discrete model problem

Consider the two-dimensional advection-diffusion problem, with uniform flow field:

$$
(8) \qquad \frac{\partial c}{\partial t} + u \frac{\partial c}{\partial x} + v \frac{\partial c}{\partial y} = D\left(\frac{\partial^2}{\partial x^2} + \frac{\partial^2}{\partial y^2}\right) \quad \text{in } \Omega \times \mathcal{T} ,
$$

with $\Omega = [0,2] \times [0,1]$, time interval $\mathcal{T} = [t_0, t_0 + T]$; constants $u, v, D > 0$, and $c : \Omega \times \mathcal{T} \to R$. We restrict ourselves to cases where advection is dominant over diffusion (Pe $\gg$ 1). On Ω, a uniform cartesian grid with size ($\Delta x = \frac{2}{I_1 + I_2}$, $\Delta y = \frac{1}{J}$) and $(I_1 + I_2 + 2) \times (J + 2)$ nodes has been defined to discretize equation (8), using three-point central differences in space and Euler backward in time. Depending on the flow direction, inflow or outflow conditions are used at the physical boundaries. For domain decomposition purposes, the domain is subdivided into two vertical strips, with common interface Γ normal to the x-direction, having $I_1 \times J$ and $I_2 \times J$ internal grid points respectively. Courant and cell-Peclet numbers are defined as: $\text{CFL}_x := \frac{u \Delta t}{\Delta x}$ and $\text{Pe}_x^c := \frac{u \Delta x}{D}$. Similar expressions are defined for the y-direction.

Conforming to the theory of the previous section, $(I_k + 2) \times (J + 2)$ unknowns are defined for each subdomain k, $k = 1, 2$, thus effectively using extended grids with $2J + 4$ additional unknowns. These unknowns are denoted by $c_k^{i,j}$, $i = 0, ..., I_k + 1$, $j = 0, ..., J + 1$, $k = 1, 2$. On the virtual overlap consisting of two adjacent vertical grid lines, the subvectors $c_I^l = (c_1^{I_1,0}, ..., c_1^{I_1,J+1})^T$ and $c_I^r = (c_2^{1,0}, ..., c_2^{1,J})^T$ are defined, together with the enhancement vectors $\tilde{c}_I^r = (c_1^{I_1+1,0}, ..., c_1^{I_1+1,J+1})^T$ and $\tilde{c}_I^l = (c_2^{0,0}, ..., c_2^{0,J+1})^T$ (cf. (6)). So to close the system of equations, $2J + 4$ coupling equations have to be added. For our advection-dominated model problem (8) they have been chosen as:

$$
(9) \quad
\begin{aligned}
(\mu_x(\tfrac{1}{2} + \tfrac{\beta}{2\Delta y}\mu_y \delta_y) + \delta_x(\tfrac{\alpha}{\Delta x} + \tfrac{\beta}{2\Delta y}\mu_y \delta_y)) c_1^{I_1 + \frac{1}{2}, j} &= \\
(\mu_x(\tfrac{1}{2} + \tfrac{\beta}{2\Delta y}\mu_y \delta_y) + \delta_x(\tfrac{\alpha}{\Delta x} + \tfrac{\beta}{2\Delta y}\mu_y \delta_y)) c_2^{\frac{1}{2}, j} &\quad , j = 1, ..., J , \\
(\mu_x(\tfrac{1}{2} + \tfrac{\delta}{2\Delta y}\mu_y \delta_y) + \delta_x(\tfrac{\gamma}{\Delta x} - \tfrac{\delta}{2\Delta y}\mu_y \delta_y)) c_2^{\frac{1}{2}, j} &= \\
(\mu_x(\tfrac{1}{2} + \tfrac{\delta}{2\Delta y}\mu_y \delta_y) + \delta_x(\tfrac{\gamma}{\Delta x} - \tfrac{\delta}{2\Delta y}\mu_y \delta_y)) c_1^{I_1 + \frac{1}{2}, j} &\quad , j = 1, ..., J ,
\end{aligned}
$$

where μ_x and δ_x are defined as: $\delta_x c^i \equiv c^{i+\frac{1}{2}} - c^{i-\frac{1}{2}}$, $\mu_x c^i \equiv c^{i+\frac{1}{2}} + c^{i-\frac{1}{2}}$. Operators μ_y and δ_y are defined analogously. Note that equations (9) are the discrete equivalent of (2) and (4) involving c, $\frac{\partial c}{\partial x}$ at the interface, and $\frac{\partial c}{\partial y}$ at the boundaries of the extended subdomains. The 4 remaining coupling equations (at the intersection of the interface with the physical boundaries) are defined in a similar way, except for the number of unknowns involved (8 instead of 12), see [2].

4. Convergence analysis

The optimal value of the coupling parameters α, β, γ and δ is estimated by means of a simple Fourier analysis. We assume that the convergence error e^m of the mth iterand, $e^m = c^m - c$, can be written as:

$$
(10) \quad e_k^{i,j,m} = \sum_{\ell=0}^{J-1} \left[F_{k,\ell}^m (\lambda_\ell^l)^i + G_{k,\ell}^m (\lambda_\ell^r)^{i - I_k - 1} \right] \exp(\imath \omega_\ell j) ,
$$

with $\imath = \sqrt{-1}$ and $\omega_\ell = 2\pi \ell \Delta y$; λ_ℓ^l and λ_ℓ^r are determined by the discretization while the amplitudes $F_{k,\ell}^m$ and $G_{k,\ell}^m$ are determined by (9) and by the left and right boundary conditions. Straightforward calculation shows:

PROPOSITION 1. *For sufficiently small CFL-numbers, the rate of convergence per Fourier mode can be approximated by:*

$$\rho_\ell^2 \approx \frac{[\frac{1+\lambda_\ell^l}{2} + \frac{\alpha(\lambda_\ell^l-1)}{\Delta x} + \frac{\beta\lambda_\ell^l \sin(\omega_\ell)\imath}{\Delta y}][\frac{\bar\lambda_\ell^r+1}{2} + \frac{\gamma(1-\bar\lambda_\ell^r)}{\Delta x} + \frac{\delta\bar\lambda_\ell^r \sin(\omega_\ell)\imath}{\Delta y}]}{[\frac{\bar\lambda_\ell^r+1}{2} + \frac{\alpha(1-\bar\lambda_\ell^r)}{\Delta x} + \frac{\beta \sin(\omega_\ell)\imath}{\Delta y}][\frac{1+\lambda_\ell^l}{2} + \frac{\gamma(\lambda_\ell^l-1)}{\Delta x} + \frac{\delta \sin(\omega_\ell)\imath}{\Delta y}]}$$

Here $\bar\lambda_\ell^r$ denotes $\frac{1}{\lambda_\ell^r}$. In our numerical experiments, we approximately solved the minimax problem: $\min_{\alpha,\beta,\gamma,\delta} \max_\ell \rho_\ell$ to estimate the optimal coupling parameters for (9). (The coupling parameters in the 4 remaining coupling equations were determined experimentally.)

5. Numerical results

In this section we report on some numerical experiments, where we have used results of the convergence analysis. In all cases we used $I_1 = I_2 = J = 10$, and $\mathrm{Pe}_x^c = \mathrm{Pe}_y^c = 10^5$.

To assess the importance of coupling parameters and the benefits of the inclusion of parameterized tangential derivatives in the coupling conditions, we varied the coupling parameters for a skew velocity field $\mathrm{CFL}_x = 1$, $\mathrm{CFL}_y = 2$ and computed the rate of convergence by determining the spectral radius of the error amplification matrix of the block-Jacobi method. The results are shown in figure 1. In the first figure α and β were varied, keeping γ and δ fixed at their (experimentally determined) optimal values. In the second figure, the opposite case was considered. For this example the convergence analysis predicts approximate optimal coupling

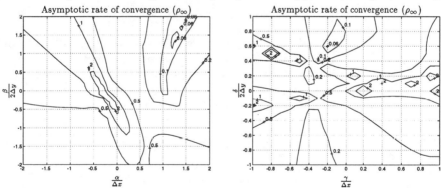

FIGURE 1. 2-D: varying coupling at interface, $\mathrm{CFL}_x = 1, \mathrm{CFL}_y = 2$

parameters: $\frac{\alpha}{\Delta x} \approx 1.27$, $\frac{\beta}{2\Delta y} \approx 1.44$, $\frac{\gamma}{\Delta x} \approx -0.19$, $\frac{\delta}{2\Delta y} \approx 0.56$. These are very close to the experimentally determined optimal values, both sets of (theoretically or experimentally determined) parameters resulting in convergence rates of approximately $\rho_\infty \approx 0.05$. The best convergence rate with the use of combinations of c and $\frac{\partial c}{\partial x}$ only, turns out to be $\rho_\infty \approx 0.3$, which is considerably worse.

We also compared several coupling mechanisms for a number of different flow directions. Velocities (u, v) were chosen such that $(\mathrm{CFL}_x)^2 + (\mathrm{CFL}_y)^2 = 2$. This time periodic boundary conditions in y-direction were used. In figure 2 the obtained convergence rates are shown. In the figure, DD denotes Dirichlet-Dirichlet coupling at the virtual grid points ($\alpha = \frac{\Delta x}{2}$, $\gamma = -\frac{\Delta x}{2}$, $\beta = \delta = 0$). ND1 is a

Neumann-Dirichlet coupling ($\alpha = \infty$ and $\beta = \gamma_2 = \delta_2 = 0$). ND2 is an extended Neumann-Dirichlet coupling where the 'Neumann' condition consists of a one-sided discretization of the advective part of (8) at the interface ($\frac{\alpha}{\Delta x} = \text{CFL}_x$, $\beta = \text{CFL}_y$, $\gamma = \delta = 0$). Finally, GND is our generalized Neumann-Dirichlet coupling, with parameters chosen according to the approximate optimization.

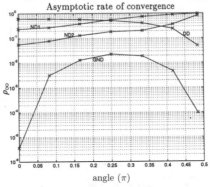

FIGURE 2. 2-D: 2 subdomains, varying angle of flow (w.r.t. x-axis)

The results indicate that for any flow angle, a suitable subdomain coupling is obtained with our approximate optimization of (9).

6. Concluding remarks

In this paper we have briefly described a flexible local coupling mechanism that allows an improvement of convergence speed of domain decomposition methods by tailoring the coupling to the discretized problem to solve. Full details can be found in [**2**]. Although in section 4 only a few coupling parameters were optimized, numerical results nevertheless show that in this way already some very good convergence rates can be obtained. The inclusion of a parameterized tangential component in the coupling algorithm appears to be essential. Recently obtained results with a slightly extended parameterized subdomain coupling by including several mixed derivatives show a further improvement of convergence rates.

The presented approach can be extended easily to variable-coefficient and non-linear problems when discretized on a curvilinear, structured grid, optimizing the coupling parameters locally in a problem-dependent way. This work is currently in progress and will be reported on in the near future.

REFERENCES

1. A. QUARTERONI, *Domain decomposition methods for systems of conservation laws: Spectral collocation approximations*, SIAM J. Sci. Stat. Comput., 11 (1990), pp. 1029–1052.
2. K. H. TAN AND M. J. A. BORSBOOM, *Problem-dependent optimization of flexible couplings in domain decomposition methods with an application to advection-dominated problems*, Tech. Report 830, University Utrecht, Department of Mathematics, Utrecht, Oct. 1993.
3. W. P. TANG, *Generalized Schwarz splittings*, SIAM J. Sci. Stat. Comput., 13 (1992), pp. 573–595.

MATHEMATICAL INSTITUTE, UTRECHT UNIVERSITY, P.O. BOX 80010, 3508 TA UTRECHT, THE NETHERLANDS.
E-mail address: tan@math.ruu.nl

DELFT HYDRAULICS, P.O. BOX 152, 8300 AD EMMELOORD, THE NETHERLANDS.
E-mail address: mart.borsboom@wldelft.nl

Contemporary Mathematics
Volume **180**, 1994

Exotic coarse spaces for Schwarz methods for lower order and spectral finite elements

OLOF B. WIDLUND

ABSTRACT. Fast domain decomposition algorithms for elliptic problems are typically two-level methods. The second level enables us to construct iterative methods with convergence rates that are independent of the number of subregions. The construction of a suitable coarse, global subspace is often the most interesting part of the design and analysis. We consider several coarse spaces for lower order finite element methods and we then show that one of them has a useful analog in the case of continuous, piecewise Q_p elements defined on cubic elements. For all these methods, the condition number of the relevant iteration operator grows only logarithmically or polylogarithmically with the number of degrees of freedom associated with a subregion, even if there are arbitrarily large jumps in the coefficients across subregional interfaces.

1. Introduction

Fast domain decomposition algorithms for elliptic problems are typically two-level methods. The second level enables us to construct iterative methods with convergence rates independent of the number of subregions. The construction of a suitable coarse, global subspace is the most interesting part of the design, analysis, and practice. In this paper, we will briefly survey recent results obtained jointly with Maksymilian Dryja, Luca Pavarino, and Barry Smith; see [**6, 7, 10**]. For related work on multilevel methods; cf. Dryja, Sarkis, and Widlund [**4**]. See also the papers by Dryja, Pavarino, and Sarkis in these proceedings.

We first consider several coarse spaces for lower order, *h-version* finite element methods. There are polylogarithmic bounds for the condition number $\kappa(T_h)$ of the

1991 *Mathematics Subject Classification.* Primary 41A10, 65N30; Secondary 65N35, 65N55.

This work was supported in part by the National Science Foundation under Grant NSF-CCR-9204255 and, in part, by the U. S. Department of Energy under contracts DE-FG02-92ER25127 and DE-FG02-88ER25053.

This paper is in final form and no version of it will be submitted for publication elsewhere.

relevant iteration operator T_h based on such a coarse space and good local spaces. For several new methods, we have

$$\kappa(T_h) \leq C(1 + \log(H/h));$$

cf. Dryja, Smith, and Widlund [7]. C is independent of the mesh parameters and the jumps in the coefficient, but depend on the aspect ratios of the subdomains.

We then consider a Galerkin method using *spectral* elements and obtain

$$\kappa(T_p) \leq C(1 + \log(p))^2.$$

The bounds are uniform in the number of subproblems. The number of degrees of freedom for each subproblem is on the order of $(H/h)^3$ and p^3, respectively, and the bounds of the condition numbers are thus polylogarithmic functions of these quantities. Each subproblem can be handled by a processor of a parallel or distributed computing system; the size of the local problem will be limited primarily by the amount of fast memory available to the individual processors.

In this paper, we consider only *iterative substructuring methods* and only three dimensional problems. The local problems of an iterative substructuring method communicate with their neighbors only through the boundary values of the iterates; this is different from the Schwarz methods that use overlapping subregions.

Earlier important work on iterative substructuring methods for the h-version finite element methods was carried out by Bramble, Pasciak, and Schatz, [2], Dryja [5], and Smith [12, 13]. Work on p-version finite element methods is described in Mandel [8]. Important work on the p-version for two dimensions is described in Babuška, Craig, Mandel, and Pitkäranta [1]. Results on iterative substructuring for higher order elements methods for two dimensions are included in Pavarino's thesis [9], which also contains results on several domain decomposition algorithms based on overlapping subregions.

2. The elliptic problems

We consider linear, elliptic problems on a bounded Lipschitz domain Ω in R^3, formulated as: Find u such that

$$a(u, v) = \int_\Omega \rho(x) \nabla u \cdot \nabla v \, dx = f_\Omega(v), \, \forall \, v \in V.$$

$\rho(x) > 0$ can be discontinuous, with very different values for different substructures, but $\rho(x) = \rho_i$ on the substructure Ω_i. $\bar{\Omega} = \cup \bar{\Omega}_i$, $\Omega_i \cap \Omega_j = \emptyset, i \neq j$. V is a subspace of $H^1(\Omega)$ chosen appropriately.

The finite element space is either the space $V^h \subset V$ of continuous, piecewise linear functions on the elements, into which the substructures have been partitioned, or $V^p \subset V$, the space of continuous piecewise Q_p functions.

For the *lower order case*, we can, if so desired, use quite general substructures, which determine a coarse "triangulation". We can think of substructures as pieces of a three-dimensional jigsaw puzzle. Each such piece is a union of whole elements.

For the *spectral case,* we assume that the region is a union of elements that are cubes or images of a reference cube under reasonable smooth mappings. No element should be "too distorted".

The finite element problems are obtained by restricting u and the test functions to the space $\tilde{V} = V^h$ or V^p.

3. Excerpts from additive Schwarz analysis

In this section, we give a brief overview of a useful method of analyzing domain decomposition methods; see Dryja et al. [**6, 7**] for many further details.

We decompose the finite element space $\tilde{V}$ into subspaces V_i. We need not use a direct sum. For each subspace V_i, we introduce an orthogonal projection P_i onto V_i:

$$a(P_i u, v) = a(u, v), \ \forall v \in V_i, \ u \in \tilde{V},$$

or use an approximate solver defined by

$$\tilde{a}_i(T_i u, v) = a(u, v), \ \forall v \in V_i, \ u \in \tilde{V}.$$

T_i defines the inner product $\tilde{a}_i(\cdot, \cdot)$ and vice versa.

The major effort in proving a good bound for the condition number of the relevant operator $T = \sum T_i$, goes into proving that there exist C_0 and ω, of modest size, such that there exists a decomposition

$$u = \sum_{i=0}^{N} u_i, \ u_i \in V_i \ \forall u \in \tilde{V}, \ \text{with}$$

$$\sum_{i=0}^{N} \tilde{a}_i(u_i, u_i) \leq C_0^2 a(u, u), \ \text{and}$$

$$a(u_i, u_i) \leq \omega \tilde{a}_i(u_i, u_i) \ \forall u_i \in V_i.$$

There is also a well established theory for multiplicative, Gauss-Seidel-type methods that involves the same parameters; see Bramble et al. [**3**] and Dryja et al. [**6, 7**].

For many iterative substructuring methods, a local analysis is possible. We can then compare $a^{(j)}(u, u)$, the contribution of the substructure Ω_j to the strain energy, to the contributions $\tilde{a}_i^{(j)}(u_i, u_i)$ of Ω_j to the preconditioner; we derive upper and lower bounds for

$$\frac{\sum_{i=0}^{N} \tilde{a}_i^{(j)}(u_i, u_i)}{a^{(j)}(u, u)}.$$

This greatly simplifies the derivation of bounds independent of the jumps of the coefficients; cf. [**7**]. Exactly one of the local problems, the one which corresponds to the global subspace, must have the same null space as the local finite element problem to be able to obtain bounds independent of the number of subregions.

4. Choosing a good coarse space V_0

The obvious choice for V_0 is V^H and Q_1 for the lower order and spectral case, respectively. However, if all the elements of the other subspaces vanish at the vertices of the substructures, then $\kappa(T)$ must grow linearly with H/h and p^2, respectively, for three dimensional problems.

We therefore turn to more exotic spaces. As previously pointed out, we also gain greater freedom in choosing the shape of the subregions.

There are a number of important geometric objects: *interiors, faces, edges,* and *vertices.* We can merge edges and vertices, creating a *wire basket* basket W_i for each substructure. The subspaces are directly related to these objects. A set of counting functions describes it all. These functions also define a good coarse space.

The counting functions ν_i are defined on Γ_h, the set of nodes on all interfaces between the substructures, by

$$\nu_i(x) = \text{number of } \partial\Omega_{j,h} \quad \text{to which } x \in \partial\Omega_{i,h} \text{ belongs}$$
$$\nu_i(x) = 0, \ x \in \Gamma_h \setminus \partial\Omega_{i,h}$$

Just one basis function, μ_i, is related to each Ω_i. It is given by the pseudo inverse of $\nu_i(x)$, in the Laplace case, and by a similar formula involving the ρ_i, in the general case. (Special rules are used for boundary substructures.) The $\mu_i(x)$ are discrete harmonic in each substructure and their span defines our first good coarse space, V_C^h; see Dryja et al. [**6, 7**].

The coarse component of the decomposition is selected as

$$u_0 = \sum_i \bar{u}_i \mu_i.$$

$\bar{u}_i$ is the average of $u \in V^h$ over $\partial\Omega_{i,h}$. This is an interpolation formula, which reproduces constants.

5. Two additional good coarse spaces

Another good coarse space, V_W^h, was introduced in Barry Smith's thesis; cf. [**12, 13**]. It can be regarded as being *wire basket based.* The elements of this coarse space are piecewise discrete harmonic functions. Their values on all of $\partial\Omega_i$ are defined by the values on the wire basket W_i; the value on each face F^k is constant and equals the average $\bar{u}_{\partial F^k}^h$ of the nodal values on the boundary of the face. The related interpolation formula is given, on Ω_i, by

$$I_W^h u^h = \sum_{x_l \in W_i} u^h(x_l)\varphi_l + \sum_{\partial F^k \subset \partial\Omega_i} \bar{u}_{\partial F^k}^h \theta_{F^k}.$$

Here $\theta_{F^k} = 1, x \in F^k$, and $= 0, x \in \Gamma_h \setminus F^k$, and φ_l is the discrete harmonic extension of the standard nodal basis function of x_l. A simple bilinear form is used

$$\tilde{a}_0(u^h, u^h) = (1 + \log(H/h))h \sum_i \rho_i \min_{c_i} \|\underline{u} - c_i \underline{z}^{(i)}\|_{l_2(W_i)}^2,$$

where $z_j^{(i)} = 1, \forall j$. A linear system results with only one essentially global degree of freedom, c_i, per substructure. Additionally, there is a linear system with a diagonal matrix that determines the nodal values on the wire baskets.

A different coarse space, V_M^h, for which a bound that is linear in $\log(H/h)$ has been established, is obtained by assigning independent degrees of freedom to the functions θ_{F^k}. This is a larger *face based* space; cf. also Sarkis [11]. We also need to use high quality local spaces to obtain such a strong bound.

6. A good coarse space for spectral elements

We have designed a wire basket based coarse space for the spectral elements; see Pavarino and Widlund [10] and additional forthcoming papers.

The values on the wire basket W_i of a given Q_p polynomial are extended to the faces, and then to the entire cube, by using formulas based on separation of variables. There are also additional terms similar to those involving the θ_{F^k}; one special function, $\kappa_k(x, y, z)$, is constructed for each face of the cube. We begin with the values of the function 1 on the wire basket and extend them elsewhere by the formula just suggested. We then subtract the resulting function from 1 and obtain $\kappa(x, y, z)$. We write

$$\kappa(x, y, z) = \sum_{k=1}^{6} \kappa_k(x, y, z),$$

where each $\kappa_k(x, y, z) = 0$ for five of the six faces of the cube.

The weights of the κ_k terms can be chosen, as for V_W^h, as the averages over the boundaries of the faces. The resulting space, by construction, contains the constant functions. A simple auxiliary bilinear form $\tilde{a}_0(\cdot, \cdot)$ can be constructed with only one essentially global degree of freedom per substructure.

Acknowledgement: Our results have been obtained jointly with Maksymilian Dryja, Luca F. Pavarino, and Barry F. Smith.

References

1. Ivo Babuška, Alan Craig, Jan Mandel, and Juhani Pitkäranta, *Efficient Preconditioning for the p-Version Finite Element Method in Two Dimensions*, SIAM J. Numer. Anal., 28, 624–661, 1991.

2. James H. Bramble, Joseph E. Pasciak, and Alfred H. Schatz, *The construction of preconditioners for elliptic problems by substructuring, IV*, Math. Comp., 53, 1–24, 1989.

3. James H. Bramble, Joseph E. Pasciak, Junping Wang, and Jinchao Xu, *Convergence Estimates for Product Iterative Methods with Applications to Domain Decomposition*, Math. Comp., 57, 1–21, 1991.

4. Maksymilian Dryja, Marcus Sarkis, and Olof B. Widlund, *Multilevel Schwarz Methods for Elliptic Problems with Discontinuous Coefficients in Three Dimensions*, TR #662, CS Department, Courant Institute, March 1994.

5. Maksymilian Dryja, *A Method of Domain Decomposition for 3-D Finite Element Problems*, in First International Symposium on Domain Decomposition Methods for Partial Differential Equations, Roland Glowinski, Gene H. Golub, Gérard A. Meurant, and Jacques Périaux, editors, SIAM, Philadelphia, PA, 1988.

6. Maksymilian Dryja and Olof B. Widlund, *Schwarz Methods of Neumann-Neumann Type for Three-Dimensional Elliptic Finite Element Problems*, TR #626, CS Department, Courant Institute, March 1993. To appear in Comm. Pure Appl. Math.

7. Maksymilian Dryja, Barry F. Smith, and Olof B. Widlund, *Schwarz Analysis of Iterative Substructuring Algorithms for Elliptic Problems in Three Dimensions*, TR #638, CS Department, Courant Institute, May 1993. To appear in SIAM J. Numer. Anal.

8. Jan Mandel, *Solving Large Scale Problems in Mechanics. The Development and Application of Computational Solution Methods*, in Adaptive Iterative Solvers in Finite Elements, M. Papadrakakis, editor, John Wiley & Sons, London, 1993.

9. Luca F. Pavarino, *Domain Decomposition Algorithms for the p-version Finite Element Method for Elliptic Problems*, NYU PhD thesis, TR #616, CS Department, Courant Institute, September 1992.

10. Luca F. Pavarino and Olof B. Widlund, *A Polyalgorithmic Bound for an Iterative Substructuring Method for Spectral Elements in Three Dimensions*, TR #661, CS Department, Courant Institute, March 1994.

11. Marcus Sarkis, *Two-Level Schwarz Methods for Nonconforming Finite Elements and Discontinuous Coefficients*, TR #629, CS Department, Courant Institute, March 1993.

12. Barry F. Smith, *A Domain Decomposition Algorithm for Elliptic Problems in Three Dimensions*, Numer. Math. 60, 219–234, 1991.

13. Barry F. Smith, *A Parallel Implementation of an Iterative Substructuring Algorithm for Problems in Three Dimensions*, SIAM J. Sci. Comput., 14, 406–423, 1993.

COURANT INSTITUTE OF MATHEMATICAL SCIENCES, 251 MERCER STREET, NEW YORK, NY 10012
E-mail address: widlund@widlund.cs.nyu.edu

PART II

Algorithms

Contemporary Mathematics
Volume **180**, 1994

Preconditioning Via Asymptotically-defined Domain Decomposition

S.F. Ashby, C.T. Kelley, P.E. Saylor, and J.S. Scroggs

ABSTRACT. Asymptotic analysis is used to derive preconditioners based on operator splitting and domain decomposition for the numerical solution of the advection-diffusion equation. Specifically, asymptotics is used to identify subdomains in which the solution is dominated by a certain operator, and this information is used to construct an effective preconditioner. We analyze the one-dimensional case in a function space setting and present numerical results for both one and two dimensions.

1. Introduction

In this paper, we construct and apply a preconditioning technique for the matrices arising in the numerical solution of differential equations. The preconditioning strategy is based on asymptotic analysis and uses the interaction of solution phenomena with the differential operator to create an approximate inverse that depends on both the operator and on the right hand side. This preconditioning differs from the usual approach that considers only the discrete or differential operator, and not how that operator interacts with the solution. Our preconditioning strategy is based on decomposing the computational domain in such a way that each subdomain

1991 *Mathematics Subject Classification*. Primary 65F10, 65M06, 76M20.

The work of the first and third authors was supported in part by the Applied Mathematical Sciences subprogram of the Office of Energy Research, Department of Energy, by Lawrence Livermore National Laboratory under contract W-7405-ENG-48. The work of the second author was supported by Air Force Office of Scientific Research grant #AFOSR-FQ8671-9101094 and National Science Foundation grant #DMS-9024622. The work of the fourth author was supported by National Science Foundation grant #DMS-9201252.

This paper is in final form and no version of it will be submitted for publication elsewhere.

isolates a fundamentally different type of physical behavior. We then associate a simple "partial" operator with each subdomain, the choice being made according to the dominant physics within that subdomain. To identify these partial operators, we use a uniform asymptotic expansion. The partial operators are then discretized and combined to obtain a preconditioner for use in an iterative method.

Our goal is to approximate the solution of a singular-perturbation problem governed by a differential operator where the singular behavior of the solution is governed by a small (but non-zero) parameter. Toward this end, we discretize the operator so that exact solution to the discrete problem has an error no larger than some specified tolerance. We then obtain an approximate solution to the discrete problem to within the same tolerance. We use asymptotic analysis to define a *preconditioner* for accelerating convergence of an iterative method that is applied to the original problem. We remark that the idea of using a uniform asymptotic expansion to precondition the problem was first proposed by Chin, Hedstrom, and Howes [9]. This idea has been applied to canonical problems in fluid flows [10, 12, 6] and isentropic gas dynamics [4].

2. Model Problem

Our target application is the numerical simulation of fluid flow. In certain high-speed regimes, shock layers or boundary layers can arise, and the flow is governed by the compressible viscous gas dynamics equation,

$$\nabla \cdot F'(\mathbf{u}) - \nabla \cdot G(\nabla \times \mathbf{u}) = 0,$$

with appropriate boundary conditions. The components of the vector $\mathbf{u}$ represent density, velocity, and energy, F is the flux, and G is the stress. In this paper we consider problems in which the "ratio" of $\nabla \cdot F(\mathbf{u})$ to $\nabla \cdot G(\nabla \times \mathbf{u})$ changes throughout the computational domain. For our purposes, it is sufficient to study a canonical problem whose solution is smooth except for the presence of a boundary layer. We will develop the technique for the linear one-dimensional singularly-perturbed advection-diffusion equation, and then demonstrate the effectiveness of the preconditioner on examples in one and two dimensions. Specifically, we will consider

(1) $$(L_c - \epsilon L_d)[u] = f$$

on the domain $\Omega = [0, 1]$, with homogeneous Dirichlet boundary conditions

(2) $$u(0) = 0, \quad u(1) = 0.$$

Here the operators $L_c = \nabla$ and $L_d = \Delta$ represent the effects of advection and diffusion, respectively. We are interested in solutions to this problem when $1 \gg \epsilon > 0$ is small, which gives rise to a boundary layer. The forcing function $f(x)$ may depend on ϵ, but we assume that f and its derivatives of all orders are bounded independently of ϵ. The analysis of this simple problem will illustrate the fundamental ideas.

3. Quantitative Analysis

The solution to (1-2) may be decomposed into two components, $u = u_{null} + u_f$. One component, u_f, has derivatives bounded independent of ϵ for the entire domain, and the other component, u_{null}, has derivatives bounded independently of ϵ except for a region that we can locate. Component u_f satisfies the non-homogeneous equation

$$(3) \qquad (L_c - \epsilon L_d)[u_f] = f$$

with boundary conditions that yield a regular perturbation problem. The singular nature of the solution is captured in u_{null}, which satisfies

$$(L_c - \epsilon L_d)[u_{null}] = 0.$$

Thus, u_{null} is in the null space of the operator, and it is chosen so that the sum $u_{null} + u_f$ satisfies (2). In general, u_{null} is nonzero. More specifically, let

$$(4) \qquad u_f(x) = F(x) + \sum_{n=1}^{\infty} \epsilon^n (f^{(n-1)}(x) - f^{(n-1)}(0))$$

where $F(x) = \int_0^x f(y)dy$. This is a regular asymptotic expansion for u_f that satisfies the differential Eqn. (3). For example, using expansion (4) for u_f, we find

$$(5) \qquad u_{null}(x) = K_0 e^{\frac{x-1}{\epsilon}} + K_1 + O(\epsilon)$$

where the constants K_0 and K_1 are determined by applying the boundary conditions

$$K_0 e^{-\frac{1}{\epsilon}} + K_1 = 0 \quad \text{and} \quad K_0 + K_1 = -F(1).$$

4. Qualitative Analysis

In this section, the qualitative behavior of the solution is obtained by deriving a uniform asymptotic expansion. The expansion is not new [7], but is repeated here as motivation for our preconditioner.

Consider Eqn. (1), and let

$$(6) \qquad u_{as}^{outer}(t, x, \epsilon) = U_0(t, x) + \epsilon U_1(t, x) + \epsilon^2 U_2(t, x) + \cdots$$

be the regular asymptotic expansion of u in the (outer) region of smooth behavior $\Omega_{outer} = [0, \tau]$, for some τ yet to be determined. Since ϵ is small, the dominant behavior is given by U_0, which satisfies

$$(7) \qquad L_c[U] = 0.$$

In other words, the operator in (1) is advection dominated in this subdomain. Next let

$$(8) \qquad u_{as}^{inner}(t,x,\epsilon) = \hat{u}_0(t,x) + \epsilon \hat{u}_1(t,x) + \epsilon^2 \hat{u}_2(t,x) + \cdots$$

be the asymptotic expansion for the solution in the (inner) boundary layer region $\Omega_{inner} = (\tau, 1]$. Here the behavior is given by $\hat{u}_0$, which satisfies

$$(9) \qquad (L_c - \epsilon L_d)[\hat{u}] = 0.$$

Thus, in the boundary layer region, the solution is neither advection nor diffusion dominated. By applying appropriate conditions at the interface between the subdomains (i.e., $U_0(\tau) = \hat{u}_0(\tau)$), the first term in the uniform asymptotic expansion for the solution to our model problem is

$$(10) \qquad u(x,y) \approx u_0^{uniform} = \begin{cases} U_0(x), & x \in \Omega_{outer} \\ \hat{u}_0(x), & x \in \Omega_{inner} \end{cases}.$$

We discuss how to choose τ in §6.

5. Preconditioning

In this section, we describe how asymptotics can be used to define a preconditioner for our problem. The goal is to find an operator M such that the problem

$$(11) \qquad M^{-1}(L_c - \epsilon L_d)[u] = M^{-1}f$$

is easier to solve than the original problem. It would be natural to choose the preconditioner

$$(12) \qquad M = (A_{outer})\chi_{[0,\tau)} + (A_{inner})\chi_{[\tau,1]}.$$

based on the portions of the operator that are dominant in the various subdomains. Here, χ_I denotes the characteristic function of the interval I. Next, we define each of the operators on the right-hand-side of (12).

Preconditioner in the Outer Region. In the outer region, the action of L_c dominates the differential operator, hence we define $V = A_{outer}[v]$ to be the solution of

$$L_c[V] = v, \quad \text{subject to} \quad V(0) = 0.$$

This preconditioner is applied to the equation $(L_c - \epsilon L_d)[v] = f$ for $x \in \Omega_{outer}$ to obtain

$$v(x) - \epsilon v'(x) - (v(0) - \epsilon v'(0)) = F(x).$$

Preconditioner in the Boundary Layer Region. In the boundary layer, neither the action of L_c nor the action of ϵL_d dominates the differential operator, and so the full operator needs to be used in this (smaller) region. The boundary condition at $x = \tau$ is provided by the preconditioner A_{outer}. To summarize, the preconditioner for the boundary layer region $A_{inner}[v]$ is defined as the solution V of

$$(L_c - \epsilon L_d)[V] = v$$

$$\text{subject to} \quad V(\tau) = A_{outer}[v' - \epsilon v''](\tau) \quad \text{and} \quad V(1) = 0.$$

Note that A_{outer} must be applied before A_{inner}. We apply this preconditioner to both sides of the equation $(L_c - \epsilon L_d)[v] = f$ for $x \in \Omega_{inner}$ to obtain $v(x) - \mu(x) = \hat{F}(x)$, where $\hat{F} = A_{inner}[f]$, and μ is an element of the null space of $(L_c - \epsilon L_d)$ that satisfies the boundary conditions imposed by our preconditioner, namely, $v(1) + \mu(1) = 0$ and $v(\tau) + \mu(\tau) = A_{outer}[v' - \epsilon v''](\tau)$.

Preconditioned Problem. After applying the preconditioner on each subdomain, Eqn. (11) becomes

$$A_{outer}[u]\chi[0, \tau] + A_{inner}[u]\chi[\tau, 1] = G$$

where $G(x) = [F(x)]\chi[0, \tau] + \left[\hat{F}(x)\right]\chi[\tau, 1]$. Since the family of functions $J_1 e^{\frac{x-1}{\epsilon}} + J_2$ is in the null space of $(L_c - \epsilon L_d)$, the above equation may be written

$$
\begin{aligned}
u(x) + \epsilon \Bigg(& \left[u'(0) - u'(x)\right]\chi[0, \tau] \\
(13) \qquad & + \left[(u'(0) - u'(\tau))\left(\frac{e^{\frac{x-1}{\epsilon}} - 1}{e^{\frac{\tau-1}{\epsilon}} - 1}\right)\right]\chi[\tau, 1] \Bigg) = G(x).
\end{aligned}
$$

Using the homogeneous Dirichlet boundary conditions, we write our preconditioned problem more generally as: Find a constant α and a function $u(x)$ such that

$$\left[u(x) - \epsilon u'(x) + \epsilon\alpha\right]\chi[0, \tau] + \left[u(x) + \mu(x)\right]\chi[\tau, 1] = G(x)$$

where u satisfies the boundary conditions $u(0) = 0$ and $u(1) = 0$. Notice that this is a first-order equation for u with two boundary conditions. The parameter α reflects the use of (the inverse of) a first-order operator to precondition a second-order problem.

6. Analysis of the Iterative Method

In this section, we analyze the convergence of the Richardson and GM-RES iterative methods in a function space setting. This analysis makes clear the dependence of the L_2 convergence properties on bounds for high derivatives of the solution in a way that an analysis of a discretized problem could not. We emphasize that this analysis has not been extended to the discrete case, but we present supporting numerical evidence in the next two sections.

Let $A = L_c - \epsilon L_d$ and apply homogeneous Dirichlet boundary conditions on $[0, 1]$. We will solve $Au = f$ via the GMRES method [11] using M as a left preconditioner. Let the functions $\{v_k(x)\}$ be the GMRES iterates. Denote the error by $e_k(x) = u(x) - v_k(x)$. In the L^2 norm we have [11, 8]

$$(14) \qquad \|e_k\|_2 = \|P_{opt}(M^{-1}A)[e_0]\|_2,$$

where P_{opt} is a kth degree *residual* polynomial, that is, $P_{opt}(0) = 1$. GMRES chooses the kth degree polynomial P_{opt} that is optimum in the *residual error* norm. That is, P_{opt} is a residual polynomial that minimizes the L_2 norm of the residual error,

$$(15) \qquad \|r_k\|_2 = \|P_{opt}(M^{-1}A)[r_0]\|_2 \leq \|P(M^{-1}A)[r_0]\|_2$$

over all k^{th} degree residual polynomials. To demonstrate the convergence of the GMRES iterates, we may choose any k^{th} degree residual polynomial, and any initial iterate that makes the left side of (15) easy to estimate. Toward that end, we choose the polynomial $P(z) = (1 - z)^k$, which corresponds to a stationary Richardson iteration. In addition, we assume that the initial iterate is zero, that is, $v_0(x) = 0$. This implies that the initial error is $e_0 = u$, which allows us to exploit the smoothness properties of u. We emphasize that this choice of initial iterate is crucial to the analysis that follows. Of course, in practice, one may wish to use a different v_0, but one should ensure that it is sufficiently smooth.

To show that a specified tolerance δ in the L_2 norm is obtained when $k = k_\delta$, we show that τ can be chosen as a function of ϵ and δ so that

$$(16) \qquad \|r_k\|_2 \leq \|(I - M^{-1}A)^k M^{-1}A[e_0]\|_2 \leq \delta.$$

Let v be a general function satisfying homogeneous Dirichlet boundary conditions, then

$$
(17) \qquad \begin{aligned}
(I - M^{-1}A)^k[v] = \epsilon^k &\left(\left[v^{(k)}(x) - v^{(k)}(0) \right] \chi[0, \tau] \right. \\
&+ \left. \left[\left(v^{(k)}(\tau) - v^{(k)}(0) \right) \left(\frac{e^{\frac{x-1}{\epsilon}} - 1}{e^{\frac{\tau-1}{\epsilon}} - 1} \right) \right] \chi[\tau, 1] \right).
\end{aligned}
$$

Thus, since $r_0 = M^{-1}A[u]$, the bound in the L_2 norm of the residual error at the kth iterate is

$$(18) \qquad \|r_k\|_2^2 \leq \epsilon^{2k} C_k^2 + \epsilon^{2k+1} C_k C_{k+1} + \epsilon^{2k+1} C_{k+1}^2.$$

Here, C_k is an upper bound on the magnitude of $u^{(k)}$ for $x \in \Omega_{outer}$. Inequality (18) does not imply that the GMRES iterates converge in the limit as $k \to \infty$ because the sequence $\{C_k\}$ is unbounded (see below). However, given a δ, we show that τ can be chosen so that (16) holds for $k = k_\delta$.

To choose τ, we use the analysis in §3 to obtain

$$(19) \qquad C_k = \max_{\Omega_{outer}} u^{(k)}(x) = \frac{K}{\epsilon^k} e^{\frac{\tau-1}{\epsilon}} + \gamma_k.$$

Here, $\gamma_k = \max_{\Omega_{outer}} f^{(k)}(x) + O(\epsilon)$ is a constant based on (4). The constant

$$(20) \qquad K = K_0 + O(\epsilon)$$

is determined by (5). From Eqn. (18), we must satisfy $\epsilon^k \left(\Gamma + \frac{2K}{\epsilon^k} e^{\frac{\tau-1}{\epsilon}} \right) (1 + O(\epsilon)) \leq \delta$ where Γ is an upper bound on γ_k and γ_{k+1}. Assuming $\Gamma > \epsilon$, the desired tolerance on the residual error is satisfied for $k = k_\delta$ if $\epsilon^k \left(2\Gamma + \frac{2K}{\epsilon^k} e^{\frac{\tau-1}{\epsilon}} \right) \leq \delta$. Next, we will show that it is possible to choose τ so that the above relation holds. Suppose we wish to converge in k_δ iterations. Then the above relation is satisfied if $\tau < 1 + \epsilon \ln \left(\frac{\delta}{4K} - \frac{\epsilon^{k_\delta} \Gamma}{K} \right)$.

We assume that k_δ is large enough so that $\frac{\delta}{4K} > \Gamma \epsilon^{k_\delta}$. (This provides a lower bound on the number of iterations given δ and ϵ). We choose τ so that $\tau < 1 + \epsilon \ln \left(\frac{\delta}{8K} \right)$. This expression is the basis for the heuristic $\tau = 1 + c_\tau \epsilon \ln(\delta)$, for choosing τ, where the constant c_τ is near unity.

Error Analysis. In general, a small residual error $||r_k||_2$ does not guarantee a small error $||e_k||_2$. We can establish a bound on the error based on the residual. This bound follows from an examination of the solution to $(M^{-1}A)[e_k] = r_k$, where $||r_k||_2 \leq \delta$. For $x \in \Omega_{outer}$, it has been shown [5] that $e_k = r_k + O(\epsilon)$. In addition, for $x \in \Omega_{inner}$ we have that $e_k = r_k + O(\epsilon) + O(e^{\frac{\tau-1}{\epsilon}})$. Thus, the norm of the error is related to the norm of the residual by $||e_k||_2 = ||r_k||_2 + O(\epsilon^{1/2})$. A tighter bound on the norm of the error might be possible; however, with simple analysis, we have shown that a small residual norm implies a small error norm, assuming a sufficiently smooth initial error.

7. One-Dimensional Implementation and Demonstration

We demonstrate the convergence properties by approximating the solution of the advection-diffusion problem (1-2) with $f(x) = -1$ so that the error is no larger than a tolerance TOL. The error tolerance applies to both the discretization and to the iterative process.

The discretization of the advection operator is a second-order strictly upwind scheme, and the diffusion operator is approximated using a second-

Table 1: Experiments varying TOL ($\epsilon = .01$, $c_\tau = 1.0$)

TOL	Δx_{outer}	Number of grid points	Number of GMRES steps
.1	.33	138	1
.01	.11	144	1
.001	.032	166	2

Table 2: Experiments varying ϵ ($TOL = .025$, $c_\tau = 1.0$, $\Delta x_{outer} = .17$)

ϵ	Number of grid points	Number of GMRES steps
.1	26	1
.01	141	1
.005	270	1

order centered scheme. Based on known behavior of the solution [5], the boundary layer is in a neighborhood of $x = 1$ of size $O(\epsilon \ln(\epsilon))$. The nonuniform grid is chosen to resolve this boundary layer. In the outer region, we use $\Delta x_{outer} \approx TOL^{1/2}$. This grid spacing is halved for successive mesh points in the interface region between the boundary layer and the outer region. The boundary layer region uses $\Delta x_{inner} \approx \epsilon \Delta x_{outer}$. This will provide a uniformly second-order accurate method. At the interface we use the second-order upwind method for the advection operator for all points $x_i < \tau$, and the discretization of the full operator for all points $x_i > \tau$ (here, we assume τ is between two points in our grid.)

The experiments were performed in Matlab. The GMRES iteration was halted when $||r_k||_2/||r_0||_2 < TOL$. These experiments suggest that the conclusions from our analysis of the continuous problem also apply to the discretized problem (e.g., implementation in a finite-dimensional space).

Table 3: Experiments varying c_τ

c_τ	Number of outer-region points	Number of GMRES steps
$1.2 - 1.0$	30	2
.8	31	2
.6	32	3
.4	57	27

The results summarized in Table 1 show that the method has slow growth in the number of iterations as TOL is varied. The experiments reported in Table 2 show that the number of iterations is (nearly) independent of ϵ.

However, total work per iteration increases. Next, in Table 3, only c_τ is varied and the discretization fixed. With $\epsilon = .01$ and $TOL = .001$, there are 166 points in the grid. The number of iterations is sensitive to c_τ, as seen in the dramatic increase in the number of GMRES steps for $c_\tau = 0.4$. Other experiments indicated that there is a strong interaction between the choice of c_τ and the location of the interface between the *outer* and *inner* subdomains. That is, τ should be located outside or barely inside the region that is refined. When τ is well inside the refined region, the number of GMRES iterations increases dramatically, regardless of whether τ is inside the boundary layer or not.

8. Two-Dimensional Results

Discretizations. Consider a tensor-product mesh $P_{i,j} = (x_i, y_j)$, where $i = 0, \ldots, N_x$ and $j = 0, \ldots, N_y$. Also let $u_{i,j}$ represent the numerical approximation to the solution at $P_{i,j}$, that is, $u_{i,j} \approx u(x_i, y_j)$. The points need not be uniformly distributed. In practice, the coarse mesh would be used in the "smooth" subdomain Ω_c, and a fine mesh in the "boundary layer" subdomain Ω_d. We use a first order upwind approximation for the first order spatial derivatives, and centered differences are used for the second order spatial derivatives. Consequently, the discretization of (1) can be formulated as the system of linear equations,

$$(21) \qquad Au = f$$

where $A = A_c - \epsilon A_d$, with A_c, A_d, and f being the discretizations of L_c, L_d, and the boundary conditions, respectively.

In this section, we introduce and compare several preconditioning matrices, including two based on asymptotics and domain decomposition. The description of the preconditioning matrices is most easily accomplished if we introduce a block partitioning of A (as in [1]). Let $u = [u_c^T \; u_d^T]^T$, where u_c are the unknowns corresponding to the smooth subdomain Ω_c and u_d are the unknowns corresponding to the boundary layer subdomain Ω_d. Moreover, let the unknowns within a subdomain be ordered according to the direction of the convection [2, 3]. Since we are using a strictly upwind differencing scheme for the convection terms, this means A_c is lower triangular. Given this ordering, we can partition the matrices as follows:

$$(22) \qquad A_c = \begin{bmatrix} A_{11} & \\ A_{21} & A_{22} \end{bmatrix} \quad \text{and} \quad A_d = \begin{bmatrix} D_{11} & D_{12} \\ D_{21} & D_{22} \end{bmatrix}$$

where A_{11} and A_{22} are lower tridiagonal (we have assumed a and b are positive), D_{11} and D_{22} are pentadiagonal, and A_{21}, D_{21}, and D_{12} contain one nonzero sub- or super-diagonal. The off-diagonal blocks A_{21}, D_{21}, and D_{12} represent the portions of the difference stencils that couple the two subdomains, and the diagonal blocks contain the portions of the difference stencils that couple points entirely within one subdomain or the other. That is, we solve the equivalent linear system $M^{-1}Au = M^{-1}f$ where M is chosen to approximate A in some sense and is easy to invert.

We present several preconditioners derived from asymptotic analysis and compare them to "standard" preconditioners. Our baseline "preconditioner" is the identity matrix (i.e., no preconditioning), $M_{id} = I$. Since the diagonal of A is constant in our test problems, diagonal preconditioning has no effect on GMRES convergence. We also consider the Gauss-Seidel (lower triangular part of A) preconditioner,

$$(23) \qquad M_{gs} = \begin{bmatrix} A_{11} - \epsilon\tilde{D}_{11} & 0 \\ A_{21} - \epsilon D_{21} & A_{22} - \epsilon\tilde{D}_{22} \end{bmatrix},$$

where $\tilde{D}_{ii}$ is the lower triangular part of D_{ii}. The next preconditioner is the matrix representation of the discrete convection operator,

$$(24) \qquad M_{co} = A_c = \begin{bmatrix} A_{11} & 0 \\ A_{21} & A_{22} \end{bmatrix}.$$

This preconditioner should be effective in the smooth subdomain, but it neglects the importance of the diffusion term in the boundary layer subdomain. To compensate for this, we might consider the block diagonal preconditioner

$$(25) \qquad M_{bd} = \begin{bmatrix} A_{11} & 0 \\ 0 & A_{22} - \epsilon D_{22} \end{bmatrix}.$$

This is our first asymptotics-motivated domain decomposition preconditioner. It is equivalent to solving the convection equation (7) in Ω_c and the full equation (9) in Ω_d, and ignoring the coupling of unknowns across the subdomain interface. Thus, this preconditioner corresponds to solving the original problem with certain nonphysical conditions imposed at the interface. The advantage of this preconditioner is its inherent large-grain parallelism: the two subproblems can be solved independently. We rely on the iterative method to "glue" the two pieces together.

If we wish to obtain a more physically realistic preconditioner, we can include the coupling of the Ω_c and Ω_d unknowns. This yields the following block lower triangular matrix

$$(26) \qquad M_{dd} = \begin{bmatrix} A_{11} & 0 \\ A_{21} - \epsilon D_{21} & A_{22} - \epsilon D_{22} \end{bmatrix}.$$

Finally, note that the physically motivated domain decomposition preconditioners M_{bd} and M_{dd} require the solution of the original equation (1),

Table 4: Condition Number

Grid Size	10 × 10		20 × 20		30 × 30	
	Actual	Estimate	Actual	Estimate	Actual	Estimate
$M = I$	33.	23.	110	74.	240	150
M_{gs}	8.0	4.7	29	16.	66	43.
M_{co}	4.8	2.3	12	3.8	23	4.9
M_{bd}	10.	4.5	33	8.9	81	15.
M_{dd}	8.9	2.4	35	4.1	97	6.1

Table 5: Number of iterations

Grid Size	10 × 10	20 × 20	30 × 30	40 × 40
$M = I$	48	90	112	133
M_{gs}	15	38	79	87
M_{co}	15	27	39	52
M_{bd}	19	35	58	89
M_{dd}	15	27	39	53

but on a smaller domain. Since the boundary layer region is a fraction of the total computational domain, this subproblem is small compared to the full problem.

The condition number of the upper Hessenberg matrix that is generated during the GMRES iterations is an estimate for the condition number of the preconditioned matrix, and is an indication of the *stiffness* of the problem. The estimates in Table 4 were obtained after the method had obtained a relative residual of 10^{-4}. By comparing the results presented in Table 4 with those in Table 5, we see that this estimate is a better predictor of the total number of iterations than the actual condition number. The results in Table 5 indicate that the most efficient preconditioner would be M_{co}; however, this might change if a nonuniform mesh were used.

References

[1] S. F. ASHBY, P. S. SAYLOR, AND J. S. SCROGGS, *Physically motivated domain decomposition preconditioners*, in Proc. Copper Mountain Conf. on Iterative Methods, vol. 1, Copper Mountain, CO, April 1992.

[2] R. C. CHIN, T. A. MANTEUFFEL, AND J. DE PILLIS, *ADI as a preconditioning for solving the convection-diffusion equation*, SIAM J. Sci. Statist. Comput., 5 (1984), pp. 281–299.

[3] H. C. ELMAN AND G. H. GOLUB, *Line iterative methods for cyclically reduced discrete convection-diffusion problems*, SIAM J. Sci. Statist. Comput., 13 (1992), pp. 339–363.

[4] M. GARBEY AND J. S. SCROGGS, *Asymptotic-induced method for conservation laws*, in Asymptotic Analysis and the Numerical Solution of PDEs, H. Kaper and M. Garbey, eds., New York, New York, 1990, Marcel Dekker, Inc., pp. 75–98.

[5] F. A. HOWES, *Multi-dimensional reaction-convection-diffusion equations*, in Ordinary and Partial Differential Equations, B. Sleeman and R. Jarvis, eds., vol. 1151, New York, 1984, Springer-Verlag, pp. 217–223.

[6] J. S. SCROGGS AND D. C. SORENSEN, *An asymptotic induced numerical method for the convection-diffusion-reaction equation*, in Math. for Large Scale Comp., J. Diaz, ed., Marcel Decker, 1989, pp. 81–114.

[7] J. KEVORKIAN AND J. D. COLE, *Perturbation Methods in Applied Mathematics*, Springer-Verlag, New York, 1981.

[8] N. M. NACHTIGAL, S. C. REDDY, AND L. N. TREFETHEN, *How fast are nonsymmetric matrix iterations?*, SIAM J. Matrix Anal. Appl., 13 (1992), pp. 778–792.

[9] R. C. Y. CHIN, G. W. HEDSTROM, AND F. A. HOWES, *Considerations on solving problems with multiple scales*, in Multiple Time Scales, J. Brackbill and B. Cohen, eds., Academic Press, Orlando, Florida, 1985, pp. 1–27.

[10] R. C. Y. CHIN, G. W. HEDSTROM, J. S. SCROGGS, AND D. C. SORENSEN, *Parallel computation of a domain decomposition method*, in Advances in Computer Methods for PDEs 6, 1987, pp. 375–381.

[11] Y. SAAD AND M. SCHULTZ, *GMRES: a generalized minimal residual algorithm for solving nonsymmetric linear systems*, SIAM J. Sci. Stat. Comp., 7 (1986), pp. 856–869.

[12] J. S. SCROGGS, *A parallel algorithm for nonlinear convection-diffusion equations*, in Proc of the 3rd Int Symp. on Domain Decomposition for PDEs, T. Chan, R. Glowinski, J. Periaux, and O. Widlund, eds., SIAM, 1989, pp. 373–384.

CENTER FOR COMPUTATIONAL SCIENCES AND ENGINEERING, LAWRENCE LIVERMORE NATIONAL LABORATORY, P.O. BOX 808, L-316, LIVERMORE, CA 94551
E-mail address: sfashby@llnl.gov

CENTER FOR RESEARCH IN SCIENTIFIC COMPUTATION, MATHEMATICS DEPARTMENT, NORTH CAROLINA STATE UNIV., RALEIGH, NC 27502
E-mail address: Tim_Kelley@ncsu.edu

DEPARTMENT OF COMPUTER SCIENCE, UNIVERSITY OF ILLINOIS, URBANA, IL 61801
E-mail address: saylor@cs.uiuc.edu

CENTER FOR RESEARCH IN SCIENTIFIC COMPUTATION, MATHEMATICS DEPARTMENT, NORTH CAROLINA STATE UNIVERSITY, RALEIGH, NC 27502
E-mail address: scroggs@wave.math.ncsu.edu

Contemporary Mathematics
Volume **180**, 1994

A Spectral Stokes Solver in
Domain Decomposition Methods

M. AZAIEZ AND A. QUARTERONI

ABSTRACT. We present a domain decomposition solver for the linear Stokes problem. We use a technique based on an equivalence between the single-domain and multidomain formulations of the Stokes problem in which the transmission conditions at subdomain interfaces are properly taken in account. The discrete problem is solved using the Uzawa algorithm.

1. Introduction

The aim of this note is to present a domain decomposition method for the spectral solution of Stokes equations. After recalling the spectral collocation method in its single–domain version, we introduce our multidomain approach which is based on the concept of transmission interface conditions. The resulting problem is handled by an Uzawa solution algorithm which requires at each iteration the solution of a Poisson boundary value problem for each velocity component. At this stage we apply the projection decomposition method introduced in [5]. We conclude by a numerical investigation that shows the efficiency and accuracy of our approach.

2. The Stokes Problem

The Stokes problem reads : find u and p such that

$$-\nu \Lambda u + \nabla p = f \quad \text{in} \quad \Omega ,$$

(1)
$$\nabla \cdot u = 0 \quad \text{in} \quad \Omega ,$$

$$u = 0 \quad \text{on} \quad \partial\Omega ,$$

1991 *Mathematics Subject Classification.* Primary 35Q30, 65M55; Secondary 65M70.

The work was partially supported by Sardinian Regional Authorities and Fondi MURST 40 %. The first author thanks Mr F. Ben Belgacem for several discussions.

This paper is in final form and no version of it will be submitted for publication elsewhere.

where $\Omega =]-1, 1[^2$, f is the density force and ν the viscosity. This problem admits the following variational formulation:

Find $u \in X$ and $p \in M$ such that

$$(2) \qquad \nu\,(\nabla u, \nabla v) - (\nabla \cdot v, p) = <f, v>, \quad \forall\, v \in X$$

$$(3) \qquad (\nabla \cdot u, q) = 0 \ \forall\, q \in M,$$

where $X = \left(H_0^1(\Omega)\right)^2$ and $M = L_0^2(\Omega) = \left\{v \in L^2(\Omega), \int_\Omega v\,dx = 0\right\}$. Here $(\cdot, \cdot)$ denotes the scalar product of $L^2(\Omega)$ and $<\cdot, \cdot>$ the duality between X' and X (for notation see, e.g., Lions and Magenes [1]).

This problem is well posed and has a unique solution in $X \times M$.

3. Spectral single-domain discretization

We need first to introduce the polynomial spaces of approximation. For the velocity we choose $X_N = (P_N^0(\Omega))^2$ and for the pressure $M_N = P_N(\Omega) \bigcap L_0^2(\Omega)$, where $P_N(\Omega)$ denotes the space of algebraic polynomials of degree not greater than N with respect to each coordinate.

Let L_N be the Legendre polynomial of degree N, the solution of the Sturm-Liouville equation

$$\left((1-x^2)L_N'(x)\right)' + N(N+1)L_N(x) = 0, \qquad -1 < x < 1.$$

We introduce the following discrete inner product:

$$(4) \qquad (u, v)_N = \sum_{i,j=0}^{N} u(\xi_i, \xi_j) v(\xi_i, \xi_j) \rho_i \rho_j, \quad \forall (u, v) \in (C^0(\bar{\Omega}))^2,$$

where $\{\xi_j, 0 \leq j \leq N\}$ are the $N+1$ Gauss-Lobatto-Legendre points, the roots of the polynomial: $(1-x^2)L_N'(x)$, and $\{\rho_j, 0 \leq j \leq N\}$ are the associated weights (see, e.g., [2]). Then the approximate problem based on the Legendre collocation method reads : Find (u_N, p_N) in $X_N \times M_N$ such that for all $v_N \in X_N$ and all $q_N \in M_N$

$$(5) \qquad \mathcal{A}_N\,(u_N, v_N) + b_N\,(v_N, p_N) = (f, v_N)_N,$$

$$(6) \qquad b_N\,(u_N, q_N) = 0,$$

where $\mathcal{A}_N(u, v) = \nu\,(\nabla u, \nabla v)_N$ and $b_N(v, q) = -\,(\nabla.v, q)_N$.

Both velocity and pressure will be computed on the Gauss-Lobatto grid Ξ_N:

$$\Xi_N = \{x = (\xi_i, \xi_j), 0 \leq i \leq N, 0 \leq j \leq N\}.$$

As proven by Bernardi and Maday [3], the drawback of the previous formulation is that it is affected by spurious modes. These are nonzero polynomials $q^* \in P_N(\Omega) \bigcap L_0^2(\Omega)$ such that $b_N(v, q^*) = 0$, $\forall v \in X_N$. If such a q^* exists then each couple $(u_N, p_N + q^*)$ would also be a solution, yielding a potential instability in the pressure approximation.

The subspace $\mathcal{Z}_N$ of $P_N(\Omega) \bigcap L_0^2(\Omega)$ made of spurious modes has dimension 7 and is spanned by

$$L_N(y),\ L_N(x),\ L_N(x)L_N(y),\ L_N'(x)L_N'(y),$$
$$L_N'(x)yL_N'(y),\ xL_N'(x)L_N'(y),\ xL_N'(x)yL_N'(y).$$

To get rid of these undesirable modes and to get satisfactory approximation, the pressure space M_N^o can be chosen such that it is a supplementary space of $\mathcal{Z}_N$, i.e.

$$(7) \qquad\qquad P_N(\Omega) \cap L_0^2(\Omega) = M_N \oplus \mathcal{Z}_N.$$

In this case, one can prove the existence of an Inf-Sup condition with a constant behaving as $\mathcal{O}(N^{-1})$ [3], i.e.,

$$\inf_{q \in M_N^o} \sup_{v \in X_N} \frac{b_N(v, q)}{\|v\|_X \|q\|_M} = \mathcal{O}(N^{-1}).$$

The discrete problem $(5) - (6)$ can be solved using the following Uzawa algorithm [2]. Formally we write

$$(8) \qquad\qquad u_N = S_N^{-1}\left(f - \nabla_N p_N\right)$$

where S_N and ∇_N are the algebraic representations of $\mathcal{A}_N$ and b_N, respectively. Inserting u_N in the continuity equation we get:

$$(9) \qquad\qquad \nabla_N.S_N^{-1}\nabla_N p_N = \nabla_N.S_N^{-1}f.$$

The Uzawa operator is then

$$(10) \qquad\qquad \nabla_N.S_N^{-1}\nabla_N$$

and it is positive, symmetric with a condition number of $O(N^2)$. This enables the use of conjugate gradient iterations to solve (9).

4. Non-overlapping domain decomposition

Our non-overlapping domain decomposition solver will be based on Uzawa's algorithm.

The solution of such a system is accomplished via a global inner/outer iterative scheme where at each iteration only elliptic problems need to be solved. An equivalence principle between single and multidomain formulations of the Helmholtz problem is used in this case, and the same iterative algorithm is applied to the resolution of the inner iteration yielding now a sequence of single domain elliptic problems.

Let Ω be a bounded domain (that we assume to be a rectangle for the sake of simplicity) of $I\!\!R^2$ partitioned into Ω_1 and Ω_2 with $\Gamma = \partial\Omega_1 \bigcap \partial\Omega_2$.

For the velocity space we take

$$(11) \qquad X_h = \left\{ v_h \in \left(\mathcal{C}^0(\bar{\Omega})\right)^2 \mid v_h^i = v_h|_{\Omega_i} \in \left(P_N(\Omega_i)\right)^2, v_h|_{\partial\Omega} = 0 \right\}.$$

and for the pressure

(12) $M_h = \left\{ q_h \in C^0(\bar{\Omega}) \mid q_h^i = q_h|_{\Omega_i} \in P_N(\Omega_i), \int_{\Omega} q_h = 0 \right\}.$

Note that the pressure is required to be continuous over Ω a priori.

The multidomain problem is defined as follows.

Find (u_h, p_h) in $X_h \times M_h$ such that for all v_h in X_h and for all q_h in M_h

(13) $\mathcal{A}_h(u_h, v_h) + b_h(v_h, p_h) = (f, v_h)_h$

(14) $b_h(u_h, q_h) = 0,$

where $\mathcal{A}_h(u_h, v_h) = \sum_{i=1,2} \nu \left(\nabla u_h^i, \nabla v_h^i \right)_{N,i}$, $b_h(v_h, q_h) = -\sum_{i=1,2} \left(\nabla . u_h^i, q_h^i \right)_{N,i}$, and where $(.,.)_{N,i}$ denotes the generalization of the discrete inner product $(.,.)_N$ on Ω_i based on transformed nodes ξ_j^i and weights ρ_j^i. Also in this case the problem is well-posed when the pressure is taken in a space complementary to the one of the spurious modes. Concerning the latter, taking for instance the domain $\Omega =] - 2, 2[\times]0, 2[$, it can be proven (see [4]) that the space of spurious modes

$$\mathcal{Z}_h = \{q_h \in M_h : b_h(v, q_h) = 0 \; \forall v \in X_h\},$$

has dimension 7. A set of generators is provided by the following three functions

(15)
$$(L_N(x+1), (-1)^N L_N(x-1)),$$
$$(L_N(y), L_N(y)),$$
$$(L_N(x+1)L_N(y), (-1)^N L_N(x-1)L_N(y))$$

(for each couple, the first component denotes the value attained in Ω_1, the second one is in Ω_2) plus four other functions which are the characteristic polynomials associated with the four corner points of Ω. For a domain Ω of more general shape (i.e., made by a union of rectangles), we still have the three spurious modes (15), plus as many characteristic functions as the number of corners of $\partial\Omega$. We point out that the existence of internal cross points does not introduce additional spurious modes. The latter property would not be true if we used a space of discountinuous pressures.

The pressure subspace M_h^0 that we are going to use is a complementary space of $\mathcal{Z}_h$, i.e. it satisfies $M_h = M_h^0 \oplus \mathcal{Z}_h$, and we propose a subdomain iteration method to solve the discrete problem.

5. A solver based on subdomain iterations

Using the spectral collocation method, the formulation (13) − (14) of the multidomain method can be given the following pointwise interpretation:

Find $(u_h, p_h) \in X_h \times M_h$ such that

$$-\nu \Delta u_h^1 + \nabla p_h^1 = f_h^1 \quad \text{in } \Xi_N^1 \cap \Omega_1$$
$$\nabla . u_h^1 = 0, \quad \text{in } \Xi_N^1 \backslash \Gamma$$
$$u_h^1 = 0 \quad \text{on } \Xi_N^1 \cap \partial\Omega;$$

and

$$-\nu\Delta u_h^2 + \nabla p_h^2 = f_h^2, \qquad \text{in } \Xi_N^2 \bigcap \Omega_2$$
$$\nabla.u_h^2 = 0, \qquad \text{in } \Xi_N^2 \backslash \Gamma$$
$$u_h^2 = 0 \qquad \text{on } \Xi_N^2 \bigcap \partial\Omega.$$

The interface conditions (according to [6]) are:

(16) $$u_h^1 = u_h^2 \text{ on } \Xi_N^\Gamma,$$

(17) $$\left(\nu\frac{\partial u_h^2}{\partial n_\Gamma} - p_h^2 n_\Gamma\right) - \left(\nu\frac{\partial u_h^1}{\partial n_\Gamma} - p_h^1 n_\Gamma\right) = -w^2 R_h^2 - w^1 R_h^1 \text{ on } \Xi_N^\Gamma,$$

(18) $$p_h^1 = p_h^2 \text{ on } \Xi_N^\Gamma,$$

(19) $$w^1\nabla \cdot u_h^1 + w^2\nabla \cdot u_h^2 = 0 \text{ on } \Xi_N^\Gamma.$$

Here, n_Γ is the normal unit vector on Γ directed outward with respect to Ω_2, $w^1 = \rho_N^1$ and $w^2 = \rho_0^2$, $R_h^i = -\nu\Delta u_h^i + \nabla p_h^i - f_h^i$, $i = 1, 2$, and Ξ_N^i is the set of Legendre Gauss-Lobatto nodes in $\bar{\Omega}_i$ while

$$\Xi_N^\Gamma = \Xi_N^1 \bigcap \Gamma = \Xi_N^2 \bigcap \Gamma.$$

Conditions (16) yield the continuity of the velocity, whereas (17) and (18) enforce the continuity of the normal derivative in a weak form, i.e.

(20) $$\nu\frac{\partial u_h^2}{\partial n_\Gamma} - \nu\frac{\partial u_h^1}{\partial n_\Gamma} = -w^2 R_h^2 - w^1 R_h^1 \text{ on } \Xi_N^\Gamma.$$

This global system is solved by using the Uzawa algorithm, consisting of deducing a global problem for the pressure field (p_h^1, p_h^2) similar to (9) and then solving it by a conjugate gradient procedure. At any step one Poisson problem for each velocity component is solved. At this stage we apply a domain decomposition technique which enforces both (16) and (20) on Γ. Among all available domain decomposition spectral solvers, we apply the one based on the Spectral Projection Method (see [5]). This algorithm is effective and allows spectral accuracy to be achieved.

6. Some numerical results

Our test case concerns the approximation of (1) corresponding to the exact solution: $u_{ex} = (\sin(\pi x)\sin(\pi y), \sin(\pi x)\sin(\pi y))$ and $p_{ex} = \sin(\pi x)\cos(\pi y)$, defined on the domain $\Omega =]0, 1[^2$.

Let $\|u_h\|_h = \left(\sum_{i=1}^{ndom}(\nabla.u_h, \nabla.u_h)_{N,i}\right)^{\frac{1}{2}}$ be the discrete norm and $ndom$ is the number of subdomains (of equal measure) into which the domain Ω has been subdivided.

We recall that the numerical divergence is zero at all the collocation nodes of $\bar{\Omega}_1$ and $\bar{\Omega}_2$, except those lying on the interface Γ where, however, (19) holds.

We are also interested in the number of Uzawa iterations using a suitable mass diagonal matrix preconditioner (see [7]). We report the number of Uzawa iterations, for several numbers of subdomains, taking the polynomial degree N equal to 8 within each subdomain (Table 1)

Table 1

Subdomains	1x1	2x2	3x3	4x4	5x5	6x6	7x7	8x8	9x9	10x10
Iterations	7	16	16	16	18	18	16	16	16	16

We observe that the iteration number is independent of the number of subdomains. In Table 2 we report the value of the divergence norm $\|u_h\|_h$ for several number of subdomains (left) and different value of the polynomial degree N on each subdomain (top). Spectral approximation is verified.

Table 2

	6	8	10	12	14	16
1x1	$0.2e-1$	$0.5e-3$	$0.4e-5$	$0.3e-7$	$0.1e-9$	$0.9e-12$
2x2	$0.3e-3$	$0.3e-5$	$0.7e-7$	$0.3e-10$	$0.2e-12$	$0.2e-11$
4x4	$0.3e-4$	$0.3e-7$	$0.2e-10$	$0.1e-11$	$0.4e-12$	$0.4e-11$
8x8	$0.9e-6$	$0.2e-9$	$0.8e-12$	$0.2e-11$	$0.9e-11$	$0.8e-11$
10x10	$0.3e-7$	$0.5e-10$	$0.1e-11$	$0.3e-11$	$0.8e-11$	$0.9e-11$

In these tables, we have indeed used the full space M_h defined in (12). The spurious modes that are intrinsically produced by our scheme $(13)-(14)$ are easy to filter out along the Uzawa iteration process and the pressure obtained is then filtered by using the method described in [8]

REFERENCES

1. J.-L. Lions, E. Magenes, *Problèmes aux Limites non Homogènes et Applications*, Dunod, Paris,1968.
2. C. Canuto, M.Y. Hussaini, A. Quarteroni and T. A. Zang, *Spectral Methods in Fluid Dynamics*, Springer Verlag, New York, 1988.
3. C. Bernardi and Y. Maday, *Spectral Methods*, in *Handbook of Numerical Analysis*, P.G. Ciarlet & J.-L. Lions eds, North-Holland, Amsterdam, 1994.
4. G. Sacchi Landriani and H. Vandeven, *A Multidomain Spectral Collocation Method for the Stokes Problems*, Numer. Math. 58 (1990), pp. 441–464.
5. P. Gervasio, E. Ovchinnikov and A. Quarteroni, *The Spectral Projection Decomposition Method of Elliptic Equations*, submitted to SIAM J. Numer. Anal.
6. A. Quarteroni, *Domain Decomposition and Parallel Processing for the Numerical Solution of Partial Differential Equations*, Surv. Math. Ind. 1 (1991), pp.75–118.
7. B. Métivet, *Résolution des Equations de Navier-Stokes par Méthodes Spectrales*. Thèse, Université Pierre-et-Marie-Curie (1987).
8. E. M. Ronquist, *Optimal Spectral Element Methods for the Unsteady 3-Dimensional Incompressible Navier-Stokes Equations*, Thesis, Massachusetts Institute of Technology (1988).

UNIVERSITÉ PAUL SABATIER, LABORATOIRE MMF, 31062, TOUL OUSE. FRANCE
E-mail address: azaiez@soleil.ups-tlse.fr

DIPARTIMENTO DI MATEMATICA, POLITECNICO DI MILANO, AND CRS4, CAGLIARI
E-mail address: alfio@crs4.it

Contemporary Mathematics
Volume **180**, 1994

Preconditioned Iterative Methods in a Subspace for Linear Algebraic Equations with Large Jumps in the Coefficients

NIKOLAI S. BAKHVALOV AND ANDREW V. KNYAZEV

ABSTRACT. We consider a family of symmetric matrices $A_\omega = A_0 + \omega B$, with a nonnegative definite matrix A_0, a positive definite matrix B, and a nonnegative parameter $\omega \leq 1$. Small ω leads to a poor conditioned matrix A_ω with jumps in the coefficients. For solving linear algebraic equations with the matrix A_ω, we use standard preconditioned iterative methods with the matrix B as a preconditioner. We show that a proper choice of the initial guess makes possible keeping all residuals in the subspace $Im(A_0)$. Using this property we estimate, uniformly in ω, the convergence rate of the methods.

Algebraic equations of this type arise naturally as finite element discretizations of boundary value problems for PDE with large jumps of coefficients. For such problems the rate of convergence does not decrease when the mesh gets finer and/or ω tends to zero; each iteration has only a modest cost. The case $\omega = 0$ corresponds to the fictitious component/capacitance matrix method.

1. Introduction

In recent years, the study of preconditioners for iterative methods for solving large linear systems of equations, arising from discretizations of stationary boundary value problems of mathematical physics, has become a major focus of numerical analysts and engineers. In each iteration step of such a method, only a linear system with a special matrix, the preconditioner, has to be solved; the given system matrix has to be available only in terms of a matrix-vector multiplication. The basic theory of convergence of these methods is very well developed for the symmetric, positive definite case. It is well known that the preconditioner

---

1991 *Mathematics Subject Classification.* Primary 65F35.

The work of the second author was supported in part by the National Science Foundation under the supplement to Grant NSF-CCR-9204255.

This paper is in final form and no version of it will be submitted for publication elsewhere.

must approximate the inverse of the matrix of the original system well in order to obtain rapid convergence properties. For finite element/difference problems rapid convergence typically means that the rate of convergence is independent of the mesh size. It is common to use a conjugate gradient type method as an accelerator in such iterations.

There are several methods of constructing preconditioners, which allow the use of efficient methods, such as fast direct solvers, for solving the related linear systems. Many such methods with asymptotically optimal *a priori* estimates of the computational cost [11] are known, and some of these preconditioned iterative methods are among the best for solving mesh problems when the mesh parameter is small enough.

A particularly challenging class of problems arises with models described by partial differential equations (PDE) with discontinuous coefficients. Many important physical problems are of this nature. Such difficult problems arise in the design and study of numerical methods for composite materials built from essentially different components. Composites (or composite materials) are media with a large number of non-homogeneous inclusions of small sizes in at least one direction. Stationary states of such media are described by elliptic PDE with highly oscillating coefficients. Homogenization is the process of finding a set of constant coefficients such that the solution of the original PDE can be approximated well by that of the much simpler problem. The computation of homogenized coefficients for a composite with periodic structure reduces to solving a series of periodic boundary value problem for the original PDE in a domain of periodicity, see [8]. For composites with essentially different components, the coefficients of the PDE in the domain of periodicity are discontinuous and have large jumps.

Fictitious domain/embedding method is another source of PDE with jumps of coefficients, cf. e.g. [19, 2, 10, 14]. In this method, the original boundary value problem for the PDE is changed into a new boundary value problem with a domain in which the original one is embedded. In the new fictitious part of the domain the coefficients of PDE are chosen close to zero, if the original boundary condition is of Neumann type, or very large in the Dirichlet case. The solutions of these new problems approximate, or might even coincide with, the desired solutions. Therefore, the use of the fictitious domain method improves the shape of the boundary, but leads to a PDE with very large jumps of coefficients.

Similar problems occur in the semi-conductor device modelling.

There are several difficulties associated with the numerical solution of these problems using preconditioned iterative methods. For a number of methods, the larger the jumps of the coefficients, the slower the convergence. However, it has recently been shown [4, 5, 6, 7] that for continuous models, and with a special initial guess, the rate of convergence does not depend on the size of the jumps.

In the present paper, we explain the main idea of [4, 5, 6, 7] in the simplest form for algebraic systems of linear equations with a symmetric coefficient matrix.

We consider a parametric family of symmetric matrices $A_\omega = A_0 + \omega B$, with a nonnegative definite matrix A_0, a positive definite matrix B, and a parameter $\omega, 0 \leq \omega \leq 1$. Small ω leads to a large drop of coefficients of A_ω. For solving linear algebraic equations with matrix A_ω, we use standard preconditioned iterative methods with matrix B as a preconditioner. We show that a proper choice of the initial guess makes it possible to keep all residuals in the subspace $Im(A_0)$ and the difference between all iteration vectors and the solution in the subspace $Im(B^{-1}A_0)$. Using this property we estimate, uniformly in ω, the convergence rate of the methods.

A similar method, based on the same idea, was proposed in [1]. This method can only be applied for solving PDE with piece wise constant coefficients. We have to note that the proof of the mesh extension theorem in [1] is not correct.

In the case $\omega = 0$, these methods are closely related to the *capacitance matrix methods* [18, 16, 17, 9, 15].

The importance of the idea of *iterative methods in a subspace* is widely recognized in the theory of the *domain decomposition methods*, e.g. [13].

2. Preconditioned Iterative Methods for the symmetric positive definite case

We first consider a linear algebraic system $Au = f$ with a symmetric positive definite matrix A. In practice, a direct solution of that system often requires very considerable computational work, so iterative methods of the form

$$(1) \qquad u^{n+1} = u^n - \gamma^n B^{-1}(Au^n - f)$$

with a symmetric positive definite matrix B, the preconditioner, are of great importance.

There is an equivalent form of the method for residuals $r^n = Au^n - f$:

$$(2) \qquad \begin{aligned} r^{n+1} &= r^n - \gamma^n AB^{-1}r^n, r^0 = Au^0 - f, \\ \sigma^{n+1} &= \sigma^n + r^{n+1}, \sigma^0 = r^0, \\ u^{n+1} &= u^0 - \gamma^n B^{-1}\sigma^n. \end{aligned}$$

Preconditioned iterative methods of this kind can be very effective for the solution of systems of equations arising from discretizations of elliptic operators and their effectiveness can, of course, be further enhanced by the use of conjugate gradient type methods. For an appropriate choice of the preconditioner B, the convergence does not slow down when the mesh is refined and each iteration has a small cost; see e.g. [11].

Let

$$(3) \qquad 0 < \delta_0 (Bv, v) \leq (Av, v) \leq \delta_1 (Bv, v), v \neq 0.$$

The importance of choosing the preconditioner B such that δ_1/δ_0, the condition number of $B^{-1}L$, either is independent of N, the size of the system, or depends weakly, e.g. polylogarithmically in N, is widely recognized. At the same time, the numerical solution of the system with the matrix B should ideally require on the order of N, or $N \ln N$ arithmetic operations for single processor computers.

It is well known that, typically, the smaller the ratio δ_0/δ_1, the slower the convergence.

We cite here the simplest convergence estimate. Let $\gamma_k = \delta_1$; then

$$(4) \qquad (B(u^n - u), u^n - u) \le q^n (B(u^0 - u), u^0 - u), \quad q \equiv 1 - \delta_0/\delta_1,$$

for an arbitrary initial guess u^0.

3. The symmetric nonnegative definite case

We now consider the case $A = A_0$ where A_0 is a symmetric nonnegative definite matrix. The matrix B may also be symmetric nonnegative definite with the kernel $Ker B \subseteq Ker A_0$. Such matrices B and A_0 arise, for example, when considering periodic boundary value problems. In the present paper, however, we, for simplicity, consider only positive definite B.

Inequality (3) can not be true any longer, because the matrix A_0 has a kernel. Let, instead of (3), the following analogous inequalities hold:

$$(5) \qquad 0 < \delta_0(Bv, v) \le (A_0 v, v) \le \delta_1(Bv, v), v \ne 0, v \in Im(B^{-1}A_0),$$

and, thereby, δ's *have been redefined*. The subspace $Im(B^{-1}A_0)$ will play the key role and we denote it by $Im = Im(B^{-1}A_0)$.

LEMMA 1. *The subspace Im consists of all B-normal, i.e. normal in the scalar product $(B\star, \star)$, solutions of the system $A_0 u = f$ for all possible right-hand sides f.*

The following theorem is well known, cf. [2].

THEOREM 1. *Let u be a B-normal solution of $A_0 u = f$ for a given $f \in Im A_0$ and let the initial guess u^0 be chosen such that $u^0 \in Im$. Then the iterative method (1) with $A = A_0$ and $\gamma_k = \delta_1$ converges to this B-normal solution and convergence estimate (4) holds with δ's from (5).*

The proof is based on the fact that the subspace Im is invariant with respect to the operator $B^{-1}A_0$.

4. The symmetric positive definite case

We finally consider a parametric family of symmetric positive definite matrices $A = A_\omega = A_0 + \omega B$, with a parameter $\omega, 0 < \omega \le 1$. The condition number of the matrix A_ω as well as that of the preconditioned matrix $B^{-1}A_\omega$ tends to infinity as ω tends to zero, and the common convergence theory of the method (1) for the positive definite case becomes useless.

We can improve the convergence by using a special initial guess, however.

THEOREM 2. *Let the initial guess u^0 be chosen such that $u^0 - B^{-1}f/\omega \in Im$. Then the iterative method (1) with $A = A_\omega$ and $\gamma_k = \delta_1 + 1$ converges and the convergence estimate (4) holds with $q = 1 - \delta_0/(\delta_1 + 1)$ and δ's from (5).*

The proof is very similar to the proof of the previous theorem. The subspace Im is invariant with respect to the operator $B^{-1}A_\omega = B^{-1}A_0 + \omega I$ and the initial error $u^0 - u$ is in this subspace.

We also note that the initial residual and, therefore, all residuals of the method (2) lie in the subspace ImA_0.

Analogous convergence results may be, evidently, obtained for preconditioned conjugate gradient methods with this choice of the initial guess.

A dual approach to the numerical solution of problems with large jumps of coefficients, using a mixed variational formulation, is described in [12]. For a general saddle point problem

$$(6) \qquad \begin{pmatrix} F_\omega & G \\ G^* & 0 \end{pmatrix} \begin{pmatrix} u \\ v \end{pmatrix} = \begin{pmatrix} 0 \\ f \end{pmatrix},$$

where $F_\omega = F_0 + \omega I$, F_0 symmetric nonnegative definite, and ω is a parameter, $0 \le \omega \le 1$, we obtain

$$\begin{cases} P^\perp F_\omega v = 0, & P = G(G^*G)^{-1}G^*, P^\perp = I - P, \\ Pv = p, & p = G(G^*G)^{-1}f. \end{cases}$$

P is an orthoprojector and the system is further reformulated as a single matrix equation with a symmetric, nonnegative definite matrix $P^\perp F_\omega P^\perp$. This matrix, as well as our preconditioned matrix $B^{-1}A_\omega$, is an ωI perturbation in the subspace $Im(P^\perp)$ of a symmetric nonnegative definite matrix. Therefore, our trick with the special initial guess for a standard iterative method is useful for this matrix, too; see [12].

This approach was applied in [3] to the diffusion equation taking A^{-1} as the diffusion coefficient and $-G$ as the gradient operator. The method is most efficient when multiplication by P can be calculated using a fast direct method, e.g. for a problem with a periodic boundary condition, which arises naturally in homogenization and fictitious domain methods.

In the near future we are planning to consider a case of a nonsymmetric family $A_\omega = A_0 + \omega B$, by using symmetrization in the form $(B^{-1}A_\omega)^*B^{-1}A_\omega$.

The authors are indebted to Olof Widlund and to the Organizing Committee of the Seventh International Conference on Domain Decomposition Methods for their support.

REFERENCES

1. N. N. Abramov and V. V. Kucherenko. An algorithm for the numerical solution of the Dirichlet problem. *Soviet Math. Dokl.*, 39(2):302–305, 1989.
2. G. P. Astrakhatsev. Method of fictitious domains for a second-order elliptic equation with natural boundary conditions. *U.S.S.R. Computational Math. and Math. Phys.*, 18:114–121, 1978.
3. N. S. Bakhvalov and A. V. Knyazev. A new iterative algorithm for solving problems of the fictitious flow method for elliptic equations. *Soviet Math. Doklady*, 41(3):481–485, 1990.
4. N. S. Bakhvalov and A. V. Knyazev. Efficient computation of averaged characteristics of composites of a periodic structure of essentially different materials. *Soviet Math. Doklady*, 42(1):57–62, 1991.
5. N. S. Bakhvalov and A. V. Knyazev. An efficient iterative method for solving the Lamé equations for almost incompressible media and Stokes' equations. *Soviet Math. Doklady*, 44(1):4–9, 1992.
6. N. S. Bakhvalov and A. V. Knyazev. Fictitious domain methods and computation of homogenized properties of composites with a periodic structure of essentially different components. In Gury I. Marchuk, editor, *Numerical Methods and Applications*, pages 221–276. CRC Press, Boca Raton, 1994.
7. N. S. Bakhvalov, A. V. Knyazev, and G. M. Kobel'kov. Iterative methods for solving equations with highly varying coefficients. In Roland Glowinski, Yuri A. Kuznetsov, Gérard A. Meurant, Jacques Périaux, and Olof Widlund, editors, *Fourth International Symposium on Domain Decomposition Methods for Partial Differential Equations*, pages 197–205, Philadelphia, PA, 1991. SIAM.
8. N. S. Bakhvalov and G. P. Panasenko. *Homogenization: Averaging of Processes in Periodic Media. Mathematical Problems in the Mechanics of Composite Materials*, volume 36 of *Math. and Its Applications. Soviet Series*. Kluwer Academic Publishers, Boston, 1989.
9. Ch. Börgers and O. Widlund. On finite element domain imbedding methods. *SIAM J. Numer. Anal.*, 27(4):963–978, 1990.
10. B. L. Buzbee, F. W. Dorr, J. A. George, and G. Golub. The direct solution of the discrete Poisson equation on irregular regions. *SIAM J. Numer. Anal.*, 8:722–736, 1971.
11. E. G. D'yakonov. *Minimization of computational work. Asymptotically optimal algorithms for elliptic problems*. Nauka, Moscow, 1989. In Russian.
12. A. V. Knyazev. Iterative solution of PDE with strongly varying coefficients: algebraic version. In R. Beauwens and P. de Groen, editors, *Iterative Methods in Linear Algebra*, pages 85–89, Amsterdam, 1992. Elsevier. Proceedings IMACS Symp. Iterative Methods in Linear Algebra, Brussels, 1991.
13. Yu. A. Kuznetsov. Numerical methods in subspaces. In Gury I. Marchuk, editor, *Computational Processes and Systems, No. 3*. Nauka, Moscow, 1985. In Russian.
14. G. I. Marchuk, Yu. A. Kuznetsov, and A. M. Matsokin. Fictitious domain and domain decomposition methods. *Soviet J. Numerical Analysis and Math. Modelling*, 1:1–82, 1986.
15. D. P. O'Leary and O. Widlund. Capacitance matrix methods for the Helmholtz equation on general three dimensional regions. *Math. Comput.*, 33:849–879, 1979.
16. V. Pereyra, W. Proskurowski, and O. Widlund. High order fast Laplace solvers for the Dirichlet problem on general regions. *Math. Comput.*, 31:1–16, 1977.
17. W. Proskurowski and O. Widlund. On the numerical solution of Helmholtz's equation by the capacitance matrix method. *Math. Comput.*, 30:433–468, 1976.
18. W. Proskurowski and O. Widlund. A finite element - capacitance matrix method for the Neumann problem for Laplace's equation. *SIAM Stat. and Sci. Comput.*, 1:410–425, 1980.
19. V. K. Saul'ev. On solving boundary value problems with high performance computers by a fictitious domain method. *Siberian Math. J.*, 4(4):912, 1963. In Russian.

INSTITUTE OF NUMERICAL MATHEMATICS RUSSIAN ACADEMY OF SCIENCE, MOSCOW, RUSSIA AND COURANT INSTITUTE OF MATHEMATICAL SCIENCE, NYU, NEW YORK 10012

E-mail address: na.knyazev@na-net.ornl.gov

Contemporary Mathematics
Volume 180, 1994

The Hierarchical Basis Multigrid Method and Incomplete LU Decomposition

RANDOLPH E. BANK AND JINCHAO XU

ABSTRACT. A new multigrid technique is developed in this paper for solving large sparse algebraic systems from discretizing partial differential equations. By exploring the connection between the hierarchical basis method and incomplete LU decomposition, the resulting algorithm can be effectively applied to problems discretized on completely unstructured grids. Numerical experiments demonstrating the efficiency of the method are also reported.

1. Introduction

In this work, we explore the connection between the methods of sparse Gaussian elimination [8, 13], incomplete LU (ILU) decomposition [9, 10] and the hierarchical basis multigrid (HBMG) [16, 4].

Hierarchical basis methods have proved to be one of the more robust classes of methods for solving broad classes of elliptic partial differential equations, especially the large systems arising in conjunction with adaptive local mesh refinement techniques [5, 2], and have been shown to be strongly connected to space decomposition methods and to classical multigrid methods [14, 15, 4, 9]. As with typical multigrid methods, classical hierarchical basis methods are usually defined in terms of an underlying refinement structure of a sequence of nested meshes. In many cases this is no disadvantage, but it limits the applicability of the methods to truly unstructured meshes, which may be highly nonuniform but *not* derived from some grid refinement process. A major goal of our study is to generalize the construction of hierarchical bases to such meshes, allowing HBMG and other hierarchical basis methods to be applied. Some work on multigrid methods on non-nested meshes is reported in Bramble, Pasciak and Xu [6], Xu [14], and Zhang [17].

1991 *Mathematics Subject Classification.* Primary 65F10; Secondary 65N20.

The first author was supported by the Office of Naval Research under contract N00014-89J-1440. The second author was supported by National Science Fundation.

This paper is in final form and no version of it will be submitted for publication elsewhere.

In Section 2, we develop a simple graph elimination model for classical hierarchical basis methods on sequences of nested meshes. This elimination model can be interpreted as a particular ILU decomposition where certain fill–in edges, namely those corresponding the element edges on a coarser mesh, are allowed. This graph elimination model can be generalized in a very simple and straightforward fashion to the case of completely unstructured meshes, providing a simple mechanism for defining hierarchical bases on such meshes. The key concept is that of *vertex parents* of a given vertex v_i in the mesh. In the case of a sequence of refined meshes, the parents of v_i are just the endpoints of the triangle edge which was bisected when v_i was created. By generalizing this notion slightly, we are able to define vertex parents for vertices in an unstructured mesh, which, in effect, supplies us with a heuristic procedure for systematically unrefining the unstructured mesh.

In Section 3, we describe algebraic aspects of HBMG and its application to completely unstructured meshes. In the classical case, vertices are ordered (blocked) by refinement level and we apply symmetric block Gauss-Seidel to the linear system represented in the hierarchical basis. We note that it is the transformation of the stiffness matrix from nodal to hierarchical basis which has a strong connection to ILU, and this is completely algebraic in nature once the transformation is defined. In turn, the transformation matrix relies upon the vertex parents function to determine its sparsity structure and blocking, and upon the geometric properties of the refinement procedure to determine the numerical values of the elements. Since we define vertex parents on unstructured meshes as part of our ILU graph elimination algorithm, we can determine the structure of the transformation matrix just as in the case of nested meshes. The numerical values are selected by examining the geometry of the mesh, in a fashion similar to the nested mesh case. At the level of implementation, HBMG and other iterations based on hierarchical bases are algebraically identical for the cases of structured and unstructured meshes. Indeed, we made only slight changes to our HBMG routines to implement the new method; essentially all new coding was devoted to the graph elimination process and the hueristics for determining vertex parents.

In Section 4, we present a numerical illustration of the method and make a few concluding remarks.

2. Graph Theoretical Properties of Hierarchical Bases

In this section, we explore the connection between the HBMG method and ILU decomposition in terms of graph theory. We will consider first the standard Gaussian elimination and classical ILU factorization. We then progress to the HBMG method, first considering the triangular meshes generated through a process of grid refinement and then considering completely unstructured triangular meshes. To begin, we briefly review the process of Gaussian Elimination from a graph theoretical point of view. A more complete discussion of this point can be found in Rose [13] or George and Liu [8].

Corresponding to a sparse, symmetric, positive definite $N \times N$ matrix A, let

$\mathcal{G}(X, E)$ be the graph that consists of a set of N ordered vertices $v_i \in X$, $1 \le i \le N$, and a set of edges E such that the edge (connecting vertices v_i and v_j) $e_{ij} \in E$ if and only if $a_{ij} \neq 0$, $i \neq j$. Note that edges in the graph $\mathcal{G}$ correspond to the nonzero off-diagonal entries of A. For the case of interest here, A is the stiffness matrix for the space of continuous piecewise linear polynomials represented in the standard nodal basis. Then $\mathcal{G}$ is just the underlying triangulation of the domain (with some possible minor modifications due to the treatment of Dirichlet boundary conditions). If we view $\mathcal{G}$ as an *unordered* graph, then the graph corresponds to the class of matrices of the form $P^t A P$, where P is a permutation matrix; that is, reordering the vertices of the graph corresponds to forming the product $P^t A P$ for a suitable permutation matrix P.

For convenience, we shall need a few additional terminologies from graph theory. Let $v_i \in X$; the set of adjacent vertices $adj(v_i)$ is given by

$$adj(v_i) = \{v_j \in X \,|\, e_{ij} \in E\}.$$

Roughly speaking, the set $adj(v_i)$ corresponds to the set of column indices for the nonzero entries in row i of matrix A (or the set of row indices for nonzero entries of column i of A), with the exception of the diagonal entry a_{ii}.

A *clique* $\mathcal{C} \subseteq X$ is a set of vertices which are all pairwise connected; that is $v_i, v_j \in \mathcal{C}, i \neq j \Rightarrow e_{ij} \in E$. If A is a dense $N \times N$ matrix, then its graph is a clique on N vertices. More generally, with a proper ordering of the vertices, cliques correspond to dense submatrices of A.

With these definitions, we can define the graph theoretic equivalent of Gaussian elimination of A. First, in terms of matrices let

$$A = \begin{bmatrix} a_{11} & r^t \\ r & B \end{bmatrix},$$

where r is an $N - 1$-vector and B is an $(N-1) \times (N-1)$ matrix. The first step of Gaussian elimination consists of the factorization

$$
\begin{aligned}
A &= \begin{bmatrix} 1 & 0 \\ r/a_{11} & I \end{bmatrix} \begin{bmatrix} a_{11} & 0 \\ 0 & B - rr^t/a_{11} \end{bmatrix} \begin{bmatrix} 1 & r^t/u_{11} \\ 0 & I \end{bmatrix} \\
&= L_1 D_1 L_1^t.
\end{aligned}
$$

The matrix $A' = B - rr^t/a_{11}$ is a symmetric, positive definite matrix of order $N - 1$ to which the factorization can be inductively applied. Note that A' may be less sparse than B due to the *fill–in* caused by the outer product rr^t/a_{11}.

In graph theoretic terms, eliminating vertex v_1 from $\mathcal{G}$ transforms $\mathcal{G}(X, E)$ to a new graph $\mathcal{G}'(X', E')$, corresponding to matrix A', as follows

   (i) Eliminate vertex v_1 and all its incident edges from $\mathcal{G}$. Set $X' = X - \{v_1\}$. Denote the resulting set of edges $E_1 \subseteq E$.

  (ii) Create the set F of *fill–in* edges as follows: For each distinct pair $v_j, v_k \in adj(v_1)$ in $\mathcal{G}$, add the edge e_{jk} to F if it is not already present in E_1. Set $E' = E_1 \cup F$.

Note that the set $adj(v_1)$ in $\mathcal{G}$ becomes a clique in $\mathcal{G}'$.

Within this framework, an ILU factorization is one in which all the fill–in called for in step 2 above is not allowed. The classical form of ILU is to allow *no* fill–in, that is, no new edges are added in step 2 ($E' = E_1$). That forces the resulting matrix A', (which is now not necessarily equal to $B - rr^t/a_{11}$) to have the same sparsity pattern as B. The effect of the neglected fill–in elements in terms of the numerical values of entries in A' varies, and is not considered here; at the moment, our concern is with the sparsity pattern itself.

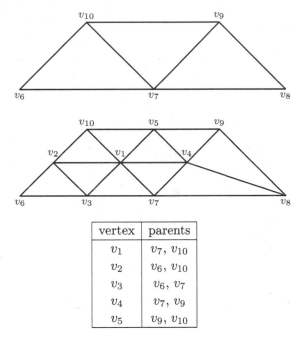

FIGURE 1. The coarse grid $\mathcal{T}_c$ and the fine grid $\mathcal{T}_f$. Vertices $\{v_i\}_{i=6}^{10} = X_c$, while $\{v_i\}_{i=1}^{5} = X_f$.

Now let us view HBMG on a set of nested meshes in terms of ILU. For convenience, we will restrict consideration to the case of only two levels. Let $\mathcal{T}_c$ be the coarse triangulation, and $\mathcal{T}_f$ be the fine triangulation, where some elements $t \in \mathcal{T}_c$ are refined into four elements in $\mathcal{T}_f$ by pairwise connection of the midpoints of the three edges of t (regular refinement). Some elements near the boundary of the refined region can be bisected (*green* refinement), while others are left unrefined. The details of such a refinement algorithm can be found in [**5, 2**] and are not of great interest to the current discussion. An example is shown in Figure 1. Let X be the set of vertices in $\mathcal{T}_f$, and $X_c \subset X$ be the set of vertices in $\mathcal{T}_c$. Denote by X_f the set of fine grid vertices not in X_c ($X_f = X \setminus X_c$).

For each vertex $v_i \in X_f$, there are a (unique) pair of vertices $v_j, v_k \in X_c$ such that v_i is the midpoint of the edge connecting v_j and v_k in the coarse grid $\mathcal{T}_c$. This pair of vertices is called the *vertex parents* of v_i. The vertex parents for the set X_f for our example are given in Figure 1.

Suppose now that we choose an ordering in which all the vertices in X_f are ordered first, followed by those in X_c. We then consider eliminating the vertices in X_f as follows:

(i) Eliminate vertex v_1 and all its incident edges from $\mathcal{G}$. Set $X' = X - \{v_1\}$. Denote the resulting set of edges $E_1 \subseteq E$.

(ii) Add *one* fill–in edge connecting the vertex parents of v_i, say $v_j, v_k \in X_c$. Set $E' = E_1 \cup \{e_{jk}\}$.

It is easy to see this is an *ILU* algorithm in that only selected fill–in edges are allowed, namely those connecting vertex parents. It also is important to note that the triangulation $\mathcal{T}_f$ is the graph for the original stiffness matrix A represented in the standard nodal basis. After all the vertices in X_f are eliminated, the resulting graph is just the triangulation $\mathcal{T}_c$; that is, the sparity structure of the coarse grid matrix corresponds to the coarse grid triangulation. One of the important properties of HBMG is that the corresponding coarse grid matrix is just the stiffness matrix with respect to the nodal basis of the coarse grid (e.g., the hierarchical basis). For this to occur requires a particular (but natural) choice of numerical values for the multipliers used in computing the *ILU*. This is a topic for the next section. A comparison of the elimination graphs for vertex v_1, using regular Gaussian elimination, classical *ILU* and HBMG is shown in Figure 2.

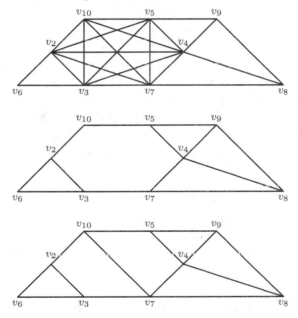

FIGURE 2. The elimination graphs $\mathcal{G}'$ generated by eliminating vertex v_1. Standard Gaussian elimination is shown at the top, classical *ILU* in the center, and HBMG at the bottom.

This brings us to HBMG for completely unstructured, nonnested meshes. So far we have considered two different algorithms. We begin with the simpler of the two.

Suppose that for a vertex $v_i \in X$, we denote a pair of distinct vertices $v_j, v_k \in adj(v_i)$ as the *tentative vertex parents* of v_i. Generally, in selecting tentative vertex parents, we try to emulate the case of nested HBMG as closely as possible; that is, we want v_i to be "close" in some sense to the midpoint of the straight line connecting v_j and v_k. Not all vertices can be assigned as reasonable parents; those that cannot will be called "corners" of the region. Among those vertices that do have tentative parents, we chose the vertex v_i which best satisfies the heuristic criterion used in selecting tentative parents, and eliminate it as follows:

(i) Eliminate v_i and all its incident edges. Set $X' = X - \{v_i\}$.

(ii) If not already present, add one fill-in edge connecting the tentative vertex parents of v_i, say v_j and v_k. These become the *permanent* vertex parents of v_i. Set $E' = E_i \cup \{e_{jk}\}$.

(iii) For each $v_\ell \in adj(v_i)$ in $\mathcal{G}$, compute new tentative vertex parents based on the new graph $\mathcal{G}'$.

The third step is essential, since v_i could have been a tentative parent in $\mathcal{G}$ of any the vertices in $adj(v_i)$, including its permanent vertex parents v_j and v_k. Furthermore, the additional edge connecting v_j and v_k allows them to be considered as tentative vertex parents of each other in $\mathcal{G}'$. This algorithm continues inductively until no more vertices can be eliminated (all remaining vertices are corners).

After the first vertex is eliminated, the remaining elimination graphs are not necessarily triangulations of the domain, but typically contain polygonal elements of various orders. The second variant of the algorithm addresses this issue. In this algorithm, a vertex v_i can have either two or three tentative vertex parents, again selected from among the vertices in $adj(v_i)$. The first two are chosen as in the first variant. If v_i is "too far" from being colinear with its tentative parents v_j and v_k, then a third tentative parent, say $v_m \in adj(v_i)$ is chosen such that the resulting triangle with vertices v_j, v_k and v_m is optimized with respect to shape regularity. Once again, not all vertices may have tentative vertex parents, although in practice, allowing for the possibility of three parents tends to reduce the number of corners.

In any event, the vertex v_i which best optimizes the criteria, is then eliminated as follows:

(i) Eliminate v_i and all its incident edges. Set $X' = X - \{v_i\}$.

(ii) If not already present, add a fill-in edge connecting the two principal vertex parents of v_i. Add additional fill-in edges as required, such that the resulting graph remains a triangulation of the domain. Denote this set of edges by F and set $E' = E_i \cup F$. The two or three tentative vertex parents of v_i become permanent.

(iii) For each $v_\ell \in adj(v_i)$ in $\mathcal{G}$, compute new tentative vertex parents based on the new graph $\mathcal{G}'$.

This algorithm is aesthetically more pleasing than the first. Since all the elimination graphs are triangulations, it seems easier to handle in terms of the mathematical analysis. However, it generally allows for more fill-in than the first algorithm, and requires more complex data structures based on linked lists to represent the graphs. The additional fill-in and the possibility of having vertices with three vertex parents makes the resulting HBMG algorithm more expensive (see next section). Of course, this would be justified if the resulting HBMG algorithm performed significantly better, but so far in our experience, both algorithms perform comparably well in terms of convergence rate. Thus, at present we have no justification for the more expensive version.

As a final point in this section, we consider the assignment of vertex levels. Each vertex in the mesh has a unique level; this level is used to partition the stiffness matrix in HBMG and other hierarchical basis iterations. This is not an important point for the current discussion, since we have assumed only two levels, but it is very important for the case of more than two levels. In the classical HBMG using a sequence of nested meshes, the level ℓ_i of vertex v_i is defined as follows. All vertices on the coarse grid are assigned $\ell_i = 1$. Thereafter, the remaining vertices are assigned levels in terms of the levels of their parents, according to

$$(1) \qquad \ell_i = \max(\ell_k, \ell_j) + 1,$$

where v_k and v_j are the parents of vertex v_i. In the case of unstructured meshes, equation (1) can still be used, modified appropriately for the case of vertices with three vertex parents. All vertices without parents (corners) are assigned $\ell_i = 1$, and then (1) uniquely determines the level of the remaining vertices. In computing vertex levels, one should process the vertices in the *reverse* order of elimination, so that the level of all parents of vertex v_i will be defined prior to the processing of v_i.

3. Algebraic HBMG and ILU

In this section we consider the algebraic aspects of the HBMG method, and its relation to Gaussian elimination. Again for convenience we will consider the case of only two levels. Let A denote the stiffness matrix for the fine grid, and consider the block partitioning

$$(2) \qquad A = \begin{bmatrix} A_f & C \\ C^t & A_c \end{bmatrix},$$

where A_f corresponds to the nodal basis functions in X_f, A_c corresponds to the (fine grid) nodal basis functions in X_c, and C corresponds to the coupling between the two sets of basis functions. We consider transformations of the form $A' = SAS^t$, where S has the block structure

$$(3) \qquad S = \begin{bmatrix} I & 0 \\ R & I \end{bmatrix}.$$

From (2)-(3), we obtain

$$(4) \qquad SAS^t = \begin{bmatrix} A_f & A_f R^t + C \\ RA_f + C^t & \hat{A}_c \end{bmatrix},$$

where

$$(5) \qquad \hat{A}_c = RA_f R^t + C^t R^t + RC + A_c.$$

Different algorithms can be characterized by different choices of R. For example, in classical block Gaussian elimination $R = -C^t A_f^{-1}$, and $\hat{A}_c = A_c - C^t A_f^{-1} C$ is the Schur complement. In this case, the off diagonal blocks are reduced to zero, but at the cost of having fairly dense matrices R and $\hat{A}_c$.

In the case of classical HBMG, the matrix R^t is sparse, and contains information about changing from the nodal to hierarchical basis [4, 1]. Each row of R^t is zero except for two entries which are equal to $1/2$. For the row corresponding to vertex $v_i \in X_f$, the two nonzero entries are in the columns corresponding to $v_j \in X_c$ and $v_k \in X_c$, where v_j and v_k are the vertex parents of v_i. In this case, the matrix $\hat{A}_c$ is just the stiffness matrix for the coarse grid represented in the *coarse* grid nodal basis. Although we know a priori that the graph for $\hat{A}_c$ is just the coarse grid triangulation $\mathcal{T}_c$, we can formally compute this graph by applying the symbolic *ILU* elimination process. The matrix $RA_f + C^t$ is not zero as in the case of Gaussian elimination; indeed it is less sparse than C^t. However, the matrix is small in some sense; the usual Cauchy inequality estimate [3, 1, 7] written in this notation is:

$$|x^t(RA_f + C^t)y| \leq \gamma (x^t \hat{A}_c x)^{1/2} (y^t A_f y)^{1/2},$$

where $\gamma < 1$ is the constant in the strengthened Cauchy inequality. It is worth commenting that in implementation, the matrix $RA_f + C^t$ is never formed explicitly; all that is required to implement HBMG and other iterations using hierarchical bases (either additive or multiplicative variants) is the set of matrices A_f, C, $\hat{A}_c$ and R. For our current discussion, R is the critical matrix. A_f and C are, of course, just parts of the nodal matrix, and $\hat{A}_c$ is explicitly computed from (5) onece R is known.

The sparsity pattern of R is completely determined by the vertex parents function; in the case of classical HBMG on nested meshes, it follows from the refinement structure of the mesh. The numerical values of the coefficients (the "weights" or the "multipliers") are all equal to $1/2$ for the nested case; the $1/2$ arises naturally from the geometry of the refinement process, stating that a vertex created at the midpoint of an edge of a coarse grid triangle is between its vertex parents.

Now let us consider HBMG on nonnested meshes. From the algebraic point of view, nothing changes from the nested case once the matrix is defined, and to define R we need only two things: the vertex parents function to define the sparsity pattern, and the weights to define the numerical values. From this point of view, it should be clear that this process will (implicitly) construct linear combinations of the fine grid nodal basis functions, whose energy inner products appear as matrix elements in the matrix $\hat{A}_c$, just as in the nested case. The difference is that in the nested case, these complicated linear combinations reduce to simple nodal basis functions for the coarse mesh. For nonnested meshes, they remain complicated linear combinations of the fine grid basis functions. On the other hand, one never need explicitly deal with these basis functions (except in the mathematical analysis), since the iteration itself is completely algebraic. The critical issue for HBMG is *not* that one obtains simple coarse grid nodal basis functions, but rather that the *support* of the basis functions which are obtained is increasing at the proper rate, and as long as the complicated basis functions have that property, one should see the expected convergence rates. The use of modified hierarchical basis functions appears in the work of Hoppe and Kornhuber [11] and Kornhuber [12] in connection with the solution of obstacle problems.

We now consider the construction of R for the two heuristic algorithms for unstructured meshes described in the last section. In both cases, the vertex parents function (sparsity pattern for R) is determined by the graph elimination algorithms discussed in the last section. Each column of R^t will have either two or three nonzeroes, corresponding to its permanent vertex parents.

As for the weights, for the algorithm which allows just two vertex parents, let vertex v_i have parents v_j and v_k, and let $v_i' = \theta v_j + (1 - \theta)v_k$ be the orthogonal projection of v_i onto the straight line connecting v_j and v_k. Then we take the corresponding weights (numerical values of the nonzeroes in R) to be θ and $1 - \theta$.

In the case of a vertex with three vertex parents, let v_j and v_k be the two principal parents and v_ℓ the third. In this case, we compute the barycentric coordinates of v_i with respect to the triangle with vertices v_j, v_k and v_ℓ, and these barycentric coordinates become the numerical values used in R. Since v_i was supposed to be close to the line connecting v_j v_k, the barycentric coordinate corresponding to v_ℓ should generally be small in comparison with the others, which in turn, should be close to the form described above for method using only vertex parents. Allowing for the possibility of three permanent parents means that the matrix R will be less sparse, which in turn means $\hat{A}_c$ will be less sparse, a situation which is compounded as recursion adds more levels. Overall, this leads to more work (a bigger constant, but apparently not a change in the order of magnitude), but as far as we can tell from our early experience with the algorithm, does not significantly improve the rate of convergence.

4. Numerical Illustrations and Conclusions

In this section, we present a simple example of our algorithm. This example was developed using the finite element package $PLTMG$ [2]. We consider a square domain Ω with a circular hole. We triangulate this domain using 684 triangles and 398 vertices as shown in Figure 3. The mesh is unstructured, in that it was not generated through the refinement of a coarser mesh.

We solved the equation $-\Delta u = 1$ in Ω with a combination of homogeneous Dirichlet and Neumann boundary conditions on $\partial\Omega$. Applying the algorithms outlined in Sections 2-3, we created an algebraic hierarchical basis with 10 levels, and 95 vertices on the coarsest level. Levels were determined using (1). In Figure 3 we show the convergence history of the multiplicative (symmetric Gauss-Seidel) hierarchical basis iteration, starting from a zero initial guess. The quantity plotted is

$$\sigma_k = \log\left\{\frac{\| r_k \|}{\| r_0 \|}\right\},$$

where r_k is the residual at iteration k and $\| \cdot \|$ is the ℓ^2 norm. From the data points, we estimate by least squares that the convergence rate is approximately 0.44, which

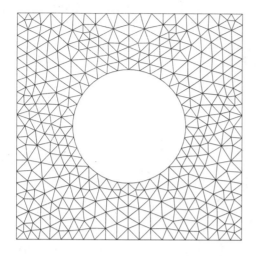 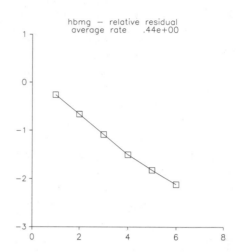

FIGURE 3. The triangulation and the convergence history for HBMG.

is fairly typical of this particular iterative method applied to a similar problem on a sequence of refined meshes.

We close with several remarks. First, the success of the method obviously depends rather crucially on the method for choosing tentative vertex parents, and the criterion which determines which vertex to eliminate next in the symbolic (graph) elimination phase of the algorithm. These are of course both heuristics, which are based on emulating case of hierarchical basis for a sequence of refined meshes. At the moment, we do not think our heuristics are optimal, and we expect them to be significantly improved as we gain further insights through the mathematical analysis of the iteration.

Second, we think that our scheme for choosing vertex parents, and that of the classical hierarchical basis multigrid method, are generally appropriate for self adjoint Laplace–like operators. We anticipate that as we gain more experience with the method, variations more suitable for anisotropic problems or convection dominated problems will be developed. For example, one can imagine adding weights and/or directions to the edges in the graph and incorporating this information into the heuristics used to select vertex parents.

REFERENCES

1. R. E. BANK, *Hierarchical preconditioners for elliptic partial differential equations*, in Proceedings of the SERC Summer Workshop in Numerical Analysis, 1992.
2. ———, *PLTMG: A Software Package for Solving Elliptic Partial Differential Equations, Users' Guide 7.0*, Frontiers in Applied Mathematics, SIAM, Philadelphia, 1994.
3. R. E. BANK AND T. F. DUPONT, *Analysis of a two level scheme for solving finite element equations*, Tech. Report CNA-159, Center for Numerical Analysis, University of Texas at Austin, 1980.
4. R. E. BANK, T. F. DUPONT, AND H. YSERENTANT, *The hierarchical basis multigrid method*, Numer. Math., 52 (1988), pp. 427–458.
5. R. E. BANK, A. H. SHERMAN, AND A. WEISER, *Refinement algorithms and data structures*

for regular local mesh refinement, in Scientific Computing (Applications of Mathematics and Computing to the Physical Sciences) (R. S. Stepleman, ed.), North Holland, 1983, pp. 3–17.

6. J. BRAMBLE, J. PASCIAK, AND J. XU, *The analysis of multigrid algorithms with non-imbedded spaces or non-inherited quadratic forms*, Math. Comp., 56 (1991), pp. 1–43.

7. V. EIJKHOUT AND P. VASSILEVSKI, *The role of the strengthened Cauchy-Buniakowskii-Schwarz inequality in multilevel methods*, SIAM Review, 33 (1991), pp. 405–419.

8. A. GEORGE AND J. LIU, *Computer Solution of Large Sparse sitive Definite Systems*, Prentice Hall, Englewood Cliffs, NJ, 1981.

9. W. HACKBUSCH, *Multigrid Methods and Applications*, Springer–Verlag, Berlin, 1985.

10. W. HACKBUSCH AND G. WITTUM, EDS, *Incomplete Decomposition Algorithms, Theory and Applications*, Notes on Numerical Fluid Mechanics, vol. 41, Vieweg, Braunschweig, 1993.

11. R. H. W. HOPPE AND R. KORNHUBER, *Adaptive multilevel methods for obstacle problems*, SIAM J. Numer. Anal., (to appear).

12. R. KORNHUBER, *Monotone multigrid methods for variational inequalities I*, Numer. Math., (to appear).

13. D. J. ROSE, *A graph theoretic study of the numerical solution of sparse positive definite systems*, in Graph Theory and Computing, Academic Press, New York, 1972.

14. J. XU, *Theory of Multilevel Methods*, PhD thesis, Cornell University Report AM-48, 1989.

15. ———, *Iterative methods by space decomposition and subspace correction*, SIAM Review, 34(1992), pp. 581–613.

16. H. YSERENTANT, *On the multi-level splitting of finite element spaces*, Numer. Math., 49 (1986), pp. 379–412.

17. S. ZHANG, *Multilevel Iterative Techniques*, PhD thesis, Pennsylvania State University, Department of Mathematics Report 88020, 1988.

DEPARTMENT OF MATHEMATICS, UNIVERSITY OF CALIFORNIA AT SAN DIEGO, LA JOLLA, CA 92093.
E-mail address: rcb@sdna1.ucsd.edu

DEPARTMENT OF MATHEMATICS, PENN STATE UNIVERSITY, UNIVERSITY PARK, PA 16802.
E-mail address: xu@math.psu.edu

Contemporary Mathematics
Volume 180, 1994

Domain Decomposition and Multigrid Algorithms
for Elliptic Problems on Unstructured Meshes

TONY F. CHAN AND BARRY F. SMITH

ABSTRACT. Multigrid and domain decomposition methods have proven to be versatile methods for the iterative solution of linear and nonlinear systems of equations arising from the discretization of PDEs. The efficiency of these methods derives from the use of a grid hierarchy. In some applications to problems on unstructured grids, however, no natural multilevel structure of the grid is available and thus must be generated as part of the solution procedure.

In this paper, we consider the problem of generating a multilevel grid hierarchy when only a fine, unstructured grid is given. We restrict attention to problems in two dimensions. Our techniques generate a sequence of coarser grids by first forming a maximal independent set of the graph of the grid or its dual and then applying a Cavendish type algorithm to form the coarser triangulation. Iterates on the different levels are combined using standard interpolation and restriction operators. Numerical tests indicate that convergence using this approach can be as fast as standard multigrid and domain decomposition methods on a structured mesh.

1. Introduction

Recently, unstructured meshes have become quite popular in large scale scientific computing [2], [10]. They have the advantage over structured meshes of the extra flexibility in adapting efficiently to complicated geometries and to rapid changes in the solution. However, this flexibility may come with a price.

1991 *Mathematics Subject Classification.* Primary 65N30; Secondary 65F10.

The authors were supported in part by the National Science Foundation under contract ASC 92-01266, the Army Research Office under contract DAAL03-91-G-0150, and subcontract DAAL03-91-C-0047, and ONR under contract ONR-N00014-92-J-1890. Part of this work was performed while the first author was visiting the Computer Science Department of the Chinese University of Hong Kong.

The final version of this paper will be submitted for publication elsewhere.

Traditional solvers which exploit the regularity of the mesh may become less efficient on an unstructured mesh. Moreover, vectorization and parallelization may become more problematic. Thus, there is a need to adapt and modify current solution techniques for structured meshes so that they can run as efficiently on unstructured meshes.

In this paper, we will present some domain decomposition (DD) and multigrid (MG) methods for solving elliptic problems on unstructured triangular meshes in two space dimensions. These are among the most efficient algorithms for solving elliptic problems. The application of multigrid methods to unstructured grid problems have received some attention; see for example [10] and references therein. There has been relatively little work on domain decomposition methods for unstructured grid problems. Cai and Saad [4] considered overlapping domain decomposition methods for general sparse matrices which in principle can be applied to the stiffness matrices arising from discretizations of elliptic problems on unstructured grids. However, if a coarse grid is to be used (often necessary for fast convergence), it cannot be deduced from the algebraic structure of the sparse matrix alone and geometric information about the coarse grid and the associated interpolation operators must be supplied.

For multigrid and domain decomposition algorithms, a hierarchy of grids, together with the associated interpolation and restriction operators, is needed. For structured meshes, this grid hierarchy is naturally available and is indeed exploited in these algorithms. For an unstructured mesh, however, the coarser grids may not be given. Thus, a procedure is needed that generates this grid hierarchy, as well as the associated interpolation and restriction operators. One approach is to generate the coarser meshes independently, using a mesh generator, possibly the one which generated the fine mesh in the first place. This approach has been used by Mavriplis [10], who constructed multigrid algorithms for the Navier-Stokes equations on unstructured meshes in two and three space dimensions. Another approach is to generate the grid hierarchy automatically and directly from the given unstructured fine grid. This approach requires less from the user because only the fine grid, on which the solution is sought, is needed.

In this paper, we will follow the second approach. Our techniques generate a sequence of coarser grids by first forming two maximal independent sets, one for the interior vertices and the other for the boundary vertices, and then applying Cavendish's algorithm [6] to form the coarser triangulation. Thus, in this approach, the coarse mesh vertices form a subset of the fine mesh vertices. We also consider a variant in which this nested property of the vertices does not hold. Iterates on the different levels are combined using standard finite element interpolation and restriction operators. The mesh can be multiply-connected. Numerical tests indicate that convergence using both coarsening approaches can be as fast as standard multigrid and domain decomposition methods on a structured mesh.

For a recently developed convergence theory for two level Schwarz domain decomposition methods using nonnested coarse grids see Cai [**5**] and Chan, Smith and Zou [**8**].

2. Domain Decomposition and Multigrid Algorithms

Thus, we are interested in solving the following elliptic problem:

$$(2.1) \qquad -\nabla \cdot \alpha(x,y)\nabla u = f(x,y), \qquad \alpha(x,y) > 0,$$

on a 2D (not necessarily simply connected) region Ω with appropriate boundary conditions. We assume that Ω is triangulated into a fine grid, which can generally be unstructured and non-quasi-uniform, and a finite element or finite difference method is applied resulting in the algebraic system $Au = b$. The DD and MG algorithms we shall construct are used as preconditioners for A and are used in conjunction with a preconditioned Krylov subspace method.

We first discuss overlapping domain decomposition algorithms (a recent survey can be found in [**7**].) The fine grid Ω is decomposed into p overlapping subdomains $\Omega_i, i = 1, \cdots, p$, either as specified by the user or automatically determined by a mesh partitioning algorithm (in this paper, we will use exclusively the *recursive spectral bisection* (RSB) method of Pothen, Simon and Liou [**11**]). Associated with each Ω_i are restriction and extension operators R_i and R_i^T ($R_i u$ extracts the components of u corresponding to Ω_i and $R_i^T u_i$ is the zero-extension of an iterate u_i on Ω_i to Ω) and the local stiffness matrix $A_i \equiv R_i A R_i^T$. In order to achieve a good convergence rate, we will also use a coarse grid Ω_0. In this paper, we shall assume that Ω_0 is a proper triangular mesh itself which is not necessarily nested to Ω. We construct the associated interpolation operator R_H^T, which maps an iterate u_0 on Ω_0 to Ω, as follows. If a fine grid node lies within a triangle of Ω_0, we use linear interpolation to obtain its value, otherwise we set its value to zero. Once R_H^T is defined, the restriction operator R_H is defined to be its transpose. Finally, we compute the coarse grid stiffness matrix A_H by applying a piecewise linear finite element method to (2.1) on Ω_0. Note that in general $A_H \neq R_H A R_H^T$ due to the non-nestedness of the grids.

With these operators defined, we can now define the additive Schwarz preconditioner (which corresponds to a generalized block Jacobi method) as follows:

$$M_{as}^{-1} = R_H^T A_H^{-1} R_H + \sum_{i=1}^{p} R_i^T A_i^{-1} R_i.$$

Thus, each application of the preconditioner involves restricting the residual vector to each subdomain and performing a subdomain solve. In addition, a weighted restriction of the residual vector is computed on the coarse grid and inverted by a coarse grid solve. These local and coarse solutions are then mapped back onto the fine mesh and added together to obtain the desired result. Multiplicative versions (i.e. Gauss-Seidel) can also be defined analogously given an ordering of the subdomains.

Multilevel preconditioners (including classical multigrid methods) are closely related to domain decomposition methods and their implementations can be treated in the same framework. A grid hierarchy is needed and the associated interpolation and restriction operators can be defined in an analogous way. For example, let the fine grid be level 1 and the coarsest grid level l. Let R_i denote the restriction operator from level 1 to level i and the transpose R_i^T the corresponding interpolation operator. Then an additive multilevel preconditioner can be written in the following form:

$$M_{aml}^{-1} = \sum_{i=1}^{l} R_i^T S_i R_i,$$

where S_i is a "smoother" on level i. For instance, for multi-level diagonal scaling, S_i is simply the inverse of the diagonal of the stiffness matrix on level i. Classical V-cycle MG methods can be viewed as symmetrized multiplicative versions of the above preconditioner. Note that in practice the action of R_i and R_i^T are computed via a recursion using mappings between *adjacent* grid levels.

In our implementation of the domain decomposition algorithms, the coarse grid Ω_0 is obtained by a sequence of recursively applied coarsening steps (see next section), and hence the grid hierarchy is naturally defined for performing the multigrid iteration as well. Of course, this is not the only way to construct the coarse grid for a DD method. For example, one can use the subdomains to directly construct a coarse grid without going through a grid hierarchy. We shall not pursue these other possibilities in this paper.

3. Construction of the Grid Hierarchy

In this section, we will describe our techniques for constructing the coarse grid hierarchy, as well as the associated interpolation and restriction operators, directly from the given unstructured fine mesh. It suffices to describe this for one coarse level because the procedure can be recursively applied to obtain all the coarse meshes.

We shall need the notion of a *maximal independent set* of the vertices of a graph. A subset of vertices V of a graph G is said to be *independent* if no two vertices of V are connected by an edge. V is said to be *maximally independent* if adding any additional vertex to it makes it dependent. Note that maximal independent sets of vertices of a graph are generally not unique.

The procedure has four steps:

(i) Form a maximally independent set of the boundary vertices and from these construct a set of coarse boundary edges,

(ii) Form a maximally independent set of the interior vertices,

(iii) Apply a Cavendish type algorithm [6] to triangulate the resulting collection of coarse boundary edges and coarse interior vertices,

(iv) Construct the interpolation and restriction operators.

Step (i) is fairly straightforward. For each disjoint boundary segment, the boundary vertices are ordered say in a clockwise direction, starting with a random vertex. Then every other vertex is thrown out and the remaining ones are connected with new coarse boundary edges. This forms a coarse representation of the boundary segment. After several coarsenings, one may find that the boundary is no longer qualitatively similar to the original boundary. This may be prevented by simply retaining some of the vertices in the coarse grid boundary that would normally be dropped.

Step (ii) uses a greedy wavefront type algorithm. A random interior vertex is selected for inclusion in the maximally independent set. Then every interior vertex connected to it is eliminated from consideration for inclusion in the maximally independent set. Next, one of the interior vertices connected to the newly eliminated vertices is selected for inclusion and the procedure repeats until all interior vertices have been considered. An algorithm similar to this has been used by Barnard and Simon [1] in designing graph partitioning algorithms. This procedure can be implemented in linear time, i.e. proportional to the total number of interior vertices.

The input to Step (iii) is thus a collection of coarse boundary edges and coarse interior vertices. A version of Cavendish's algorithm [6] is then applied to triangulate this collection. This algorithm is an advancing front technique and "grows" new triangles from those already built by selecting an interior vertex to be "mated" to an existing edge. In doing so, it tries to optimize the aspect ratio of the new triangle formed, preferring those that are close to being equilateral. It is possible to implement this algorithm in linear time, i.e. proportional to the number of interior vertices, but our current implementation is not optimal.

Finally, in Step (iv), the interpolation operator is constructed in the form of a sparse matrix and stored. To determine the entries of this interpolation matrix, the coarse triangles are taken in sequence and the entries corresponding to all the fine grid vertices within the coarse triangle are then computed using the standard piecewise linear interpolation. This procedure can also be implemented in linear time because the fine grid triangles close to the vertices of the coarse triangle (which are also fine grid vertices as well) can be found by a local search. We emphasize that this is not possible if the coarser grids are generated completely independently. The restriction matrix is then just the transpose of the interpolation matrix. Clearly higher order finite elements may also be used; then the interpolation would be piecewise polynomial and could still be calculated in a local manner and hence remain a linear time algorithm.

In the alternative variant only Step (ii) is changed. We consider a candidate coarse vertex at the center of each element. Those that are adjacent to

the selected boundary nodes are then eliminated. We then construct a maximal
independent set of the remaining candidate vertices using the same approach as
indicated above. This procedure is equivalent to calculating a maximal independent set of the *dual* graph of the mesh. It is important to note that the coarse
grid vertices generated in this manner will not lie in the same location as any
fine grid nodes. In our experiments on a few sample grids, the number of coarse
grid nodes generated using this alternative approach is slightly more than that
generated using the first approach outlined.

4. Numerical Results

All the numerical experiments were performed using the Portable, Extensible
Toolkit for Scientific Computation (PETSc) of Gropp and Smith, [9] running on
a Sun SPARC 10.

We will report numerical results for solving the Poisson equation on three
different unstructured triangular meshes. All meshes are enclosed in the unit
square. Two kinds of boundary conditions are used: (1) homogeneous Dirichlet,
or (2) a *mixed* condition: if $x > .2$, a homogeneous Neumann boundary condition
is imposed, otherwise homogeneous Dirichlet is imposed. We use piecewise linear
finite elements for the discretization and we solve the resulting systems of linear
equations by either a V-cycle multigrid method (with a pointwise Gauss Seidel
smoother, using 2 pre and 2 post smoothing sweeps per level) or an overlapping
Schwarz domain decomposition method. In all cases, the discrete right hand side
is chosen to be a vector of all 1's and the initial iterate set to zero. Both the
MG and DD methods are used as preconditioners, with full GMRES [12] as an
outer accelerator. The iteration is stopped when the l_2 norm of the residual has
been reduced by a factor of 10^{-6}. Our goal is to compare the performance of
our versions of domain decomposition and multigrid algorithms on unstructured
meshes to that of the same algorithms on similar *structured* meshes. For this
purpose, we also use two structured meshes in our experiments (a uniform mesh
on a square and on an annulus).

Table 1 shows the number of MG iterations for the *Eppstein* mesh, [3], shown
in Figure 1, a relative small quasi-uniform unstructured mesh on the unit square.
Figures 2 and 3 show the coarser meshes using regular and dual graph coarsening.
These results should be compared to those for a uniform square mesh on the unit
square in Table 2, because the two meshes are topologically similar. We see that
although the performance of our MG algorithm on the unstructured mesh is
slightly higher than that on the structured square mesh, its performance is quite
satisfactory, for both types of boundary conditions.

Next, we look at a more realistic mesh, the *Airfoil* mesh, (from T. Barth
and D. Jesperson of NASA Ames), shown in Figure 4. The coarse meshes are
shown in Figures 5, 6 and 7. One may note that several poorly shaped triangles

FIGURE 1. The *Eppstein* mesh: 547 nodes

FIGURE 2. The *Eppstein* mesh: level 2. regular coarsening (left) and dual graph coarsening (right)

FIGURE 3. The *Eppstein* mesh: level 3. regular coarsening (left) and dual graph coarsening (right)

TABLE 1. MG iterations for the *Eppstein* mesh, 547 nodes

	Regular coarsening		Dual graph coarsening	
MG Levels	Dir. B.C.	Mixed B.C.	Dir. B.C.	Mixed B.C.
2	4	4	3	4
3	4	5	4	6

TABLE 2. MG iterations for the uniform *Square* mesh, 4225 nodes

MG Levels	Nodes	Dir. B.C.	Mixed B.C.
2	1089	3	3
3	289	3	3
4	81	3	3

are generated on the coarsest grid. These do not seem to seriously effect the convergence rate. In theory, these bad elements could be adjusted during a "cleanup" pass over the mesh, after the Cavendish algorithm was applied. We see that the performance for the Dirichlet boundary condition cases, given in Table 3, are quite comparable to that for the *Eppstein* mesh but is noticeably worse for the mixed boundary conditions. Note there is no difference in the performance for both types of coarsening. The deterioration in performance for the mixed boundary condition case is due to the fact that the coarser grid domains do not completely cover the portion of the fine grid boundary on which Neumann boundary conditions are applied. In Chan, Smith and Zou [8] it is shown that to obtain an optimal convergence rate the coarser grid must completely cover the Neumann part of the fine grid boundary. We also compare the performance to that on a quasi-uniform mesh on an annulus region; see Figure 4 and Table 4. Due to our treatment of the curved boundary, the fine grid boundary is not covered by the coarser grids and hence poor convergence results. Overall, the performance of our MG algorithm on the unstructured *Airfoil* mesh is comparable or better than on the *Annulus* mesh.

In Table 5, we show results for a larger unstructured mesh around an airfoil, namely the *Barth* mesh, (from T. Barth of NASA Ames) shown in Figure 9. The levels 2 and 4 coarse meshes are shown in Figure 10 (not all the grid points are shown). We only include the Dirichlet boundary condition results. We can observe that the MG performance is quite comparable to the other unstructured meshes (i.e. *Airfoil, Eppstein*) and the structured meshes (i.e. *Square, Annulus*).

TABLE 3. MG iterations for the *Airfoil* mesh, 4253 nodes

	Regular coarsening			Dual graph coarsening		
MG Levels	Nodes	Dir. B.C.	Mixed B.C.	Nodes	Dir. B.C.	Mixed B.C.
2	1180	4	8	1507	4	8
3	518	4	9	328	4	9
4	89	4	10	171	5	10

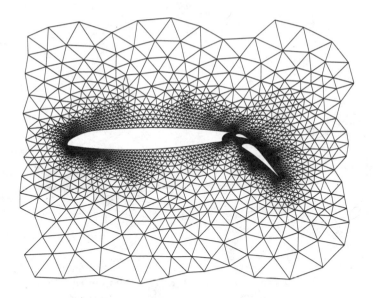

FIGURE 4. The *Airfoil* mesh: 4253 nodes

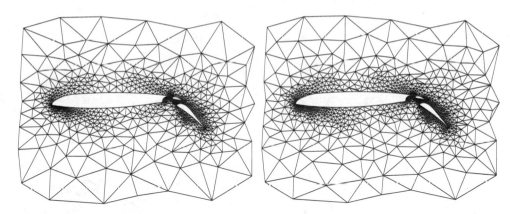

FIGURE 5. The *Airfoil* mesh: Level 2. Regular (left) and dual graph coarsening (right).

TABLE 4. MG iterations for the *Annulus* mesh, 2176 nodes

MG Levels	Nodes	Dir. B.C.	Mixed B.C.
2	576	4	18
3	160	5	18
4	48	5	18

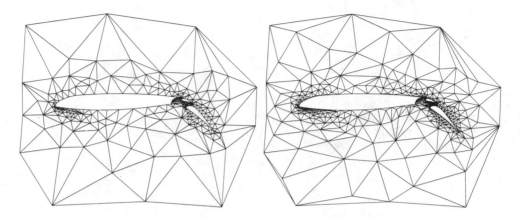

FIGURE 6. The *Airfoil* mesh: Level 3. Regular (left) and dual graph coarsening (right).

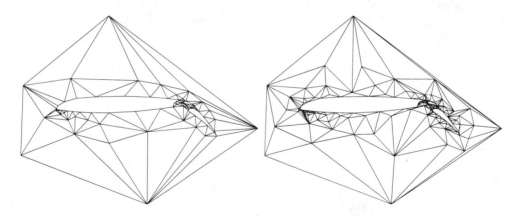

FIGURE 7. The *Airfoil* mesh: Level 4. Regular (left) and dual graph coarsening (right).

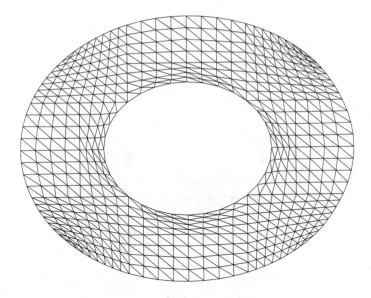

FIGURE 8. The *Annulus* mesh: 2176 nodes

TABLE 5. MG iterations for the *Barth* mesh, 6691 nodes

MG Levels	Regular coarsening		Dual graph coarsening	
	Nodes	Dir. B.C.	Nodes	Dir. B.C.
2	1614	5	1810	5
3	405	6	574	6
4	112	7	189	6

Again, both coarsening strategies work equally well.

Finally, we show in Table 6 the results for the multiplicative version of our domain decomposition algorithm on the *Airfoil* mesh. The column labeled *overlap* refers to the number of fine grid elements that are extended from each subdomain into the interior of its neighbors. Thus, an overlap of 0 means there is no overlap at all. The column labeled *Level of coarse grid* refers to which level of the grid hierarchy is used as the coarse grid in the DD algorithm. The 16 subdomains computed by the Recursive Spectral Bisection method are shown in Figure 11. We can make several observations from the results. First, the use of a coarse grid reduces the number of iterations significantly. Second, the use of some overlap is very cost effective but the number of iterations levels off quickly as the overlap increases. These results for overlapping Schwarz are also very similar to those obtained for structured grids.

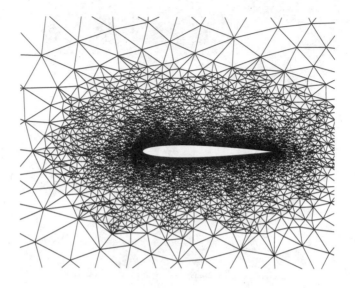

FIGURE 9. The *Barth* mesh: 6691 nodes

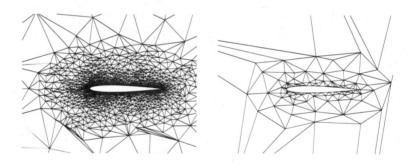

FIGURE 10. The *Barth* mesh: Level 2, 1614 nodes (left) Level 4, 112 nodes (right)

TABLE 6. Multiplicative DD iterations for the *Airfoil* mesh. 16 Subdomains

Overlap (no. elements)	Level of coarse grid	Regular coarsening	Dual graph coarsening
0	None	56	56
0	4	21	22
0	3	15	12
1	None	16	16
1	4	10	10
1	3	7	7
2	None	14	14
2	4	8	8
2	3	5	5

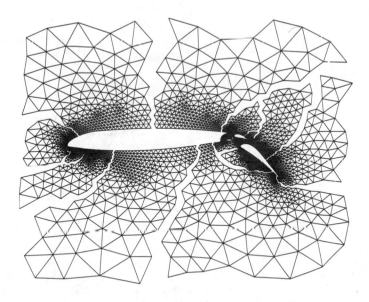

FIGURE 11. The *Airfoil* mesh: 16 subdomains computed by RSB

5. Summary

In summary, we have constructed domain decomposition and multigrid algorithms for solving elliptic problems on general unstructured meshes, which in our limited experience perform nearly as well as these algorithms would perform on similar structured meshes. Only the fine mesh is needed and all auxiliary components of the algorithms, such as the coarse grid hierarchy, the interpolation operators, and the domain partitioning, are computed automatically. The algorithms can in principle be extended to three space dimensions and to indefinite, non-self-adjoint and higher order problems.

Acknowledgements: We thank Mr. Nip Chun-kit of the Chinese University of Hong Kong for providing us with a Matlab implementation of the Cavendish algorithm which helped us in developing the C version in PETSc. We also thank Horst Simon and John Gilbert for providing several of our test meshes. The first author also thanks Prof. Jiachang Sun of the Computing Center of Academia Sinica for many helpful discussions on the subject.

REFERENCES

1. S.T. Barnard and H.D. Simon. A fast multilevel implementation of recursive spectral bisection. *Proc. 6th SIAM Conf. Parallel Proc. for Sci. Comp.*, pp. 711–718, 1993.
2. H. Deconinck and T.J. Barth. Special course on unstructured grid methods for advection dominated flows. *Lecture Series, Von Karman Inst., Belgium*, AGARD-R-787, March 1992.
3. M. Bern, D. Eppstein and J. Gilbert. Provably good mesh generation. *31st Annual Symp. on Foundations of Computer Science*, IEEE, pp. 231-241, 1990.
4. X.-C. Cai and Y. Saad. Overlapping domain decomposition algorithms for general sparse matrices. Technical Report 93-027, Univ. of Minn., AHPCRC, March 1993.
5. X.-C. Cai. A non-nested coarse space for Schwarz type domain decomposition methods. Technical Report: Department of Computer Science, University of Colorado at Boulder, November 1993.
6. J. C. Cavendish. Automatic triangulation of arbitrary planar domains for the finite element method. *Int'l J. Numer. Meth. Engineering*, 8:679–696, 1974.
7. T. F. Chan and T. Mathew. Domain decomposition algorithms. *Acta Numerica*, pages 1–83, 1994.
8. T. F. Chan and B. F. Smith and J. Zou. Overlapping Schwarz methods on unstructured grids using non-matching coarse grids. Technical Report: CAM 94-08, UCLA Department of Mathematics, Febuary 1994.
9. W. D. Gropp and Barry F. Smith. Portable, Extensible Toolkit for Scientific Computation (PETSc), Available via anonymous ftp at `info.mcs.anl.gov` in the directory `pub/pdetools`.
10. D. J. Mavriplis. Unstructured mesh algorithms for aerodynamic calculations. Technical Report 92-35, ICASE, NASA Langley, Virginia, July 1992.
11. A. Pothen, H. Simon, and K. P. Liou. Partitioning sparse matrices with eigenvectors of graphs. *SIAM J. Mat. Anal. Appl.*, 11:430–452, 1990.
12. Y. Saad and M. H. Schultz. GMRES: a generalized minimal residual algorithm for solving nonsymmetric linear systems. *SIAM J. Sci. Statist. Comput.*, 7:856–869, 1986.

DEPARTMENT OF MATHEMATICS, UNIVERSITY OF CALIFORNIA, LOS ANGELES, CA 90024-1555

E-mail address: chan@math.ucla.edu

DEPARTMENT OF MATHEMATICS, UNIVERSITY OF CALIFORNIA, LOS ANGELES, CA 90024-1555

E-mail address: bsmith@math.ucla.edu

Contemporary Mathematics
Volume 180, 1994

Two-Grid Methods for Mixed Finite Element Approximations of Nonlinear Parabolic Equations

CLINT N. DAWSON AND MARY F. WHEELER

ABSTRACT. Mixed finite element approximation of nonlinear parabolic equations is discussed. The equation considered is a prototype of a model that arises in flow through porous media. A two-grid approximation scheme is developed and analyzed for implicit time discretizations. In this approach, the full nonlinear system is solved on a "coarse" grid of size H. The nonlinearities are expanded about the coarse grid solution, and the resulting linear but nonsymmetric system is solved on a "fine" grid of size h. Error estimates are derived which demonstrate that the error is $\mathcal{O}(h^{k+1} + H^{2(k+1)-d/2} + \triangle t)$, where k is the degree of the approximating space for the primary variable and d is spatial dimension, with $k \geq 1$ for $d \geq 2$. For the RT_0 space ($k = 0$) on rectangular domains, we present a modified scheme for treating the coarse grid problem. Here we show that the error is $\mathcal{O}(h + H^{3-d/2} + \triangle t)$. The above estimates are useful for determining an appropriate H for the coarse grid problem.

1. Introduction

In this paper, we consider the mixed finite element approximation of ρ satisfying

$$(1) \qquad \frac{\partial \rho}{\partial t} - \nabla \cdot (K(\rho)\nabla \rho) = f(x,t), \quad \text{on } \Omega \times (0, T],$$

with initial condition

$$(2) \qquad \rho(x, 0) = \rho^0(x), \quad \text{on } \Omega,$$

1991 *Mathematics Subject Classification.* Primary 65M60, 65P05 .
Key words and phrases. Parabolic equations, mixed finite element methods.
This work supported by grants from the Department of Energy and the State of Texas.
This paper will not be submitted for publication elsewhere.

and boundary condition

(3) $$(K(\rho)\nabla\rho) \cdot \nu = 0, \text{ on } \partial\Omega \times (0, T],$$

where $\Omega \subset IR^d$, $d \leq 3$, is a bounded, convex domain with C^1 boundary $\partial\Omega$, ν is the unit exterior normal to $\partial\Omega$, and K is a symmetric positive definite tensor and $K : \Omega \times IR \to IR^{d \times d}$. We further assume that there exist positive constants K_* and K^* such that for $z \in IR^d$

(4) $$K_* \parallel z \parallel^2 \leq z^t K(x, s) z \leq K^* \parallel z \parallel^2, \text{ for } x \in \Omega \text{ and } s \in IR,$$

and that each element of K is twice continuously differentiable with derivatives up to second order bounded above by K^*.

Equation (1) is a simplification of a nonlinear parabolic equation (Richard's Equation) which arises in the modeling of two-phase flow in porous media, where the two phases are air and water [3]. The major difference between Richards' Equation and (1) is in the time derivative term, where ρ is replaced by a nonlinear function $\theta(\rho)$. The function $\theta'(\rho)$ may be zero and this complicates the analysis; however, this difficulty may be handled by techniques given in [1]. For brevity, we do not consider this generalization here.

Let $L^p(\Omega)$ for $p \geq 1$ denote the standard Banach space defined on Ω, with norm $\parallel \cdot \parallel_p$. We shall denote by $W^{m,p}(\Omega)$ the standard Sobolev space (m partial derivatives in L^p) with norm $\parallel \cdot \parallel_{m,p}$. The $W^{m,p}$ norm for vectors will be understood to be for each component. When $p = 2$ we omit the subscript on the norms.

Let $(\cdot, \cdot)$ denote the $L^2(\Omega)$ inner product, scalar and vector. Let $V = H(div; \Omega) = \{v \in (L^2(\Omega))^d : \nabla \cdot v \in L^2(\Omega)\}$, $\tilde{V} = V \cap \{v \cdot \nu = 0\}$, and $W = L^2(\Omega)$.

As in [2] we define the mixed weak form of (1) as $(\rho, \Gamma, \Psi) \in W \times V \times \tilde{V}$ satisfying

(5) $$(\frac{\partial\rho}{\partial t}, w) + (\nabla \cdot \Psi, w) = (f, w), \quad w \in W,$$

(6) $$(\Gamma, v) - (\rho, \nabla \cdot v) = 0, \quad v \in \tilde{V},$$

(7) $$(\Psi, v) = (K(\rho)\Gamma, v), \quad v \in V.$$

We consider finite-dimensional subspaces W_h^k, $\tilde{V}_h^k$, and V_h^k of W, $\tilde{V}$, and V, respectively (they may be Raviart-Thomas-Nedelec spaces of index k RT_k [9, 10] or Brezzi-Douglas-Marini spaces of index k [4], for instance) associated with a quasi-uniform family of polygonal decompositions of Ω by triangles (tetrahedra) or bricks with diameter h. For simplicity in the discussion below, we will drop the superscript k.

Let $N > 0$ and $\Delta t = T/N$, $t^n = n\Delta t$, and $\varphi^n = \varphi(\cdot, t^n)$. In our analysis we shall use the following norms: for X a normed vector space defined on Ω,

$$\|\varphi\|_{l^2((0,T);X)} \equiv (\sum_{n=1}^{N} \Delta t \|\varphi^n\|_X^2)^{\frac{1}{2}},$$

$$\|\varphi\|_{l^\infty((0,T);X)} \equiv \max_{1 \leq n \leq N} \|\varphi^n\|_X,$$

and

$$\|\varphi\|_{L^2((0,T);X)} \equiv (\int_0^T \|\varphi^n(\cdot, t)\|_X^2 dt)^{\frac{1}{2}}.$$

The discrete time mixed finite element approximation to (5)-(7) is defined as follows: Given $(\rho_h^0, \Gamma_h^0, \Psi_h^0) \in W_h \times V_h \times \tilde{V}_h$, for $n = 1, \ldots, N$, let $(\rho_h^n, \Gamma_h^n, \Psi_h^n) \in W_h \times V_h \times \tilde{V}_h$, satisfy

(8) $(\dfrac{\rho_h^n - \rho_h^{n-1}}{\Delta t}, w_h) + (\nabla \cdot \Psi_h^n, w_h) = (f^n, w_h), \quad w_h \in W_h,$

(9) $(\Gamma_h^n, v_h) - (\rho_h^n, \nabla \cdot v_h) = 0, \quad v_h \in \tilde{V}_h,$

(10) $(\Psi_h^n, v_h) = (K(\rho_h^n)\Gamma_h^n, v_h), \quad v_h \in V_h.$

This procedure is based on a modification of the standard mixed finite element procedure and was introduced for linear elliptic problems in [2]. For brevity, a proof of existence and uniqueness of the solution to the nonlinear algebraic system (8)-(10) is not given; the reader is referred to [7] since the proof closely follows the argument of Milner for standard mixed method approximations to nonlinear elliptic problems.

Instead of solving (8)–(10) each time step for a fine mesh h we shall also consider two two-level procedures, both of which involve solving the nonlinear problem only a coarse grid of diameter $H >> h$. This work has been motivated by the work of Xu [12] for Galerkin procedures applied to nonlinear elliptic equations. In the simplest scheme, the fine mesh discrete problem is linearized by evaluating $K(\rho)$ at the coarse mesh solution ρ_H. In the second scheme $K(\rho)$ is approximated by a first-order Taylor expansion about ρ_H, a correction by one Newton iteration on the fine space. We shall show that the simplest scheme yields no improvement in accuracy over the coarse grid approximation. The second scheme; however, yields substantial improvement and is a viable computational approach.

This paper is divided into five sections. Notation and approximation assumptions are given and error estimates are derived for (8)–(10) in Section 2. The two two-level schemes are defined and estimates derived for both in Section 3. A new procedure based on postprocessing the coarse grid solution and then applying a Newton correction is defined and analyzed in Section 4 for the space RT_0. In Section 5, we give conclusions and extensions.

2. Notation and Approximation Results

We shall employ several projection operators.

Denote by P the L^2 projection operator. More precisely, let $P_{W_h} : L^2(\Omega) \to W_h$ and $P_{V_h} : (L^2(\Omega))^d \to V_h$, where for $g \in L^2(\Omega)$,

$$(11) \qquad (P_{W_h} g, w_h) = (g, w_h), \quad w_h \in W_h,$$

and for $q \in (L^2(\Omega))^d$,

$$(12) \qquad (P_{V_h} q, v_h) = (q, v_h), \quad v_h \in V_h.$$

For convenience we shall set $\hat{g}_h = Pg$, where P is understood to be P_{W_h} or P_{V_h} depending on whether g is a scalar or vector quantity.

We shall use the well known Π_h projection for mixed finite element approximation spaces. We shall assume that there exists a projection operator $\Pi_h : (H^1(\Omega))^d \to V_h$ such that

$$(13) \qquad (\nabla \cdot \Pi_h g, w_h) = (\nabla \cdot g, w_h), \quad w_h \in W_h.$$

We assume that

$$\nabla \cdot \Pi_h = P_{W_h}, \quad \nabla \cdot : V \overset{\text{onto}}{\to} W_h.$$

We also assume that the following approximation properties hold.

Approximation Properties A:
- There exists a positive constant Q independent of h such that

$$(14) \qquad \|u - \Pi_h u\| \leq Q\|u\|_r h^r, \quad 1 \leq r \leq k+1,$$

$$(15) \qquad \|p - \hat{p}_h\| \leq Q\|p\|_r h^r, \quad 0 \leq r \leq k+1,$$

and

$$(16) \qquad \|p - \hat{p}_h\|_\infty \leq Q\|p\|_\infty h^r, \quad 0 \leq r \leq k+1.$$

- Given $\Upsilon \in (W^{1,\infty}(\Omega))^d$ there exists a $\Upsilon_h \in V_h$ such that

$$(17) \qquad \|\Upsilon - \Upsilon_h\| \leq Qh\|\Upsilon\|_{1,\infty}.$$

Finally, we assume the following inverse property on V_h holds, namely, for $v_h \in V_h$,

$$(18) \qquad \|v_h\|_\infty \leq \|v_h\| h^{-d/2}.$$

These assumptions are known to hold for the Raviart-Thomas-Nedelec and Brezzi-Douglas-Marini spaces mentioned above.

In this paper C shall denote a generic constant. For convenience we will also assume that the solution is smooth and that the maximum index of convergence $k+1$ is attained.

We rewrite (5)–(7) with $t = t^n$. Using the definition of $\hat{\rho}$, the Π_h projection, and the assumption that $\nabla \cdot V_h \subset W_h$, we deduce that

$$(19)\,(\frac{\hat{\rho}_h^n - \hat{\rho}_h^{n-1}}{\Delta t}, w_h) + (\nabla \cdot \Pi_h \Psi^n, w_h) = (f^n, w_h) + (\epsilon^n, w_h), \quad w_h \in W_h,$$

$$(20) \qquad (\hat{\Gamma}_h^n, v_h) - (\hat{\rho}_h^n, \nabla \cdot v_h) = 0, \quad v_h \in \tilde{V}_h,$$

and

$$(21) \qquad (\Pi_h \Psi^n, v_h) = (\Pi_h \Psi^n - \Psi^n, v_h) + (K(\rho^n)\Gamma^n, v_h), \quad v_h \in V_h,$$

where ϵ^n is a time truncation error of order Δt.

We now derive an error estimate for the algorithm (8)-(10).

Set $\mu^n = \hat{\rho}_h^n - \rho_h^n$, $\zeta^n = \hat{\Gamma}_h^n - \Gamma_h^n$, and $\chi^n = \Pi_h \Psi^n - \Psi_h^n$. Subtracting (8) from (19), (9) from (20), and (10) from (21) and in the resulting equations using the test functions $w_h = \mu^n$, $v_h = \chi^n$, and $v_h = \zeta^n$ in the first, second, and third equations, respectively, we obtain the error equations

$$(22) \qquad (\frac{\mu^n - \mu^{n-1}}{\Delta t}, \mu^n) + (\nabla \cdot \chi^n, \mu^n) = (\epsilon^n, \mu^n),$$

$$(23) \qquad (\zeta^n, \chi^n) = (\mu^n, \nabla \cdot \chi^n),$$

and

$$
\begin{aligned}
(\chi^n, \zeta^n) = {} & (\Pi_h \Psi^n - \Psi^n, \zeta^n) + (K(\rho_h^n)\zeta^n, \zeta^n) \\
& - ((K(\rho_h^n) - K(\rho^n))\Gamma^n, \zeta^n) \\
& + (K(\rho_h^n)(\Gamma^n - \hat{\Gamma}_h^n), \zeta^n).
\end{aligned}
$$

(24)

Combining (22) - (24), applying smoothness and boundedness assumptions on $K(\rho)$, and Hölder's inequality we obtain

$$
\begin{aligned}
\frac{1}{2\triangle t} & [\|\mu^n\|^2 - \|\mu^{n-1}\|^2] + \|K(\rho_h^n)^{1/2}\zeta^n\|^2 \\
& \leq (\frac{\mu^n - \mu^{n-1}}{\triangle t}, \mu^n) + \|K(\rho_h^n)^{1/2}\zeta^n\|^2 \\
& \leq \frac{1}{2}\|\mu^n\|^2 + \frac{1}{2}\|\epsilon^n\|^2 + (\Pi_h\Psi^n - \Psi^n, \zeta^n) \\
& \quad + ((K(\rho_h^n) - K(\rho^n))\Gamma^n, \zeta^n) + (K(\rho_h^n)(\Gamma^n - \hat{\Gamma}_h^n), \zeta^n) \\
& \leq \frac{1}{2}\|\mu^n\|^2 + \frac{1}{2}\|\epsilon^n\|^2 + \delta\|\zeta^n\|^2 \\
& \quad + C\left[\|\Pi_h\Psi^n - \Psi^n\|^2 + \|\hat{\Gamma}_h^n - \Gamma^n\|^2 + (\|\rho^n - \hat{\rho}_h^n\| \|\Gamma^n\|_\infty)^2\right],
\end{aligned}
$$

(25)

where $\delta \leq K_*$. Multiplying (26) by $\triangle t$ and summing on $n, n = 1, 2, ..., N$ and applying Gronwall's Lemma, we see that

$$
\begin{aligned}
\|\mu^N\|^2 & + \sum_{n=1}^N \triangle t \|K(\rho_h^n)^{1/2}\zeta^n\|^2 \\
& \leq C \sum_{n=1}^N \triangle t \left[\|\Pi_h\Psi^n - \Psi^n\|^2 + \|\hat{\Gamma}_h^n - \Gamma^n\|^2 + (\|\rho^n - \hat{\rho}_h^n\| \|\Gamma^n\|_\infty)^2\right] \\
& \quad + \|\mu^0\|^2 + C\triangle t^2.
\end{aligned}
$$

(26)

Using (27), Approximation Properties A, and the triangle inequality we deduce the following theroem:

THEOREM 1. *Let $V_h = V_h^k$, $\tilde{V}_h = \tilde{V}_h^k$, and $W_h = W_h^k$ and define the triplet $(\rho_h^n, \Gamma_h^n, \Psi_h^n) \in W_h^k \times V_h^k \times \tilde{V}_h^k$ by (8) - (10), for $n \geq 1$. Assume that the Approximation Properties A hold. Take $\rho_h^0 = \hat{\rho}_h(\cdot, 0)$. Then there exists a positive constant C, independent of h such that*

$$
(27) \quad \|\rho_h^N - \rho^N\| + (\sum_{n=1}^N \triangle t \|K(\rho_h^n)^{1/2}(\Gamma_h^n - \Gamma^n)\|^2)^{\frac{1}{2}} \leq C(h^{k+1} + \triangle t).
$$

3. Two-Level Schemes

We first consider a scheme based on a correction by one Newton iteration on the fine space. More precisely, we solve (8)–(10) on a coarse mesh with $h = H$

and then solve the following linear system for $(\tilde{\rho}_h^n, \tilde{\Gamma}_h^n, \tilde{\Psi}_h^n) \in W_h \times V_h \times \tilde{V}_h$, for $n \geq 1$ and $h << H$:

$$(28) \qquad (\frac{\tilde{\rho}_h^n - \tilde{\rho}_h^{n-1}}{\Delta t}, w_h) + (\nabla \cdot \tilde{\Psi}_h^n, w_h) = (f^n, w_h), \quad w_h \in W_h,$$

$$(29) \qquad (\tilde{\Gamma}_h^n, v_h) = (\tilde{\rho}_h^n, \nabla \cdot v_h), \quad v_h \in \tilde{V}_h,$$

$$(30) \quad (\tilde{\Psi}_h^n, v_h) = (K(\rho_H^n)\tilde{\Gamma}_h^n, v_h) + (K'(\rho_H^n)\Gamma_H^n(\tilde{\rho}_h^n - \rho_H^n), v_h), \quad v_h \in V_h.$$

Equation (30) is motivated by the Taylor expansion

$$K(\rho^n)\Gamma^n = K(\rho_H^n)\Gamma^n + K'(\rho_H^n)\Gamma^n(\rho^n - \rho_H^N) + \frac{K''(\alpha^n)}{2}\Gamma^n(\rho^n - \rho_H^n)^2,$$

for some α^n between ρ^n and ρ_H^n.

We now derive an estimate for this two-level scheme. Set $\xi^n = \hat{\rho}_h^n - \tilde{\rho}_h^n$, $v^n = \hat{\Gamma}_h^n - \tilde{\Gamma}_h^n$, and $\theta^n = \Pi_h \Psi^n - \tilde{\Psi}_h^n$.

Subtracting (28) from (19), (29) from (20), and (30) from (21), letting $w_h = \xi^n$, $v_h = \theta^n$, and $v_h = v^n$, in the first, second, and third equations respectively, we obtain

$$(31) \qquad (\frac{\xi^n - \xi^{n-1}}{\Delta t}, \xi^n) + (\nabla \cdot \theta^n, \xi^n) = (\epsilon^n, \xi^n)$$

$$(32) \qquad (v^n, \theta^n) = (\xi^n, \nabla \cdot \theta^n),$$

and

$$(33) \quad (\theta^n, v^n) = (\Pi_h \Psi^n - \Psi^n, v^n) - (K(\rho^n)(\hat{\Gamma}_h^n - \Gamma^n), v^n)$$
$$+ (K(\rho_H^n)v^n, v^n) + ((\xi^n + (\rho^n - \hat{\rho}_h^n))K'(\rho_H^n)\Gamma_H^n, v^n)$$
$$+ ((K'(\rho_H^n)(\rho^n - \rho_H^n)(\hat{\Gamma}_h^n - \Gamma_H^n), v^n)$$
$$+ \frac{1}{2}((\rho^n - \rho_H^n)^2 K''(\alpha^n)\hat{\Gamma}_h^n, v^n).$$

Combining (31)–(34) and using Hölder's inequality, we deduce that

$$\frac{1}{2\Delta t}[\|\xi^n\|^2 - \|\xi^{n-1}\|^2] + \|K(\rho_H^n)^{1/2}v^n\|^2$$

$$(34) \qquad \leq |(\xi^n, \epsilon^n)| + \delta\|v^n\|^2$$
$$+ C\|v^n\| \|\xi^n\| (\|\Gamma^n\|_\infty + \|(\Gamma_H^n - \Gamma^n)\|_\infty)$$
$$+ C \left[\|\Pi_h \Psi^n - \Psi^n\|^2 + \|\hat{\Gamma}_h^n - \Gamma^n\|^2 \right.$$
$$+ \|(\rho^n - \hat{\rho}_h^n)\Gamma_H^n\|^2$$
$$+ \|(\rho^n - \rho_H^n)(\hat{\Gamma}_h^n - \Gamma_H^n)\|^2$$
$$\left. + \|(\rho^n - \rho_H^n)^2\hat{\Gamma}_h^n\|^2 \right],$$

where $\delta \leq K_*/2$, and $C = C(K^*, K_*)$.

We now choose $\hat{N}$ to be the index corresponding to the maximum of $\|\xi^n\|$ for $n = 1, \ldots, N$. Multiplying (35) by $\triangle t$ and summing from $n = 1$ to $M, M \leq N$, using Approximation Properties A and Theorem 1, we see that

$$\|\xi^M\|^2 - \|\xi^0\|^2 + \sum_{n=1}^{M} \triangle t \|K(\rho_H^n)^{1/2} v^n\|^2$$

$$(35) \quad \leq C\triangle t^2 \|\rho_{tt}\|^2_{L^2((0,T),L^2(\Omega))} + \delta \sum_{n=1}^{M} \triangle t \|v^n\|^2 + T^* + T^{**}$$

$$+ C \left[\sum_{n=1}^{M} \triangle t \|\xi^n\|^2 + \|\Psi\|^2_{l^2((0,T),H^{k+1})} h^{2(k+1)} + \|\Gamma\|^2_{l^2((0,T),H^{k+1})} h^{2(k+1)} \right],$$

where T^* and T^{**} are defined as follows.

First, by Theorem 1 and the inverse assumption (18)

$$T^* = K^* \|\xi^{\hat{N}}\| \left(\sum_{n=1}^{M} \triangle t \|v^n\|^2 \right)^{\frac{1}{2}} \left(\sum_{n=1}^{M} \triangle t \|(\Gamma_H^n - \hat{\Gamma}_H^n)\|_{\infty}^2 \right)^{\frac{1}{2}}$$

$$(36) \qquad \leq K^* \|\xi^{\hat{N}}\| \left(\sum_{n=1}^{M} \triangle t \|v^n\|^2 \right)^{\frac{1}{2}} (H^{-d/2}(H^{k+1} + \triangle t)).$$

For $d \geq 2$, $k \geq 1$, H and $\triangle t$ can be chosen sufficiently small so that

$$(37) \qquad T^* \leq \frac{1}{2} \|\xi^{\hat{N}}\|^2 + \frac{K_*}{8} \sum_{n=1}^{N} \triangle t \|v^n\|^2.$$

Moreover,

$$T^{**} = C^* \sum_{n=1}^{M} (\|(\rho^n - \hat{\rho}_h^n)\Gamma_H^n\|^2 + \|(\rho^n - \rho_H^n)(\hat{\Gamma}_h^n - \Gamma_H^n)\|^2$$

$$+ \|(\rho^n - \rho_H^n)^2 \hat{\Gamma}_h^n\|^2) \triangle t$$

$$\leq C^* \sum_{n=1}^{M} \triangle t (\|\rho^n - \hat{\rho}_h^n\|^2 \|\Gamma^n\|_{\infty}^2$$

$$+ \|\rho^n - \hat{\rho}_h^n\|_{\infty}^2 \|\Gamma_H^n - \Gamma^n\|^2$$

$$+ (\|\hat{\rho}_H^N - \rho_H^N\|_{\infty} + \|\rho^N - \hat{\rho}_H^N\|_{\infty})^2 \|\hat{\Gamma}_h^n - \Gamma_H^n\|^2$$

$$+ \|\rho^n - \rho_H^n\|_{\infty}^2 \|\rho^n - \rho_H^n\|^2 \|\hat{\Gamma}_h^n\|_{\infty}^2).$$

Thus, from Theorem 1 and approximation properties,

$$T^{**} \leq C^* \sum_{n=1}^{M} \triangle t [h^{2(k+1)} + (H^{-d/2}(H^{k+1} + \triangle t) + H^{k+1})^2 (H^{k+1} + \triangle t)^2$$

$$(38) \qquad + (H^{-d/2}(H^{k+1} + \triangle t))^2 (H^{k+1} + \triangle t)^2].$$

Here $C*$ depends on $\|\rho\|_{l^2((0,T);W^{(k+1),\infty})}$, $\|\Gamma\|_{l^2((0,T);W^{1,\infty})}$, $\|\Gamma\|_{l^2((0,T);H^{(k+1)})}$, K^* and K_*.

Taking $M = \hat{N}$, and for $d \geq 2$ taking $k \geq 1$, noting that $\xi^0 = 0$, and applying Gronwall's Lemma to (36) we see that for $\triangle t$ and H sufficiently small,

$$\tag{39} \|\xi^{\hat{N}}\| \leq C(h^{k+1} + \triangle t + H^{2k+2-d/2}).$$

Combining (36) - (39), taking $M = N$, we deduce that

$$\tag{40} \|\xi^N\| + (\sum_{n=1}^{N} \triangle t \|K(\rho_H^n)^{1/2} v^n\|^2)^{\frac{1}{2}}$$
$$\leq C(C^*)(h^{k+1} + \triangle t + H^{2k+2-d/2}).$$

Applying the triangle inequality and approximation properties, we obtain the following theorem:

THEOREM 2. *Let* $V_h = V_h^k$, $\tilde{V}_h = \tilde{V}_h^k$, *and* $W_h = W_h^k$ *and define the one Newton correction triplet* $(\tilde{\rho}_h^n, \tilde{\Gamma}_h^n, \tilde{\Psi}_h^n)$, $\in W_h^k \times V_h^k \times \tilde{V}_h^k$ *by* (28) - (30). *Assume that the Approximation Properties A hold, and take* $\tilde{\rho}_h^0 = \hat{\rho}_h(\cdot, 0)$. *Then there exists a positive constant* C^{**}, *independent of* h, *such that*

$$\tag{41} \|\tilde{\rho}_h^N - \rho^N\| + (\sum_{n=1}^{N} \triangle t \|K(\rho_H^n)^{1/2}(\tilde{\Gamma}_h^n - \Gamma^n)\|^2)^{\frac{1}{2}}$$
$$\leq C^{**}(h^{k+1} + \triangle t + H^{2k+2-d/2}).$$

C^{**} *depends on* $\|\rho\|_{l^2((0,T);W^{(k+1),\infty})}$, $\|\Gamma\|_{l^2((0,T);W^{1,\infty})}$, $\|\Gamma\|_{l^2((0,T);H^{(k+1)})}$, K^* *and* K_*.

We now consider the simple two level scheme: solve(8)–(10) on a coarse mesh with $h = H$ and then solve the following linear system for $(\bar{\rho}_h^n, \bar{\Gamma}_h^n, \bar{\Psi}_h^n) \in W_h \times V_h \times \tilde{V}_h$ for $n \geq 1$,

$$\tag{42} (\frac{\bar{\rho}_h^n - \bar{\rho}_h^{n-1}}{\triangle t}, w_h) + (\nabla \cdot \bar{\Psi}_h^n, w_h) = (f^n, w_h), \quad w_h \in W_h,$$

$$\tag{43} (\bar{\Gamma}_h^n, v_h) = (\bar{\rho}_h^n, \nabla \cdot v_h), \quad v_h \in \tilde{V}_h,$$

$$\tag{44} (\bar{\Psi}_h^n, v_h) = (K(\rho_H^n)\bar{\Gamma}_h^n, v_h) \quad v_h \in V_h.$$

We now derive an estimate for this two-level scheme. Set $\alpha^n = \hat{\rho}_h^n - \bar{\rho}_h^n$, $\beta^n = \hat{\Gamma}_h^n - \bar{\Gamma}_h^n$, and $\lambda^n = \Pi_h \Psi^n - \bar{\Psi}_h^n$. Subtracting (42) from (19), (43) from (20), and (44) from (21), letting $w_h = \alpha^n$, $v_h = \lambda^n$, and $v_h = \beta^n$, in the first, second, and third equations respectively, we obtain

$$\tag{45} (\frac{\alpha^n - \alpha^{n-1}}{\triangle t}, \alpha^n) + (\nabla \cdot \lambda^n, \alpha^n) = (\epsilon^n, \alpha^n)$$

$$\tag{46} (\beta^n, \lambda^n) = (\alpha^n, \nabla \cdot \lambda^n),$$

and

$$(47) \quad \begin{aligned} (\lambda^n, \beta^n) \;=\;& (\Pi_h \Psi^n - \Psi^n, \beta^n) - (K(\rho^n)(\bar{\Gamma}_h^n - \Gamma^n), \beta^n) \\ &+ (K(\rho_H^n)\beta^n, \beta^n) + ((K(\rho^n) - K(\rho_H^n))\hat{\Gamma}_h^n, \beta^n). \end{aligned}$$

Combining (45) - (47) we obtain

$$(48) \quad \begin{aligned} \|\alpha^N\|^2 + \sum_{n=1}^{N} &\triangle t \|K(\rho_H^n)^{\frac{1}{2}}\beta^n\|^2 \\ &\leq C(\|\rho_H^n - \rho^n\|^2 + \|\Gamma^n - \bar{\Gamma}_h^n\|^2 + \triangle t^2). \end{aligned}$$

THEOREM 3. *The error bound for the simple two level scheme defined by (42) - (44) is given by*

$$(49) \quad \begin{aligned} \|\bar{\rho}^N - \rho^N\| + (\sum_{n=1}^{N} &\triangle t \|K(\rho_H^n)^{\frac{1}{2}}(\bar{\Gamma}_h^n - \Gamma^n)\|)^{\frac{1}{2}} \\ &\leq C(H^{k+1} + \triangle t + h^{k+1}). \end{aligned}$$

In the estimate (49) we observe that solving a linear problem on a fine grid with no Newton-type correction yields no asymptotic improvement over a coarse grid solution. This is different from the result obtained for Galerkin methods for nonlinear elliptic equations [12].

4. Newton Correction for RT_0 Spaces

Our previous results did not treat the case $k = 0$ for $d \geq 2$. Here we assume Ω is a rectangular parallelepiped, and we restrict our attention to the RT_0 ($k = 0$) space defined on a tensor product grid. We also assume K is a diagonal matrix or the mesh is uniform.

In this case, the space W_h is the space of discontinuous piecewise constants defined on a tensor product partition. The ith component of the velocity field V_h is a continuous piecewise linear polynomial in the ith direction and discontinuous piecewise constants in the other directions.

The mixed finite element method for the RT_0 space with special numerical quadrature rules has been shown [11] to be equivalent to cell centered finite difference methods. The results obtained here are applicable to the latter methods.

A discrete time mixed finite element approximation for the RT_0 space is defined as follows: Given $(\rho_h^0, \Gamma_h^0, \Psi_h^0) \in W_h \times V_h \times \tilde{V}_h$, for $n = 1, \ldots, N$, let $(\rho_h^n, \Gamma_h^n, \Psi_h^n) \in W_h \times V_h \times \tilde{V}_h$, satisfy

$$(50) \quad (\frac{\rho_h^n - \rho_h^{n-1}}{\triangle t}, w_h) + (\nabla \cdot \Psi_h^n, w_h) = (f^n, w_h), \quad w_h \in W_h,$$

$$(51) \quad (\Gamma_h^n, v_h) - (\rho_h^n, \nabla \cdot v_h) = 0, \quad v_h \in \tilde{V}_h,$$

$$(52) \quad (\Psi_h^n, v_h) = (K(\mathcal{P}(\rho_h^n))\Gamma_h^n, v_h), \quad v_h \in V_h.$$

Here $\mathcal{P} : W_h \to M_h$ is a "postprocessing" operator, where M_h denotes the space of continuous bilinears $(d = 2)$ or trilinears $(d = 3)$ defined on the same partition as the RT_0 space. For $d = 2$, $\mathcal{P}$ linearly interpolates the four adjacent cell centered values to the vertices of the tensor product grid. A similar procedure can be defined for $d = 3$. This postprocessing operator is motivated by well-known superconvergence results for the scalar variable ρ at the center of each grid block [**5, 8**].

In deriving an error estimate for this algorithm we proceed as in Section 2; however here we will employ superconvergence results for the RT_k spaces on rectangular elements. It is known for example that the Π_h and the P_{V_h} projections are super close, namely $O(h^{k+2})$ in L^2 [**8, 6**]. It is also known that two weighted L^2 projections are $O(h^{k+2})$ provided the weight functions are $\mathcal{C}^1$. We shall use these results in the analysis given below.

Set $\mu^n = \hat{\rho}_h^n - \rho_h^n$, $\zeta^n = \hat{\Gamma}_h^n - \Gamma_h^n$, and $\chi^n = \Pi_h \Psi^n - \Psi_h^n$. Subtracting (50) from (19), (51) from (20), and (52) from (21) and in the resulting equations using the test functions $w_h = \mu^n$, $v_h = \chi^n$, and $v_h = \zeta^n$ in the first, second, and third equations, respectively, we obtain the error equations

$$\text{(53)} \qquad (\frac{\mu^n - \mu^{n-1}}{\Delta t}, \mu^n) + (\nabla \cdot \chi^n, \mu^n) = (\epsilon^n, \mu^n),$$

$$\text{(54)} \qquad (\zeta^n, \chi^n) = (\mu^n, \nabla \cdot \chi^n),$$

and

$$
\begin{aligned}
\text{(55)} \qquad (\chi^n, \zeta^n) \;=\; & (\Pi_h \Psi^n - \hat{\Psi}^n, \zeta^n) + (K(\mathcal{P}(\rho_h^n)\zeta^n, \zeta^n) \\
& - ((K(\mathcal{P}(\rho_h^n)) - K(\rho^n))\check{\Gamma}_h^n, \zeta^n) \\
& + (K(\mathcal{P}(\rho_h^n))(\check{\Gamma}_h^n - \hat{\Gamma}_h^n), \zeta^n).
\end{aligned}
$$

Here we have defined $\check{\Gamma}_h^n$ by

$$\text{(56)} \qquad (K(\rho^n)(\check{\Gamma}_h^n - \Gamma^n), v_h) = 0, \quad v_h \in V_h.$$

Note we have also replaced Ψ^n on the right side of (55) by its L^2 projection.

Combining (53) - (55), applying smoothness and boundedness assumptions on $K(\rho)$, superconvergence, and Holder's inequality we obtain

$$\frac{1}{2\Delta t}[\|\mu^n\|^2 - \|\mu^{n-1}\|^2] + \|K(\mathcal{P}(\rho_h^n)^{1/2}\zeta^n\|^2$$

$$\leq (\frac{\mu^n - \mu^{n-1}}{\Delta t}, \mu^n) + \|K(\mathcal{P}(\rho_h^n)^{1/2}\zeta^n\|^2$$

$$\leq \frac{1}{2}\|\mu^n\|^2 + \frac{1}{2}\|\epsilon^n\|^2 + (\Pi_h \Psi^n - \hat{\Psi}_h^n, \zeta^n)$$

$$+ ((K(\mathcal{P}(\rho_h^n)) - K(\rho^n))\check{\Gamma}_h^n, \zeta^n) + (K(\mathcal{P}(\rho_h^n))(\check{\Gamma}_h^n - \hat{\Gamma}_h^n), \zeta^n)$$

$$\text{(57)} \qquad \leq C\|\mu^n\|^2 + \delta\|\zeta^n\|^2 + C\left[h^{2(k+2)} + \Delta t^2\right],$$

where $\delta \le K_*$. Here we have also used the fact that

$$\|\mathcal{P}(\hat{\rho}_h^n) - \rho^n\| \le Ch^2.$$

Multiplying (57) by $\triangle t$ and summing on $n, n = 1, 2, ..., N$ and applying Gronwall's Lemma, we see that

$$(58) \quad \|\mu^N\|^2 + \sum_{n=1}^{N} \triangle t \|K(\rho_h^n)^{1/2} \zeta^n\|^2 \le \|\mu^0\|^2 + C(\triangle t^2 + h^{2(k+2)}).$$

Using (58), Approximation Properties A, and the triangle inequality we deduce the following theroem:

THEOREM 4. Let $V_h = V_h^0$, $\tilde{V}_h = \tilde{V}_h^0$, and $W_h = W_h^0$, that is, the RT_0 spaces defined on a tensor product grid, and define the triplet $(\rho_h^n, \Gamma_h^n, \Psi_h^n) \in W_h^k \times V_h^k \times \tilde{V}_h^k$ by (50) - (52). Take $\rho_h^0 = \hat{\rho}_h(\cdot, 0)$. Then there exists a positive constant C, independent of h such that

$$(59) \quad \|\rho_h^N - \hat{\rho}_h^N\| + (\sum_{n=1}^{N} \triangle t \|K(\rho_h^n)^{1/2}(\hat{\Gamma}_h^n - \Gamma_h^n)\|^2)^{\frac{1}{2}} \le C(h^2 + \triangle t).$$

We now proceed as in Section 3, and let ρ_H^n be defined by (50)-(52) with $h = H$. The proof is identical to that given above accept for the bounds on the terms T^* and T^{**}. These bounds can be modified by simply replacing H^{k+1} by H^{k+2}; i.e., since $k = 0$, we replace H by H^2.

We obtain the following theorem:

THEOREM 5. Let $V_h = V_h^0$, $\tilde{V}_h = \tilde{V}_h^0$, and $W_h = W_h^0$ and define the one Newton correction triplet $(\tilde{\rho}_h^n, \tilde{\Gamma}_h^n, \tilde{\Psi}_h^n), \in W_h^0 \times V_h^0 \times \tilde{V}_h^0$ by (28) - (30). Assume that the Approximation Properties A holds, and take $\tilde{\rho}_h^0 = \hat{\rho}_h(\cdot, 0)$. Then there exists a positive constant C, independent of h such that

$$\|\tilde{\rho}_h^N - \rho^N\| + (\sum_{n=1}^{N} \triangle t \|K(\rho_H^n)^{1/2}(\tilde{\Gamma}_h^n - \Gamma^n)\|^2)^{\frac{1}{2}}$$

$$(60) \qquad\qquad \le C(h + \triangle t + H^{3-d/2}).$$

5. Conclusions and extensions

We remark that we even though we only considered the case $k = 0$ in Section 4, we have established superconvergence for the RT_k spaces, $k \ge 0$, for the scheme (50)-(52). The two-level scheme with the Newton correction outlined in Section 3 could be extended to multiple levels with multiple corrections. We are currently investigating these possibilities. Computational results for the algorithms outlined here are also in progress.

We also remark that one may be able to improve the rate of convergence given in Theorem 5 for the lowest-order case by substituting $\mathcal{P}(\rho_H^n)$ and $\mathcal{P}(\tilde{\rho}_h^n)$ for ρ_H^n

and $\tilde{\rho}_h^n$. Preliminary theoretical results indicate that this may give superconvergence for $\tilde{\rho}_h^n$ at the centers of each cell of order $h^2 + H^{4-d/2} + \triangle t$.

References

1. T. Arbogast, *An error analysis for Galerkin approximations to an equation of mixed elliptic-parabolic type*, Technical Report TR90-33, Dept. of Comp. and Appl. Math., Rice University, Houston, TX, 1990.
2. T. Arbogast, M. F. Wheeler, and I. Yotov, *Mixed finite element methods on general geometry*, in preparation.
3. J. Bear, *Dynamics of fluids in porous media*, Elsevier, New York, 1972.
4. F. Brezzi, J. Douglas, Jr., and L. D. Marini, *Two families of mixed finite elements for second order elliptic problems*, Numer. Math. **47**, pp. 217-223, 1985.
5. J. Douglas, Jr., and J. E. Roberts, *Global estimates for mixed methods for second order elliptic equations*, Math. Comp. **44**, pp. 39-52, 1985.
6. R. E. Ewing, R. D. Lazarov, and J. Wang, *Superconvergence of the velocity along the Gauss lines in mixed finite element methods*, SIAM J. Numer. Anal. **28**, pp. 1015-1029, 1991.
7. F. Milner, *Mixed finite element methods for quasilinar second-order elliptic problems*, Math. Comp. **44**, pp. 303-320, 1985.
8. M. Nakata, A. Weiser, and M. F. Wheeler, *Some superconvergence results for mixed finite element methods for elliptic problems on rectangular domains*, in The Mathematics of Finite Elements and Applications V, J. R. Whiteman, ed., Academic Press, London, 1985.
9. J. C. Nedelec, *Mixed finite elements in IR^3*, Numer. Math. **35**, pp. 315-341, 1990.
10. P. A. Raviart and J. M. Thomas, *A mixed finite element method for 2nd order elliptic problems*, in Mathematical Aspects of the Finite Element Method, Rome 1975, Lecture Notes in Mathematics, Springer-Verlag, Berlin, 1977.
11. A. Weiser and M. F. Wheeler, *On convergence of block-centered finite differences for elliptic problems*, SIAM J. Numer. Anal. **25**, pp. 351-375, 1988.
12. J. Xu, *Two-grid finite element discretization for nonlinear elliptic equations*, preprint.

DEPARTMENT OF COMPUTATIONAL AND APPLIED MATHEMATICS, RICE UNIVERSITY, HOUSTON, TX 77251
E-mail address: clint@rice.edu, mfw@rice.edu

Contemporary Mathematics
Volume **180**, 1994

Domain Decomposition Algorithms for PDE Problems with Large Scale Variations

Luc Giraud*   Ray S. Tuminaro[†]

Abstract
We consider a Schur complement BPS-like domain decomposition algorithm for the 2D drift-diffusion equations arising from semiconductor modeling. In particular, we focus on two problems: anisotropic phenomena and large changes in the PDE coefficients as one moves spatially within the domain. The preconditioners that we discuss are essentially BPS preconditioners [2] where the interface coupling is approximated using band matrices generated by the probing technique [3]. To cope with anisotropic phenomena, we introduce additional band matrices (in the context of the probe preconditioner) to approximate the coupling between neighboring interfaces. To address coefficient variations over the domain, we make use of the close connection between domain decomposition and multigrid and introduce specialized interpolation, projection, and averaging techniques to develop an accurate coarse grid approximation. We demonstrate the benefits of the new approach using computational experiments.

1 Introduction

We consider algorithms for the solution of the drift-diffusion equations in two dimensions. The solution of these equations is of great importance for semiconductor device modeling. In this paper, we focus on efficiently solving the linear systems that arise from Gummel's method [8] via Schur complement domain decomposition algorithms. The principal difficulties presented by our formulation of the drift-diffusion equations are anisotropic behavior introduced by the discretization and large variations in the PDE coefficients. To solve the resulting linear systems, the conjugate gradient method is used in conjunction with a BPS-like [2] preconditioner:

$$M^{-1} = M_E^{-1} + I_H^h A_H^{-1} I_h^H,$$

where M_E^{-1} is the interface approximation, $I_H^h A_H^{-1} I_h^H$ is the coarse grid problem, A_H corresponds to the discretization matrix of the original problem on

---

1991 *Mathematics Subject Classification*. Primary 65N55; Secondary 65C20.

a coarse grid whose elements are defined by the non-overlapping subdomains, and $I_h^H = [I_H^h]^T$ denotes the transfer operator between the original fine grid and the coarse grid. In Section 2, we describe briefly the drift-diffusion equations and show how anisotropic behavior and large variations in the PDE coefficients arise from our formulation. Next in Section 3, we present new methods for M_E^{-1} suitable for anisotropic problems and specifically designed for situations when the coupling between neighboring interfaces is stronger than the coupling within an interface [6]. Then, in Section 4 we propose grid transfer operators suitable for highly variable coefficient problems. These operator-dependent transfers correspond to extensions of operator-dependent prolongation and restriction operators used in standard multigrid methods [7]. Lastly, in Section 5 computational results are given to illustrate the different approaches.

2   The Drift-Diffusion Equations

A drift-diffusion model is used to approximate the behavior of a single semiconductor device [9]. This model consists of a potential equation

$$\varepsilon \nabla^2 \psi + q[n_{ie} e^{-q\psi/kT} v - n_{ie} e^{q\psi/kT} u + N_D - N_A] = 0,$$

and two continuity equations

$$\nabla \cdot [\frac{kT}{q} \mu_n n_{ie} e^{q\psi/kT} \nabla u] + R = 0 \quad \text{and} \quad \nabla \cdot [\frac{kT}{q} \mu_p n_{ie} e^{-q\psi/kT} \nabla v] - R = 0,$$

where ε is the scalar permittivity of the semiconductor, n_{ie} is the effective intrinsic carrier concentration, q is the elementary charge, k is the Boltzmann constant, T is the temperature in Kelvin, N_D is the density of donor impurities, N_A is the density of acceptors, and μ_n and μ_p are the electron and hole mobilities respectively. The dependent variables are the electric potential, ψ, and the two Slotboom variables u and v. The density of free electrons, n, and free holes, p, can be recovered using

$$n = n_{ie} e^{q\psi/kT} u \qquad\qquad p = n_{ie} e^{-q\psi/kT} v.$$

Notice that the drift-diffusion equations are nonlinear and that the individual equations are symmetric self-adjoint operators with highly variable coefficients (due to the exponential operators).

A discrete approximation to the drift-diffusion equations is obtained by approximating each of the derivative terms with central differences on a highly stretched grid. Unfortunately, a significant degree of anisotropic behavior is introduced as a consequence of this highly stretched grid. The resulting system of three discrete nonlinear equations is then solved using a nonlinear Gauss-Seidel method known as the Gummel iterative technique [8]. This nonlinear technique requires the solution of three symmetric linear systems (corresponding to the three discrete PDEs) within each Gummel

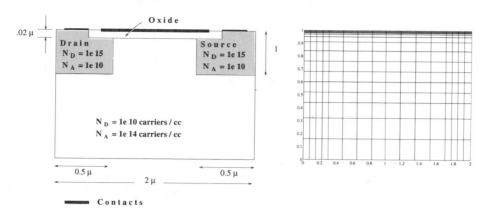

FIG. 1. *MOSFET Device*   FIG. 2. *Example of a stretched grid*

iteration. To solve these linear systems, we use a conjugate gradient algorithm in conjunction with the BPS-like preconditioner

$$M^{-1} = M_E^{-1} + I_H^h A_H^{-1} I_h^H.$$

The remainder of this paper discusses the development of the two components of this preconditioner.

3 Interface Component

To construct the interface component M_E^{-1}, we follow [3]. That is, the diagonal blocks of the Schur complement that correspond to the coupling between fine grid points on the same subdomain edge are approximated by band matrices. This choice is motivated by the following observations for a two subdomain decomposition. First, the Schur complement matrix is "close" to a band matrix. That is, the entries decay rapidly from the diagonal. Second, all entries of a band matrix having upper and lower bandwidth d can be computed by the action of this matrix on $2d+1$ carefully chosen probe vectors p^k (e.g. , $p^k = (.., 0, 0, 1, 0, 0, 1, 0, 0, 1, ...)^T$ for $d = 1$). In the rest of the paper, d denotes the upper and lower bandwidth of the approximation matrices. To generalize the probing technique to multiple domains, we must generate one band matrix for each edge. One possibility is to build all the band approximations on the vertical edges at the same time and then on the horizontal edges. That is, a composite probe vector is defined first over all the vertical edges by combining individual probe vectors for each edge. Then the same process is performed for the horizontal edges. This variant gives rise to the preconditioner referred to as $M_E^{vh(d)}$. The second variant further subdivides both horizontal and vertical edges into 'red' and 'black' sets to minimize the approximation errors arising from coupling between vertical (or horizontal) interfaces. The resulting preconditioner is denoted by $M_E^{rb(d)}$.

TABLE 1

# of iterations for $u_{xx} + \epsilon u_{yy} = f$ on a 128×128 grid

$\varepsilon = 1.0$					
	# domains				
M^{-1}	2×2	4×4	8×8	16×16	32×32
$M_E^{sim(1)}$	15	24	32	48	81
$M_E^{rb(1)}$	15	24	31	48	81
$M_E^{full(1)}$	-	21	32	50	69

$\varepsilon = 10^{-3}$					
	# domains				
M^{-1}	2×2	4×4	8×8	16×16	32×32
$M_E^{sim(1)}$	9	56	200	485	–
$M_E^{rb(1)}$	9	26	54	71	120
$M_E^{rb(2)}$	8	26	54	71	120
$M_E^{full(1)}$	-	14	16	17	17

For anisotropic problems, the off-diagonal blocks of the Schur complement can have large entries. In such situation, the previous block Jacobi
type preconditioners can be inefficient. In order to capture the off-diagonal
block coupling, we consider a new preconditioner which approximates the
Schur complement coupling along each line of the original grid by a separate
band matrix. The band matrix for each line differs depending on whether
a line is aligned with an interface or not. On a uniform grid with $n_x \times n_y$
grid points partitioned among $N_x \times N_y$ equi-sized rectangles, the $N_y - 1$
horizontal grid lines that are aligned with the interfaces have n_x points in
the Schur complement operator and those lines between the interfaces have
$N_x - 1$ points. The probe idea can be applied to compute an approximation
of the Schur complement restricted to the lines. For detailed information on
these full probing approximations, we refer to [6] and simply state that the
new preconditioner looks like an additive version of alternating line relaxation method or an additive ADI method applied on the Schur complement
operator. In this paper the resulting preconditioner is referred to as $M_E^{full(d)}$.

In Table 1 we display the number of iterations required using various
interface preconditioners without a coarse grid preconditioner for a Poisson
problem defined on the unit square with Dirichlet boundary conditions.
Convergence is attained when the Euclidean norm of the residual is reduced
by a factor 10^5. These results show that while all the preconditioners are
equivalent for the nonanisotropic case, there is a great deal of variation

for anisotropic problems. In the anisotropic case, the red-black probing improves the convergence noticeably and with full probing we obtain very fast convergence for the anisotropic case. When the problem is very anisotropic and constant coefficient, a coarse grid preconditioner is not needed as the problem is essentially one dimensional and thus preconditioned accurately by the band solves.

4   Coarse Grid Component

The coarse grid component of the preconditioner is defined by $I_H^h A_H^{-1} I_h^H$ where the restriction operator I_h^H is the transpose of the prolongation operator I_H^h. The definition of the grid transfer operators is crucial to develop an accurate coarse grid approximation. When the coefficients are highly variable, the close relationship between the BPS-type preconditioner and multigrid can be exploited. In the context of multigrid for regular meshes, where the fine grid contains one point in between all adjacent coarse grid points, the definition of those operators is relatively well understood [1], and [4]. In our work we have generalized these results to the case when the number of fine grid points between two adjacent coarse grid points is greater than one and not necessarily constant. These operator dependent transfers correspond to an extension of standard operator dependent prolongation and restriction operators used in standard multigrid methods. It can be shown that for certain coarse grids the 1D version of this prolongation is equivalent to recursive use of standard operator dependent prolongation on a hierarchy of grids in a multigrid method where a harmonic average is used to average the PDE coefficients [7]. For a detailed description of this grid transfer operator in the context of multigrid, we refer to [7] where averaging techniques and Galerkin formulations are also discussed for obtaining the operator A_H.

5   Experimental results

Results are given corresponding to a simple MOSFET device simulation. In Fig. 1, we illustrate the characteristics of this device. The contacts correspond to Dirichlet boundary conditions while the other boundaries are Neumman conditions. For the potential equation the solution is solved over the whole domain. For the u and v equations, we solve on the whole domain excluding the oxide where Neumman conditions are used on the oxide interface. A scaled down version of the grid used in this paper is given in Fig. 2. Table 2 displays the average number of conjugate gradient iterations for each linear solve corresponding to the two continuity equations for a MOSFET simulation discretized on a 129×129 grid. In the table, we show only the results using the full probing technique as the other probing techniques required many more iterations. From the table, we can see the importance of the coarse grid component in the preconditioner. That is,

TABLE 2

average # of iterations for the n and p equations.

Preconditioner	# domains			
	4×4	8×8	16×16	32×32
without coarse grid	33	58	103	156
standard grid transfers	41	58	65	60
op. dep. transfers	38	45	40	31

without the coarse grid the number of iterations grows significantly as the number of domains is increased. Further, we can see the effect of using carefully chosen operator-dependent grid transfers, compared with using simple bilinear interpolation. In particular, for the problem using 32×32 domains there is a factor of two difference in the number of iterations between the two while the work per iteration is approximately the same. For more details on the probing and anisotropic phenomena we refer the reader to [6]. For more details on the grid transfers we refer the reader to [7]. Finally, more extensive numerical experiments for several semiconductor devices using a few different domain decomposition algorithms will be presented in [5].

References

[1] R. Alcouffe, A. Brandt, J. Dendy, and J. Painter, *The multigrid method for diffusion equations with strongly discontinuous coefficients*, SIAM J. Sci. Stat. Comput., 2 (1981), pp. 430–454.

[2] J. Bramble, J. Pasciak, and J. Schatz, *The construction of preconditioners for elliptic problems by substructuring I.*, Math. Comp., 47 (1986), pp. 103–134.

[3] T. F. Chan and T. P. Mathew, *The interface probing technique in domain decomposition*, SIAM J. Matrix Anal. Appl., 13 (1992), pp. 212–238.

[4] J. Dendy, *Black box multigrid*, J. Comp. Phys., 48 (1982), pp. 366–386.

[5] L. Giraud and R. Tuminaro, *A domain decomposition method for the drift-diffusion equations*, Tech. Rep. in preparation, CERFACS, Toulouse, France.

[6] ——, *A domain decomposition probing variant suitable for anisotropic problems*, Tech. Rep. TR/PA/93/36, CERFACS, Toulouse, France, 1993.

[7] ——, *Grid transfer operators for highly variable coefficient problems*, Tech. Rep. TR/PA/93/37, CERFACS, Toulouse, France, 1993.

[8] H.K.Gummel, *A self-consistent iterative scheme for one-dimensional steady state transistor calculations*, IEEE Transactions on Electron Devices, ED-11 (1964), pp. 455–465.

[9] S. Selberherr, *Analysis and Simulation of Semiconductor Devices*, Springer-Verlag, New York, 1984.

* CERFACS, Toulouse, France
† Sandia National Laboratory, Albuquerque, NM

Contemporary Mathematics
Volume **180**, 1994

A One Shot Domain Decomposition/Fictitious Domain method for the Navier–Stokes Equations

ROLAND GLOWINSKI, TSORNG–WHAY PAN AND JACQUES PÉRIAUX

ABSTRACT. In this paper which is motivated by computation on parallel MIMD machines, we address the numerical solution of some class of elliptic problems by a combination of domain decomposition and fictitious domain methods. We take advantage of the fact that the Steklov–Poincare operators associated with the subdomain interfaces and with the fictitious domain treatment of internal boundaries have very similar properties. We use these properties to derive fast solution methods of conjugate gradient type with good parallelization properties which force simultaneously the matching at the subdomain interfaces and the actual boundary conditions. Preliminary results obtained on a KSR machine are presented. A similar methodology has been applied to simulate viscous flows around obstacles modelled by Navier–Stokes equations.

1. Introduction

Fictitious domain methods for Partial Differential Equations have shown recently a most interesting potential for solving complicated problems from Science and Engineering (see, e.g., [**1, 2**] for some impressive illustrations of the above statement). The main reason of popularity of fictitious domain methods (sometime called *domain imbedding methods*; cf. [**3**]) is that they allow the use of fairly structured meshes on a simple shape auxiliary domain containing the actual one, allowing therefore the use of fast solvers. In [**4, 5**], we have used Lagrange multiplier and finite element methods combined with fictitious domain techniques to compute the numerical solutions of elliptic problems with Dirichlet boundary conditions and simulate some nonlinear time dependent problems, namely the flow of a viscous–plastic medium in a cylindrical pipe and time dependent external incompressible viscous flow modelled by the Navier–Stokes equations.

1991 *Mathematics Subject Classification*. Primary 65M60; Secondary 76D05.
The final version of this paper will be submitted for publication elsewhere.

In this paper motivated by computation on parallel MIMD machines, we address the numerical solution of a class of elliptic problems by a combination of domain decomposition and fictitious domain methods. From the fact that the Steklov–Poincare operators associated with the subdomain interfaces and with the fictitious domain treatment of internal boundaries have very similar properties; we derive fast solution methods of conjugate gradient type with good parallelization properties which enforce simultaneously the matching at the subdomain interfaces and the actual boundary conditions. A similar methodology has been applied to simulate viscous flows around obstacles modelled by Navier–Stokes equations. In Section 2, we describe the formulation of a family of Dirichlet problems and discuss an equivalent formulation which is at the basis of the domain decomposition/fictitious domain methods. Preliminary results obtained on a KSR1 machine are presented. In Section 3 we apply a similar methodology to simulate external incompressible viscous flow modelled by Navier–Stokes equations.

2. One Shot DD/FD Method for the Dirichlet problem

2.1. Formulation of the Dirichlet problem

We consider the following elliptic problem:

$$(2.1) \qquad \alpha u - \nu \Delta u = f \ in \ \Omega \setminus \bar{\omega},$$

$$(2.2) \qquad u = g_0 \ on \ \gamma, \ u = g_1 \ on \ \Gamma,$$

where Ω is a "box" domain in $\mathbb{R}^d (d \geq 1)$, ω is a bounded domain in $\mathbb{R}^d (d \geq 1)$ such that $\omega \subset\subset \Omega$ (e.g., see Fig. 2.1 in which $\Omega = \Omega_1 \cup \Omega_2$ and $\omega = \omega_1 \cup \omega_2$), Γ (resp., γ) is the boundary $\partial \Omega$ (resp., $\partial \omega$), $\alpha \geq 0$, and $\nu > 0$; finally, f, g_0, and g_1 are given functions defined over $\Omega \setminus \bar{\omega}$, γ and Γ, respectively. If f, g_0, and g_1 are smooth enough, problem (2.1)–(2.2) has a unique solution. The equivalent variational formulation of problem (2.1)–(2.2) is

$$(2.3) \qquad \int_{\Omega_0} (\alpha u v + \nu \nabla u \cdot \nabla v) \, dx = \int_{\Omega_0} f v \, dx, \ \forall v \in V_0; \ u \in V_g,$$

where $\Omega_0 = \Omega \setminus \bar{\omega}$, $V_g = \{v | v \in H^1(\Omega_0), v = g_0 \ on \ \gamma, v = g_1 \ on \ \Gamma\}$ and $V_0 = \{v | v \in H^1(\Omega_0), v = 0 \ on \ \gamma \cup \Gamma\}$.

2.2. Domain decomposition/fictitious domain approach

For simplicity we consider the case where ω is the union of two disjoint bounded domains, ω_1 and ω_2, and Ω is the union of two subdomains Ω_1 and Ω_2 (see Fig. 2.1); we denote by γ_0 the interface between Ω_1 and Ω_2, by γ_1 (resp., γ_2) the boundary of ω_1 (resp., ω_2), and let $\Gamma_1 = \Gamma \cap \partial \Omega_1$ and $\Gamma_2 \doteq \Gamma \cap \partial \Omega_2$. Combining the fictitious domain method discussed in [4] to a domain decomposition method (see, e.g., [6]) and applying to the solution of problem (2.1)–(2.2), (2.3), we obtain the following equivalent problem:

$$
(2.4) \quad \begin{cases}
Find\ u_i \in V_g^i,\ \lambda_i \in L^2(\gamma_i),\ \lambda_d \in L^2(\gamma_0)\ such\ that \\[2mm]
\displaystyle\int_{\Omega_i} (\alpha u_i v_i + \nu \nabla u_i \cdot \nabla v_i)\, dx = \int_{\Omega_i} \tilde{f} v_i\, dx \\[4mm]
\displaystyle + \int_{\gamma_i} \lambda_i v_i\, d\gamma + (-1)^i \int_{\gamma_0} \lambda_d v_i\, d\gamma,\ \forall v_i \in V_0^i,\ for\ i = 1, 2,
\end{cases}
$$

$$
(2.5) \quad \int_{\gamma_i} \mu_i (u_i - g_0)\, d\gamma = 0,\ \forall \mu_i \in L^2(\gamma_i),\ for\ i = 1, 2,
$$

$$
(2.6) \quad \int_{\gamma_0} \mu_d (u_2 - u_1)\, d\gamma = 0,\ \forall \mu_d \in L^2(\gamma_0),
$$

where $\tilde{f}$ is a $L^2(\Omega)$ extension of f, $V_g^i = \{v | v \in H^1(\Omega_i), v = g_1\ on\ \Gamma_i\}$ and $V_0^i = \{v | v \in H^1(\Omega_0), v = 0\ on\ \Gamma_i\}$ for $i = 1, 2$. We have equivalence in the sense that if relations (2.4)–(2.6) hold then $u_i = u|_{\Omega_i}$, for $i = 1, 2$, and conversely.

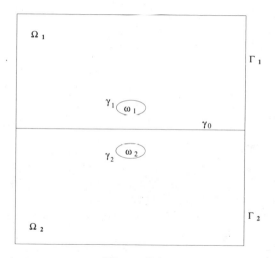

Figure 2.1.

In (2.4)–(2.6), the function λ_i which is a *Lagrange multiplier* associated with the *boundary condition* $u = g_0$ on γ_i is essentially the jump of $\nu \dfrac{\partial u}{\partial n}$ at γ_i for $i = 1, 2$ and the function λ_d which can be viewed as a *Lagrange multiplier* associated with the *interface boundary condition* $u_1 = u_2$ on γ_0 is nothing but the function $\nu \dfrac{\partial u}{\partial n_2}|_{\gamma_0} = -\nu \dfrac{\partial u}{\partial n_1}|_{\gamma_0}$, where n_i is the normal unit vector at γ_0, outward to Ω_i.

Due to the combination of the two methods, there are *two Lagrange multipliers* associated with the *boundary conditions* and with the matching of solution at the *subdomain interfaces*, respectively. We can solve the saddle–point system (2.4)–(2.6) by a conjugate gradient algorithm driven by the multiplier associated with the boundary conditions, the one driven by the multiplier associated with the matching at the subdomain interfaces, or by the one called the *one shot method* driven by the *two multipliers* at the same time [7]. These methods have different parallelization properties and can be parallelized on MIMD machines. The one driven by the multiplier associated with the boundary conditions was discussed in [7] and its speed on a KSR1 is much slower than that of the one shot method for the two subdomains decomposition shown in Fig. 2.1. In Section 2.3 we are would like to test the one shot method with more subdomains.

2.3. Performance on a KSR1 machine

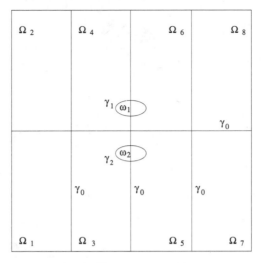

Figure 2.2.

We consider problem (2.1)–(2.2) with $\alpha = \nu = 1$ as test problem and let $u(x, y) = x^2 + y^2$ be the solution of the test problem. Then $f(x, y) = x^2 + y^2 - 4$. Let $\omega = \omega_1 \cup \omega_2$ where $\omega_i = \{(x, y) | \frac{(x - 1.0)^2}{(1/8)^2} + \frac{(y - c_i)^2}{(1/16)^2} < 1\}$, for $i = 1, 2$, $c_1 = 1.1875$, $c_2 = 0.8125$; take $\Omega = (0, 2) \times (0, 2)$.

In the numerical experiments, we consider the case where Ω is the union of eight subdomains $\Omega_1, \ldots,$ to Ω_8 (see Fig. 2.2). The finite dimensional spaces V_{gh}^i and V_{0h}^i of V_g^i and V_0^i, respectively for $i = 1, \ldots, 8$ are as follows:

$$V_{gh}^i = \{v_h | v_h \in V_g^i \cap C^0(\bar{\Omega}_i), v_h = g_h \text{ on } \Gamma_i, v_h|_T \in P_1 \ \forall T \in \mathcal{T}_h^i\},$$
$$V_{0h}^i = \{v_h | v_h \in V_0^i \cap C^0(\bar{\Omega}_i), v_h = 0 \text{ on } \Gamma_i, v_h|_T \in P_1 \ \forall T \in \mathcal{T}_h^i\},$$

where g_h is an approximation of g, $\mathcal{T}_h^i$ is a triangulations of Ω_i for $i = 1, \ldots, 8$ and P_1 is the space of the polynomials in x, y of degree ≤ 1. For $i = 0, 1, 2$, the finite dimensional space Λ_h^i of $L^2(\gamma_i)$ is defined as follows:

$$\Lambda_h^i = \{\mu_h | \mu_h \in L^\infty(\gamma_i), \mu_h \text{ is constant on the}$$
$$\text{segment joining 2 consecutive mesh points on } \gamma_i\}.$$

The choice of mesh points on γ_1 and γ_2 are shown on Fig. 2.3. For stability reason –the so called LBB inf–sup condition– the length of each segment on γ_1 and γ_2 has to be chosen greater than the meshsize h. The obvious choice for the mesh points on γ_0 are the midpoints of the edges located on γ_0 (see Fig. 2.3).

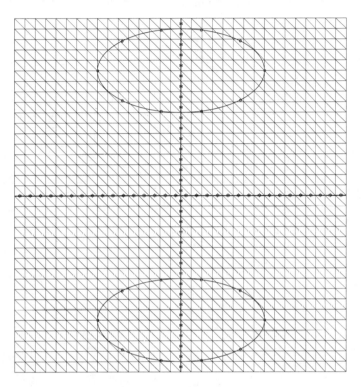

FIGURE 2.3. MESH POINTS MARKED BY "*" ON γ_1 AND γ_2 AND PART OF MESH POINTS MARKED BY "*" ON γ_0 WITH $h = 1/64$.

In the one shot method, the elliptic problems have been solved on each subdomain by a *Fast Elliptic Solver* based on *cyclic reduction* [**3, 8–10**]. Concerning implementation of the one shot method on the KSR1 machine, eight discrete elliptic problems can be solved simultaneously. For meshsize $h = 1/32$, $1/64$, $1/128$, and $1/256$, the number of iterations of the one shot method is 78, 91, 116, and 151 respectively and the number of iterations for the preconditioned one shot method is 68, 75, 88, and 92 respectively. Thus the preconditioner for two dimensional problems works very well. The CPU time per iteration of the one shot method with or without preconditioner is about the same. In Tables 2.1 and 2.2 we have shown the elapsed time and speedup per iteration of the discrete analogues of the preconditioned one shot method for different meshsizes where N_p is the number of processors used in computation. The speedup per iteration in Table 2.2 is better as the size of problem is larger.

Table 2.1. CPU per iteration on a KSR1				
N_p	h=1/32	h=1/64	h=1/128	h=1/256
1	0.291 sec.	1.130 sec.	5.251 sec.	22.608 sec.
2	0.189 sec.	0.649 sec.	2.828 sec.	11.756 sec.
4	0.125 sec.	0.394 sec.	1.543 sec.	6.177 sec.
8	0.099 sec.	0.231 sec.	0.883 sec.	3.334 sec.

Table 2.2. Speedup per iteration on a KSR1				
N_p	h=1/32	h=1/64	h=1/128	h=1/256
1	1.00	1.00	1.00	1.00
2	1.54	1.74	1.86	1.92
4	2.33	2.87	3.40	2.66
8	2.94	4.89	5.95	6.78

3. External incompressible viscous flow

3.1. Navier–Stokes equations

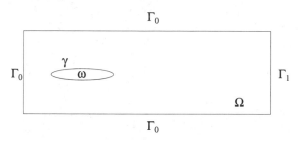

Figure 3.1.

In [**5**], we have used Lagrange multiplier/fictitious domain methods with finite element methods to simulate external incompressible viscous flow modelled by the Navier–Stokes equations. Here we would like to consider the same flow problems with a one shot method. The Navier–Stokes equations are the following:

$$(3.1) \qquad \frac{\partial \mathbf{u}}{\partial t} - \nu \Delta \mathbf{u} + (\mathbf{u} \cdot \nabla)\mathbf{u} + \nabla p = \mathbf{f} \; in \; \Omega \setminus \bar{\omega},$$

$$(3.2) \qquad \nabla \cdot \mathbf{u} = 0 \; in \; \Omega \setminus \bar{\omega},$$

$$(3.3) \qquad \mathbf{u}(\mathbf{x}, 0) = \mathbf{u}_0(\mathbf{x}), \quad \mathbf{x} \in \Omega \setminus \bar{\omega}, (with \; \nabla \cdot \mathbf{u}_0 = 0),$$

$$(3.4) \qquad \mathbf{u} = \mathbf{g}_0 \; on \; \Gamma_0, \; \nu \frac{\partial \mathbf{u}}{\partial \mathbf{n}} - \mathbf{n}p = \mathbf{g}_1 \; on \; \Gamma_1, \; \mathbf{u} = \mathbf{g}_2 \; on \; \gamma.$$

In (3.1)–(3.4), Ω and ω are bounded domains in $\mathbb{R}^d (d \geq 2)$ (see Fig. 3.1), Γ (resp., γ) is the boundary of Ω (resp., ω) with $\Gamma = \Gamma_0 \cup \Gamma_1$, $\Gamma_0 \cap \Gamma_1 = \emptyset$ and $\int_{\Gamma_1} d\Gamma > 0$, $\mathbf{n}$ is the outer normal unit vector at Γ_1, $\mathbf{u} = \{u_i\}_{i=1}^{i=d}$ is the flow velocity, p is the presure, $\mathbf{f}$ is a density of external forces, $\nu (> 0)$ is a viscosity parameter, and

$(\mathbf{v} \cdot \nabla)\mathbf{w} = \{\sum_{j=1}^{j=d} v_j \frac{\partial w_i}{\partial x_j}\}_{i=1}^{i=d}$. For the fictitious domain formulation, we imbed $\Omega \setminus \bar{\omega}$ in Ω and define

$$(3.5) \qquad \mathbf{V}_{\mathbf{g}_0} = \{\mathbf{v}|\mathbf{v} \in (\mathbf{H}^1(\Omega))^d, \mathbf{v} = \mathbf{g}_0 \ on \ \Gamma_0\},$$

$$(3.6) \qquad \mathbf{V}_0 = \{\mathbf{v}|\mathbf{v} \in (\mathbf{H}^1(\Omega))^d, \mathbf{v} = \mathbf{0} \ on \ \Gamma_0\},$$

$$(3.7) \qquad \Lambda = (\mathbf{L}^2(\gamma))^d.$$

Let $\mathbf{U}_0$ be an extension of $\mathbf{u}_0$ with $\nabla \cdot \mathbf{U}_0 = 0$ in Ω and $\tilde{\mathbf{f}}$ an extension of $\mathbf{f}$. Then we have equivalence between (3.1)–(3.4) and the following problem

For $t > 0$, find $\mathbf{U}(t) \in \mathbf{V}_{\mathbf{g}_0}$, $P(t) \in \mathbf{L}^2(\Omega)$, $\lambda(t) \in \Lambda$ such that

$$(3.8) \qquad \left\{ \begin{array}{l} \int_\Omega \frac{\partial \mathbf{U}}{\partial t} \cdot \mathbf{v} \, d\mathbf{x} + \nu \int_\Omega \nabla \mathbf{U} \cdot \nabla \mathbf{v} \, d\mathbf{x} + \int_\Omega (\mathbf{U} \cdot \nabla)\mathbf{U} \cdot \mathbf{v} \, d\mathbf{x} \\[2ex] \quad - \int_\Omega P\nabla \cdot \mathbf{v} \, d\mathbf{x} = \int_\Omega \tilde{\mathbf{f}} \cdot \mathbf{v} \, d\mathbf{x} + \int_{\Gamma_1} \mathbf{g}_1 \cdot \mathbf{v} \, d\Gamma \\[2ex] \quad + \int_\gamma \lambda \cdot \mathbf{v} \, d\gamma, \ \forall \mathbf{v} \in \mathbf{V}_0, \ a.e. \ t > 0, \end{array} \right.$$

$$(3.9) \qquad \nabla \cdot \mathbf{U}(t) = 0 \ in \ \Omega, \ \mathbf{U}(\mathbf{x}, 0) = \mathbf{U}_0(\mathbf{x}), \ \mathbf{x} \in \Omega,$$

$$(3.10) \qquad \mathbf{U}(t) = \mathbf{g}_2(t) \ on \ \gamma,$$

in the sense that $\mathbf{U}|_{\Omega \setminus \bar{\omega}} = \mathbf{u}$, $P|_{\Omega \setminus \bar{\omega}} = p$. The multiplier λ is the jump of $\nu \frac{\partial \mathbf{U}}{\partial \mathbf{n}} - \mathbf{n}P$ at γ and the effect of the actual geometry is concentrated on $\int_\gamma \lambda \cdot \mathbf{v} \, d\gamma$ in the right–hand–side of (3.8), and on (3.10).

To solve (3.8)–(3.10), we shall consider a time discretization by an operator splitting method, like the ones in, e.g., [11–14]. With these methods we can de couple the nonlinearity and the incompressibility in the Navier–Stokes/fictitious domain problem (3.8)–(3.10). Applying the θ–*scheme* (cf. [14]) to (3.8)–(3.10), we obtain *quasi–Stokes/fictitious domain subproblems* and nonlinear *advection– diffusion subproblems* (e.g., see [5]). In Section 3.2, a one shot method for the quasi–Stokes/FD subproblems shall be discussed. Due to the fictitious domain method and the operator splitting method, advection–diffusion subproblems may be solved in a least–squares formulation by a conjugate gradient alogrithm [14] in a simple shape auxiliary domain Ω without concern for the constraint $\mathbf{u} = \mathbf{g}$ at γ. Thus, advection–diffusion subproblems can be solved with domain decom- position methods.

3.2. The one shot method for the quasi–Stokes/FD subproblems

The quasi–Stokes/fictitious domain subproblem is the following:

$$
(3.11) \quad
\begin{cases}
\textit{Find } \mathbf{U} \in \mathbf{V_{g_0}}, \ P \in \mathbf{L}^2(\Omega), \ \lambda \in \Lambda \textit{ such that} \\[4pt]
\displaystyle \int_\Omega (\alpha \mathbf{U} \cdot \mathbf{v} + \nu \nabla \mathbf{U} \cdot \nabla \mathbf{v}) \, d\mathbf{x} - \int_\Omega P \nabla \cdot \mathbf{v} \, d\mathbf{x} \\[10pt]
\displaystyle - \int_\gamma \lambda \cdot \mathbf{v} \, d\gamma = \int_\Omega \mathbf{f} \cdot \mathbf{v} \, d\mathbf{x} + \int_{\Gamma_1} \mathbf{g}_1 \cdot \mathbf{v} \, d\Gamma, \ \forall \mathbf{v} \in \mathbf{V}_0,
\end{cases}
$$

$$(3.12) \qquad \nabla \cdot \mathbf{U} = 0 \ in \ \Omega,$$

$$(3.13) \qquad \mathbf{U} = \mathbf{g}_2 \ on \ \gamma,$$

The one shot methodology has been used for solving problem (3.11)–(3.13) in which the *two Lagrange multipliers* are *the pressure P* for (3.12) and λ for *the actual boundary condition* (3.13). Here we consider a bilinear form $b(\cdot, \cdot)$ which is symmetric and elliptic over Λ. We may choose $b(\lambda, \mu) = \int_\gamma \lambda \cdot \mu \, d\gamma, \ \forall \lambda, \ \mu \in \Lambda.$

The one shot algorithm is the following:

(3.14) $\{P^0, \lambda^0\} \in \mathbf{L}^2(\Omega) \times \Lambda$ *given; solve the following Dirichlet problem:*

$$
(3.15) \quad
\begin{aligned}
&\int_\Omega (\alpha \mathbf{U}^0 \cdot \mathbf{v} + \nu \nabla \mathbf{U}^0 \cdot \nabla \mathbf{v}) \, d\mathbf{x} = \int_\Omega \mathbf{f} \cdot \mathbf{v} \, d\mathbf{x} + \\[6pt]
&\int_\gamma \lambda^0 \cdot \mathbf{v} \, d\gamma + \int_\Omega P^0 \nabla \cdot \mathbf{v} \, d\mathbf{x} + \int_{\Gamma_1} \mathbf{g}_1 \cdot \mathbf{v} \, d\Gamma, \ \forall \mathbf{v} \in \mathbf{V}_0; \ \mathbf{U}^0 \in \mathbf{V_{g_0}},
\end{aligned}
$$

set $r_1^0 = \nabla \cdot \mathbf{U}^0$, $\mathbf{r}_2^0 = (\mathbf{U}^0 - \mathbf{g}_2)|_\gamma$, *and define* $\mathbf{g}^0 = \{g_1^0, \mathbf{g}_2^0\}$ *as follows:*

$$(3.16) \qquad\qquad\qquad g_1^0 = \alpha \phi^0 + \nu r_1^0,$$

with ϕ^0 *the solution of*

$$
(3.17) \quad
\begin{cases}
-\Delta \phi^0 = r_1^0 \ in \ \Omega, \\[6pt]
\dfrac{\partial \phi^0}{\partial \mathbf{n}} = 0 \ on \ \Gamma_0; \ \phi^0 = 0 \ on \ \Gamma_1,
\end{cases}
$$

$$(3.18) \qquad\qquad b(\mathbf{g}_2^0, \mu) = \int_\gamma \mathbf{r}_2^0 \cdot \mu \, d\gamma \ \forall \mu \in \Lambda; \ \mathbf{g}_2^0 \in \Lambda.$$

We take $\mathbf{w}^0 = \{w_1^0, \mathbf{w}_2^0\} = \{g_1^0, \mathbf{g}_2^0\}.$

　　Then for $n \geq 0$, *assuming that* P^n, λ^n, $\mathbf{U}^n$, r_1^n, $\mathbf{r}_2^n$, $\mathbf{w}^n$, $\mathbf{g}^n$ *are known, compute* P^{n+1}, λ^{n+1}, $\mathbf{U}^{n+1}$, $\mathbf{w}^{n+1}$, r_1^{n+1}, $\mathbf{r}_2^{n+1}$, $\mathbf{g}^{n+1}$ *as follows: solve the intermediate Dirichlet problem:*

$$
(3.19) \quad
\begin{aligned}
&\int_\Omega (\alpha \bar{\mathbf{U}}^n \cdot \mathbf{v} + \nu \nabla \bar{\mathbf{U}}^n \cdot \nabla \mathbf{v}) \, d\mathbf{x} \\[6pt]
&= \int_\gamma \mathbf{w}_2^n \cdot \mathbf{v} \, d\gamma + \int_\Omega w_1^n \nabla \cdot \mathbf{v} \, d\mathbf{x}, \ \forall \mathbf{v} \in \mathbf{V}_0; \ \bar{\mathbf{U}}^n \in \mathbf{V}_0,
\end{aligned}
$$

set $\bar{r}_1^n = \nabla \cdot \bar{\mathbf{U}}^n$, $\bar{\mathbf{r}}_2^n = \bar{\mathbf{U}}^n|_\gamma$, and define $\bar{\mathbf{g}}^n = \{\bar{g}_1^n, \bar{\mathbf{g}}_2^n\}$ as follows:

$$(3.20) \qquad \bar{g}_1^n = \alpha\bar{\phi}^n + \nu\bar{r}_1^n,$$

with $\bar{\phi}^n$ the solution of

$$(3.21) \qquad \begin{cases} -\Delta\bar{\phi}^n = \bar{r}_1^n \text{ in } \Omega, \\ \dfrac{\partial\bar{\phi}^n}{\partial\mathbf{n}} = 0 \text{ on } \Gamma_0; \ \bar{\phi}^n = 0 \text{ on } \Gamma_1, \end{cases}$$

$$(3.22) \qquad b(\bar{\mathbf{g}}_2^n, \mu) = \int_\gamma \bar{\mathbf{r}}_2^n \cdot \mu\, d\gamma \ \forall \mu \in \Lambda; \ \bar{\mathbf{g}}_2^n \in \Lambda.$$

We compute then $\rho_n = \int_\Omega r_1^n g_1^n\, d\mathbf{x} + \int_\gamma \mathbf{r}_2^n \cdot \mathbf{g}_2^n\, d\gamma / \int_\Omega \bar{r}_1^n w_1^n\, d\mathbf{x} + \int_\gamma \bar{\mathbf{r}}_2^n \cdot \mathbf{w}_2^n\, d\gamma$, and set

$$(3.23) \qquad P^{n+1} = P^n - \rho_n w_1^n, \ \mathbf{U}^{n+1} = \mathbf{U}^n - \rho_n \bar{\mathbf{U}}^n,$$

$$(3.24) \qquad \lambda^{n+1} = \lambda^n - \rho_n \mathbf{w}_2^n, \ \mathbf{g}^{n+1} = \mathbf{g}^n - \rho_n \bar{\mathbf{g}}^n,$$

$$(3.25) \qquad r_1^{n+1} = r_1^n - \rho_n \bar{r}_1^n, \ \mathbf{r}_2^{n+1} = \mathbf{r}_2^n - \rho_n \bar{\mathbf{r}}_2^n.$$

If $\int_\Omega r_1^{n+1} g_1^{n+1}\, d\mathbf{x} + \int_\gamma \mathbf{r}_2^{n+1} \cdot \mathbf{g}_2^{n+1}\, d\gamma / \int_\Omega r_1^0 g_1^0\, d\mathbf{x} + \int_\gamma \mathbf{r}_2^0 \cdot \mathbf{g}_2^0\, d\gamma \leq \epsilon$, take $\lambda = \lambda^{n+1}$, $P = P^{n+1}$, $\mathbf{U} = \mathbf{U}^{n+1}$. If not, compute

$$(3.26) \quad \gamma_n = \int_\Omega r_1^{n+1} g_1^{n+1}\, d\mathbf{x} + \int_\gamma \mathbf{r}_2^{n+1} \cdot \mathbf{g}_2^{n+1}\, d\gamma / \int_\Omega r_1^n g_1^n\, d\mathbf{x} + \int_\gamma \mathbf{r}_2^n \cdot \mathbf{g}_2^n\, d\gamma,$$

and set $\mathbf{w}^{n+1} = \mathbf{g}^n + \gamma_n \mathbf{w}^n$.
Do $n = n + 1$ and go back to (3.19).

3.3. Numerical experiments

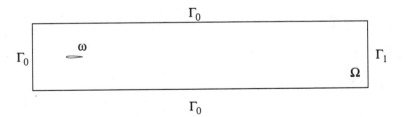

Figure 3.2.

We consider the test problem where ω is a NACA0012 airfoil with zero degree angle of attack centered at $(0, 0)$ and Ω is $(-0.625, 4.375) \times (-0.5, 0.5)$ (see Fig. 3.2). The boundary conditions are defined as follows:

$$(3.27) \qquad \mathbf{u} = \begin{cases} (1 - e^{-ct}) \begin{pmatrix} 1 \\ 0 \end{pmatrix} \text{ on } \Gamma_0, \\ \mathbf{0} \text{ on } \gamma, \end{cases}$$

where c is a positive constant and $\nu\dfrac{\partial \mathbf{u}}{\partial \mathbf{n}} - \mathbf{n}p = \mathbf{0}$ on Γ_1.

As a finite dimensional subspace of $\mathbf{V}$, we choose $\mathbf{V}_h = \{\mathbf{v}_h | \mathbf{v}_h \in H^1_{0h} \times H^1_{0h}\}$ where

$$H^1_{0h} = \{\phi_h | \phi_h \in C^0(\bar{\Omega}), \phi_h|_T \in P_1, \forall T \in \mathcal{T}_h, \phi_h = 0 \text{ on } \Gamma_0\},$$

where $\mathcal{T}_h$ is a triangulation of Ω (see, e.g, Fig. 3.3), P_1 being the space of the polynomials in x_1, x_2 of degree ≤ 1. A traditional way of approximating the pressure is to take it in the space

$$H^1_{2h} = \{\phi_h | \phi_h \in C^0(\bar{\Omega}), \phi_h|_T \in P_1, \forall T \in \mathcal{T}_{2h}\},$$

where $\mathcal{T}_{2h}$ is a triangulation twice coarser than $\mathcal{T}_h$. Concerning the space Λ_h approximating Λ, we define it by

$$\Lambda_h = \{\mu_h | \mu_h \in (L^\infty(\partial\omega))^2, \mu_h \text{ is constant on the segment joining}$$
$$2 \text{ consecutive mesh points on } \partial\omega\}.$$

A particular choice for mesh points on γ is visualized on Fig. 3.3.

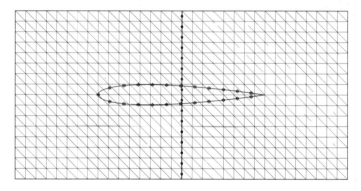

FIGURE 3.3. MESH POINTS MARKED BY "*" ON γ, PART OF MESH POINTS MARKED BY "•" ON THE INTERFACE BETWEEN Ω_1 AND Ω_2 AND PART OF THE TRIANGULATION OF Ω WITH $h = 1/64$.

Using the θ–scheme, we solve at each time step two quasi–Stokes subproblems by the one shot method (3.14)–(3.26) and one advection–diffusion subproblem in a least–squares formulation by a conjugate gradient algorithm. We divide Ω into two subdomains $\Omega_1 = (-0.625, 0.0) \times (-0.5, 0.5)$ and $\Omega_2 = (0.0, 4.375) \times (-0.5, 0.5)$ (see Fig. 3.4) and use domain decomposition methods introduced in Section 2 to solve the elliptic problems arising in the one shot method and in the conjugate gradient algorithm for the least–square problems. The mesh points on the interface between Ω_1 and Ω_2 are shown in Fig. 3.3.

Here we have chosen meshsizes $h_v = 1/64$ for velocity and $h_p = 1/32$ for pressure, time step $\triangle t = 0.01$ and $c = 20$ in (3.29). The number of iterations for the one shot method is from 40 to 60 except the first several time steps.

The number of iterations of the conjugate gradient method for the least–squares method is from 1 to 2. In Fig. 3.5, we observe a *Kàrmàn vortex shedding* (here, the Reynolds number is 1000).

Figure 3.4

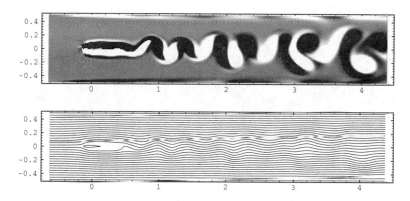

FIGURE 3.5. VORTICITY DENSITY (TOP) AND STREAM LINES (BOT-TOM) FOR THE FLOW PASSING AROUND NACA0012 WITH ZERO DEGREE ANGLE OF ATTACK. FLOW DIRECTION IS FROM THE LEFT TO THE RIGHT, THE REYNOLDS NUMBER IS 1000, DIMENSIONLESS TIME IS 6.

4. Conclusion

Domain decomposition methods combined to fictitious domain methods seem to provide an efficient alternative to conventional solution methods for the solution of Poisson and Navier–Stokes equations on parallel MIMD computer.

This new methodology looks also promising for the simulation of time dependent solution of viscous flow problems around moving rigid bodies. However, further experiments are needed for very large problems to explore parallelization properties of one shot algorithm for 3-D flows, turbulent flow with one point (Baldwin–Lomax) or two point (k–ε) closure models and also local higher order approximations for higher values of the Reynolds number.

Acknowledgements

We would like to acknowledge the helpful comments and suggestions of the following individuals: L. C. Cowsar, E. J. Dean, G. H. Golub, Y. Kuznetsov, W. Lawton, P. Le Tallec, J. Pasciak, M. Ravachol, H. Resnikoff, J. Singer., J. Weiss, R. O. Wells, M. F. Wheeler, O. B. Widlund, and X. Zhou.

The support of the following corporations and institutions is acknowledged: AWARE, Dassault Aviation, INRIA, Texas Center for Advanced Molecular Computation, University of Houston, Université P. et M. Curie. We also benefited from the support of NSF (Grants DMS 8822522, DMS 9112847, INT 8612680 and DMS 9217374), the Texas Board of Higher Education (Grants 003652156ARP and 003652146ATP), DRET (Grant 89424) and DARPA (Contracts AFOSR F49620–89–C–0125 and AFOSR –90–0334).

References

1. Young, D. P., Melvin, R. G., Bieterman, M. B., Johnson, F. T., Samanth, S. S., Bussoletti, J. E., 1991, *A locally refined finite rectangular grid finite element method. Application to Computational Physics*, J. Comp. Physics **92** (1991), pp. 1–66.
2. Bussoletti, J. E., Johnson, F. T., Samanth, S. S., Young, D. P., Burkhart, R. H.,, *EMTRANAIR: Steps toward solution of general 3D Maxwell's equations*, in Computer Methods in Applied Sciences and Engineering, R. Glowinski ed., Nova Science, Commack, NY, 1992, pp. 49–72.
3. Buzbee, B. L., Dorr, F. W., George, J. A., Golub, G. H., 1971, *The direct solution of the discrete Poisson equation on irregular regions*, SIAM J. Num. Anal. **8** (1971), pp. 722–736.
4. Glowinski, R., Pan, T. W., Periaux, J., *A fictitious domain method for Dirichlet problem and applications*, Comp. Meth. Appl. Mech. Eng. **111** (1994), pp. 283–303.
5. Glowinski, R., Pan, T. W., Periaux, J., *A fictitious domain method for external incompressible viscous flow modeled by Navier–Stokes equations*, Comp. Meth. Appl. Mech. Eng. **112** (1994), pp. 133–148.
6. R. Glowinski, M. F. Wheeler, *Domain decomposition and mixed finite element methods for elliptic problems*, in Domain decomposition methods for partial differential equations, R. Glowinski, G. H. Golub, G. Meurant, J. Periaux eds., SIAM, Philadelphia, 1988, 144–172.
7. Glowinski, R., Pan, T. W., Periaux, J., *A one shot domain decomposition/fictitious domain method for the solution of elliptic equations*, in the proceeding of Parallel CFD'93, Paris, France, May, 1993 (to appear).
8. O. Buneman, *A compact non-iterative Poisson solver*, Report 294, Standford University Institute for Plasma Research (1969), Stanford, Cal., 1969.
9. G. H. Golub, C. F. Van Loan, *Matrix computations*, Johns Hopkin University Press, Baltimore, 1983.
10. R. A. Sweet, *a cyclic reduction algorithm for solving block tridiagonal systems of arbitrary dimension*, SIAM J. Num. Anal. **14** (1977), 706–720.
11. R. Glowinski, *Numerical methods for nonlinear variational problems*, Springer–Verlag, New York, 1984.
12. R. Glowinski, *Viscous flow simulation by finite element methods and related numerical techniques*, Progress and Supercomputing in Computational Fluid Dynamics, E. M. Murman and S. S. Abarbanel eds., Birkhauser, Boston, 1985, 173–210.
13. M. O. Bristeau, R. Glowinski and J. Periaux, *Numerical methods for the Navier–Stokes equations*, Comp. Phys. Rep. **6** (1987), 73–187.
14. R. Glowinski, *Finite element methods for the numerical simulation of incompressible viscous flow. Introduction to the control of the Navier–Stokes equations*, Lectures in Applied Mathematics **28**, AMS, Providence, R. I., 1991, 219–301.

DEPARTMENT OF MATHEMATICS, UNIVERSITY OF HOUSTON, HOUSTON, TEXAS 77204, UNIVERSITÉ P. ET M. CURIE, PARIS, AND CERFACS, TOULOUSE, FRANCE

DEPARTMENT OF MATHEMATICS, UNIVERSITY OF HOUSTON, HOUSTON, TEXAS 77204
E-mail address: : pan@math.uh.edu

DASSAULT AVIATION, 92214 SAINT–CLOUD, FRANCE
E-mail address: : periaux@menusin.inria.fr

Contemporary Mathematics
Volume **180**, 1994

Domain-oriented multilevel methods

M. GRIEBEL

ABSTRACT. For the discretization of elliptic linear PDE's, instead of the usual nodal basis, we use a generating system that contains the nodal basis functions of the finest and all coarser levels. The Galerkin approach now results in an enlarged semidefinite linear system to be solved. Traditional iterative methods for that system turn out to be equivalent to modern multilevel methods for the fine grid system. Besides level-oriented iterative methods that lead to multilevel algorithms, other orderings of the unknowns of the enlarged system can be considered as well. A domain-wise block Gauss-Seidel iteration for the enlarged system results in a certain domain decomposition method with convergence rates independent of the mesh width of the fine grid. Furthermore, this approach directly leads to a $O(1)$-preconditioner for the Schur complement that arises in conventional domain decomposition methods.

1. The Generating System

Consider a partial differential equation with linear, symmetric and elliptic operator $Lu = f$ in Ω, with Dirichlet boundary conditions and associated weak formulation $a(u, v) = f(v), \forall v \in V$. For the discretization on some grid Ω_k with uniform mesh width $h_k = 2^{-k}$ usually a basis $B_k = \{\phi_i^{(k)}, i = 1, .., n_k\}$ with nodal basis functions $\phi_i^{(k)}$ is used, that span the corresponding space $V_k = \text{span}\{\phi_i^{(k)}, i = 1, .., n_k\}$. Here, n_k denotes the number of interior grid points and thus the dimension of V_k. Any function $u \in V_k$ can be denoted by

$$u = \sum_{i=1}^{n_k} u_{k,i} \cdot \phi_i^{(k)}$$

with corresponding coefficient vector $u_k^B = (u_{k,i})_{i=1,..,n_k}$ of nodal values. Now, the Galerkin approach leads to the linear system

$$(1) \qquad L_k^B u_k^B = f_k^B$$

1991 *Mathematics Subject Classification*. 65F10, 65N30, 65N55, 65N99.

This work is supported by the Bayerische Forschungsstiftung via FORTWIHR — The Bavarian Consortium for High Performance Scientific Computing.

This paper is in final form, and no version of it will be submitted for publication elsewhere.

with the vector u_k^B of unknowns.

In the context of multilevel methods for the iterative solution of (1), a sequence $\Omega_1, \Omega_2, ..., \Omega_2$ of grids, with associated sequence $B_1, B_2, ..., B_k$ of nodal bases and corresponding spaces $V_1, V_2, ..., V_k$ with dimensions $n_1, n_2, ..., n_k$ is employed. Inspired by that, we now will use directly the generating system

$$E_k = B_1 \cup B_2 \cup ... \cup B_k = \bigcup_{l=1}^{k} B_k$$

for the representation of functions in V_k and for the discretization process. Compare also [2, 4]. This corresponds to the level-wise splitting $V_k = \sum_{l=1}^{k} V_l$ of the underlying discretization space V_k. Since E_k is only a generating system and not a basis, the representation of any function $u \in V_k$ by

$$u = \sum_{l=1}^{k} \sum_{i=1}^{n_l} u_{i,l} \cdot \phi_i^{(l)}$$

with the enlarged vector $u_k^E = (u_1^{B^T}, u_2^{B^T}, ..., u_k^{B^T})^T$ is not unique any more.

For the simple 1D case, Figure 1 shows the functions contained in E_3 and one example for a multilevel representation of a function.

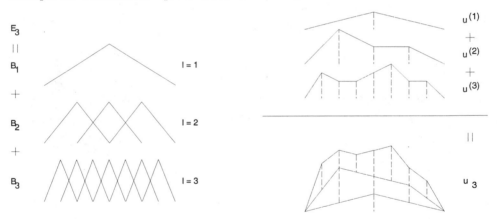

FIGURE 1. The generating system E_3 in 1D (left) and the multilevel representation of u_3 by E_3 (right).

2. The Semidefinite System

Now, we use the generating system E_k directly in the Galerkin discretization process. Then, we obtain the enlarged linear system

(2) $$L_k^E u_k^E = f_k^E$$

with semidefinite matrix L_k^E where for $i_1 = 1, .., n_{l_1}$, $i_2 = 1, .., n_{l_2}$ and $l_1, l_2 = 1, .., k$

$$(L_k^E)_{i_1, i_2, l_1, l_2} = a(\phi_{i_1}^{(l_1)}, \phi_{i_2}^{(l_2)}) \quad \text{and} \quad (f_k^E)_{i_2, l_2} = f(\phi_{i_2}^{(l_2)}).$$

This linear system is of size $n_k^E = \sum_{l=1}^k n_l$, which is in 1D about 2 times, in 2D about 4/3 times and in 3D about 8/7 times larger than n_k, i.e. the size of (1).

Assuming a level-oriented ordering of the unknowns, we obtain the following structure for L_k^E (here for the simple example of $k = 3$):

$$L_k^E = \begin{pmatrix} R_3^1 L_3^B P_1^3 & R_3^1 L_3^B P_2^3 & R_3^1 L_3^B \\ R_3^2 L_3^B P_1^3 & R_3^2 L_3^B P_2^3 & R_3^2 L_3^B \\ L_3^B P_1^3 & L_3^B P_2^3 & L_3^B \end{pmatrix} = \begin{pmatrix} R_3^1 \\ R_3^2 \\ I_3 \end{pmatrix} \cdot L_3^B \cdot \begin{pmatrix} P_1^3 & P_2^3 & I_3 \end{pmatrix}$$

where $R_i^j = {P_j^i}^T$ and P_j^i denotes the interpolation/prolongation from V_j to V_i, $j < i$, and I_i denotes the identity in V_i. Note that $P_j^i = \prod_{q=1}^{i-j} P_{i-q}^{i-q+1}$. Thus, we see that with help of the matrix

$$S_k = \begin{pmatrix} P_1^k & P_2^k & \cdots & P_{k-1}^k & I_k \end{pmatrix}$$

our enlarged system (2) can be written as

$$S_k^T L_k^B S_k u_k^E = S_k^T f_k^B.$$

Now, we see that the discrete Galerkin operators $L_l^B = R_k^l L_k^B P_l^k, l = 1, .., k$, i.e. the stiffness matrices of every level of discretization, are contained as diagonal blocks. The couplings between different levels are contained in the outer diagonal blocks.

Note that our enlarged system is consistent, i.e. $\text{rank}(L_k^E) = \text{rank}(L_k^E, f_k^E)$, and therefore solvable. There exist many different solutions due to the semidefiniteness of L_k^E. Since the unique solution u_k^B of (1) can be obtained from *any* solution u_k^E of (2) by $u_k^B = S_k u_k^E$, the idea is now to produce some u_k^E for (2) by a traditional iterative method and to apply S_k. This will be studied in the following sections.

3. Level-Oriented Methods

In the previous example we employed a level-wise ordering of the unknowns u_k^E that resulted in a level-block partitioning of the matrix L_k^E and the system (2) and was associated with the splitting $V_k = \sum_{l=1}^k V_l = \sum_{l=1}^k \sum_{i=1}^{n_l} V_{l,x_i}$, where $V_{l,x_i} = \text{span}\{\phi_i^{(l)}\}$.

It can be seen easily that traditional iterative methods for (2) are equivalent to modern multilevel methods for (1), c.f. [**2, 4**]. Note that this has also been shown in [**10**] (in a slightly different but equivalent language). For instance, the simple Jacobi-preconditioner for (2) is equivalent to the BPX-preconditioner [**1**] for (1). The BPX-preconditioner can be written as $BPX_k = S_k D_k^{E-1} S_k^T$, where $D_k^E = \text{diag}(L_k^E)$. Now, if we define the generalized condition number κ of a positive semidefinite matrix to be the quotient of the largest and non-vanishing smallest eigenvalue, we obtain directly

$$\kappa(BPX_k L_k^B) = \kappa(S_k D_k^{E-1} S_k^T L_k^B) = \kappa(D_k^{E-1} S_k^T L_k^B S_k) = \kappa(D_k^{E-1} L_k^E)$$

and since $\kappa(BPX_k L_k^B) = O(1)$ (see [**6, 7, 10, 13**]) we have

$$(3) \qquad \kappa(D_k^{E-1} L_k^E) = O(1).$$

Thus, the Jacobi-preconditioned CG-method for (2) converges to some solution within a number of iterations that is independent of k.

Furthermore, we can consider Gauss-Seidel-type methods for (2). They are equivalent to multigrid methods with a Gauss-Seidel smoother, c.f. [2, 4]. For example, the simple Gauss-Seidel iteration on (2) with level-wise ordering $l = 1, .., k$ corresponds to the multigrid (0,1)-V-cycle with one post-smoothing step by Gauss-Seidel. The symmetric Gauss-Seidel-iteration corresponds to the (1,1)-V-cycle. Here, an outer iteration switches from level to level and an inner iteration operates on the specific grids.

The convergence rate of the Gauss-Seidel iteration on (2) can be estimated by

$$\rho = \sqrt{1 - K_0/(1 + K_1)^2} = 1 - O(1).$$

Here, $K_0 := \lambda_{\min \neq 0}(\tilde{L}_k^E) \geq c_0 > 0$ (c.f. (3)), where $\tilde{L}_k^E = D_k^{E-1/2} L_k^E D_k^{E-1/2}$ and c_0 is some constant that is independent of k.

Furthermore, $K_1 := \|\tilde{F}_k^E\|_2$ where $\tilde{F}_k^E$ is given as the lower triangle part of $\tilde{L}_k^E$. With help of the Cauchy-Schwarz inequality (see, e.g. [4, 10, 12, 13]), K_1 can be estimated from above by some constant independent of k. We get $K_1 \leq \lambda_{max}(|\tilde{L}_k^E|) \leq c_1 < \infty$, where $|\tilde{L}_k^E|$ denotes the matrix that is produced from L_k^E by taking the absolute value of each entry. Note that with $\lambda_{\max}(\tilde{L}_k^E) \leq \lambda_{\max}(|L_k^E|)$, c_1 is an upper bound for the largest eigenvalue of the Jacobi-preconditioned matrix, i.e. $\tilde{L}_k^E$, as well.

In addition, $K_1 \leq c_1$ holds for *all* possible Gauss-Seidel traversal orderings, c.f. [4]. Therefore, we obtain a k-independent convergence rate not only for the Gauss-Seidel method for (2) with some level-weise traversal ordering that corresponds to a multigrid method, but also for *any other* traversal ordering as well. This will be exploited in the sequel.

4. Domain-Oriented Methods

Now, we consider domain-oriented Gauss-Seidel iterations for (2). We assume a decomposition on Ω in J non-overlapping subdomains $\Omega_j, j = 1, .., J$, i.e. $\Omega = \bigcup_{j=1}^{J} \Omega^j$ with mutually disjoint interiors so that no grid point lies on an internal boundary and split the grid points $\Omega_l = \bigcup_{j=1}^{J} \Omega_l^j$ on each level $l = 1, .., k$ accordingly. Then, we group together the associated functions of E_k and the unknowns of u_k^E that belong to the same subdomain. The system (2) is partitioned analogously. This corresponds to the splitting

$$V_k = \sum_{j=1}^{J} \left(\sum_{l=1}^{k} \sum_{x \in \Omega_l^j} V_{l,x} \right)$$

where j runs over the domains, l runs over the levels and x runs over the respective grid points. Note that in comparison to the level-wise splitting of the previous section the order of summation is exchanged. The term in parenthesis corresponds now to the block structuring of system (2). Figure 2 shows a 2D example with four subdomains and $k = 3$.

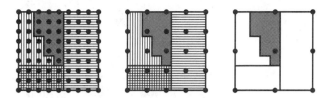

FIGURE 2. Domain-oriented regrouping of E_3 in 2D.

Now, we perform a block Gauss-Seidel iteration for the block partitioned system (2). Then, an outer iteration switches from subdomain to subdomain. If we treat the arising subdomain problems by one inner Gauss-Seidel iteration where within each block a level-wise ordering of the unknowns is applied, we obtain in a natural way a *local* multigrid method (i.e. a local (0,1)-V-cycle). Altogether, this results in a Gauss-Seidel iteration for the overall system (2) with just a domain-wise traversal ordering. Since we have seen in the last section that the upper bound $c_1 < \infty$ of K_1 is independent of the traversal ordering, we directly obtain that the convergence rate of our domain oriented Gauss-Seidel method is independent of k as well.

For the simple 1D case, Figure 3 shows the methodological difference between the level- and domain-oriented methods.

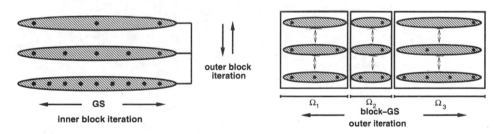

FIGURE 3. The level-oriented (left) and domain-oriented (right) GS-methods for (2) in 1D, $k = 3$.

Note that for the inner block iteration some alternatives to the level-wise traversal ordering exist. If we apply the domain decomposition principle *recursively* until in every domain only one grid point is contained, we obtain the so called point-block method as described in [**3, 4**]. Furthermore, if we restrict ourselves in each subdomain to the subsystem belonging to B_k (and keep the unknowns that belong to $B_l, l < k$ fixed) we can apply exact solvers as well. Note however, that for the outer block iteration all degrees of freedom of E_k take part in the residual computations. In contrast to the conventional domain decomposition method this allows information to travel over long distances as well and maintains fast multigrid-like convergence rates.

5. Schur Complement Preconditioning

Now, we use the generating system approach to derive a simple preconditioner for the Schur complement problem arising in conventional domain decomposition methods. This preconditioner results in a condition number that is independent of k.

 For the ease of explanation only, we restrict ourselves to the simple situation depicted left. There, the grid points are split by the the middle line separator into the set of points Ω_k^2 situated on the separator and the set of remaining grid points $\Omega_k^1 = \Omega_k \backslash \Omega_k^2$ that belong to the interior to the two resulting subdomains.

The nodal basis system (1) is partitioned correspondingly, i.e.

$$\begin{pmatrix} L_{11} & L_{12} \\ L_{21} & L_{22} \end{pmatrix} \begin{pmatrix} u_1 \\ u_2 \end{pmatrix} = \begin{pmatrix} f_1 \\ f_2 \end{pmatrix}$$

where u_2 belongs to the separator and u_1 to the interior of the left and right subdomain. Then, the Schur complement reads $K_{22} = L_{22} - L_{21} L_{11}^{-1} L_{12}$.

Now, we will use the corresponding part B_k^1 of the nodal basis B_k for the grid points Ω_k^1 but the corresponding part E_k^2 of the generating system E_k for the separator. This results in a smaller generating system $\hat{E}_k = B_k^1 \cup E_k^2$. Figure 4 shows the center points and the supports of the contained functions.

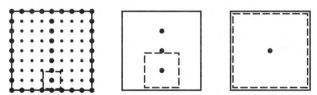

FIGURE 4. Support of the functions of $\hat{E}_k$.

Now, using $\hat{E}_k$, the Galerkin approach results in the semidefinite system

$$\begin{pmatrix} L_{11} & L_{12}^E \\ L_{21}^E & L_{22}^E \end{pmatrix} \begin{pmatrix} u_1 \\ u_2^E \end{pmatrix} = \begin{pmatrix} f_1 \\ f_2^E \end{pmatrix}$$

and, since L_{11} involves only B_k^1 and is therefore invertible, we obtain the enlarged semidefinite Schur complement

$$K_{22}^E = L_{22}^E - L_{21}^E L_{11}^{-1} L_{12}^E.$$

Now, we could work directly with this semidefinite enlarged Schur complement K_{22}^E and the associated linear system $K_{22}^E u_2^E = f_2^E - L_{21}^E L_{11}^{-1} f_1$ like previously with L_k^E and (2). Moreover, we could apply level- or domain-oriented Gauss-Seidel methods that would give us a non-unique solution of the Schur complement system, etc.

Here, however, we want to derive a preconditioner for K_{22}. A short calculation gives

$$K_{22}^E = \hat{S}_k^T \cdot (L_{22} - L_{21} L_{11}^{-1} L_{12}) \cdot \hat{S}_k^E = \hat{S}_k^T \cdot K_{22} \cdot \hat{S}_k^E$$

with $\hat{S}_k = S_k|_{E_k^2} : E_k^2 \to B_k^2$, i.e. the application of S_k to the separator only. Analogously to the BPX-preconditioner we can derive a preconditioner for K_{22} from K_{22}^E. We obtain

$$C_{22} = \hat{S}_k \hat{D}_k^{-1} \hat{S}_k^T \quad \text{with} \quad \hat{D}_k = \text{diag}(L_{22}^E).$$

Note that this construction principle (use the nodal basis in the interior of the subdomains but the subpart of the generating system E_k on the separator) works in the higher dimensional case as well. An analysis in [5, 8] shows $\kappa(C_{22}K_{22}) = O(1)$. A similar construction is given in [9]. See also [11], remark 10.3.

For our simple 2D example above and $L = \Delta$, Table 1 gives the condition numbers of K_{22} and $C_{22}K_{22}$ for different values of k. We clearly see that $\kappa(K_{22})$ behaves like $O(h_k^{-1})$ whereas $\kappa(C_{22}K_{22})$ behaves for sufficiently large k practically like $O(1)$. (The slight increase of the condition number is similar to that observed for the BPX-preconditioner in 2D.)

TABLE 1. Condition numbers of K_{22} and $C_{22}K_{22}$ for different k.

k	2	3	4	5	6	7	8
$\kappa(K_{22})$	1.95	3.81	7.64	15.26	30.51	61.02	122.04
$\kappa(C_{22}K_{22})$	1.65	2.08	2.42	2.67	2.85	2.99	3.09

REFERENCES

1. J. Bramble, and J. Pasciak and J. Xu, *Parallel multilevel preconditioners*, Math. Comp., 31, (1990), 333-390
2. M. Griebel, *Multilevel algorithms considered as iterative methods on indefinite systems*, SFB-Report 342/29/91 A, TUM-I9143, Institut für Informatik, TU München, 1991, also SIAM J. Sci. Comput., 15(3), (1994)
3. M. Griebel, *Grid- and point-oriented multilevel algorithms*, in Notes on Numerical Fluid Mechanics, 41, (1993), 32-46, Vieweg Verlag, Braunschweig
4. M. Griebel, *Multilevelmethoden als Iterationsverfahren über Erzeugendensystemen*, Teubner Skripten zur Numerik, (1994), Teubner Verlag, Braunschweig
5. P. Oswald, *Two remarks on multilevel preconditioners*, report Nr. Math/91/1, (1991), Mathematische Fakultät, FSU Jena
6. P. Oswald, *Norm equivalencies and multilevel Schwarz preconditioning for variational problems*, report Math/92/1, (1992), Mathematische Fakultät, FSU Jena
7. P. Oswald, *On discrete norm estimates related to multilevel preconditioners in the finite element method*, Proc. Int. Conf. Constr. Theory of Functions, (1991), Varna
8. P. Oswald, *Stable splittings of Sobolev spaces and fast solution of variational problems*, report Math/92/5, (1992), Mathematische Fakultät, FSU Jena
9. C. Tong, and T. Chan and C. Kuo, *A domain decomposition preconditioner based on a change to the multilevel nodal basis*, SIAM J. Sci. Stat. Comput., 12, (1991), 1486-1495
10. J. Xu, *Iterative methods by space decomposition and subspace correction: a unifying approach*, SIAM Review, 34(4), (1992), 581-613
11. J. Xu, *Theory of multilevel methods*, Report AM48, (1989), Department of Mathematics, Penn State Univ.
12. H. Yserentant, *Old and new convergence proofs for multigrid methods*, Acta Numerica, (1993)
13. X. Zhang, *Multilevel Schwarz methods*, Numer. Math., 63, (1992), 521-539

INSTITUT FÜR INFORMATIK DER TU MÜNCHEN, D-80290 MÜNCHEN, GERMANY
E-mail address: griebel@informatik.tu-muenchen.de

Contemporary Mathematics
Volume **180**, 1994

Multigrid and Domain Decomposition Methods for Electrostatics Problems

MICHAEL HOLST AND FAISAL SAIED

ABSTRACT. We consider multigrid and domain decomposition methods for the numerical solution of electrostatics problems arising in biophysics. We compare multigrid methods designed for discontinuous coefficients with domain decomposition methods, including comparisons of standard multigrid methods, algebraic multigrid methods, additive and multiplicative Schwarz domain decomposition methods, and acceleration of multigrid and domain decomposition methods with conjugate gradient methods. As a test problem, we consider a linearization of the Poisson-Boltzmann equation, which describes the electrostatic potential of a large complex biomolecule lying in an ionic solvent.

1. Introduction

In recent years, multigrid (MG) and domain decomposition (DD) methods have been used extensively as tools for obtaining approximations to solutions of partial differential equations (see, for example, the references in [**15**]). In this paper, we consider MG and DD methods for the numerical solution of the Poisson-Boltzmann equation, which describes the electrostatic potential of a large complex biomolecule lying in an ionic solvent (see, for example, [**3, 14**] for an overview). We compare MG methods designed for discontinuous coefficients with DD methods, when applied to a two-dimensional, linearized Poisson-Boltzmann equation. Several approaches are considered, including standard MG methods, algebraic MG methods, additive and multiplicative Schwarz methods, and the acceleration of MG and DD methods with conjugate gradient (CG) methods.

1991 *Mathematics Subject Classification.* Primary 65N30; Secondary 65F10.

The first author was supported in part by DOE Grant No. DOE DE-FG02-91ER25099.

The second author was supported in part by NSF Grant No. NSF ASC 92 09502 RIA.

This paper is in final form and no version of it will be submitted for publication elsewhere.

2. Background Material

The nonlinear Poisson-Boltzmann equation (PBE) for the dimensionless electrostatic potential $u(\mathbf{r}) = e_c \phi(\mathbf{r}) k_B^{-1} T^{-1}$ has the form:

$$-\nabla \cdot (\epsilon(\mathbf{r}) \nabla u(\mathbf{r})) + \bar{\kappa}^2 \sinh(u(\mathbf{r})) = \left(\frac{4\pi e_c^2}{k_B T} \right) \sum_{i=1}^{N_m} z_i \delta(\mathbf{r} - \mathbf{r}_i), \quad \mathbf{r} \in \mathbb{R}^3, \quad \Phi(\infty) = 0,$$

where $\phi(\mathbf{r})$ denotes the electrostatic potential at field position $\mathbf{r}$. The coefficients appearing in the equation are necessarily discontinuous by several orders of magnitude, describing both the molecular surface ($\epsilon(\mathbf{r})$) and an ion-exclusion layer ($\bar{\kappa}(\mathbf{r})$) around the molecule. The placement and magnitude of atomic charges are represented by the source terms involving the delta-functions.

Using known analytical solutions for special situations, approximate boundary conditions are obtained for a finite domain $\Omega \subset \mathbb{R}^3$ containing the molecule and some of the surrounding solvent; the problem is then solved as a finite-boundary problem. A linearized form of the equation is often solved as an approximation to the full nonlinear problem [5, 9, 12]. Damped-inexact-Newton methods combined with algebraic MG methods have been shown to be efficient and robust for the full nonlinear problem [7, 8].

3. Multigrid Methods

Consider a nested sequence of finite-dimensional Hilbert spaces

$$\mathcal{H}_1 \subset \mathcal{H}_2 \subset \cdots \subset \mathcal{H}_J = \mathcal{H},$$

each with an associated inner-product $(\cdot, \cdot)_k$ inducing the norm $\| \cdot \|_k = (\cdot, \cdot)_k^{1/2}$. Also associated with each $\mathcal{H}_k$ is an operator A_k, assumed to be SPD with respect to $(\cdot, \cdot)_k$. The spaces $\mathcal{H}_k$, which may be finite element function spaces or simply $\mathbf{R}^{n_k}$ (where $n_k = dim(\mathcal{H}_k)$), are connected by prolongation operators $I_{k-1}^k \in \mathbf{L}(\mathcal{H}_{k-1}, \mathcal{H}_k)$, and restriction operators $I_k^{k-1} \in \mathbf{L}(\mathcal{H}_k, \mathcal{H}_{k-1})$. It is assumed that the operators satisfy *variational conditions*:

(3.1) $$A_{k-1} = I_k^{k-1} A_k I_{k-1}^k, \qquad I_k^{k-1} = (I_{k-1}^k)^T.$$

These conditions hold naturally in the finite element setting, and are imposed directly in algebraic MG methods.

Given $B \approx A^{-1}$ in the space $\mathcal{H}$, the *basic linear method* constructed from the preconditioned system $BAu = Bf$ has the form:

(3.2) $$u^{n+1} = u^n - BAu^n + Bf = (I - BA)u^n + Bf.$$

Now, given some B, or some procedure for applying B, we can either formulate a linear method using $E = I - BA$, or employ a CG method for $BAu = Bf$ if B is SPD.

The recursive formulation of MG methods has been well-known for more than fifteen years; mathematically equivalent forms of the method involving product error propagators have been recognized and exploited theoretically only very recently. In particular, it can be shown [**2, 11**] that if the conditions (3.1) hold, then the MG error propagator can be factored as:

$$E_J = I - B_J A_J = (I - T_{J;J})(I - T_{J;J-1}) \cdots (I - T_{J;1}),$$

where:

$$I_{k-i}^k = I_{k-1}^k I_{k-2}^{k-1} \cdots I_{k-i+1}^{k-i+2} I_{k-i}^{k-i+1}, \quad I_k^{k-i} = I_{k-i+1}^{k-i} I_{k-i+2}^{k-i+1} \cdots I_{k-1}^{k-2} I_k^{k-1}, \quad I_k^k = I,$$

$$T_{J;1} = I_1^J A_1^{-1} I_J^1 A_J, \quad T_{J;k} = I_k^J R_k I_J^k A_J, \quad k = 2, \ldots, J,$$

where $R_k \approx A_k^{-1}$ is the "smoothing" operator employed in each space $\mathcal{H}_k$. We make this remark simply to stress the similarities between MG methods and certain DD methods discussed in the next section.

For problems such as the Poisson-Boltzmann equation, the coefficient discontinuities are complex, and they may not lie on coarse mesh element boundaries as required for accurate finite element approximation (and as required for validity of finite element error estimates). MG methods typically perform badly, and even the regularity-free MG convergence theory [**2**] is invalid.

Possible approaches include coefficient averaging methods (cf. [**1**]) and the explicit enforcement of the conditions (3.1) (cf. [**1, 6, 13**]). By introducing a symbolic stencil calculus and employing Maple or Mathematica, the conditions (3.1) can be enforced algebraically in an efficient way for certain types of sparse matrices; details may be found for example in [**7**].

4. Domain Decomposition Methods

DD methods were first proposed by H.A. Schwarz as a theoretical tool for studying elliptic problems on complicated domains, constructed as the union of simple domains. An interesting early reference not often mentioned is [**10**], containing both analysis and numerical examples, and references to the original work by Schwarz.

Given a domain Ω and coarse triangulation by N regions $\{\Omega_k\}$ of mesh size H, we refine (several times) to obtain a fine mesh of size h. The regions defined by the initial triangulation Ω_k are then extended by δ_k to form the "overlapping subdomains" Ω_k'. Now, let V and V_0 denote the finite element spaces associated with the h and H triangulation of Ω, respectively. The variational problem in V has the form:

$$\text{Find } u \in V \text{ such that } a(u, v) = f(v), \quad \forall v \in V.$$

The form $a(\cdot, \cdot)$ is bilinear, symmetric, coercive, and bounded, whereas $f(\cdot)$ is linear and bounded. Therefore, through the Riesz representation theorem we can associate with the above problem an abstract operator equation $Au = f$, where A is SPD.

DD methods can be seen as iterative methods for solving the above operator equation, involving approximate projections of the error onto subspaces of V associated with the overlapping subdomains Ω_k'. To be more specific, let $V_k = H_0^1(\Omega_k') \cap V$, $k = 1, \ldots, N$; it is not difficult to show that $V = V_1 + \cdots + V_N$, where a coarse space V_0 may also be included in the sum.

As with MG methods, we denote A_k as the restriction of the operator A to the space V_k. Algebraically, it can be shown that $A_k = N_k^T A N_k$, where N_k is the natural inclusion in $\mathbb{R}^{n_k}$, and N_k^T is the corresponding projection. In other words, DD methods automatically satisfy the variation conditions (3.1) in the subspaces V_k, $k \neq 0$. Now, if $R_k \approx A_k^{-1}$, we can define the approximate A-orthogonal projector from V onto V_k as $T_k = N_k R_k N_k^T A$. An overlapping DD method can be written as a basic linear method as in equation (3.2), where the *multiplicative Schwarz* error propagator E is:

$$E = I - BA = (I - T_N)(I - T_{N-1}) \cdots (I - T_0).$$

The *additive Schwarz* error propagator E is:

$$E = I - BA = I - \omega(T_0 + T_1 + \cdots + T_N).$$

An additive-multiplicative variant has been proposed in [4], which takes only the coarse space projection an additive term in the following way:

$$E = I - BA = (I - T_N)(I - T_{N-1}) \cdots (I - T_1) - \omega T_0.$$

This approach decouples the coarse problem in V_0, allowing it to be solved in parallel with the other subproblems.

5. An empirical comparison of MG and DD for a 2D PBE

We now compare several MG and DD methods for a two-dimensional, linearized Poisson-Boltzmann equation. The numerical solution proceeds as follows for two test problems.

In each case, we begin with a simple "triangular" molecule with three point charges. In the first case, we force the molecule surface to align with the coarsest mesh in the MG methods, and to align with the non-overlapped subdomains in the DD methods. In the second case, the discontinuities do not align with the coarsest mesh or the subdomain boundaries (the "non-aligned" case).

Beginning with an initial mesh size H, we uniformly refine the mesh five times, yielding a mesh of size h. Subdomains are then given a small overlap (one fine mesh triangle, $\delta_k = h_k$). Piecewise linear finite elements are used to discretize the problem in all subdomains for the DD methods, and on all levels for the MG methods; the DD methods employ a coarse space. Figure 1 shows the initial triangulation and a sample overlapped subdomain, and a sample solution. Table 1 gives a key to the remaining figures.

Figure 2 shows the performance of the methods, as a function of CPU time on a SPARC 10, for the aligned problem. The MG methods appear to be the most

TABLE 1. Various multigrid and domain decomposition methods.

Method	Description
MG	FEM-based MG, weighted Jacobi smoothing
M	multiplicative Schwarz
A	additive Schwarz (with a damping parameter)
AM	multiplicative Schwarz with additive coarse term
CGMG	CG preconditioned with MG
CGM	CG preconditioned with M
CGA	CG preconditioned with A
CGAM	CG preconditioned with AM
MGG	Algebraic MG, weighted Jacobi smoothing
CGMGG	CG preconditioned with algebraic MG

efficient methods; however, it should be noted that inexact subdomain solvers often lead to improved DD solve times (we employed a sparse direct method). Also, when viewed as error reduction per iteration rather than time in Figure 3, multiplicative Schwarz and multigrid have strikingly similar behavior.

The non-aligned case is illustrated in Figure 4. As expected, the standard MG method fails when the conditions (3.1) are strongly violated. The DD methods remain robust for this problem, whereas the algebraic MG methods appear to be the most efficient. However, note that setup time for the algebraic MG methods (although negligible for this problem) can be quite substantial for some problems.

6. Summary and Conclusions

Convergence theorems for MG and DD methods, applicable in the presence of discontinuous coefficients, rely heavily on the conditions (3.1). Although additional assumptions must be employed to prove that the convergence rate is independent of the meshsize, number of levels, or number of subdomains, very general proofs (although with no rate information) can be given using essentially only (3.1), demonstrating the robustness of this approach.

While the conditions (3.1) are enforced for the algebraic MG methods, they also hold automatically for DD methods, independent of the location of discontinuities in the coefficients. This is not true for the coarse space, which is identical to the MG coarse grid problem; the DD methods appearing in the plots here include a coarse space, but do not explicitly enforce the conditions (3.1) for the coarse problem. If the discontinuities were made worse, the DD methods presented here might also have difficulty with the non-aligned case.

MG and DD methods are comparable sequentially for two-dimensional electrostatics problems. DD methods seem to be naturally more robust, although MG can be made robust and efficient by enforcing the conditions (3.1) explicitly. While the MG methods were generally more efficient, the DD methods offer advantages, such as ease of implementation, as well as parallel implementations.

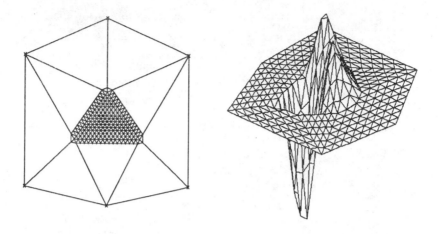

FIGURE 1. An overlapping subdomain and a sample solution.

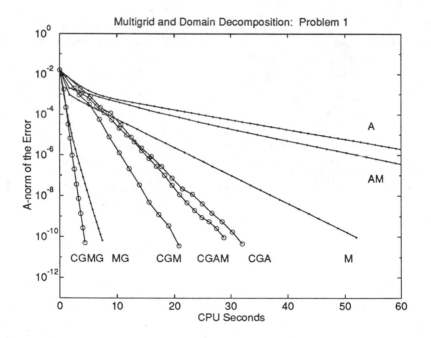

FIGURE 2. CPU seconds for Case 1.

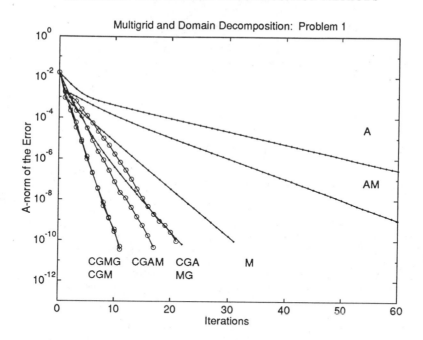

FIGURE 3. Iterations for Case 1.

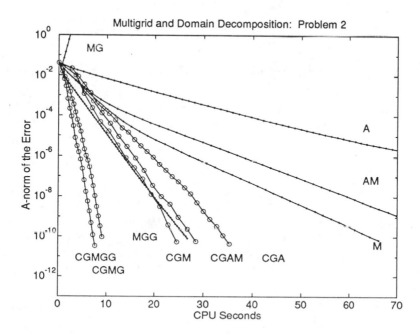

FIGURE 4. CPU seconds for Case 2.

REFERENCES

1. R. E. Alcouffe, A. Brandt, J. E. Dendy, Jr., and J. W. Painter, *The multi-grid method for the diffusion equation with strongly discontinuous coefficients*, SIAM J. Sci. Statist. Comput. **2** (1981), 430–454.
2. J. H. Bramble, J. E. Pasciak, J. Wang, and J. Xu, *Convergence estimates for multigrid algorithms without regularity assumptions*, Math. Comp. **57** (1991), 23–45.
3. J. M. Briggs and J. A. McCammon, *Computation unravels mysteries of molecular biophysics*, Computers in Physics **6** (1990), 238–243.
4. X.-C. Cai, *An optimal two-level overlapping domain decomposition method for elliptic problems in two and three dimensions*, SIAM J. Sci. Statist. Comput. **14** (1993), 239–247.
5. M. E. Davis and J. A. McCammon, *Solving the finite difference linearized Poisson-Boltzmann equation: A comparison of relaxation and conjugate gradient methods*, J. Comput. Chem. **10** (1989), 386–391.
6. J. E. Dendy, Jr., *Two multigrid methods for three-dimensional problems with discontinuous and anisotropic coefficients*, SIAM J. Sci. Statist. Comput. **8** (1987), 673–685.
7. M. Holst, *Multilevel methods for the Poisson-Boltzmann equation*, Ph.D. thesis, Numerical Computing Group, Department of Computer Science, University of Illinois at Urbana-Champaign, 1993, Also published as Tech. Rep. UIUCDCS-R-03-1821.
8. M. Holst, R. Kozack, F. Saied, and S. Subramaniam, *Treatment of electrostatic effects in proteins: Multigrid-based-Newton iterative method for solution of the full nonlinear Poisson-Boltzmann equation*, Proteins: Structure, Function, and Genetics **18** (1994), 231–245.
9. M. Holst and F. Saied, *Multigrid solution of the Poisson-Boltzmann equation*, J. Comput. Chem. **14** (1993), 105–113.
10. L. V. Kantorovich and V. I. Krylov, *Approximate methods of higher analysis*, P. Noordhoff, Ltd, Groningen, The Netherlands, 1958.
11. S. F. McCormick and J. W. Ruge, *Unigrid for multigrid simulation*, Math. Comp. **41** (1983), 43–62.
12. A. Nicholls and B. Honig, *A rapid finite difference algorithm, utilizing successive over-relaxation to solve the Poisson-Boltzmann equation*, J. Comput. Chem. **12** (1991), 435–445.
13. J. W. Ruge and K. Stüben, *Algebraic multigrid*, Multigrid Methods (S. McCormick, ed.), SIAM, 1987, pp. 73–130.
14. K. A. Sharp and B. Honig, *Electrostatic interactions in macromolecules: Theory and applications*, Annu. Rev. Biophys. Biophys. Chem. **19** (1990), 301–332.
15. J. Xu, *Iterative methods by space decomposition and subspace correction*, SIAM Review **34** (1992), 581–613.

DEPARTMENT OF APPLIED MATHEMATICS AND CRPC, CALIFORNIA INSTITUTE OF TECHNOLOGY 217-50, PASADENA, CA 91125
E-mail address: holst@ama.caltech.edu

DEPARTMENT OF COMPUTER SCIENCE, UNIVERSITY OF ILLINOIS AT URBANA-CHAMPAIGN, URBANA, IL 61801
E-mail address: saied@cs.uiuc.edu

Contemporary Mathematics
Volume **180**, 1994

A Parallel Subspace Decomposition
Method for Hyperbolic Equations

EDGAR KATZER

July 5, 1994

ABSTRACT. A parallel subspace decomposition method for solving hyperbolic equations is presented. For a linear model problem, the conservation law is discretized by a cell vertex finite volume method on a triangular grid. Additive and multiplicative Schwarz methods together with a new sequence of subspace corrections are used for solving the normal equations, in a modification of the frequency decomposition approach [**5**]. Uniformly bounded convergence rates for all characteristic directions are obtained. The present approach may be extended to linear hyperbolic and elliptic systems [**9**].

1. Introduction

The numerical solution of flow problems is still a challenging task (see e. g. [**1**], [**2**], [**6**], [**7**], [**10**]). The governing equations, e. g. the Euler equations of steady flow, are a nonlinear system of composite elliptic/hyperbolic character. The efficient solution of hyperbolic equations is a prerequisite for fast solutions of the Euler equations.

Here we solve a simple hyperbolic model problem with constant coefficients:

$$(1.1) \qquad a_1 \, \partial_x u + a_2 \, \partial_y u = 0 \, , \ a_2 > 0 \, ,$$

1991 *Mathematics Subject Classification.* Primary 65M55; Secondary 76M25.

Many fruitful and inspiring discussions with Prof. Hackbusch, Kiel, are gratefully acknowledged.

The present work was supported by the "Deutsche Forschungsgemeinschaft" (German Research Society) and the "Land Sachsen-Anhalt" (Saxony Anhalt) within the program: Graduiertenkolleg "Modellierung, Berechnung und Identifikation mechanischer Systeme" (Modelization, Calculation and Identification of Mechanical Systems).

This paper is in final form and no version of it will be submitted for publication elsewhere.

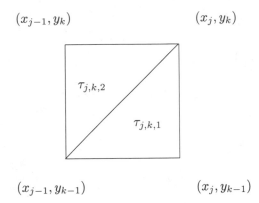

(x_{j-1}, y_k) (x_j, y_k)

$\tau_{j,k,2}$

$\tau_{j,k,1}$

(x_{j-1}, y_{k-1}) (x_j, y_{k-1})

FIGURE 1. Subdivision of a square into two triangles

and x-periodic inflow and boundary conditions on the unit square. A conservative integral formulation is given by Stokes' theorem:

$$(1.2) \qquad \int_{\partial \tau} a_1 \, u \, dy - a_2 \, u \, dx = 0 \,,$$

for sufficiently smooth control elements τ with boundary $\partial \tau$.

A finite volume discretization on a triangular grid and a minimization problem are introduced in section 2. In section 3 we present a subspace decomposition approach and a parallel method based on an additive Schwarz iteration. Section 4 shows numerical results for additive and multiplicative Schwarz iteration. Details are given in [8] and [9].

2. Discrete minimization problem

A regular triangular grid is introduced by subdividing the cells of an equidistant grid according to Figure 1. This grid is equivalent to a triangular grid shown in Figure 2. The conservation equation (1.2) is approximated by the trapezoidal rule and yields for a triangle of type $\tau_{j,k,1}$:

$$(2.1) \qquad Lu\left(\tau_{j,k,1}\right) = \frac{1}{h}\left(a_2 \, u_{j,k} - (a_2 - a_1)u_{j,k-1} - a_1 \, u_{j-1,k-1}\right)$$

and for a triangle of type $\tau_{j,k,2}$:

$$(2.2) \qquad Lu\left(\tau_{j,k,2}\right) = \frac{1}{h}\left(a_1 \, u_{j,k} + (a_2 - a_1)u_{j-1,k} - a_2 \, u_{j-1,k-1}\right)\,.$$

As there are twice as many triangles as grid points, the discrete system $Lu = f$ is overspecified and no solution of the flux equations exists in the general case. Therefore a discrete minimization problem based on the squared Euclidean norm $E(u) = \|Lu - f\|_2^2$ is defined:

$$(2.3) \qquad E(u^*) \le E(u) \,, \ \forall \, u \in U \,.$$

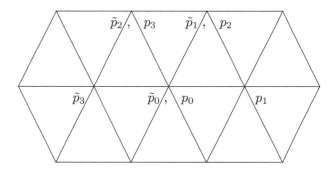

FIGURE 2. Location of coarse grid prolongations

The well known solution of this problem is given by the normal equations:

$$(2.4) \qquad Au^* = L^*Lu^* = L^*f = b \ .$$

Matrix A is positive definite but not an M-matrix. The system is stable and the solution is unique. The discrete solution is second-order accurate on an equidistant grid (see [8]). This is rather surprising, because we do not fulfill the flux equations (2.1, 2.2) exactly but only in the mean. Apparently, the minimization procedure does not deteriorate the accuracy.

3. Subspace decomposition method

The discrete system (2.4) is solved by a subspace decomposition method. The additive Schwarz variant is perfectly suited for a parallel algorithm. We introduce several subspaces, $U_\kappa = \text{Range}\ (p_\kappa)$, of the fine grid space U by prolongations $p_\kappa : V \to U$ on a coarse grid space V. The prolongations are then given in stencil notation:

$$p_0 = \tfrac{1}{2} \begin{bmatrix} 0 & 1 & 1 \\ 1 & 2 & 1 \\ 1 & 1 & 0 \end{bmatrix} \ , \ p_1 = \tfrac{1}{2} \begin{bmatrix} 0 & 1 & -1 \\ -1 & 2 & -1 \\ -1 & 1 & 0 \end{bmatrix} \ ,$$

$$p_2 = \tfrac{1}{2} \begin{bmatrix} 0 & -1 & -1 \\ 1 & 2 & 1 \\ -1 & -1 & 0 \end{bmatrix} \ , \ p_3 = \tfrac{1}{2} \begin{bmatrix} 0 & -1 & 1 \\ -1 & 2 & -1 \\ 1 & -1 & 0 \end{bmatrix} \ .$$

Here, p_0 is the well known seven-point interpolation [3]. The other prolongations, defined on shifted coarse grids, are no more interpolations. The negative signs represent oscillating components transverse to selected characteristic directions. These prolongations are modifications of the frequency decomposition approach [4, 5]. As the stencil notation gives no information on the location of the coarse grid, Figure 2 shows the location of the center of the prolongations in the triangular grid.

Numerical tests showed that these four prolongations are not sufficient for a robust method. It was necessary to introduce four additional prolongations $\tilde{p}_0 \ldots \tilde{p}_3$ with the same stencils as $p_0 \ldots p_3$ but located at shifted coarse grid positions given in Figure 2.

On each subspace U_κ, $1 \leq \kappa \leq K$, smaller minimization problems are given by:

$$(3.1) \qquad\qquad E(u + v_\kappa) \leq E(u + v) , \ \forall \, v \in U_\kappa .$$

This defines coarse grid corrections $G_\kappa(u) = u + v_\kappa$. We obtain the multiplicative Schwarz method:

$$(3.2) \qquad\qquad \Phi^{MS} = \tilde{G}_3 \tilde{G}_2 \tilde{G}_1 \tilde{G}_0 G_3 G_2 G_1 G_0 .$$

With additional smoothing steps given by one iteration of the gradient method S, we obtain a twogrid method:

$$(3.3) \qquad\qquad \Phi^{TG} = \tilde{G}_3 S \tilde{G}_2 S \tilde{G}_1 S \tilde{G}_0 S G_3 S G_2 S G_1 S G_0 S .$$

A parallel algorithm is obtained by an additive Schwarz method:

$$(3.4) \qquad\qquad \Phi^{AS}(u) = u + \sum_{\kappa=1}^{K} \alpha_\kappa v_\kappa .$$

Where the coefficients α_κ are optimized by a small minimization problem:

$$(3.5) \qquad\qquad E(\Phi^{AS}(u)) \leq E(u + v) , \ \forall \, v = \sum_{\kappa=1}^{K} \tilde{\alpha}_\kappa v_\kappa .$$

These corrections can be calculated in parallel on K processors and require the solution of coarse grid systems with different condition numbers. Standard iterative methods need different iteration counts on the processors which leads to load imbalance. Therefore a solution algorithm with time complexity independent of the condition, as e. g. a multigrid or a noniterative approach, is required for solving the coarse grid systems. The determination of the α_κ and the update of U is done sequentially on a single processor and needs communication. Although the solution of (3.5) is a small problem compared with the solution of all coarse grid corrections v_κ, it needs almost the same time as one correction. At the moment the small problem is solved sequentially which causes some load imbalance and reduces the parallel efficiency. The small problem should be solved parallel on several processors at the cost of increased communication. Results for parallel efficiencies are presented in [9].

TABLE 1. Convergence rates for subspace decomposition methods

q	-4.0	-2.0	-1.0	-0.5	0.0	0.25
$\rho(\Phi^{AS})$	0.87	0.88	0.88	0.87	0.74	0.87
$\rho(\Phi^{MS})$	0.63	0.64	0.64	0.64	0.47	0.65
$\rho(\Phi^{TG})$	0.64	0.54	0.55	0.53	0.44	0.50
q	0.50	0.75	1.0	1.5	2.0	4.0
$\rho(\Phi^{AS})$	0.88	0.86	0.73	0.88	0.87	0.86
$\rho(\Phi^{MS})$	0.67	0.65	0.47	0.63	0.64	0.62
$\rho(\Phi^{TG})$	0.50	0.52	0.44	0.54	0.56	0.59

4. Numerical results

The accuracy of the minimization solution is discussed in [8]. Here we present convergence rates for additive and multiplicative Schwarz iterations.

All results in Table 1 are asymptotic error reduction rates obtained on a 32×32 grid. The robustness of the two-grid iteration is analyzed. The characteristic direction, represented by the parameter $q = a_1/a_2$, has only minor influence on the convergence. Convergence rates are uniformly bounded and thus all presented methods are robust. As expected, the multiplicative Schwarz iteration is faster than the additive variant. With smoothing we obtain error reduction rates of approximately 0.5.

5. Conclusion

A new parallel algorithm for solving hyperbolic equations is presented. The linear advection equation is used as a model problem. The conservation form of the equations is discretized on a triangular grid by a cell vertex scheme. The overspecified system is transformed into a minimization problem which is uniformly stable for all characteristic directions. The solution is second order accurate on an equidistant grid.

The normal equations are solved with a subspace decomposition technique. Subspaces are defined by prolongations on a coarse grid. It is a modification of the frequency decomposition approach of Hackbusch [4, 5].

The multiplicative Schwarz iteration together with smoothing iterations shows good convergence rates of approximately 0.5. For the additive Schwarz iteration we obtain slower convergence rates of approximately 0.9, but the algorithm may easily be parallelized.

In all cases, the convergence rates are independent of the characteristic direction; thus the algorithm is robust. This is essential for future applications on flow problems with varying flow directions.

The present approach can be extended to linear and nonlinear systems. Results for linear systems are presented in [9]. The extension to Euler and Navier-Stokes equations is planned for future work.

REFERENCES

1. K. M. J. de Cock, *Multigrid convergence acceleration for the 2D Euler equations applied to high-lift systems*, Contributions to Multigrid, Fourth European Multigrid Conference EMG '93 (P. W. Hemker and P. Wesseling, eds.), CWI Tract 103, Centrum voor Wiskunde en Informatica, Amsterdam, 1994, pp. 25–40.
2. E. Dick, *Multigrid formulation of polynomial flux-difference splitting for steady Euler equations*, J. Comput. Phys. **91** (1990), 161–173.
3. W. Hackbusch, *Multi-Grid Methods and Applications*, Springer-Verlag, Berlin, Heidelberg and New York, 1985.
4. _____, *A new approach to robust multi-grid solvers*, First International Conference on Industrial and Applied Mathematics ICIAM '87 (J. McKenna and R. Temam, eds.), SIAM, Philadelphia, PA., 1988, pp. 114–126.
5. _____, *The frequency decomposition multi-grid method. Part I: Application to anisotropic equations*, Numer. Math. **56** (1990), 229–245.
6. P. W. Hemker and S. P. Spekreijse, *Multiple grid and Osher's scheme for the efficient solution of the Euler equations*, Appl. Numer. Math. **2** (1986), 475–493.
7. A. Jameson, *Solution of the Euler equations for two dimensional transonic flow by a multigrid method*, Appl. Math. Comput. **13** (1983), 327–356.
8. E. Katzer, *A subspace decomposition twogrid method for hyperbolic equations*, Bericht Nr. 9218, Institut für Informatik und Praktische Mathematik, Universität Kiel, Kiel, Germany, 1992.
9. _____, *A Parallel subspace decomposition method for elliptic and hyperbolic systems*, Tenth GAMM-Seminar Kiel on Fast Solvers for Flow Problems, Notes on Numerical Fluid Mechanics, Vieweg, Braunschweig and Wiesbaden (to appear).
10. W. A. Mulder, *A high resolution Euler solver based on multigrid, semi-coarsening, and defect correction*, J. Comput. Phys. **100** (1992), 91–104.

INSTITUTE FOR ANALYSIS AND NUMERICAL MATHEMATICS, OTTO-VON-GUERICKE-UNIVERSITÄT, 39016 MAGDEBURG, GERMANY

Contemporary Mathematics
Volume **180**, 1994

Numerical Treatments for the Helmholtz Problem by Domain Decomposition Techniques

Seongjai Kim

Abstract. A parallelizable iterative procedure based on nonoverlapping domain decomposition techniques for numerical solution of the Helmholtz problem in a bounded domain is discussed. An automatic efficient strategy for choosing the algorithm parameter is demonstrated. Numerical results are reported.

1. Introduction

Consider the (complex–valued) scalar Helmholtz problem

(1.1)
$$-\frac{\omega^2}{c(x)^2}u - \Delta u = f(x), \quad x \in \Omega,$$
$$i\frac{\omega}{c(x)}u + \frac{\partial u}{\partial \nu} = 0, \quad x \in \Gamma,$$

where $\Omega \subset \mathbb{R}^d$, $d \leq 3$, is a bounded domain with a Lipschitz boundary $\Gamma = \partial\Omega$, the coefficient $c(x)$ denotes the wave speed and is bounded below and above by positive constants c_0 and c_1, respectively, ν is the outer unit normal to Γ, and the angular frequency $\omega > 0$. The second equation of (1.1) represents first–order absorbing boundary condition that allows normally incident waves to pass out of Ω transparently.

The Helmholtz problem appears difficult to solve. In addition to having a complex–valued solution, the problem (1.1) is neither Hermitian symmetric nor coercive; as a consequence, most standard iterative methods either fail to converge or converge so slowly as to be impractical. The question to be treated in this paper is that of finding the numerical solution of (1.1) in an effective and computationally efficient fashion. We shall define a parallelizable domain decomposition iterative procedure and indicate an efficient strategy of choosing iteration parameters.

1991 *Mathematics Subject Classification.* Primary 65N55, 65F10; Secondary 65N06.

This paper is in final form and no version of it will be submitted for publication elsewhere.

Concerning the iterative numerical solvers for the Helmholtz problem, we refer to Bayliss, Goldstein and Turkel [1] for the preconditioned conjugate–gradient algorithms applied to the normal equations, and Douglas, Hensley and Roberts [2] for an ADI algorithm. The convergency of the strip–type domain decomposition algorithms has been tested by the author [3].

An outline of the paper is as follows. In §2 the domain decomposition algorithm is defined and the iterative procedure using the Robin–type interface condition is illustrated for finite difference approximate solution of the problem (1.1). In §3, an automatic efficient strategy for finding the algorithm parameter is presented. Some experimental results are reported in §4. The last section indicates the conclusions and possible applications.

2. Domain decomposition iterative procedure

Let $\{\Omega_j,\ j = 1, \ldots, M\}$ be a partition of Ω:

$$\overline{\Omega} = \cup_{j=1}^{M} \overline{\Omega}_j; \quad \Omega_j \cap \Omega_k = \emptyset, \quad j \neq k.$$

Assume that $\Omega_j,\ j = 1, 2, \cdots, M$, is convex. In practice, each Ω_j would be a rectangular or cubic region. Let

$$\Gamma_j = \Gamma \cap \partial\Omega_j, \quad \Gamma_{jk} = \Gamma_{kj} = \partial\Omega_j \cap \partial\Omega_k, \quad \Sigma = \cup_{j,k=1}^{M} \Gamma_{jk}.$$

Let us consider the decomposition of the problem (1.1) over $\{\Omega_j\}$. The problem (1.1) is equivalent to the following: Find $u_j,\ j = 1, \ldots, M$, such that

(2.1)
$$
\begin{aligned}
-\frac{\omega^2}{c^2}u_j - \Delta u_j &= f(x), \quad x \in \Omega_j, \\
i\frac{\omega}{c}u_j + \frac{\partial u_j}{\partial \nu} &= 0, \quad x \in \Gamma_j, \\
\frac{\partial u_j}{\partial \nu_j} + i\beta u_j &= -\frac{\partial u_k}{\partial \nu_k} + i\beta u_k, \quad x \in \Gamma_{jk},
\end{aligned}
$$

where ν_j is the outwards normal to Ω_j, the consistency conditions are replaced by the Robin interface boundary condition, see [6, 3]. The algorithm parameter β is a complex function on Σ with $\mathrm{Re}(\beta) > 0$, for which the subproblems in (2.1) are well–posed. The problem (2.1) is very interesting from a computational point of view; we do not know of any convergence analysis for the problem.

Let $\Omega \subset \mathbb{R}^2$ be composed of M nonoverlapping rectangular regions with the interface edges parallel to either coordinate axes. Let δ_x^2 denote the centered second order difference with respect to x, and ∂_ν, ∂_f and ∂_b be the centered, forward and backward first differences, respectively, in the direction of the outer normal (here, an exterior bordering of the domain is assumed). Let $\Delta_h = \delta_x^2 + \delta_y^2$. Then, one proper finite difference approximation to (2.1) for

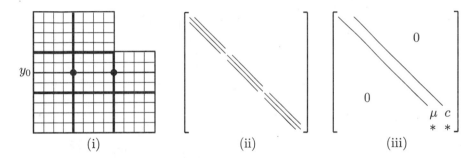

Fig. 3.1: (i). *A decomposition of the domain Ω with mesh lines.* (ii). *The tridiagonal system for the restricted one–dimensional problem on Ω^{y_0}.* (iii). *The matrix U of the LU–factorization performed up to the $(m-1)$–th row.*

two–dimensional problems can be defined by

$$
\begin{aligned}
-\omega^2 u_j^n - \Delta_h u_j^n &= f, \quad x \in \Omega_j, \\
i\omega u_j^n + \partial_\nu u_j^n &= 0, \quad x \in \Gamma_j, \\
\partial_\mathsf{f} u_j^n + i\beta u_j^n &= \partial_\mathsf{b} u_k^{n-1} + i\beta u_k^{n-1}, \quad x \in \Gamma_{jk}.
\end{aligned}
$$

(2.2)

Note that the Robin boundary condition is approximated by a combination of forward–backward differences. This combination is very necessary for both convergence and efficience, and the second–order approximation of the (centered) five point finite difference scheme would not be destroyed. For each subdomain Ω_j, only the subdomains sharing an edge as an interior interface boundary are considered as the adjacent subdomains Ω_k.

3. ADOP: *Alternating Direction Optimal Procedure*

In this section we present a heuristic, automatic method of finding efficient algorithm parameter β for general coefficient problems. Consider an L–shape domain and its domain decomposition depicted in Fig. 3.1 (i). There the bold lines denote the interfaces. We shall determine β, line by line, by using horizontal or vertical mesh lines.

Let us find the values of β on the dotted points along the line $y = y_0$. Ignoring the term u_{yy}, we restrict our problem to the one–dimensional subspace $\Omega^{y_0} := \{(x, y) : y = y_0\}$ decomposed into three subdomains with two dotted points being the interfaces. If the points in Ω^{y_0} are ordered from left to right, one iteration for the restricted problem can be performed by inverting a block diagonal matrix (three blocks), where each block is a tridiagonal matrix of dimension, say, m, see Fig. 3.1 (ii). First, consider the tridiagonal block corresponding to the left–end subdomain. When the LU–factorization in which the diagonal elements of L are 1 is performed up to the $(m-1)$–th row, the matrix U can be expressed like in Fig. 3.1 (iii). This factorization is possible even if the block has not been completely assembled on the interface point. We choose β satisfying

(3.1) $$-\frac{c}{\mu} = 1 - i\beta h,$$

on the first interface point. By using this β, one can complete not only the last row of the first block but also the first row of the next block corresponding to the mid–subdomain. Now, we consider the second tridiagonal block. After performing LU–factorization up to the $(m-1)$–th row, choose β for the second interface point as in the first case. For each mesh line having interface points, including both the horizontal and the vertical mesh lines, this searching can be continued. This procedure is readily applicable to the multiple decompositions of more general domains, clearly. This procedure is developed in [4] and referred to ADOP (alternating direction optimal procedure):

LEMMA 3.1 ([4]). *Let G be the iteration matrix of the one–dimensional problem of (2.2) restricted on Ω^{y_0}, with the parameter β found by ADOP. Then the spectral radius of G is zero, i.e., $\rho(G) = 0$.*

The above lemma implies ADOP seeks the parameter β in such a way that the spectral radii of the iteration matrices of the one–dimensional alternating direction problems are zero. It should be noticed that ADOP is automatic and non–expensive. In the next section, efficiency of ADOP will be numerically checked.

4. Numerical results

This section reports some experimental data for the algorithm (2.2) with the parameter β founded by ADOP. In [4], it is numerically checked that the ADOP parameters introduce a faster convergence than any other constant parameters. The computation is performed in complex double precision on an IBM RS/6000, a serial machine. Let $\Omega = (0,1)^2$. For the results reported in this section, the source function f is selected such that the true solution $u(x,y) = \dfrac{\phi(x) \cdot \phi(y)}{\omega^2}$, where $\phi(x) = e^{i\omega(x-1)} + e^{-i\omega x} - 2$. Zero initial values are assumed. Each subproblem in the algorithm (2.2) is solved directly. The errors are estimated on the relative L^∞-error $r_\infty^n = \dfrac{\|U^n - u\|_{L^\infty(\Omega)}}{\|u\|_{L^\infty(\Omega)}}$, where U^n is the approximate solution of the n–th iteration. For the stopping criterion, $\dfrac{\|U^n - U^{n-1}\|_{L^\infty(\Omega)}}{\|U^n\|_{L^\infty(\Omega)}} \leq 10^{-4}$ is used. We choose three different typical functions for $c(x,y)$:

(4.1)
$$
\begin{aligned}
c_1(x,y) &= 1, \\
c_2(x,y) &= 1 + 2x^3 + y, \\
c_3(x,y) &= e^{xy}(2 - \sin(2\pi x))(2 + \sin(4\pi y)).
\end{aligned}
$$

In Table 4.1, iteration counts n and the error r_∞^n are presented for various

$M_x \times M_y$	1×1	2×1	4×1	8×1	16×1	32×1	64×1
n	–	10	10	21	68	147	200
r_∞^n	.048	.048	.048	.049	.048	.049	.048

Table 4.1: *Strip decompositions, when $\omega = 25$, $c = c_1$ and $h = 1/64$.*

$M_x \times M_y$	1×1	2×1	5×1	10×1	15×1	20×1	30×1
n	–	21	34	80	75	95	123
r_∞^n	.026	.026	.026	.026	.026	.026	.026
time(sec)	88.0	46.0	26.6	40.3	35.2	42.5	54.1

Table 4.2: *Strip decompositions, when $\omega = 40$, $c = c_2$ and $h = 1/120$.*

$M_x \times M_y$	4×1	4×4	8×1	8×8	16×1	16×4	16×16
N	343	419	544	742	864	861	1167
$r_{\infty\infty}$	.0265	.0265	.0265	.0264	.0264	.0263	.0264

Table 4.3: *Total iteration numbers N for solving 100 time steps of the problem (4.2) and the error $r_{\infty\infty}$, when $\omega = 25$, $c = c_3$, $h = 1/64$ and $\Delta t = 1/200$.*

strip domain decompositions $M_x \times M_y$, when $\omega = 25$, $c = c_1$, and $h = 1/64$.

Table 4.2 shows iteration counts n, the error r_∞^n and the CPU–time(second), when $\omega = 40$, $c = c_2$ and $h = 1/120$. For these two examples, constant parameters cannot be used, and the standard iterative methods (relaxations and extrapolations) do not converge.

Next, we consider the following time–discretized Schrödinger problem:

$$
\begin{aligned}
i\frac{u^m - u^{m-1}}{\Delta t} - \frac{\omega^2}{c(\mathbf{x})^2}u^m - \Delta u^m &= f(\mathbf{x}, t^m), & \mathbf{x} \in \Omega, \\
i\frac{\omega}{c(\mathbf{x})}u^m + \frac{\partial u^m}{\partial \nu} &= 0, & \mathbf{x} \in \Gamma, \\
u^0(\mathbf{x}) &= u_0(\mathbf{x}), & \mathbf{x} \in \Omega,
\end{aligned}
$$

(4.2)

where $\mathbf{x} = (x, y)$ and $t^m = m\,\Delta t$ for some $\Delta t > 0$. To check the error propagation, we choose the true solution $u(x, y, t) = (1.5 + \sin(\pi t))\,\psi(x)\,\phi(y)/\omega^2$.

Table 4.3 indicates the total iteration counts N to solve the first 100 time steps of the problem (4.2) by using the domain decomposition method presented in §§2–3 and and the errors

$$
r_{\infty\infty} = \max_{1 \leq m \leq 100} \frac{\|U^m - u^m\|_\infty}{\|u^m\|_\infty},
$$

where U^m is the approximate solution at $t = t^m$, when $\omega = 25$, $c = c_3$, $h = 1/64$ and $\Delta t = 1/200$. The average iteration counts for solving the problem of one time step can be obtained by dividing by 100. When on each time step an one–domain direct solver is used, we have the error $r_{\infty\infty} = .0265$.

5. Conclusions

We have defined a nonoverlapping domain decomposition iterative procedure for the (complex–valued) Helmholtz problem in the finite differences framework. By a combination of forward–backward finite differences, the Robin interface boundary condition is approximated. An effective strategy for finding the algorithm parameters ADOP is introduced and the effectiveness is numerically tested. In addition to being effective, ADOP is automatic as a preprocessor and its cost is never expensive.

When an iterative (domain decomposition) algorithm is designed, iteration parameters are often introduced to accelerate the convergence speed of the iteration. For certain model problems, the parameters can be selected easily and effectively. However, the problem of choosing iteration parameters for a realistic problem may not be so simple. ADOP is proposed as an answer to the problem of choosing iteration parameters.

For the problem (1.1), it is numerically checked that if $\max(\frac{\omega}{c})h \leq \frac{1}{4}$ and $M_x \leq \frac{1}{4h}$, the procedure ADOP leads to convergence for strip domain decompositions. A numerical example for Schrödinger equation is added. From the example, one can expect ADOP will be more useful for singularly perturbed problems such as, e.g., second order time–dependent partial differential equations. When the wave speed c in (1.1) is complex–valued with $\text{Re}(c) > 0$ and $\text{Im}(c) > 0$, the convergence of the iterative algorithm (2.2) can be analyzed [5].

References

[1] A. Bayliss, C. Goldstein and E. Turkel, *An iterative method for the Helmholtz equation*, J. Comput. Phys. 49 (1983), 443–457.

[2] J. Douglas, Jr., J. L. Hensley, and J. E. Roberts, *An alternating–direction iteration method for Helmholtz problems* (Technical Report #214, Mathematics Department, Purdue University, W. Lafayette, IN 47907, 1993).

[3] S. Kim, *A parallelizable iterative procedure for the Helmholtz problem*, to appear in Appl. Numer. Math. (1994).

[4] S. Kim, *ADOP: an automatic strategy for finding efficient iteration parameters in domain decomposition algorithms* , submitted for publication.

[5] S. Kim, *Domain decomposition method and parallel computing: finite elements with mass lumping for scalar waves*, submitted for publication.

[6] P.L. Lions, *On the Schwarz alternating method III: a variant for nonoverlapping subdomains*, in: T.F. Chan, R. Glowinski, J. Periaux and O. B. Widlund, Ed., *Third International Symposium on Domain Decomposition Method for Partial Differential Equations*, SIAM, Philadelphia (1990), 202–223.

DEPARTMENT OF MATHEMATICS
PURDUE UNIVERSITY, W. LAFAYETTE, IN 47907
E–mail address: skim@math.purdue.edu

Contemporary Mathematics
Volume **180**, 1994

Schwarz Methods for Obstacle Problems with Convection-Diffusion Operators

YU.A. KUZNETSOV, P. NEITTAANMÄKI AND P. TARVAINEN

ABSTRACT. Multiplicative and additive Schwarz methods are applied to the algebraic problems arising from finite element or finite difference approximations of obstacle problems with convection-diffusion operators. We show that the methods are monotonically convergent in the subset of supersolutions. Moreover, we present a new technique, by which we obtain two-sided approximations for the mesh contact domain. Numerical experiments are included to illustrate the theoretical results.

1. Introduction

In this paper we consider the numerical solution of obstacle problems with convection-diffusion operators by Schwarz-type overlapping domain decomposition methods. We present here some theoretical and experimental results reported earlier in [7].

The motivation for studying the solution of obstacle-type variational inequalities by methods based on the ideas of domain decomposition is natural because of their complementarity property [10]: the solution of the obstacle problem decomposes the domain into two (possibly overlapping) subdomains: one, where the solution equals to a given obstacle function, and the other, where the solution satisfies linear equations. Several papers have been issued about overlapping domain decomposition methods for the obstacle problems, e.g. [2], [6], [9], but in those papers there are no considerations about taking advantage of the complementarity property in order to construct reasonable domain partitions. Moreover, the results of those papers are valid only for self-adjoint operators.

1991 *Mathematics Subject Classification.* 65K10, 65N30, 65N55.
Key words and phrases. Obstacle problems, convection-diffusion operators, Schwarz domain decomposition methods, two-sided approximations.
The third author was supported by Academy of Finland.
This paper is in final form and no version of it will be submitted for publication elsewhere.

It is clear that the multiplicative and additive Schwarz methods being applied to the mesh systems arising from finite difference or finite element discretizations of the differential problems are particular cases of block relaxation methods with overlapping groups of unknowns. The convergence of the block relaxation methods without overlapping applied to the algebraic obstacle problems was studied for the self-adjoint case in [3] and for the case of M-matrices in [1], for instance.

This paper consists of two parts: In the first, we formulate the problem and give the convergence results for the multiplicative and additive Schwarz methods. We can show that methods are monotonically convergent in the subset of super-solutions. In the second part, we introduce a new technique to obtain two-sided approximations for the mesh contact domain. The technique can be used within the Schwarz methods to decompose the computational domain into overlapping subdomains with linear and obstacle subproblems. We include some numerical experiments to illustrate this technique.

Let Ω be an open bounded polygon in $\mathbb{R}^2$. We consider the following obstacle problem: Find $u \in K$ such that

$$(1) \qquad a(u, v - u) \geq (f, v - u) \quad \forall v \in K,$$

where K is a closed, convex subset of $H_0^1(\Omega)$:

$$(2) \qquad K = \{v \in H_0^1(\Omega) | \ v \geq \psi \text{ a.e. in } \Omega\},$$

$\psi \in H^2(\Omega)$ is an obstacle function such that $\psi|_{\partial\Omega} \leq 0$, $f \in L^2(\Omega)$,

$$(3) \qquad (f, v) = \int_\Omega fv \, d\Omega, \quad v \in L^2(\Omega),$$

and the bilinear form corresponds to the convection-diffusion differential operator:

$$(4) \qquad a(u, v) = \int_\Omega [a \, \nabla u \cdot \nabla v + (\vec{b} \cdot \nabla u)v + c \, uv] \, d\Omega,$$

where the coefficients $a \geq \text{const} > 0$, $\vec{b} = (b_1, b_2)$ and $c \geq 0$ are piecewise smooth and bounded. Such kind of obstacle problems arise, for instance, in mathematical modelling of the continuous casting process [10] and some problems in mathematical economics [1].

By using discretization techniques like finite element method with upwinding [5], we obtain an algebraic problem, which can be represented either in terms of variational inequalities: find $u \in K = \{v \in \mathbb{R}^N | \ v \geq \psi\}$ such that

$$(5) \qquad (Au, v - u) \geq (f, v - u) \quad \forall v \in K,$$

or in the complementarity form: find $u \in S = \{v \in K | \ (Av - f)_j \geq 0, j = 1, \ldots, N\}$ such that

$$(6) \qquad (u - \psi)_j \cdot (Au - f)_j = 0, \quad j = 1, \ldots, N,$$

where f, $\psi \in \mathbb{R}^N$. The subset S is called the subset of supersolutions to the problem (5) or (6).

We assume that the matrix A is an M-matrix [11], not necessarily symmetric. Under the assumptions made it can be shown that the problem (5) has a unique solution [4].

2. Schwarz overlapping domain decomposition methods

Let Ω_h – the set of mesh nodes, or the mesh domain – be decomposed into m overlapping subdomains $\Omega_h^{(i)}$ such that $\Omega_h = \bigcup_{i=1}^m \Omega_h^{(i)}$, and every mesh node $x_j \in \Omega_h$ belongs to at least one subdomain $\Omega_h^{(i)}$. For given $w \in K$ and $\Omega_h^{(i)}$, $i = 1, \ldots, m$, we define the subset $K_i(w)$ of K by

$$(7) \qquad K_i(w) = \{v \in K | \ v_j = w_j, \ x_j \notin \Omega_h^{(i)}\},$$

and the operator $T_i : K \to K$ such that the solution of the subdomain problem: find $z \in K_i(w)$ such that

$$(8) \qquad (Az, v - z) \geq (f, v - z) \quad \forall v \in K_i(w),$$

is given by

$$(9) \qquad z = T_i(w), \quad i = 1, \ldots, m.$$

The multiplicative Schwarz method can be formulated in terms of the operators (9) in the following way: Let $u^0 \in S$ be given. Then for $k \geq 0$

$$(10) \qquad u^{k+1} = T_m(T_{m-1}(\cdots(T_1(u^k))\cdots)).$$

Similarly, we can formulate the additive Schwarz method: Let $u^0 \in S$ be given, and choose parameters ω_i, $i = 1, \ldots, m$, such that

$$(11) \qquad \sum_{i=1}^m \omega_i = 1.$$

Then, for $k \geq 0$,

$$(12) \qquad u^{k+1} = \sum_{i=1}^m \omega_i T_i(u^k).$$

In [7] we have shown the following convergence results:

THEOREM 1. *The multiplicative and additive Schwarz methods are monotonically convergent in* S.

Here, we mean by the monotonic convergence in the subset S, that for all $u^0 \in S$ the algorithms generate a monotonically decreasing sequence $\{u^k\}$, $u^k \in S$, which converges to the solution of the problem (5).

3. Two-sided approximations for the mesh contact domain

The aim of this section is to construct two-sided approximations within the domain decomposition methods of Section 2 for the mesh contact domain G_h,

$$(13) \qquad G_h = \{x_j \in \Omega_h| \ u_j^* = \psi_j\},$$

where u^* is the solution of the problem (5).

Let a monotonically decreasing sequence $\{u^k\}$ be given such that $u^k \xrightarrow[k\to\infty]{} u^*$ in S. If we define the mesh domains G_h^k, $k \geq 0$, by

$$(14) \qquad G_h^k = \{x_j \in \Omega_h| \ u_j^k = \psi_j\},$$

then it follows immediately that

$$(15) \qquad G_h^k \subseteq G_h^{k+1}, \ k \geq 0.$$

To obtain outer approximations within the domain decomposition procedures, we have to solve some additional linear problems of the following form: Assume that G_h^k is given for some $k \geq 0$. Find $w^k \in \mathbb{R}^N$ such that

$$(16) \qquad \begin{cases} w_j^k = \psi_j, \ \text{if } x_j \in G_h^k, \\ (Aw^k - f)_j = 0, \quad \text{otherwise.} \end{cases}$$

It can be shown [7] that

$$(17) \qquad w^k \leq u^* \leq u^k,$$

and if we define the mesh domains $\widehat{G}_h^k$, $k \geq 0$, by

$$(18) \qquad \widehat{G}_h^k = \{x_j \in \Omega_h| \ w_j^k \leq \psi_j\},$$

we can state the following conclusion [7]:

THEOREM 2. *Under the assumptions made*

$$(19) \qquad G_h^k \subseteq G_h \subseteq \widehat{G}_h^k, \quad k = 0, 1, \ldots.$$

As a consequence, we notice that in each iteration step k the mesh domain can be divided into three subdomains with respect to the two-sided approximations: the contact subdomain G_h^k, the linear subdomain $\Omega_h \setminus \widehat{G}_h^k$ and the problematic subdomain $\widehat{G}_h^k \setminus G_h^k$. It follows from the above theory and the complementarity property, that the solution u^* satisfies:

$$(20) \qquad \begin{aligned} u_j^* &= \psi_j, \quad \text{if } x_j \in G_h^k \ \text{(the contact subdomain)}, \\ (Au^* - f)_j &= 0, \quad \text{if } x_j \in \Omega_h \setminus \widehat{G}_h^k \ \text{(the linear subdomain)}, \end{aligned}$$

and only in $\widehat{G}_h^k \setminus G_h^k$ do we not know which condition of (20) the solution satisfies. Naturally, this information can be used to construct reasonable domain partitions for the Schwarz methods. Furthermore, these partitions can be modified within the domain decomposition procedure.

4. Numerical experiments

This section consists of two examples: in the first example we illustrate by means of figures the technique of two-sided approximations, and in the second example we compare execution times of an iterative procedure based on our approach and a traditional solution algorithm, the SOR method with projection.

Let Ω be the unit square and consider the obstacle problem (1) with the following data: $f \equiv -6$, $\psi(x) = -\text{dist}(x, \partial\Omega)$, $a \equiv 1$, $\vec{b} \equiv (5,5)$, $c \equiv 0$, where the function $\text{dist}(\cdot, \partial\Omega)$ means the distance from the boundary of Ω. Thus, we consider the obstacle problem with the nonself-adjoint operator. We have solved the problem by a multiplicative procedure, and Figure 1 demonstrates the behaviour of the algorithm. We have denoted by dots ($\cdot$) the contact subdomains, by bullets ($\bullet$) the problematic subdomains, and white regions denote the linear subdomains of each iteration step. It can be seen that the problematic subdomains are efficiently reduced by the two-sided approximations. Hence, Schwarz methods can be applied such that in the main part of the domain linear subdomain solvers are used.

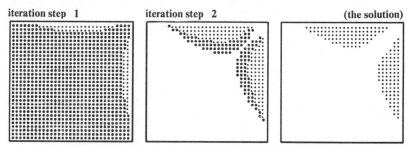

iteration step 1   iteration step 2   (the solution)

FIGURE 1. TWO-SIDED APPROXIMATIONS.

In the second example we consider the obstacle problem (1) in the unit square with the data: $a \equiv 1$, $\vec{b} \equiv (0,0)$, $c \equiv 0$, $\psi \equiv 0$, and

$$(21) \qquad f(x) = \begin{cases} -2, & x \in (3/8, 5/8) \times (3/8, 5/8), \\ 1, & \text{otherwise.} \end{cases}$$

We have implemented the multiplicative Schwarz procedure, which makes use of information from the two-sided approximations in such a way that we decompose the computational domain into rectangular subdomains such that the problematic subdomain of each iteration step is included in one of the subdomains, and others are linear subdomains. Then we apply the multiplicative Schwarz method such that in the linear subdomains we use fast direct solvers based on the fast Fourier transform, and in the nonlinear subdomains we apply the line Gauss-Seidel method. The additional linear problems needed to construct two-sided approximations are solved by the fictitious domain method.

In Table 1 we see the execution times of the projected SOR-method [5] with the acceleration parameter $\omega = 1.7$ (PSOR) and the algorithm described above

(DDM). We notice that the domain decomposition algorithm based on the two-sided approximations works faster in all cases above. We emphasize, that this algorithm is only the simplest possible implementation, without any acceleration.

Execution times (sec.) – HP 9000/735		
$n \times n$	PSOR	DDM
15×15	0.05	0.05
31×31	0.33	0.17
63×63	3.37	1.53
127×127	42.35	19.23

TABLE 1. COMPARISON OF TWO SOLUTION ALGORITHMS.

REFERENCES

1. B. Ahn, *Solution of Nonsymmetric Linear Complementarity Problems by Iterative Methods*, J. Opt. Th. Appl. **33** (1981), 175–185.
2. L. Badea, *On the Schwarz Alternating Method with More than Two Subdomains for Nonlinear Monotone Problems*, SIAM J. Num. Anal. **28** (1991), 179–204.
3. R. Cottle, G. Golub, and R. Sacher, *On the Solution of Large, Structured Linear Complementarity Problems: The Block Partitioned Case*, Appl. Math. Opt. **4** (1978), 347–363.
4. R. Cottle and A. Veinott, Jr., *Polyhedral Sets Having a Least Element*, Math. Prog. **3** (1972), 238–249.
5. R. Glowinski, *Numerical Methods for Nonlinear Variational Problems*, Springer-Verlag, New York, 1984.
6. K.-H. Hoffmann and J. Zou, *Parallel Algorithms of Schwarz Variant for Variational Inequalities*, Num. Funct. Anal. Opt. **13** (1992), 449–462.
7. Yu. Kuznetsov, P. Neittaanmäki and P. Tarvainen, *Overlapping Block Relaxation and Schwarz Methods for the Obstacle Problem with a Convection-Diffusion Operator*, Laboratory of Scientific Computing, Report 4/1993 (1993), University of Jyväskylä.
8. Yu. Kuznetsov, P. Neittaanmäki and P. Tarvainen, *Overlapping Domain Decomposition Methods for the Obstacle Problem*, Domain Decomposition Methods in Science and Engineering (Yu. Kuznetsov, J. Periaux, A. Quarteroni and O. Widlund, eds.), AMS, Providence, 1994, pp. 271–277.
9. P.L. Lions, *On the Schwarz Alternating Method* I, Domain Decomposition Methods for Partial Differential Equations (R. Glowinski, G. Golub, G. Meurant and J. Periaux, eds.), SIAM, Philadelphia, 1988, pp. 2–42.
10. J.-F. Rodrigues, *Obstacle Problems in Mathematical Physics*, North-Holland, Amsterdam, 1987.
11. R. Varga, *Matrix Iterative Analysis*, Prentice-Hall, New Jersey, 1962.

INSTITUTE OF NUMERICAL MATHEMATICS, RUSSIAN ACADEMY OF SCIENCES, LENINSKI PROSPECT 32-A, MOSCOW 117334, RUSSIA
E-mail address: labnumat@node.ias.msk.su

LABORATORY OF SCIENTIFIC COMPUTING, DEPARTMENT OF MATHEMATICS, UNIVERSITY OF JYVÄSKYLÄ, P.O. BOX 35, FIN-40351, JYVÄSKYLÄ, FINLAND
E-mail address: neittaanmaki@finjyu.bitnet

LABORATORY OF SCIENTIFIC COMPUTING, DEPARTMENT OF MATHEMATICS, UNIVERSITY OF JYVÄSKYLÄ, P.O. BOX 35, FIN-40351, JYVÄSKYLÄ, FINLAND
E-mail address: pht@math.jyu.fi

Contemporary Mathematics
Volume **180**, 1994

On Domain Decomposition
and Shooting Methods
for Two-Point Boundary Value Problems

C.-H. LAI

ABSTRACT. The fundamental properties of a shooting method for two-point boundary value problems are examined and its relation to a nonoverlapped domain decomposition technique is discussed. We address and tackle the instability problem. We extend the nonoverlapped domain decomposition to multisubdomain cases and derive an efficient parallel shooting algorithm. Numerical examples include linear problems in convection-diffusion and nonlinear problems in semiconductor device modelling.

1. Introduction

This paper intends to review some of the fundamental properties of a shooting method for two-point boundary value problems and the connection of these properties to domain decomposition methods. Idea related to this subject can be found in [**5**] [**6**]. Similar approach for two-point boundary value problems was reported in [**4**]. Attention is restricted to the following class of second order ordinary differential equations

$$(1) \qquad \frac{d^2\phi}{dx^2} + Q(x, \phi, \frac{d\phi}{dx}) = 0 \in \Omega = \{x : a < x < b\},$$

subject to boundary conditions of either

$$(2) \qquad \phi(a) = \phi_a, \ \frac{d\phi(b)}{dx} = \phi_b',$$

or

$$(3) \qquad \phi(a) = \phi_a, \ \phi(b) = \phi_b,$$

1991 *Mathematics Subject Classification.* Primary 65M55; Secondary 65N55, 65Y10.
This research was supported by SERC under grant GR/J58312.
This paper is in final form and no version of it will be submitted for publication elsewhere.
This paper was typeset using $\mathcal{AMS}$-TEX Version 2.1.

where ϕ_a, ϕ_b' and ϕ_b are given constants. Assume that the function $Q(x, \phi, d\phi/dx)$ is continuous and is Lipschitz bounded so that unique solution of (1) exists for each of the above two types of boundary conditions.

2. Elementary Properties of a Shooting Method

A shooting method for (1) subject to boundary conditions (2) can be obtained by choosing a trial value λ_a and then by solving the initial value problem

$$(4) \qquad \frac{d^2u}{dx^2} = -Q(x, u, \frac{du}{dx}) , \; u(a) = \phi_a , \; u'(a) = \lambda_a .$$

We assume that the solution $u(x; \lambda)$ of (4) does not suffer from any instability. A family of solutions $F\{u(x; \lambda)\}$ is obtained subject to the variation of the parameter λ_a. This parameter is adjusted and the initial value problem (4) is solved with the adjusted λ_a until the derivative of the solution of (4) at $x = b$ is sufficiently close to ϕ_b'. It is clear at this stage that finding the correct value of λ_a is equivalent to finding a root of the nonlinear function [1] [3],

$$(5) \qquad f(\lambda_a) \equiv \frac{\partial u(b; \lambda_a)}{\partial x} - \phi_b' = 0 .$$

To approximate the root of $f(\lambda_a) = 0$, we write the equation as

$$(6) \qquad \lambda_a = G(\lambda_a) \equiv \lambda_a - \alpha f(\lambda_a) ,$$

and consider the fixed point iteration scheme

$$(7) \qquad \lambda_a^{(n+1)} = \lambda_a^{(n)} - \alpha f(\lambda_a^{(n)}) , \; n = 0, 1, 2, \ldots .$$

Assuming G satisfies a Lipschitz condition on λ_a and that a suitable choice of α is being used so that the sequence $\{\lambda_a^{(n)}\}$ is a converging sequence which converges to a root of (5). A treatise on the choice of α can be found in [1] [2].

One disadvantage of the above shooting method lies in the fact that an initial value problem has to be solved at each step of the iteration scheme (7). The flexibility of using parallel architecture is very restricted because of the nature of the initial value problem. Since the initial value problem (4) has a unique solution for each value of λ_a and if the iteration scheme (7) converges to a root of (5), then at each step of the iteration there exists a one-to-one correspondence between $\lambda_a^{(n)}$ and $\lambda_b^{(n)} \equiv u(b; \lambda_a^{(n)})$. Hence we have,

PROPOSITION 1. *For any converging sequence $\{\lambda_a^{(n)}\}$ which converges to a root of the nonlinear equation $f(\lambda_a) \equiv \frac{\partial u(b; \lambda_a)}{\partial x} - \phi_b' = 0$ such that $u(x; \lambda_a^{(n)})$ is the solution of the intial value problem,*

$$(8) \qquad \frac{d^2u}{dx^2} = -Q(x, u, \frac{du}{dx}) , \; u(a) = \phi_a , \; u_a'(a) = \lambda_a^{(n)} ,$$

there exists a corresponding converging sequence $\{\lambda_b^{(n)} = u(b; \lambda_a^{(n)})\}$ *such that the solution* $v(x; \lambda_b^{(n)})$ *of the boundary value problem*

(9) $$\frac{d^2v}{dx^2} + Q(x, v, \frac{dv}{dx}) = 0 , \ v(a) = \phi_a , \ v(b) = \lambda_b^{(n)} ,$$

is equal to the solution $u(x; \lambda_a^{(n)})$ *of the initial value problem.*

COROLLARY. *For any converging sequence* $\{\lambda_b^{(n)} := \lambda_b^{(n)} - \beta g(\lambda_b^{(n-1)})\}$ *which converges to a root of the nonlinear equation* $g(\lambda_b) \equiv \frac{\partial v(b;\lambda_b)}{\partial x} - \phi_b' = 0$ *where* $v(x; \lambda_b^{(n)})$ *is the solution of* (9), *then there exists an* $\alpha = \alpha(\beta)$ *such that the solution of the boundary value problem* (9) *is equal to the solution of the initial value problem* (8) *with* $\lambda_a^{(n)} = \frac{\partial v(a;\lambda_b^{(n)})}{\partial x}$.

EXAMPLE 1. The initial value problem

$$\frac{d^2u}{dx^2} = \gamma\frac{du}{dx} , \ u(a) = \phi_a , \ u'(a) = \lambda_a^{(n)} ,$$

has solution

$$u(x; \lambda_a^{(n)}) = \phi_a + \frac{\lambda_a^{(n)}}{\gamma}(\frac{e^{\gamma x} - e^{\gamma a}}{e^{\gamma a}}) .$$

The boundary value problem

$$\frac{d^2v}{dx^2} - \gamma\frac{dv}{dx} = 0 , \ v(a) = \phi_a , \ v(b) = \lambda_b^{(n)} ,$$

has solution

$$v(x; \lambda_b^{(n)}) = \phi_a + \frac{\lambda_b^{(n)} - \phi_a}{e^{\gamma b} - e^{\gamma a}}(e^{\gamma x} - e^{\gamma a}) .$$

Substituting $\lambda_a^{(n)} = \frac{\partial v(a;\lambda_b^{(n)})}{\partial x}$ into the expression for $u(x; \lambda_a^{(n)})$, we obtain $v(x) = u(x)$. From the expression $\lambda_b^{(n+1)} = \lambda_b^{(n)} - \beta g(\lambda_b^{(n)})$, we can deduce that $\lambda_a^{(n+1)} = \lambda_a^{(n)} - \alpha f(\lambda_a^{(n)})$, where $\alpha = \beta\frac{\gamma e^{\gamma a}}{e^{\gamma b} - e^{\gamma a}}$.

Since the stability of (9) is easier to control than that of (8) therefore it seems better to work with (9). It follows from the Corollary that we can establish a variant of the above shooting method as following,

PROPOSITION 2. *The solution of* (1) *subject to boundary conditions* (2) *can be obtained by finding a root of the nonlinear function* $g(\lambda) \equiv \frac{\partial v(b;\lambda_b)}{\partial x} - \phi_b' = 0$, *where* $v(x; \lambda_b)$ *is the solution of the boundary value problem*

(10) $$\frac{d^2v}{dx^2} + Q(x, v, \frac{dv}{dx}) = 0 , \ v(a) = \phi_a , \ v(b) = \lambda_b .$$

Proposition 2 complicates the solution process of (1), but is used in the context of a domain decomposition. The advantage of the present variant is that (10) involves solutions of boundary value problems and the instability introduced by initial value problems can be eliminated.

3. A Nonoverlapped Domain Decomposition Method

For simplicity, the following two-point boundary value problem is considered,

$$(11) \qquad \frac{d^2\phi}{dx^2} + Q(x, \phi, \frac{d\phi}{dx}) = 0 \in \Omega = \{x : a < x < c\},$$

subject to Dirichlet boundary conditions $\phi(a) = \phi_a$ and $\phi(c) = \phi_c$. We construct two nonoverlapped subdomains $\Omega_1 = \{x : a < x < b\}$ and $\Omega_2 = \{x : b < x < c\}$ and the interface of Ω_1 and Ω_2 is $\Gamma_{12} = \{x : x = b\}$. For the problem given by (11), the coupling at the interface Γ_{12} is well known to be (a) the continuity of the function and (b) the continuity of the derivative of the function at that point.

It is obvious that one can use Proposition 2 to solve the two subproblems provided we know the value of ϕ'_b. The situation now is analogous to a shooting exercise in which we require the two shells fired from two artillery men based at $x = a$ and $x = c$ to collide at the interface Γ_{12}. We can interpret the above coupling conditions as the height and the slope of the trajectory at the point of collision. Application of Proposition 2 is achieved by simply reducing the number of variational parameters to one, in which case we have a one parameter nonlinear equation of which the solution is required. This discussion is summarised in the following Proposition.

PROPOSITION 3. *The boundary value problem* (11) *is replaced by the following two subproblems,*

$$(12) \qquad \frac{d^2u_1}{dx^2} + Q(x, u_1, \frac{du_1}{dx}) = 0 \in \Omega_1 , \; u_1(a) = \phi(a), \; u_1(b) = \lambda ,$$

and

$$(13) \qquad \frac{d^2u_2}{dx^2} + Q(x, u_2, \frac{du_2}{dx}) = 0 \in \Omega_2 , \; u_2(b) = \lambda, \; u_2(c) = \phi(c) ,$$

and the nonlinear function

$$(14) \qquad D(\lambda) \equiv \frac{\partial u_1(b; \lambda)}{\partial x} - \frac{\partial u_2(b; \lambda)}{\partial x} = 0 .$$

The two subproblems together with the nonlinear function is a variant of shooting method where the matching is chosen at $x = b$.

One advantage of the current method compared with the previous shooting method is that the two subproblems can be decomposed and indepently computed, thus ensuring intrinsic parallelism. The other advantage is that the subproblems are now boundary value problems which can avoid instability caused by solving initial value problems. Furthermore each of the subproblems is smaller than the original problem and is easier to solve. In order to solve (14), we use the fixed point iteration scheme, $\lambda^{(n+1)} = \lambda^{(n)} - \alpha_n D(\lambda^{(n)})$, where $\alpha_n := \alpha_{n-1}|D(\lambda^{(n-1)})|/|D(\lambda^{(n)}) - D(\lambda^{(n-1)})|$, details of which can be found in [6].

EXAMPLE 2. Consider $\frac{d^2\phi}{dx^2} - \gamma\frac{d\phi}{dx} = 0$, subject to $\phi(0) = 0$, $\phi(1) = 1$. Here $\gamma \gg 1$ and $b = 1 - \frac{1}{\gamma}$, "exact subsolver" means an analytic solution is obtained in the corresponding subdomain, and h is the mesh size in a finite difference scheme. In the case of finite difference method, we use the same h in both of the subdomains.

γ	n	$\lambda^{(n)}$	$\phi(b)$	$\|\phi(b) - \lambda^{(n)}\|$
10	2	0.36785	0.36785	0.37×10^{-7}
30	2	0.36788	0.36788	0.81×10^{-7}
50	2	0.36788	0.36788	0.37×10^{-6}

Table 1 : Convergence results using an exact solver.

h	n	$\lambda^{(n)}$	$\|\phi(b) - \lambda^{(n)}\|$
0.02	3	0.36663	0.12×10^{-2}
0.01	3	0.36754	0.31×10^{-3}
0.005	3	0.36777	076×10^{-4}
0.0025	3	0.36783	0.17×10^{-4}

Table 2: Convergence results for $\gamma = 10$ using a 2nd order difference scheme.

4. A Multiple Shooting Method

If the process in Proposition 3 is carried out recursively for every subsequent subdomain, then we have $\Omega_k = \{x : b_{k-1} < x < b_k\}$, $k = 1, 2, \ldots, s+1$, where $b_0 = a$ and $b_{s+1} = c$. Here Ω_k's denote a set of non-overlapped subdomains. The interfaces are located at $x = b_k$, $k = 1, 2, \ldots, s$ where the solutions at these interfaces are required. Since each pair of neighbouring subdomains is exactly the same as that given by Proposition 2, the multi-subdomain case derived from the above discussion can be considered as a multiple shooting technique.

PROPOSITION 4. *The boundary value problem given by (11) is replaced by the following $s+1$ subproblems,*

$$(15) \quad \frac{d^2 u_k}{dx^2} + Q(x, u_k, \frac{du_k}{dx}) = 0 \in \Omega_k \; , \; u_k(b_{k-1}) = \lambda_{k-1} \; , \; u_k(b_k) = \lambda_k \; ,$$

for $k = 1, 2, \ldots, s+1$, with $u_1(a) = \phi(a)$ and $u_{s+1}(c) = \phi(c)$, and the nonlinear vector function

$$(16) \quad D(\lambda) = [D_k(\lambda)] \equiv [\frac{\partial u_k(b_k; \lambda)}{\partial x} - \frac{\partial u_{k+1}(b_k; \lambda)}{\partial x}] = 0$$

where $\lambda = [\lambda_1 \; \lambda_2 \; \ldots \lambda_s]$ is an s-vector. These subproblems together with the nonlinear vector function is a multiple shooting algorithm where the matching is done at the interfaces.

In order to solve $D(\lambda) = 0$, a fixed point iteration scheme similar to that given above is used. Here α_n can be chosen either as the matrix $[J(\lambda^{(0)})]^{-1}$

where $J(\lambda^{(0)}) = D'(\lambda^{(0)})$ which reduces the scheme to Newton's iteration, or as a scalar adaptive parameter given by

(17) $$\alpha_n := \alpha_{n-1} \frac{\| D(\lambda^{(n-1)}) \|_2}{\| D(\lambda^{(n)}) - D(\lambda^{(n-1)}) \|_2}$$

Details of the choices can be found in [6].

EXAMPLE 3. We solve the same problem as that given by Example 2 using an analytic subproblem solver and present the number of iterations, n, required to update the interfaces which are evenly distributed across the physical domain.

$s \backslash \gamma$	10	20	30	40	50
3	14	15	15	13	11
7	23	21	18	20	20
15	37	35	34	37	31
31	96	62	58	51	52

Table 3: Convergence results using α_n defined in (17).

First, we construct the Jacobian matrix $J(\lambda^{(0)})$ which requires $2s$ subproblem solves. Each iteration involves $s + 1$ subproblem solves in order to compute $D(\lambda^{(n)})$. By taking $\alpha_n = [J(\lambda^{(0)})]^{-1}$, we require $n = 2$ iterations to update the interfaces. We achieve an efficient multiple shooting algorithm in a coarse-grain parallel computing environment provided we can invert $J(\lambda^{(0)})$ efficiently. Second, we use the scalar adaptive α_n in (17) and record n in Table 3. Here we do not invert $J(\lambda^{(0)})$ but the penalty is an increase in n. However the simple communication which involves only exchanging neighbouring information provides another efficient multiple shooting algorithm in a coarse-grain parallel computing environment. We observe that the number of iterations n is independent of the problem type, but increases as the number of interfaces s increases.

EXAMPLE 4. A nonlinear electrostatic problem [7] in normalised variables is tested. The problem is described by (11) with $a = 0$ and $c = 180$ and is subjected to boundary conditions $\phi(0) = 0$ and $\phi(180) = 10$. The function Q is given as

$$Q = \Gamma(x) + e^{(\phi_+ - \phi)} - e^{(\phi - \phi_-)} .$$

Here

$$\Gamma = - Ne^{-x^2/e^2} + Ne^{-(c-x)^2/e^2} , \quad N = \frac{1480}{1 - e^{-1}} ,$$

$$\phi_+ = \phi(a) + \ln \left\{ -\frac{\Gamma(a)}{2} + \sqrt{\left(\frac{\Gamma(a)}{2}\right)^2 + 1} \right\} ,$$

$$\phi_- = \phi(c) - \ln \left\{ \frac{\Gamma(c)}{2} + \sqrt{\left(\frac{\Gamma(c)}{2}\right)^2 + 1} \right\} ,$$

We evenly distribute the interfaces across the entire domain, and we use equal meshes and a second order difference scheme throughout the subdomains. We use the adaptive α_n as that given above. Table 4 records the number of iteration, n, required to update the interfaces.

$s \setminus h$	2	1	0.5	0.25	0.125
2	8	8	7	8	8
4	15	15	14	14	12
8	29	30	24	24	20
14	358	124	100	81	75

Table 4: Convergence results for the nonlinear electrostatic problem.

Note that the number of iterations n is independent of the mesh size h, but is dependent of the number of interfaces s. It is also observed that $n \approx 3s$ for small values of s and that n becomes unreasonably large for large values of s.

5. Conclusion

A framework for domain decomposition methods is built on the properties of a shooting method. A two subdomain case was presented and the convergence results are the same as the shooting method. The two subdomain case is extended to the multi-subdomain case. The multi-subdomain case provides an efficient multiple shooting algorithm on coarse-grain parallel architectures. Linear and nonlinear examples are included.

REFERENCES

1. G.H. Golub and J.M. Ortega, *Scientific Computing and Differential Equations - An Introduction to Numerical Methods*, Academic Press, New York, 1992.
2. G. Hammerlin and K.-H. Hoffmann, *Numerical Mathematics - Readings in Mathematics*, Springer-Verlag, Berlin, 1991.
3. H.B. Keller, *Numerical Methods for Two-Point Boundary Value Problem*, Blaisdell Publishing Company, London, 1968.
4. H.B. Keller, *Domain decomposition for two-point boundary value problem*, Domain Decomposition Methods vol. 4, SIAM, Philadelphia, 1991, pp. 50–57.
5. C.-H. Lai, *Shooting methods for some diffusion and convection problems*, Appl Math Modelling **16** (1992), 638–644.
6. _____, *An iteration scheme for non-symmetric interface operator*, ERCIM Research Report ERCIM-92-R003, 1992.
7. C.-H. Lai and C. Greenough, *Numerical solutions of some semiconductor devices by a domain decomposition method*, RAL Technical Report RAL-92-063, 1992.

SCHOOL OF MATHEMATICS, STATISTICS, AND COMPUTING, UNIVERSITY OF GREENWICH, WOOLWICH CAMPUS, WELLINGTON STREET, WOOLWICH, LONDON SE18 6PF, U.K.

E-mail address: c.h.lai@greenwich.ac.uk

Contemporary Mathematics
Volume **180**, 1994

ROBUST METHODS FOR HIGHLY NONSYMMETRIC PROBLEMS

W. Layton, J. Maubach and P. Rabier

ABSTRACT. We consider the solution of linear systems arising from discretizations of problems which are highly convection dominated in some subregions and diffusion dominated in others. For such problems, *robustness*, meaning uniformity in the problem parameters for consistent discretizations, rather than asymptotic optimality in h, is of paramount importance. This report presents a particularly simple method which is provably robust for highly nonsymmetric and anisotropic problems.

1. Introduction

A large number of interesting problems in scientific applications involve operators that are both highly nonsymmetric and highly anisotropic. There continues to be a great deal of research into adapting optimal order solution methods from symmetric positive definite problems to these interesting ones so that their "robustness" is incrementally improved (see other articles in this proceedings). The authors have been attacking these problems from the opposite point of view, namely: developing optimally robust methods for the problems' singular limits and then attempting to decrease their serial or parallel complexity for (simpler) problems. This report gives an overview of some results to date, the basic algorithms developed and one interesting computational example.

Semiconductor problems and fluid dynamics problems lead to (linearized) convection diffusion equations which are highly convection dominated in some subregions and diffusion dominated in others. The involved velocity field typically has stagnation points and closed loops which do not permit "streamwise" solution strategies. With these difficulties in mind, consider the problem: seek $u(x, y)$ satisfying:

$$(1.1) \qquad -\nabla \cdot (k(x,y)\nabla u) + \mathbf{b}(x,y) \cdot \nabla u + g(x,y)u = f(x,y) \quad \text{in } \Omega,$$

subject to $u = 0$ at $\partial\Omega$. The coefficients are assumed to satisfy:

$$(1.2) \qquad 0 \leq k_{\min} \leq k(x,y) \leq k_{\max} \leq 1, \; g - \frac{1}{2}\nabla \cdot \mathbf{b} \geq g_{\min} > 0,$$

and Ω is a planar polygonal domain. Accurate approximation of (1.1) usually requires adaptivity [14,15], nonlinear discretizations [9] or subgridscale modelling [4],

---

1991 *Mathematics Subject Classification*. AMS-MOS numbers: Primary 65N55, Secondary 65F10 .

This paper is in final form and no version of it will be submitted for publication elsewhere and in all these approaches, (re)assembly can often dominate linear system solution costs [16]. Therefore, we consider an elementwise data-parallel finite element solution procedure from [11,12]. Massive parallelism [8, 17, 18], point adaptivity [7,8] and nonlinear discretizations [4,6] are handled trivially by the present procedure which uses *elements* as the logical units. The new results presented herein state that the method introduced in [11, 12] is *optimally robust* for (1.1), (1.2). Specifically, the solution procedure converges uniformly in $k(x, y)$ (the cell Reynolds or Péclet number), even for centered discretizations! In Theorem 1.1 the parameter δ interpolates between the usual (centered) Galerkin method ($\delta = 0$) and the streamline diffusion finite element method ($\delta = h$).

Let us postpone the presentation of the algorithm until Section 2. The following theorem, proven in [7] shows that the Parallel FEM algorithm converges uniformly in both the anisotropy ($k_{\max}/k_{\min}$) and the Péclet number ($k_{\max}$). The technical condition upon $k(x, y)$ states that $k(x, y)$ should not change from near zero to near 1 *inside* of elements, i.e., such drastic material discontinuities should be followed by meshlines. This is certainly good computational practice but we do not know if it is necessary or an artifact of our method of proof.

Theorem 1.1. *Assume that the finite element triangulation satisfies the usual minimum angle condition, and that in addition inside each triangle e either: $0 \leq k_{\min,e} \leq k(x, y) \leq k_{\max,e} \leq C_1 h$ or $C_2 h \leq k_{\min,e} \leq k(x, y) \leq k_{\max,e} \leq 1$. Let T denote the iteration operator of the Parallel FEM algorithm, $|\cdot|_2$ be the Euclidean matrix norm and let ρ be the algorithm acceleration parameter. Then, provided ρ is scaled as $\rho = \rho_0 h$,*

$$\sup_{\substack{0 \leq k_{\min} \leq k(x,y) \leq k_{\max} \leq 1 \\ 0 \leq k_{\max}/k_{\min} \leq \infty, 0 \leq \delta \leq h}} |T|_2 \leq 1 - ch.$$

$\square$

We emphasize that this holds even for centered discretizations so that robustness is not an artifact of a special discretization which implicitly increases the size of the matrices' symmetric part. Further, the basic methods and results have recently been extended to finite difference and finite volume discretizations in [1,2]. The parallel FEM algorithm also is attractive as a combination pre- and post-conditioner. This is described in Section 2 and developed fully in [5].

2. The Algorithm

Let $\Pi^h(\Omega)$ denote an edge-to-edge triangulation of the polygonal domain Ω, in our tests usually generated self-adaptively as in [14, 15]. Without loss of generalization, let S^h denote the span of the $3-$node nonconforming linear element; all the results below extend directly to higher order elements as no M-matrix structure is used. Now, focus on a one parameter family of discretization methods ($k_{\max} \leq O(h)$ and $\delta = O(h)$ gives streamline diffusion; $0 \leq k(x, y) \leq 1$ and $\delta = 0$ gives the usual Galerkin method), given by: seek $u^h \in S^h$ satisfying

$$(2.1) \qquad \sum_{e \in \Pi^h(\Omega)} [a_e(u^h, v) - (f, v + \delta \mathbf{b} \cdot \nabla v)_e] = 0, \text{ for all } v \in S^h,$$

where $(f, v)_e = \int_e f v dx$, and $a_e(u, v) := \int_e \{k\nabla u \cdot \nabla v + \delta \mathbf{b} \cdot \nabla u \mathbf{b} \cdot \nabla v + (g - \frac{1}{2}\nabla \cdot (\mathbf{b} + \delta g \mathbf{b})uv + \frac{1}{2}(1 - \delta g)[\mathbf{b} \cdot \nabla uv - \mathbf{b} \cdot \nabla vu]\}dx$ is the usual elemental bilinear form, explicitly skew-symmetrized. Let N_j denote the nodes in the mesh (situated at the mid-edges) and ϕ_j the associated nodal basis functions (which have support on two triangles only). Color the triangles in $\Pi^h(\Omega)$ with two colors (Red and Black, traditionally). If the graph-degree of each interior vertex of $\Pi^h(\Omega)$ is even it is trivially possible to separate [2,3,10,11] the triangles (no two same-color triangles share an edge), otherwise the coloring algorithm presented in [3] is used; see [3] for details. We focus on the case of even degree interior vertices for compactness of presentation. Split the stiffness matrix A and right hand side vector f into $A = A_R + A_B$ and $f = f_R + f_B$, as follows:

$$[A_{R/B}]_{i,j} := \sum_{e \text{Red/Black}} a_e(\phi_j, \phi_i) , \quad [f_{R/B}]_j = \sum_{e \text{Red/Black}} (f, \phi_j + \delta \mathbf{b} \cdot \nabla \phi_j)_e.$$

One key observation is that the nodes can be ordered such that the 3 nodes in each Red (resp., Black) triangle are consecutively numbered causing A_R (resp., A_B) to be block diagonal with blocks of the 3×3 element matrices corresponding to Red (resp., Black) triangles. Let these two orderings be called the Red (resp., Black) ordering. Reordering corresponds to a local communication on a mesh connected array of processors [8, 17, 18].

Algorithm: Parallel FEM. *Given c_B^0 and $\rho > 0$ compute until satisfied*

   I. Calculate $d_B := (\rho I + A_B)^{-1}[f_B - (A_B - \rho I)c_B^k]$

   II. reorder $c_R^k \leftarrow d_B$

   III. Calculate $d_R := (\rho I + A_R)^{-1}[f_R - (A_R - \rho I)c_R^k]$

   IV. reorder $c_B^{k+1} \leftarrow d_R$

Upon convergence, set in consistent node ordering, $c = c_B + c_R$, then $Ac = f$. Note that as each $(\rho I \pm A_{R/B})$ is block 3×3 diagonal each step is embarassingly parallel. A good choice of ρ is $\rho = (|\lambda|_{\max}(A_e) \cdot |\lambda|_{\min}(A_e))^{1/2}$. This algorithm is precisely the D'Yakunov operator splitting method [13], with the new splitting of [11, 12].

The above algorithm is of the domain decomposition type in that each block in $(\rho I \pm A_{R/B})$ is associated with a subdomain (i.e., a single Red or Black element). There are neither overlap of subdomains nor interface conditions. Further, in cost per iteration, ability to parallelize and case of implementation it is more akin to simple relaxation methods. The next result follows immediately from the Algorithm.

Proposition 2.1. *The per iteration complexity of the Parallel FEM algorithm is as follows. (a) Computational complexity: per triangle one 3×3 matrix vector multiply, one 3×3 system solve and one 3-vector addition. (b) Communication complexity: two local communications on the physical mesh.* $\square$

Discretization (2.1) leads to the usual linear system $Ac = f$ for the nodal degrees of freedom. The previous solution algorithm reduces local errors very quickly so it is natural to study it both in combination with Krylov subspace methods and as a smoother for multilevel methods, especially on serial machines. Since (2.1) may yield a highly non-normal system, reduction of the spectral radius of A is not sufficient to reduce $\text{cond}_2(A) = |A|_2|A^{-1}|_2$. This reduction is accomplished through a judicious combination of preconditioning with $(\rho I + A_R)^{-1}$ and postconditioning with $(\rho I + A_B)^{-1}$, leading to the system $Kd = g$, where

$$(3.1)\quad K := (\rho I + A_R)^{-1}\, A(\rho I + A_B)^{-1},\ d := (\rho I + A_B)c,\ g := (\rho I + A_R)^{-1}f.$$

The following basic result is given in [5] along with extensive experiments.

Theorem 3.1. *Under the assumption of Theorem 1.1, the pre- and postconditioned system (3.1) has condition number $O(h^{-1})$ uniformly in $k(x,y), k_{\max}/k_{\min}$ and δ:*

$$\sup_{\substack{0 \le k \le 1 \\ 0 \le k_{\max}/k_{\min} \le \infty \\ 0 \le \delta \le O(h)}} (|K|_2|K^{-1}|_2) \le Ch^{-1}.$$

☐

To date we have no experience using the parallel FEM algorithm as a smoother in a multilevel method.

4. An Illustrative Example

We have attempted to construct a "model" problem which captures some interesting features of internal flow problems. To this end, let $\Omega := (-1,1)^2$, $r^2 := x^2 + y^2$, and define $k(x,y) = 1$ if $r \ge 1$ while $k(x,y) = k_f$ for $r < 1$. Let the velocity field $\mathbf{b}$ be given by $\mathbf{b}(x,y) := [-y\phi(r), x\phi(r)]^T$ with ϕ a smooth function satisfying $\phi(r) \equiv 0$ for $r \ge 1$, here chosen to be $\phi(r) := 1 - r^2$ for $r \le 1$. Note that $\text{div } \mathbf{b}(x,y) \equiv 0$. We chose $g(x,y) \equiv 2$ and $f(x,y) = 1$ if $r \le 1/2$ and 0 if $r > 1/2$. The true solution is a rotational pulse $u \cong 1/2$ if $r < 1/2$ and $u \cong 0$ if $r > 1/2$ with an $O(\sqrt{k_f})$ transition layer, thus at least an asymptotic solution was available for comparisons. Note that this problem has: closed loops and stagnation points in the convection field, characteristic layers, neither inflow nor outflow boundaries and conductivities k varying from $O(1)$ to quite small. In [7], we have tracked the numbers of iterations as $k_f \to 0$, keeping h fixed, giving the result predicted by Theorem 1.1. Here, we present a slightly more interesting test, which also indirectly verifies robustness. For elements with polynomial degree r the range $O(h) \le k_f \le O(h^{r+1})$ is the critical one. Using nonconformity linear elements we fix $k_f = h^2/10$ and solve the linear system $Ac = f$ with varying meshwidths, taking $\rho = h$. Theorem 1.1 predicts $O(h^{-1})$ iterations exactly as observed in Table 1. If the method were *not* robust, for example if the number of iterations varied as $O(k_f^{-\beta}h^{-1})$ then with $k_f = h^2, O(h^{-(1+2\beta)})$ iterations would be observed. Thus Table 1 verifies both robustness and the complexity bound of $O(h^{-1})$ iterations.

h^{-1}	Centered Galerkin $\delta = 0$ case	Streamline Upwind $\delta = h$ case
64	167	111
32	104	72
16	60	46
8	30	23

Table 1. Number of Iterations of the PFEM Algorithm.

References

1. V. Ervin and W. Layton, *Parallel algebraic splittings and the Peaceman Rachford iteration*, K.U.N. Report 9320, Univ. of Nijmegen, The Netherlands (1992).
2. V. Ervin, W. Layton and J. Maubach, *Some graph coloring questions arising in parallel numerical methods*, Algorithms in de Algebra (A.H.M. Levelt, ed.), Nijmegen (1993).
3. R. Jeurissen and W. Layton, *Load balancing by graph coloring: an algorithm*, Comput. and Math. W. Appls. **27** (1994), 27-32.
4. W. Layton, *A nonlinear subgridscale model for the Navier Stokes equations*, submitted to: SIAM J.S.S.C. (1994).
5. W. Layton, J. Maubach and B. Polman, report in preparation (1994).
6. W. Layton, J. Maubach and P. Rabier, *Parallel algorithms for monotone operators of local type*, submitted to: Numerische Math. (1994).
7. W. Layton, J. Maubach and P. Rabier, *Robustness of an elementwise parallel finite element method for convection diffusion problems*, I.C.M.A. Report 93-185 (1993).
8. W. Layton, J. Maubach, P. Rabier and A. Sunmonu, *Parallel Finite Element Methods*, Proc. Fifth I.S.M.M. Conf. on Parallel and Dist. Comput. and Systems (1992), 299-304.
9. W. Layton and B. Polman, *Oscillation absorption finite element methods for convection diffusion problems*, submitted to: SIAM J.S.S.C. (1993).
10. W. Layton and P. Rabier, *The element separation property and parallel finite element methods for the Navier-Stokes Equations*, to appear in: Appl. Math. Lett. (1994).
11. W. Layton and P. Rabier, *Domain decomposition via operator splitting for highly nonsymmetric problems*, Appl. Math. Lett. **5** (1992), 67-70.
12. W. Layton and P. Rabier, *Peaceman-Rachford procedure and domain decomposition for finite element problems*, to appear: Num. Lin. Algb. and Applications (1994).
13. G.I. Marchuk, *Splitting and alternating direction methods*, Handbook of Numerical Analysis (P.G. Ciarlet and J.L. Lions, eds.), North Holland, Amsterdam, 1990.
14. J. Maubach, *Local bisection refinement for simplical grids generated by reflection*, I.C.M.A. Report 92-170, to appear in SIAM Journal on Scientific Computing.
15. J. Maubach, *Iterative Methods for Nonlinear Partial Differential Equations*, C.W.I. Press, Amsterdam (1993).
16. D.C. O'Neal, *Optimization of finite element codes*, J. Sci. Comp. **10** (1989), 36-52.
17. A. Sunmonu, *Multitasking an element-by-element parallel finite element method on the CRAY YMP*, I.C.M.A. Report 92-168, Univ. of Pittsburgh (1992).
18. A. Sunmonu, *Implementation and Analysis of a Massively Parallel Domain Decomposition Method on Parallel Computers*, Computer Science thesis, Univ. of Pittsburgh (1993).

DEPARTMENT OF MATHEMATICS AND STATISTICS UNIVERSITY OF PITTSBURGH PITTSBURGH, PA 15260, U.S.A.

E-mail address: WJL@vms.cis.pitt.edu

DEPARTMENT OF MATHEMATICS AND STATISTICS UNIVERSITY OF PITTSBURGH PITTSBURGH, PA 15260, U.S.A.

E-mail address: Joseph@garfield.math.pitt.edu

DEPARTMENT OF MATHEMATICS AND STATISTICS UNIVERSITY OF PITTSBURGH PITTSBURGH, PA 15260, U.S.A.
E-mail address: Rabier@vms.cis.pitt.edu

Contemporary Mathematics
Volume **180**, 1994

Domain Decomposition Via the Sinc-Galerkin Method for Second Order Differential Equations

NANCY J. LYBECK AND KENNETH L. BOWERS

ABSTRACT. The solution of elliptic problems using domain decomposition techniques has been of great interest in recent years. Sinc basis functions form a desirable basis to use in approaching domain decomposition for elliptic problems because they are especially well-suited for problems with boundary singularities, and both the Sinc-Galerkin and sinc-collocation methods converge exponentially. This paper deals with overlapping and patching domain decomposition used in conjunction with the Sinc-Galerkin method for both the two-point boundary value problem and Poisson's equation on a rectangle.

1. Introduction

Sinc methods for differential equations were originally introduced in [**9**]. Since then they have become increasingly popular, and have been well-studied. Both the Sinc-Galerkin and the sinc-collocation methods converge exponentially, even in the presence of boundary singularities, as shown in [**1**], [**6**], and [**9**]. Both methods perform equally well in domain decomposition for the two-point boundary value problem, as seen in [**8**]. For Poisson's equation, the Sinc-Galerkin and sinc-collocation methods are identical. For this reason only the Sinc-Galerkin results will be presented.

Although elliptic problems are generally approached with patching domain decomposition methods, certain characteristics of the sinc basis functions make overlapping domain decomposition desirable. Because of this, both methods have been explored. See [**2**] for further details on these methods. Numerical results are presented for both overlapping and non-overlapping methods. These results exhibit nearly identical errors achieved with each method. A brief introduction to the Sinc-Galerkin method is given in §2. Domain decomposition for the two-point boundary value problem via the Sinc-Galerkin method is presented in §3. Similarly, domain decomposition for Poisson's equation via the Sinc-Galerkin method is discussed in

1991 *Mathematics Subject Classification.* 65L10, 65N35; Secondary 65L50, 65N50.
Research supported in part by NCSA grant TRA940014N.
This paper is in final form and no version of it will be submitted for publication elsewhere.

§4. Both sections §3 and §4 include one example designed to highlight the Sinc-Galerkin method's ability to deal with boundary singularities. In each case, results for both the patching and overlapping methods are presented.

2. The Sinc-Galerkin Method

The two-point boundary value problem on the finite interval (a, b) is given by

(2.1)
$$\begin{aligned} \mathcal{L}u(x) &\equiv -u''(x) + p(x)u'(x) + q(x)u(x) \\ &= f(x), \quad a < x < b \\ u(a) &= u(b) = 0 . \end{aligned}$$

The classical Sinc-Galerkin method for problems of this type is discussed in detail in [3–7],[9], and [12].

The sinc basis functions used in solving (2.1) are given by

$$S_j(x) \equiv S(j, h) \circ \phi(x) \equiv \text{sinc}\left(\frac{\phi(x) - jh}{h}\right)$$

where $h > 0$, j is an integer, $x \in (a, b)$, and

$$\text{sinc}(y) = \begin{cases} \frac{\sin(\pi y)}{\pi y}, & y \neq 0 \\ 1, & y = 0 \end{cases} .$$

The conformal map

$$\phi(z) = \ln\left(\frac{z - a}{b - z}\right)$$

is used to define the basis functions on the finite interval (a, b). The sinc nodes x_k are chosen so that

$$x_k \equiv \psi(kh) = \phi^{-1}(kh) = \frac{a + be^{kh}}{e^{kh} + 1} .$$

The approximate solution is then given by

(2.2)
$$u_m(x) = \sum_{k=-M}^{N} u_k S_k(x), \quad m = M + N + 1 .$$

Orthogonalization of the residual against each basis function

$$(\mathcal{L}u - f, S_j) = 0, \quad -M \leq j \leq N$$

uses the weighted inner product

$$(f, g) = \int_a^b f(x)g(x)(\phi'(x))^{-1/2}dx .$$

Integration by parts is used to remove all derivatives from u, and applying the sinc quadrature rule (found in [6] or [11]) yields the discrete Sinc-Galerkin system. The following theorem for the convergence of this method in the case $p(x) \equiv 0$ is proven in [9].

THEOREM 2.1. *Let the numbers u_k ($k = -M, \ldots, N$) be determined from the discrete Sinc-Galerkin system, and let $u_m(x)$ be as defined in (2.2). Then under appropriate assumptions (see [9]), with $h = \left(\pi/\sqrt{2M}\right)$, and $N = M$, the estimate*

$$\|u_m - u\|_\infty \leq CM^2 e^{-\sqrt{\pi d\alpha M}}$$

holds where u is the solution of (2.1) with $p(x) \equiv 0$. The parameters d and α depend on the analyticity and rate of decay of the solution u.

3. One-Dimensional Domain Decomposition

Both the overlapping and the patching methods of domain decomposition introduce matching conditions at the interface of the domains. This requires the addition of a boundary basis function to the approximation on each subdomain. In either case, appropriate fourth degree polynomial boundary functions are given in [8]. Similar situations arise in the solution of boundary value problems with non-homogeneous Dirichlet boundary conditions. See [8] and [10] for more information.

The following examples have been chosen to show the rapid convergence achieved by using the Sinc-Galerkin method in conjunction with both patching and overlapping domain decomposition methods.

EXAMPLE 3.1. Consider

$$-u''(x) + u'(x) + u(x) = f(x) , \quad -1 < x < 4$$

$$u(-1) = u(4) = 0 .$$

In this example, $f(x)$ was chosen so that the true solution is given by

$$u(x) = \frac{\sqrt{x+1}(x-4)^2}{16} .$$

Split the domain $\Omega = [-1, 4]$ into two subdomains given by either $\Omega^1 = [-1, 1]$ and $\Omega^2 = [.9, 4]$ for the overlapping method, or $\Omega^1 = [-1, 1]$ and $\Omega^2 = [1, 4]$ for the patching method. The approximate solutions are shown in Fig. 1. The error results are given in Table 1. Here, the sinc error is given by

$$\|E_S\| = \max_{x \in S} |u(x) - u_m(x)| ,$$

where S is the set of all grid points x_k generated from the Sinc-Galerkin method in both subdomains. The uniform error is found by letting

$$\|E_U\| = \max_{y \in U} |u(y) - u_m(y)|$$

where $U = \{y_j = -1 + 5j/100 : 0 \leq j \leq 100\}$ is a uniform grid of mesh size 0.05.

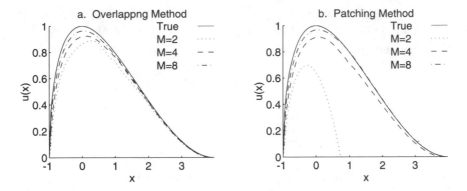

FIGURE 1. True vs. Approximate Solutions for Example 3.1. Fig. 1a shows the solution from the overlapping method. Fig. 1b shows the solution from the patching method.

TABLE 1. Error in overlapping and patching methods for Example 3.1.

$M = N$	Overlapping		Patching	
	$\|E_U\|$	$\|E_S\|$	$\|E_U\|$	$\|E_S\|$
2	$2.3394e - 01$	$2.2852e - 01$	$1.2803e + 00$	$1.2811e + 00$
4	$1.4258e - 01$	$1.4880e - 01$	$1.4278e - 01$	$1.4888e - 01$
8	$6.6033e - 02$	$6.8459e - 02$	$6.5930e - 02$	$6.8455e - 02$
16	$1.9821e - 02$	$2.0531e - 02$	$1.9811e - 02$	$2.0531e - 02$
32	$3.3660e - 03$	$3.4897e - 03$	$3.3652e - 03$	$3.4897e - 03$
64	$2.6173e - 04$	$2.7141e - 04$	$2.6159e - 04$	$2.7141e - 04$

4. Domain Decomposition for Poisson's Equation

The basis used to solve Poisson's equation is a product of sinc basis functions in the x and y directions. Let

$$S_{jk}(x, y) = S_j(x)S_k(y) \ , \ -M_x \leq j \leq N_x \ , \ -M_y \leq k \leq N_y.$$

The approximate solution takes the form

$$u_{m_x, m_y}(x, y) = \sum_{j=-M_x}^{N_x} \sum_{k=-M_y}^{N_y} u_{jk}S_{jk}(x, y)$$

where $m_x = M_x + N_x + 1$ and $m_y = M_y + N_y + 1$. As in the one-dimensional case, one orthogonalizes the residual against each basis function and perform integration by parts to reach the discrete Sinc-Galerkin system. Again, both the patching and overlapping methods for domain decomposition applied to Poisson's equation require the addition of boundary basis functions. In the example given, the domain is split only in the x direction, so there is no need for boundary basis functions in the y variable. The extra basis functions in the x variable are the same ones used in §3.

EXAMPLE 4.1. Consider the problem

$$-\Delta u(x,y) \;=\; f(x,y) \,,\; (x,y) \in \Omega = (-1,4) \times (0,1)$$

$$u(x,y) \;=\; 0 \,,\; (x,y) \in \partial\Omega \,,$$

where $f(x,y)$ is chosen so that the true solution is given by

$$u(x,y) = \sqrt{y}\sqrt{x+1}(x-4)^2(1-y)^2 \,.$$

Here the domain $\Omega = [-1,4] \times [0,1]$ is split into the two overlapping subdomains $\Omega^1 = [-1,1] \times [0,1]$ and $\Omega^2 = [.9,4] \times [0,1]$ or the two non-overlappinng subdomains $\Omega^1 = [-1,1] \times [0,1]$ and $\Omega^2 = [1,4] \times [0,1]$. In this example $M \equiv M_x = M_y$ and $N \equiv N_x = N_y$. A graphical representation of the results is found in Fig. 2, while the numerical errors are reported in Table 2. The error columns are analogous to those reported in Example 3.1. In spite of the steep solution along the x and y axes the convergence is rapid.

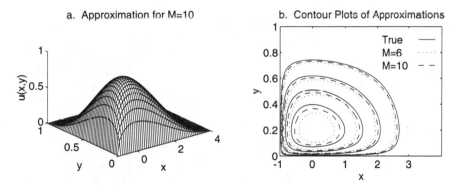

FIGURE 2. Patching results for Example 4.1. Fig. 2a shows the solution computed by the patching method for $M = 10$. Fig. 2b shows contour plots of the solutions computed by the patching method versus the true solution. The contour levels are .2, .4, .6, and .8.

TABLE 2. Error in overlapping and patching methods for Example 4.1.

$M = N$	Overlapping		Patching	
	$\|E_U\|$	$\|E_S\|$	$\|E_U\|$	$\|E_S\|$
2	$3.7729e-01$	$3.5832e-01$	$7.5272e-01$	$6.4327e-01$
4	$2.3657e-01$	$2.4320e-01$	$2.3659e-01$	$2.3423e-01$
6	$1.5294e-01$	$1.5612e-01$	$1.5282e-01$	$1.5610e-01$
8	$1.0619e-01$	$1.0813e-01$	$1.0619e-01$	$1.0813e-01$
10	$7.6184e-02$	$7.7340e-02$	$7.6181e-02$	$7.7340e-02$

5. Conclusion

The performance of the Sinc-Galerkin method in conjunction with both patching and overlapping domain decomposition methods is quite good in simple cases, as seen in Examples 3.1 and 4.1. In these test problems, the resulting systems were solved directly using MATLAB. The large size of the matrices in the two-dimensional case became prohibitive after $M = 10$. The results of Example 3.1 have been confirmed using FORTRAN on a CRAY Y-MP. Work is in progress to convert the codes for Poisson's equation to FORTRAN so that larger systems may be run for Example 4.1. An iterative method for solving these problems would be a logical next step.

REFERENCES

1. B. BIALECKI, *Sinc-collocation methods for two-point boundary value problems*, IMA J. Numer. Anal., 11 (1991), pp. 357–375.
2. C. CANUTO, M. Y. HUSSAINI, A. QUARTERONI, and T. A. ZANG, *Spectral Methods in Fluid Dynamics*, Springer-Verlag, New York, 1988.
3. N. EGGERT, M. JARRATT, and J. LUND, *Sinc function computation of the eigenvalues of Sturm-Liouville problems*, J. Comp. Phys., 69 (1987), pp. 209–229.
4. M. JARRATT, J. LUND, and K. L. BOWERS, *Galerkin schemes and the Sinc-Galerkin method for singular Sturm-Liouville problems*, J. Comp. Phys., 89 (1990), pp. 41–62.
5. J. LUND, *Symmetrization of the Sinc-Galerkin method for boundary value problems*, Math. Comp., 47 (1986), pp. 571–588.
6. J. LUND and K. L. BOWERS, *Sinc Methods for Quadrature and Differential Equations*, SIAM, Philadelphia, PA, 1992.
7. L. LUNDIN and F. STENGER, *Cardinal-type approximations of a function and its derivatives*, SIAM J. Math. Anal., 10 (1979), pp. 139–160.
8. N. J. LYBECK and K. L. BOWERS, *Sinc methods for domain decomposition*, submitted to Appl. Math. Comp.
9. F. STENGER, *A Sinc-Galerkin method of solution of boundary value problems*, Math. Comp., 33 (1979), pp. 85–109.
10. R. C. SMITH, K. L. BOWERS, and J. LUND, *A fully Sinc-Galerkin method for Euler-Bernoulli beam models*, Numer. Methods Partial Different. Equ., 8 (1992), pp. 171–202.
11. F. STENGER, *Numerical methods based on Whittaker cardinal, or sinc functions*, SIAM Rev., 23 (1981), pp. 165–224.
12. F. STENGER, *Numerical Methods Based on Sinc and Analytic Functions*, Springer-Verlag, New York, N.Y., 1993.

DEPARTMENT OF MATHEMATICAL SCIENCES, MONTANA STATE UNIVERSITY, BOZEMAN, MONTANA 59717
    E-mail address: imsgnlyb@math.montana.edu

DEPARTMENT OF MATHEMATICAL SCIENCES, MONTANA STATE UNIVERSITY, BOZEMAN, MONTANA 59717
    E-mail address: bowers@math.montana.edu

Contemporary Mathematics
Volume **180**, 1994

A Bisection Method to Find All Solutions of a System of Nonlinear Equations

PETR MEJZLÍK

June 20, 1994

ABSTRACT. This paper describes an algorithm for the solution of a system of nonlinear equations $F(x) = \theta$, where $F = (f_1, \ldots, f_n) : D \subset \mathbf{R}^m \to \mathbf{R}^n$ and D is a compact domain, given that any of the functions f_i is monotonic when restricted to any single variable at an arbitrary point. The algorithm finds an approximation of the solutions as a union of m–dimensional intervals. The computation is based on reduction of the box containing all the solutions, its bisection, and elimination of subintervals which do not contain a solution. The algorithm does not require computation of partial derivatives or their approximations. Its use is illustrated on a model case.

1. Introduction

Let $F = (f_1, \ldots, f_n) : [a, b] \subset \mathbf{R}^m \to \mathbf{R}^n$ be a continuous function. We will consider a generalized bisection method to approximate, with certainty, all solutions of a nonlinear system

$$(1.1) \qquad F(x) = \theta,$$

where $\theta = (0, \ldots, 0)$, given that

$$(1.2) \qquad \Delta_{\delta e^j} f_i(x) \cdot \Delta_{\delta e^j} f_i(y) \geq 0,$$
$$\text{for all} \quad \delta > 0, \ 1 \leq i \leq n, \ 1 \leq j \leq m, \ x, y \in [a, b],$$

where $\Delta_h f(x) = f(x + h) - f(x)$ and e^j is the j-th unit vector.

1991 *Mathematics Subject Classification.* Primary 65H10, 65H20; Secondary 65Y05.
This paper is in final form and no version of it will be submitted for publication elsewhere.

Generalized bisection methods to solve (1.1) can be described using a root inclusion function T_F (defined formally in [4]) with values *true*, *false*, and *unknown*, which has the following properties [5]:

if $T_F(S)=true$ then there is a unique solution of (1.1) within S.

if $T_F(S)=false$ then there is no solution of (1.1) within S.

The template of a generalized bisection method can then be written as follows:

Input: Bounded domain $S \subset \mathbf{R}^m$.

   (i) **if** $T_F(S) = false$ **then** the result is $\emptyset$.

   (ii) (*optional*)

      Reduce the domain S to $S' \subset S$ such that
$$\{x; F(x) = \theta, x \in S\} = \{x; F(x) = \theta, x \in S'\}.$$

   (iii) **if** diam(S') or sup$\{\|F(x)\|; x \in S'\}$ are sufficiently small,

      then S' is the result,

      else

         (a) Split S' into two subdomains S_1, S_2.

         (b) Compute solutions on S_1 and S_2 separately, using this algorithm.

Current bisection methods to solve (1.1) [1–6] use the Jacobian matrix of F, or its approximations, to compute $T_F(S)$. The method presented here does not require computation of any derivatives, taking advantage of the restriction (1.2) on the class of functions considered. Our root inclusion test has only two values (*false* and *unknown*). The refinement of distinct solutions of (1.1) is done by customizations of a precision parameter of the algorithm, instead of testing whether $T_F(S) = true$.

2. The algorithm

The following is a recursive algorithm to solve problem (1.1), given (1.2). The algorithm contains a non–sequential loop ("for all"), instances of the body of which can be computed in any order. The algorithm can be efficiently run in parallel, as it decomposes the computation into independent subproblems.

DEFINITION 2.1. *We define norm*

$$\|x\|_M = \max_i |x_i| \quad \text{for all } x \in \mathbf{R}^k, k = 1, 2, \ldots$$

Algorithm D$(S, F, \varepsilon, \delta)$:

Input: m–dimensional interval $S = [a, b] \subset \mathbf{R}^m$, function $F = (f_1, \ldots, f_n) : S \to \mathbf{R}^n$, and $\varepsilon, \delta > 0$.

   (i) Reduce the m–dimensional interval $[a, b] = [(a_1, \ldots, a_m), (b_1, \ldots, b_m)]$, preserving all the roots of F on $[a, b]$:

repeat
$a^0 := a,\ b^0 := b,$

        for all $(i,j) \in \{1,\dots,n\} \times \{1,\dots,m\}$ **do**

           if f_i does not have a root in $[a,b]$

           then return $\emptyset$,

           else $a_j := \min\ \{x_j; x = (x_1,\dots,x_m) \in [a,b], f_i(x) = 0\},$

                    $b_j := \max\ \{x_j; x = (x_1,\dots,x_m) \in [a,b], f_i(x) = 0\},$

until $\|a - a^0\|_M < \delta \ \wedge\ \|b - b^0\|_M < \delta.$

(ii) Test whether $\|F(x)\|_M < \varepsilon$ on $[a,b]$:

    if for each $1 \le i \le n :\ |\ \min_{[a,b]} f_i(x)\ |, |\ \max_{[a,b]} f_i(x)\ | < \varepsilon$

    then   **return** $[a,b]$.

(iii) Divide $[a,b]$ into two parts:

    Choose $k \in \{1,\dots,m\}$ such that $|\ a_k - b_k\ | = \max_{j=1,\dots,m} |\ a_j - b_j\ |,$

$$S_0 := [a, (b_1,\dots,b_{k-1}, \frac{a_k + b_k}{2}, b_{k+1},\dots,b_m)],$$

$$S_1 := [(a_1,\dots,a_{k-1}, \frac{a_k + b_k}{2}, a_{k+1},\dots,a_m), b],$$

    return $\mathbf{D}(S_0, F, \varepsilon, \delta) \cup \mathbf{D}(S_1, F, \varepsilon, \delta).$

To complete the specification, we describe methods to compute expressions in steps (i) and (ii) of algorithm $\mathbf{D}$. The following theorem gives a recipe for computation of the minima and maxima in step (i). The problem is reduced to the calculation of the minimal root for a single–variable monotonic continuous function on an interval of finite size, which can be done by bisection. The computation of a minimum only is considered in the theorem, as the corresponding maximum can be computed as

$$\max\{x_j; x \in [a,b], f_i(x) = 0\} = -\min\{x_j; x \in [-a,-b], f_i(-x) = 0\}.$$

THEOREM 2.1. *Let $f . [u,v] \subset \mathbf{R}^m \to \mathbf{R}$ be a continuous function, which fulfills (1.2), $j \in \{1,\dots,m\}$, and let $\Delta_{\delta e^j} f(x) \ge 0$ for all $\delta > 0$. Let us define*

$$
\begin{aligned}
c_k &= b_k \quad \text{if } \Delta_{\delta e^k} f(x) \ge 0 \text{ on } [a,b], \text{ for all } \delta > 0,\\
&= a_k \quad \text{otherwise},\\
&\quad (k = 1,\dots,j-1,j+1,\dots,m),\\
\phi_j(t) &= f(c_1,\dots,c_{j-1},t,c_{j+1},\dots,c_m).
\end{aligned}
$$

Then

  (i) ϕ_j *is continuous and non–decreasing on* $[a_j, b_j]$.

  (ii) *If f has a root within* $[a, b]$, *then*

$$\min\{x_j; x \in [a, b], f(x) = 0\} = \min\{t; \phi_j(t) = 0, t \in [a_j, b_j]\},$$
$$\textit{if } \phi_j \textit{ has a root in } [a_j, b_j],$$
$$= a_j \quad \textit{otherwise.}$$

PROOF.

  (i) The continuity of ϕ_j follows directly from the continuity of f. To prove that ϕ_j is non–decreasing, it is sufficient to note that by (1.2) the term $\Delta_{\delta e^j} f(x)$ does not change sign for any $\delta > 0$, $x \in [a, b]$, and $\mathrm{sgn}(\Delta_{\delta e^j} \phi_j(x)) = \mathrm{sgn}(\Delta_{\delta e^j} f(x))$.

  (ii) From the definition of function ϕ_j we see that

$$(2.1) \qquad\qquad \phi_j(x_j) \geq f(x) \quad \text{for all } x \in [a, b].$$

Let $x_j^* = \min\{x_j; x \in [a, b], f(x) = 0\}$. Then from the definition of ϕ_j it follows that $x_j^* \leq \min\{t; \phi_j(t) = 0, t \in [a_j, b_j]\}$ (otherwise there would be some $t \in [a_j, x_j^*)$ such that $\phi_j(t) = f(c_1, \ldots, c_{j-1}, t, c_{j+1}, c_m) = 0$). If $\phi_j(x_j^*) = 0$, then clearly

$$\min\{x_j; x \in [a, b], f(x) = 0\} = \min\{t; \phi_j(t) = 0, t \in [a_j, b_j]\}.$$

If $\phi_j(x_j^*) \neq 0$, then $\phi_j(t) > 0$ for all $x_j^* \leq t \leq b_j$, which follows from part (i) of this theorem and from (2.1). There is also $\phi_j(t) > 0$ for $a_j \leq t \leq x_j^*$, because $\phi(t) \neq 0$ on $[a_j, x_j^*]$ by the definition of ϕ_j. Thus, $\{t; \phi_j(t) = 0, t \in [a_j, b_j]\} = \emptyset$.

□

    In the case when $\Delta_{\delta e^j} f(x) \leq 0$, we can compute the minimum in step (i) of the algorithm as $\min\{x_j; x \in [a, b], f_i(x) = 0\} = \min\{x_j; x \in [a, b], -f_i(x) = 0\}$. The term on the right side can be evaluated by Theorem 2.1 because $\Delta_{\delta e^j}(-f(x)) = -\Delta_{\delta e^j} f(x) \geq 0$.

    From Theorem 2.1 it is clear that in step (ii)

$$|\max_{[a,b]} f_i(x)| = \phi_1(b_1) \quad \text{if } \Delta_{\delta e^j} f(x) \geq 0 \text{ on } [a, b],$$
$$= \phi_1(a_1) \quad \text{otherwise.}$$

$|\min_{[a,b]} f_i(x)|$ can be computed similarly.

3. Convergence

THEOREM 3.1. *Let* $F : [a, b] \subset \mathbf{R}^m \to \mathbf{R}^n$ *be a continuous function, which fulfills* (1.2), *and let* $\varepsilon, \delta > 0$. *Let* S^* *be the output of algorithm* **D** *for* $F, [a, b], \varepsilon, \delta$, *i.e.,* $S^* = \mathbf{D}(F, [a, b], \varepsilon, \delta)$. *Then*

  (i) S^* *contains all the roots of* F *on* $[a, b]$.

  (ii) $\|F(x)\|_M < \varepsilon$ *for all* $x \in S^*$.

PROOF. Both parts of the theorem follow directly from the iterational part of algorihm $\mathbf{D}$ and from the definition of $\|.\|_M$. $\square$

The following theorem shows that it is always possible to distinguish between two distinct solutions of (1.1) using a sufficiently small parameter ε in algorithm $\mathbf{D}$.

THEOREM 3.2. *Let $F : [a,b] \subset \mathbf{R}^m \to \mathbf{R}^n$ be a continuous function, which fulfills (1.2), $\delta > 0$, and let $\theta \notin F([a,b])$. Then there is $\varepsilon > 0$ such that $\mathbf{D}([a,b],F,\varepsilon,\delta) = \emptyset$.*

PROOF. The interval $[a,b]$ is a compact set, which implies that $F([a,b])$ is also compact, because function F is continuous. The norm $\|.\|_M$ has a minimum on $F([a,b])$, as $\|.\|_M$ is continuous on $\mathbf{R}^n$. Let us have

$$\varepsilon = \min\{\|y\|_M ; y \in F([a,b])\} = \min\{\|F(x)\|_M ; x \in [a,b]\}.$$

Then $\varepsilon > 0$, as $F(x) = \theta$ has no solution within $[a,b]$. From Theorem 3.1(ii) we have

$$\|F(x')\|_M < \varepsilon = \min\{\|F(x)\|_M ; x \in [a,b]\} \quad \text{for all } x' \in \mathbf{D}(F,[a,b],\varepsilon,\delta).$$

Thus, $\mathbf{D}(F,[a,b],\varepsilon,\delta) = \emptyset$. $\square$

4. Example

We illustrate the method with a system of two nonlinear inequalities:

(4.1)
$$\begin{aligned} y &\leq 1/x \\ x^2 + y^2 &\leq 4 \end{aligned}$$

for $x,y \in [0.01, 2]$.

It can be seen that the set of solutions is the "lens" in Figure 1. The system (4.1) is equivalent to the following set of equations:

(4.2)
$$\begin{aligned} f_1(x,y) &= 0 \\ f_2(x,y) &= 0 \end{aligned}$$

where

$$\begin{aligned} f_1(x,y) &= 0 \quad \text{for } y \geq \frac{1}{x}, \\ &= \frac{1}{x} - y \quad \text{otherwise,} \\ f_2(x,y) &= 0 \quad \text{for } x^2 + y^2 \leq 4, \\ &= x^2 + y^2 - 4 \quad \text{otherwise.} \end{aligned}$$

The results of computations of solutions of (4.2) by algorithm **D** are shown in Fig. 1.

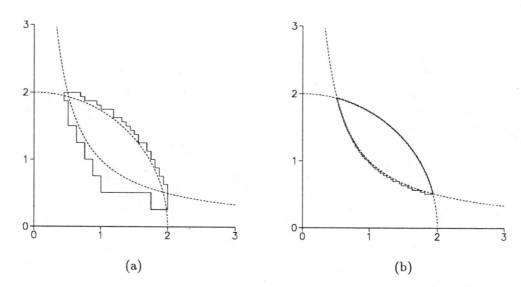

(a)   (b)

FIGURE 1. Output of algorithm **D** for for the example system.
(a) $\varepsilon = 0.5$, $\delta = \infty$, (b) $\varepsilon = 0.05$, $\delta = \infty$.

REFERENCES

1. G. Alefeld and L. Platzöder, *A quadratically convergent Krawczyk-like algorithm*, SIAM J. Numer. Anal. **20** (1983), 210–219.
2. E. R. Hansen, *A globally convergent interval method for computing and bounding real roots*, BIT **18** (1978), 415–424.
3. R. B. Kearfott, *Preconditioners for the interval Gauss–Seidel method*, SIAM J. Numer. Anal. **27** (1990), 804–822.
4. _____ , *Abstract generalized bisection and a cost bound*, Math. Comput. **49** (1987), 187–202.
5. _____ , *Some tests on generalized bisection*, ACM Trans. Math. Softw. **13** (1987), 197–220.
6. R. E. Moore and S. T. Jones, *Safe starting regions for iterative methods*, SIAM J. Numer. Anal. **14** (1977), 1051–1065.

INSTITUTE OF COMPUTER SCIENCE, MASARYK UNIVERSITY, BUREŠOVA 20, 60200 BRNO, CZECH REPUBLIC
E-mail address: mejzlik@muni.cz

Contemporary Mathematics
Volume **180**, 1994

PRECONDITIONING CELL-CENTERED FINITE DIFFERENCE EQUATIONS ON GRIDS WITH LOCAL REFINEMENT

ILYA. D. MISHEV

ABSTRACT. We consider cell-centered finite difference discretizations with local refinement for nonsymmetric boundary value problems. Preconditioners with mesh independent convergence properties for corresponding matrices are constructed. The method is illustrated with numerical experiments.

1. INTRODUCTION

This paper is devoted to construction of preconditioners of Bramble-Ewing-Pasciak-Schatz (**BEPS**) type [3] for solving nonsymmetric boundary value problems discretized by finite difference schemes on cell-centered grids with local refinement. Approximation properties of cell-centered finite difference schemes are investigated in [5], [11] for the symmetric problems, and in [7] for the nonsymmetric ones (see also [4] and [9]). The theory for two-level preconditioners is developed in [3],[6] and [8]. We extend the results obtained in [6] for nonsymmetric matrices without loss of optimality of the preconditioners, i.e., convergence rate is mesh independent.

We consider the following convection-diffusion boundary value problem: find a function $u(x)$ which satisfies the following differential equation and boundary condition:

$$(1.1) \qquad \begin{cases} \mathbf{div}(-a(x)\nabla u(x) + \underline{b}(x)u(x)) &= f(x) \quad \text{in } \Omega \\ u(x) &= 0 \quad \text{on } \Gamma \end{cases}$$

where $\Omega \subset R^2$ is a bounded domain and $\Gamma = \partial\Omega$. The coefficients $a(x)$ and $\underline{b}(x) = (b_1(x), b_2(x))$ are supposed to fulfill for some constants a_0 and β_0, β_1 the conditions

(i) $a(x) \geq a_0 > 0$, $a(x) \in W^1_\infty(\Omega)$,

(ii) $\mid b_i(x) \mid \leq \beta_1$, $b_i \in W^1_\infty(\Omega)$,

and in order to obtain coercivity it is sufficient that

(iii) $(\nabla.\underline{b}(x)) \geq \beta_0 > 0$.

The function $f(x)$ is given in Ω and $f(x) \in L^2(\Omega)$.

1991 *Mathematics Subject Classification.* 65N22, 65F10.

Key words and phrases. Cell-centered finite differences, nonsymmetric elliptic problem, local refinement, optimal preconditioners.

This paper is in final form and and no version of it will be submitted for publication elsewhere.

2. Description of the Preconditioners

Suppose the domain Ω is divided in two parts Ω_1 and Ω_2, $\Omega = \Omega_1 \cup \Omega_2$, $\Omega_1^o \cap \Omega_2^o = \phi$, where Ω_1 is the nonrefined and Ω_2 is the refined subdomain. Let us consider the composite grid ω (see [7] for a detailed description) divided in the same way, i.e., $\omega = \omega_1 \cup \omega_2$, $\omega_1 \subset \Omega_1$, $\omega_2 \subset \Omega_2$. The nodes of ω can be partitioned into three groups. The first group consists of the nodes in the refined subdomain Ω_2, the second one consists of nodes in Ω_1 next to the interface boundary, denoted by γ, and in the last are the rest of the nodes from Ω_1. Correspondingly our finite difference matrix A [7] admits a three-by-three block structure, i.e., we have

$$A = \begin{bmatrix} A_{11} & A_{12} & 0 \\ A_{21} & A_{22} & A_{23} \\ 0 & A_{32} & A_{33} \end{bmatrix} \begin{matrix} \}\omega_2 \\ \}\gamma \\ \}\omega_1 \backslash \gamma \end{matrix} .$$

We need also the s.p.d. coarse-grid matrix $\tilde{C}$, which is a approximation of matrix $\tilde{A}$ derived from the nonrefined finite difference scheme. We partition $\tilde{C}$ in the same manner as A into a three by three block structure on the nonrefined mesh $\tilde{\omega}$

$$\tilde{C} = \begin{bmatrix} \tilde{C}_{11} & \tilde{C}_{12} & 0 \\ \tilde{C}_{21} & \tilde{C}_{22} & \tilde{C}_{23} \\ 0 & \tilde{C}_{32} & \tilde{C}_{33} \end{bmatrix} \begin{matrix} \}\tilde{\omega}_2 \\ \}\gamma \\ \}\omega_1 \backslash \gamma \end{matrix} ,$$

where $\tilde{\omega}_2 = \tilde{\omega} \cap \Omega_2$. Then the preconditioner (**BEPS**) is constructed as follows [3], [6].

Given a vector $v = [v_1\ v_2\ v_3]^T$, we perform the following steps
(*i*) solve in Ω_2

$$A_{11} y_1^F = v_1;$$

(*ii*) compute the defect

$$d = v - A \begin{bmatrix} y_1^F \\ 0 \\ 0 \end{bmatrix} = \begin{bmatrix} 0 \\ v_2 - A_{21} y_1^F \\ v_3 \end{bmatrix} \begin{matrix} \}\omega_2 \end{matrix} ;$$

(*iii*) approximate the coarse-grid correction

$$\tilde{C}\tilde{y} = \begin{bmatrix} 0 \\ v_2 - A_{21} y_1^F \\ v_3 \end{bmatrix} \begin{matrix} \}\tilde{\omega}_2 \end{matrix} ;$$

(*iv*) find y_1^H in ω_2 such that

$$A_{11} y_1^H + A_{12} \tilde{y}_2 = 0 .$$

Then

$$y = B^{-1} v = \begin{bmatrix} y_1^F + y_1^H \\ \tilde{y}_2 \\ \tilde{y}_3 \end{bmatrix} .$$

In matrix notation

$$B = \begin{bmatrix} \begin{bmatrix} A_{11} \\ A_{21} \\ 0 \end{bmatrix} & \tilde{S}_c \end{bmatrix} \begin{bmatrix} I & [\ A_{11}^{-1} A_{12} \ \ 0\ ] \\ 0 & I \end{bmatrix} ,$$

where

$$\tilde{S}_c = \begin{bmatrix} \tilde{A}_{22}^c - \tilde{A}_{21}^c \tilde{A}_{11}^{c(-1)} \tilde{A}_{12}^c & \tilde{A}_{23}^c \\ \tilde{A}_{32}^c & \tilde{A}_{33}^c \end{bmatrix}$$

is the Schur complement of the coarse-grid matrix $\tilde{C}$.

Using the results in [7], we can easily prove the following auxiliary result.

Lemma 2.1. *There exists two positive constants $\hat{\gamma}_1$ and $\hat{\gamma}_2$ independent of h such that*

$$\hat{\gamma}_1 v_2^T \tilde{A} v_2 \le v^T A v$$

$$v^T A y \le \hat{\gamma}_2 (v_2^T \tilde{A} v_2)^{\frac{1}{2}} (y_2^T \tilde{A} y_2)^{\frac{1}{2}},$$

where $v = \begin{bmatrix} v_1 \\ v_2 \end{bmatrix} \begin{matrix} \}\omega\backslash\tilde{\omega} \\ \}\omega \end{matrix}$.

Our task is to estimate the eigenvalues of the preconditioned matrix $B^{-1}A$. We have

$$B^{-1}A = \begin{bmatrix} I & * \\ 0 & (\tilde{S}_c)^{-1}S_2 \end{bmatrix}$$

and from

$$\lambda(B^{-1}A) = (1, \lambda((\tilde{S}_c)^{-1}S_2))$$

is clear that we have to evaluate the spectrum of $((\tilde{S}_c)^{-1}S_2)$.

We consider two cases for the matrices C and $\tilde{C}$, respectively:

1. $\tilde{C} = \frac{\tilde{A}+\tilde{A}^T}{2}$, $C = \frac{A+A^T}{2}$,

i.e., the symmetric part of $\tilde{A}$ and A, and

2. $\tilde{C} = \tilde{A}^{(1)}$, $C = A^{(1)}$,

i.e., the part arising from the approximation of the diffusion term.

For the first case we have $v^T \tilde{A} v = v^T \tilde{C} v$. Using the results in [7], we get for the second one

$$v_2^T \tilde{A}^{(1)} v_2 \le v_2^T \tilde{A} v_2 \le E v_2^T \tilde{A}^{(1)} v_2.$$

Therefore, there exist constants γ_1 and γ_2 such that the following inequalities hold:

(2.2) $\gamma_1 v_2^T \tilde{C} v_2 \le v^T A v$

(2.3) $v^T A y \le \gamma_2 (v_2^T \tilde{C} v_2)^{\frac{1}{2}} (y_2^T \tilde{C} y_2)^{\frac{1}{2}}.$

We apply the technique proposed by Vassilevski [10] for the same problem and prove the auxiliary result.

Lemma 2.2. *Let the assumptions (2.2) and (2.3) be fulfilled. Then the following spectral equivalence relations hold:*

(2.4) $\gamma_1 v_2^{(2)^T} \tilde{S}_c v_2^{(2)} \le v_2^{(2)^T} S_2 v_2^{(2)}$ *for all $v_2^{(2)}$,*

$$w_2^{(2)^T} S_2 v_2^{(2)} \le \frac{\gamma_2^2}{\gamma_1} (w_2^{(2)^T} \tilde{S}_c w_2^{(2)^T})^{\frac{1}{2}} (v_2^{(2)^T} \tilde{S}_c v_2^{(2)^T})^{\frac{1}{2}}$$ *for all $w_2^{(2)}$ and $v_2^{(2)}$.*

Now we are ready to prove our main result.

Theorem 2.1. *The spectrum of $B^{-1}A$ lies in the following rectangle*

$$\{z \; : \; Re\,z \geq \min\,(1,\gamma_1)\,, \quad Re\,z\,,\, |Im\,z| \leq \max\,(1,\gamma_2^2/\gamma_1)\}$$

Proof. We can rewrite (2.4) in the following way

$$\frac{v_2^{(2)^T} S_2 v_2^{(2)}}{v_2^{(2)^T} \tilde{S}_c v_2^{(2)}} = \frac{v_2^{(2)^T} \frac{1}{2}\left(S_2^T + S_2\right) v_2^{(2)}}{v_2^{(2)^T} \tilde{S}_c v_2^{(2)}} \geq \gamma_1\,.$$

Hence

$$\begin{aligned} Re\,\lambda[B^{-1}A] &\geq& \min\left(1, Re\,\lambda\left[\tilde{S}_c^{-1}S_2\right]\right) \\ &\geq& \min\left(1, \lambda\left[\tilde{S}_c^{-\frac{1}{2}}\frac{1}{2}\left(S_2^T + S_2\right)\tilde{S}_c^{-1\frac{1}{2}}\right]\right) \\ &\geq& \min\,(1,\gamma_1)\,. \end{aligned}$$

For the other bound we have

$$|\lambda[B^{-1}A]| \leq \max\left(1, \left|\lambda\left[\tilde{S}_c^{-1}S_2\right]\right|\right)$$

and

$$\frac{v_2^{(2)^T} S_2 v_2^{(2)}}{v_2^{(2)^T} \tilde{S}_c v_2^{(2)}} \leq \frac{\gamma_2^2}{\gamma_1}\,.$$

Then

$$\left|\lambda\left[\tilde{S}_c^{-1}S_2\right]\right| = \left|\lambda\left[\tilde{S}_c^{-\frac{1}{2}}S_2\tilde{S}_c^{-\frac{1}{2}}\right]\right| \leq \frac{\gamma_2^2}{\gamma_1}\,.$$

□

Theorem 2.1 implies the following corollary.

Corollary 2.1. *The the preconditioned **GCG-LS** from Axelsson [1], [2] for solving the composite grid system with the preconditioner B will have rate of convergence independent of h and jumps of the coefficient $a(x)$.*

3. NUMERICAL RESULTS

In this section we illustrate the convergence behavior of the two preconditioners on two model examples. We solve the problem (1.1) in the domain $\Omega = (0,1) \times (0,1)$ with the velocity field

(3.5) $\qquad b_1 = (1 + x\cos(\alpha))\cos(\alpha)\,, \quad b_2 = (1 + y\sin(\alpha))\sin(\alpha)\,,$

where $\alpha = 15^0$. The refined subdomain is $\Omega_2 = \{0.5 \leq x_1 \leq 1,\ 0.5 \leq x_2 \leq 1\}$

Problem 1. *Consider a smooth solution $u(x)$ and a smooth coefficient $a(x)$,*

$$a(x) = \left[1 + 10(x_1^2 + x_2^2)\right]^{-1}\,, \quad u(x) = \phi_1(x_1)\psi_2(x_2)\,,$$

$$\phi_i(x_i) = \begin{cases} \sin^2\left(\pi\frac{x_i - d_i}{1 - d_i}\right)\,, & x_i \in (0.875, 1)\,, \\ 0\,, & \text{otherwise.} \end{cases}$$

TABLE 1. Preconditioner with $\tilde{C} = (\tilde{A} + \tilde{A}^T)/2$

		Problem 1		Problem 2	
n_c	h_c/h_f	iter	arfac	iter	arfac
	3	4	0.0264	3	0.0038
20	5	3	0.0098	3	0.0038
	7	4	0.0229	3	0.0036
	3	3	0.0070	3	0.0032
40	5	3	0.0055	3	0.0030
	7	2	0.0009	3	0.0029
80	3	2	0.0006	2	0.0009
	5	1	0.0001	2	0.0008

TABLE 2. Preconditioner with $\tilde{C} = \tilde{A}^{(1)}$

		Problem 1		Problem 2	
n_c	h_c/h_f	iter	arfac	iter	arfac
	3	4	0.0203	3	0.0038
20	5	3	0.0093	3	0.0038
	7	3	0.0082	3	0.0037
	3	2	0.0009	3	0.0032
40	5	2	0.0006	3	0.0030
	7	2	0.0005	3	0.0029
80	3	2	0.0004	2	0.0010
	5	1	0.0001	2	0.0008

We report the numbers of iterations for the preconditioned **GCG-LS** [1], [2]. The stopping criterion is $\|r_{last}\|/\|r_{first}\| < 10^{-6}$, $r = b - Ay$ where y is the current iteration, and $arfac = (\|r_{last}\|/\|r_{first}\|)^{1/iter}$. Our initial guess is found by constant interpolation of a coarse grid solution.

Problem 2. *Consider a piecewise continuous solution $u(x)$ and piecewise constant coefficient $a(x)$:*

$$u(x) = \left[\frac{(x_1 - b_1)(x_2 - b_2)}{a(x)} \right] \phi(x),$$

where

$$\phi(x) = \sin\left(\frac{\pi}{2} x_1\right) \sin\left(\frac{\pi}{2} x_2\right), \quad a(x) = \begin{cases} 1000, & x_i > (n_c + 3)h_c/2, \\ 1, & otherwise. \end{cases}$$

The results in Tables 1 and 2 show that the convergence rate of the considered algorithms is independent of a mesh size h, jumps of the coefficient $a(x)$ and smoothness of the solution. Although each iteration is relatively expensive (it includes solution of two problems on a refined grid and one problem on a coarse grid), the overall algorithm is very efficient because we need only a few iteration.

The theoretical and numerical results are in accordance with the general theory of overlapping domain decomposition. In fact we have overlap of the whole refined subdomain and that explanes the very good numerical results we report.

4. ACKNOWLEDGMENTS

The author would like to thank Dr. P. S. Vassilevski for the helpful discussions and constant encouridgment which make this paper possible. This work was partially supported by funding from DOE, DE-AC05-840R21400, Martin Marietta, Subcontract, SK965C and SK966V. The author is also grateful to the organizing committee of *the Seventh International Conference on Domain Decomposition in Scientific and Engineering Computing* for the help to attend the conference.

REFERENCES

1. O. Axelsson, *A generalized conjugate gradient, least square method*, Numer. Math. **51** (1987), 209–227.
2. O. Axelsson, *A restarted version of a generalized preconditioned conjugate gradient method*, Comm. Appl. Numer. Methods **4** (1988), 521–530.
3. J. H. Bramble, R. E. Ewing, J. E. Pasciak, A. H. Schatz, *Preconditioning technique for the efficient solution of problems with local grid refinement*, Comp. Meth. Applied Mechanics and Eng. **67** (1988), 149–159.
4. Z. Cai, J. Mandel, S. McCormick, *The finite volume element method for diffusion equations on composite grids*, SIAM J.Numer. Anal. **28** (1991), 392–402.
5. R. E. Ewing, R. D. Lazarov, P. S. Vassilevski, *Local refinement techniques for elliptic problems on cell-centered grids.I: Error analysis* Math. Comp. **56** (1991), 437–461.
6. R. E. Ewing, R. D. Lazarov, P. S. Vassilevski, *Local refinement techniques for elliptic problems on cell-centered grids.II: Two-grids iterative methods*, J. on Numer. Linear Algebra **56** (1991), 437–461.
7. R. D. Lazarov, I. D. Mishev, P. S. Vassilevski, *Finite Volume Methods with Local Refinement Convection-Diffusion Problems*, Computing (to appear).
8. J. Mandel, S. McCormick, *Iterative solution of elliptic equations with refinement:The two-level case*, Domain Decomposition Methods (T.F. Chan, R. Glovinski, J. Periaux, and O.B. Widlund, eds.), "SIAM", Philadelphia, 1989.
9. S. McCormic, *Multilevel Adaptive Methods for Partial Differential equations*, "SIAM", 1988.
10. P. S. Vassilevski,*Preconditioning nonsymmetric and indefinite finite element matrices*, J. Numer. Linear Algebra Appl. **1** (1992), 59–76.
11. P. S. Vassilevski, S. I. Petrova, R. D. Lazarov, *Finite difference schemes on triangular cell-centered grids with local refinement*, SIAM J. Sci. Stat. Comput. **13** (1992), 1287–1313.

DEPARTMENT OF MATHEMATICS, AND INSTITUTE FOR SCIENTIFIC COMPUTATION, TEXAS A & M UNIVERSITY, COLLEGE STATION, TX 77843-3368
E-mail address: mishev@math.tamu.edu

Contemporary Mathematics
Volume **180**, 1994

Outflow Boundary Conditions and Domain Decomposition Method

F. NATAF AND F. ROGIER

ABSTRACT. We consider an advection-diffusion problem. We write a Schur type formulation by using outflow boundary conditions on the interfaces. The condensed problem is solved by either a Jacobi algorithm (equivalent to an additive Schwarz method), GMRES, or BiCGstab. The use of outflow boundary conditions and of general iterative methods gives much better results than the original Schwarz method.

1. Introduction

Let Ω be an bounded open set of $\mathbb{R}^2$, we want to solve the following convection-diffusion problem:

$$\mathcal{L}(u) = \frac{u}{\Delta t} + a(x,y)\frac{\partial u}{\partial x} + b(x,y)\frac{\partial u}{\partial y} - \nu\Delta u = f \text{ in } \Omega$$

(1.1)
$$\mathcal{C}(u) = \tilde{g} \text{ on } \partial\Omega$$

where $\vec{a} = (a,b)$ is the velocity field, ν is the viscosity, f and $\tilde{g}$ are given functions, $\mathcal{C}$ is a linear operator. Δt is a constant which could correspond for instance to a time step for a backward-Euler scheme for the time dependent convection-diffusion equation.

In [4], Hagstrom et al. write a substructuring method for the convection-diffusion equation based on the exact outflow boundary conditions. The method is thus limited to a constant coefficient operator and makes use of nonlocal boundary conditions. In [1], Despres writes a substructuring method for the Helmholtz equation based on the radiation boundary condition of order 0. In both previous works, only non overlapping domains were considered. In this paper, we consider the convection-diffusion equation with variable coefficients and a decomposition into possibly overlapping subdomains. We use local outflow boundary conditions of order 0 and 2 (see also [7]).

1991 *Mathematics Subject Classification.* Primary 65N55, 76Rxx; Secondary 47A68.

This paper is in final form and no version of it will be submitted for publication elsewhere

The paper is organized as follows: in § 2, we write the substructuring formulation at the continuous level. In § 3, we discretize this formulation and we compare three solvers of the condensed problem.

2. A substructuring formulation

Let Ω be a bounded open set of $\mathbb{R}^2$. Let $\Omega_{i,\,1\leq i\leq N}$ be a finite sequence of sets embedded in Ω such that $\bar{\Omega} = \cup_{i=1}^{N}\bar{\Omega}_i$. Let $\Gamma = \partial\Omega$, $\Gamma_i = \partial\Omega_i - \Gamma$. The outward normal from Ω_i is $\vec{n}_i$ and $\vec{\tau}_i$ is a tangential unit vector. Let $\mathcal{B}_{i,\,1\leq i\leq N}$ be a sequence of operators leading to well posed boundary value problems (see equation (2.1) below). We assign to each subdomain i an operator S_i. Let f be a function from Ω_i to $\mathbb{R}$ and g a function from Γ_i to $\mathbb{R}$. $S_i(f,g,\tilde{g})$ is the solution v of the following boundary value problem:

$$
(2.1)\qquad
\begin{aligned}
\mathcal{L}(v) &= f(x), & x \in \Omega_i \\
\mathcal{B}_i(v) &= g(x), & x \in \Gamma_i \\
\mathcal{C}(v) &= \tilde{g}, & x \in \partial\Omega_i \cap \Gamma
\end{aligned}
$$

We introduce a sequence (η_i^j), $1 \leq i \leq N$, $1 \leq j \leq N$, $i \neq j$ of functions defined on the boundaries of the subdomains which satisfy:

$$
i)\quad \eta_i^j : \partial\Omega_i \longrightarrow [0,1]
$$

$$
ii)\quad \eta_i^j = 0 \text{ on } \partial\Omega_i - \bar{\Omega}_j
$$

$$
iii)\quad \sum_{j,j\neq i} \eta_i^j(x) = 1, \qquad x \in \partial\Omega_i
$$

REMARK 1. η_i^j is zero if $\partial\Omega_i \cap \bar{\Omega}_j = \emptyset$.

It is now possible to write a substructuring formulation. Let u be the solution to (1.1) and $u_i = u_{|\Omega_i}$. We write a system for $\mathcal{B}_i(u_i)$:

$$
\begin{aligned}
\mathcal{B}_i(u_i) = \sum_{j,j\neq i} \eta_i^j \mathcal{B}_i(u_i) &= \sum_{j,j\neq i} \eta_i^j \mathcal{B}_i(u_j) \\
&= \sum_{j,j\neq i} \eta_i^j \mathcal{B}_i(S_j(f_{|\Omega_j}, \tilde{g}, \mathcal{B}_j(u_j))) \\
= \sum_{j,j\neq i} \eta_i^j \mathcal{B}_i(S_j(f_{|\Omega_j}, \tilde{g}, 0)) &+ \sum_{j,j\neq i} \eta_i^j \mathcal{B}_i(S_j(0,0,\mathcal{B}_j(u_j)))
\end{aligned}
$$

Thus, $(\mathcal{B}_i(u_i))_{1\leq i\leq N}$ solves the following linear system:

(2.2)

$$
\mathcal{B}_i(u_i) - \sum_{j,j\neq i} \eta_i^j \mathcal{B}_i(S_j(0,0,\mathcal{B}_j(u_j))) = \sum_{j,j\neq i} \eta_i^j \mathcal{B}_i(S_j(f_{|\Omega_j}, \tilde{g}, 0))\, 1 \leq i \leq N
$$

Let $U = (U_i)_{1 \leq i \leq N}$ and $G = (G_i)_{1 \leq i \leq N}$ be the vectors

$$U = \begin{bmatrix} \mathcal{B}_1(u_1) \\ \vdots \\ \mathcal{B}_N(u_N) \end{bmatrix} \text{ and } G = \begin{bmatrix} \sum_{j, j \neq 1} \eta_1^j \mathcal{B}_1(S_j(f_{|\Omega_j}, \tilde{g}, 0)) \\ \vdots \\ \sum_{j, j \neq N} \eta_N^j \mathcal{B}_N(S_j(f_{|\Omega_j}, \tilde{g}, 0)) \end{bmatrix}$$

and $\mathcal{T}$ be the linear operator defined by

$$\mathcal{T}(U) = \begin{bmatrix} \sum_{j, j \neq 1} \eta_1^j \mathcal{B}_1(S_j(0, 0, \mathcal{B}_j(u_j))) \\ \vdots \\ \sum_{j, j \neq N} \eta_N^j \mathcal{B}_N(S_j(0, 0, \mathcal{B}_j(u_j))) \end{bmatrix}$$

System (2.2) may now be written in the following compact form:

(2.3) $$(Id - \mathcal{T})(U) = G$$

We shall consider three possibilities for $\mathcal{B}_i$:

(2.4) $$\mathcal{B}_i^I = Id$$

or

(2.5) $$\mathcal{B}_i^0 = \frac{\partial}{\partial \vec{n}_i} - \frac{\vec{a}.\vec{n}_i - \sqrt{(\vec{a}.\vec{n}_i)^2 + \frac{4\nu}{\Delta t}}}{2\nu}$$

($\vec{a}$ is the velocity field (a, b)) or

$$\mathcal{B}_i^2 = \frac{\partial}{\partial \vec{n}_i} - \frac{\vec{a}.\vec{n}_i - \sqrt{(\vec{a}.\vec{n}_i)^2 + \frac{4\nu}{\Delta t}}}{2\nu} + \frac{\vec{a}.\vec{\tau}_i}{\sqrt{(\vec{a}.\vec{n}_i)^2 + \frac{4\nu}{\Delta t}}} \frac{\partial}{\partial \vec{\tau}_i}$$
$$- \frac{\nu}{\sqrt{(\vec{a}.\vec{n}_i)^2 + \frac{4\nu}{\Delta t}}} (1 + \frac{(\vec{a}.\vec{\tau}_i)^2}{(\vec{a}.\vec{n}_i)^2 + \frac{4\nu}{\Delta t}}) \frac{\partial^2}{\partial \vec{\tau}_i^2}$$

The boundary conditions $\mathcal{B}_i^0$, or $\mathcal{B}_i^2$ are far field boundary conditions (also called Outflow B.C., Absorbing B.C., Artificial B.C., Radiation B.C.,etc , see [2], [5]) of order 0 and 2.

3. Discretization and numerical results

In order to illustrate the validity of the method, a 2D test problem has been performed. We solve the following problem:

$$\begin{cases} \dfrac{u}{\Delta t} + a(x, y) \dfrac{\partial u}{\partial x} + b(x, y) \dfrac{\partial u}{\partial y} - \nu \Delta u = 0, \quad 0 \leq x \leq 1, \, 0 \leq y \leq 1 \\ u(0, y) = 1, \quad 0 < y < 1 \\ \dfrac{\partial u}{\partial y}(x, 1) = 0, \quad 0 < x < 1 \\ \dfrac{\partial u}{\partial x}(1, y) = 0, \quad 0 < y < 1 \\ u(x, 0) = 0, \quad 0 < x < 1 \end{cases}$$

The operator $\mathcal{L}$ is discretized by a standard upwind finite difference scheme of order 1 (see [3]) and $\mathcal{B}_{i,\,1\le i\le N}$ be a finite difference approximation. We used a rectangular finite difference grid. The mesh size is denoted by h. The unit square is decomposed into overlapping rectangles. This leads to a discretization of the operator $\mathcal{T}$ and thus of system (2.3):

$$(3.1) \qquad\qquad (Id - \mathcal{T}_h)(U_h) = G_h$$

REMARK 2. Any other discretization could be used as well.

From the definition of $\mathcal{T}_h$, we see that the computation of $\mathcal{T}_h$ applied to some vector U_h amounts to the solving of N independent boundary value subproblems (one subproblem in each subdomain) which can be solved in parallel. We have considered three algorithms in order to solve (3.1): GMRES(∞), BiCGStab and a Jacobi algorithm:

$$(3.2) \qquad\qquad U_h^{n+1} = \mathcal{T}_h(U_h^n) + G_h$$

The choice of the last algorithm is due to the fact that it is equivalent to an additive Schwarz method (ASM) whose convergence has been studied in [6] and [1] for Fourier interface conditions and in [8] for outflow boundary conditions. In Tables 1 and 2, we give the number of subproblems solved so that the maximum of the error is smaller than 10^{-6}. One iteration of GMRES(∞) or of ASM counts for one solution in each subdomain and one iteration of BiCGStab counts for two solutions in each subdomain. In the tables, Id corresponds to the use of the Id as interface condition, OBC0 to $\mathcal{B}^0$ (see (2.5)) and OBC2 to $\mathcal{B}^2$ (see (2.6)).

Table 1: Computational cost vs. interface conditions and solvers

Boundary Cond.	ASM	BiCGStab	GMRES
Id	> 200	88	61
OBC0	86	38	33
OBC2	46	28	24

Table 1 corresponds to the following parameters:
8×1 subdomains, 21×120 points in each subdomain, overlap $= 2h$, $\nu = 0.1$, $\Delta t = 10^{40}$, $a = y$, $b = 0$.

Table 2: Computational cost vs. interface conditions and solvers

Boundary Cond.	ASM	BiCGStab	GMRES
Id	479	64	50
OBC0	27	22	19
OBC2	18	16	16

Table 2 corresponds to the following parameters:
4×4 subdomains, 35×35 points in each subdomain, overlap $= 2h$, $\nu = 0.1$, $\Delta t = 1$, $a = y$, $b = 0$.

The use of outflow boundary conditions leads to a significant improvement, whatever iterative solver is used. BiCGStab and GMRES give similar results with an advantage to GMRES in terms of computational cost and to BiCGStab in terms of storage requirements, since only two directions have to be stored.

4. Conclusion

The interest of using outflow boundary conditions as interface conditions is clear. We have considered here the scalar convection-diffusion equation. The same strategy can be applied to systems of PDE's.

REFERENCES

1. B. Despres, *Domain Decomposition Method and the Helmholtz Problem*, Mathematical and Numerical Aspects of Wave Propagation Phenomena, SIAM (1991), 44–52, Eds G. Cohen, L. Halpern and P. Joly.
2. B. Engquist and A. Majda, *Absorbing Boundary Conditions for the Numerical Simulation of Waves*, Math. Comp. **31** (139), (1977), 629–651.
3. C.A.J. Fletcher, Computational Techniques for Fluid Dynamics, Springer Series in Computational Physics, Springer
4. T. Hagstrom, R.P. Tewarson and A. Jazcilevich, *Numerical Experiments on a Domain Decomposition Algorithm for Nonlinear Elliptic Boundary Value Problems*, Appl. Math. Lett., **1, No 3** (1988), 299–302.
5. L. Halpern, *Artificial Boundary Conditions for the Advection-Diffusion Equations*, Math. Comp., vol 174, (1986), 425–438.
6. P.L. Lions, *On the Schwarz Alternating Method III: A Variant for Nonoverlapping Subdomains*, Third International Symposium on Domain Decomposition Methods for Partial Differential Equations, SIAM (1989), 202–223, Eds T.F. Chan, R. Glowinski, J. Periaux and O.B. Widlund, Houston.
7. F. Nataf, *Méthodes de Schur généralisées pour l'équation d'advection-diffusion (Generalized Schur methods for the advection-diffusion equation)*, C.R. Acad. Sci. Paris, t. 314, Série I, (1992), 419–422.
8. F. Nataf and F. Rogier, *Factorization of the Convection-Diffusion Operator and the Schwarz Algorithm*, to appear in Math. Models & Meth. in Appl. Sci. .

CMAP, UNITÉ CNRS URA 756, ECOLE POLYTECHNIQUE, 91128, PALAISEAU CEDEX, FRANCE
E-mail address: nataf@cmapx.polytechnique.fr

DIVISION CALCUL PARALLÈLE, ONERA, 29 AV. DE LA DIVISION LECLERC, 92322 CHÂTILLON, FRANCE
E-mail address: rogier@onera.fr

Contemporary Mathematics
Volume 180, 1994

Domain Decomposition for Adaptive
hp Finite Element Methods

J. TINSLEY ODEN, ABANI PATRA and YUSHENG FENG

ABSTRACT. A highly parallelizable domain decomposition solution technique
for adaptive hp finite element methods is developed. The technique uses good
partitioning strategies and a subspace decomposition based preconditioned iterative solver. Two level orthogonalization is used to obtain a reduced system which
is preconditioned by a coarse grid operator. Numerical results show fast convergence for the iterative solver and good control of the condition number (less than
16 for meshes with spectral orders up to 8).

1. Introduction

Adaptive hp finite element methods [1], in which both the mesh size and spectral
order are independently varied over the whole mesh, produce exponential convergence rates in the discretization error even in the presence of singularities. Together with domain decomposition and parallel algorithms they have the potential
to produce dramatic improvements in finite element modeling of problems in computational mechanics. However, the complex mesh and data structures involved
in adaptive hp finite element methods raise interesting issues in domain partitioning and the parallel solution process. Domain decomposition for h–version and
p–version finite element methods have been investigated by several authors [5][6],
but there have been no studies on domain decomposition for the hp–version finite
element methods.

Two issues that immediately arise are – automatic partitioning of hp adaptive finite element meshes and efficient parallel solution of the resulting algebraic systems
using iterative solvers. In partitioning an hp mesh, difficulties may be encountered
due to non-uniform computational load across the elements, non-uniform communication patterns and constraints between partitions. The usual choice of finite

1991 *Mathematics Subject Classification.* Primary 41A10; Secondary 65N30. The authors are
supported by DARPA contract no. DABT63-92-C-0042 This paper is the final form. No revision
will be submitted elsewhere.

element basis functions will also lead to high condition number of the associated algebraic system (possibly $O(10^{p_{max}})$, where p_{max} is the highest spectral order of the approximation) and slow convergence of almost all iterative solvers.

In this study, a highly parallelizable domain decomposition method is developed for adaptive hp finite element methods. The remainder of the paper is organized as follows. In section 2, three families of partitioning algorithms are presented. In section 3, a decomposition of the hp finite element space is formulated. The algorithm is tested on a two–dimensional elliptic boundary value problem. Numerical results and conclusions are given in section 4 and section 5, respectively.

2. Automatic Partitioning Algorithms

Efficient parallel computing requires decomposition of the problem into "load balanced" sub-problems with minimal communication and synchronization among them. The twin objectives in a partitioning algorithm thus are equi-distribution of computational effort among the sub–domains, which are eventually assigned to different processors and minimization of the interfaces between the sub-domains. These two objectives are necessary to maximize utilization of the processors, minimize communication among them and reduce the size of the interface problem. However, one may often need to accept trade-off between the two goals.

Unlike h-version methods, choice of *a-priori* computational effort measure is a non-trivial issue. In adaptive hp finite element approximation, good candidate measures of computational effort appear to be 1) error distribution in a coarse mesh solution 2) degrees of freedom distribution in the mesh and 3) element conditioning estimation.

Although degrees of freedom seems to be a natural choice, it does not reflect the computational effort very accurately. The motivation for error as a partitioning measure is in the hp adaptive strategy developed in a previous study [2].

In this section, three partitioning algorithms for adaptive hp meshes are presented. Each algorithm is implemented with different choices of computational effort measure. Due to the page limit of these proceedings,only one of the algorithms is discussed in detail. More details will be provided in future publications.

2.1. Mesh Traversal Based Decomposition (MTBD). In this family of partitioning algorithms, the mesh is traversed in some fashion and then elements are accumulated into partitions based on some estimate of computational effort. In the first choice of ordering implemented, the mesh is traversed in a nearest neighbor with lowest load fashion. This ensures some amount of locality in the decomposition as each element has at least one neighboring element in the decomposition. However this ordering often results in disconnected partitions. This drawback can be somewhat overcome by using an ordering created by mapping the centroids of the elements onto a Peano–Hilbert curve[4].

2.2. Recursive Load Based Bisection of Coordinate (RLBBC) This family of algorithms uses an explicit choice of interface to create the partition. The advantage of these methods is that both objectives, load balance and minimum interface are explicitly addressed. However, the cost of doing so inhibits the method.

In principle, the methods are comprised of selecting candidate separator surfaces and then selecting a separator based on maximum load balance and minimum interface. The selection involves assigning to each candidate separator surface a number indicative of load balance and interface size associated with the resulting domain partitions. This approach is a generalization of the work of Miller, Vavasis and their coworkers [3] on partitioning two and three dimensional finite element grids for the *h* version.

2.3. Recursive Load Based Bisection of Ordering (RLBBO). In this family of algorithms, an attempt is made to combine the advantages of the mesh traversal algorithms with that of the interface partitioning. The elements are ordered using the "Peano-Hilbert curve" ordering, and then a recursive splitting is applied to the resulting one–dimensional ordering of the elements. The basic algorithm is outlined below:

Algorithm

1. Create an ordering of the elements by mapping the centroids of the elements onto a Peano-Hilbert curve.

2. Let t_K be the distance of the centroid of element K along this curve.

3. Compute maximum and minimum of t_K.

4. Compute n trial separator levels as

$$t_i = t_{min} + \frac{t_{max} - t_{min}}{n}$$

5. For each t_i compute q_i

$$q_i = abs(\frac{dof_{left}}{dof_{right}} - 1) \cdot dof_{tot} + dof_{inter}$$

Replace dof by error or other load estimate as appropriate

6. Choose as interface t_i corresponding to lowest q_i

7. Apply 1-6 recursively.

One particularly demanding *hp* mesh (*p* ranges from 1 to 7) and corresponding partitions are shown in the figure 1(a) and figure 1(b). The resulting partitions seem to have balanced load and nice interfaces.

3. Domain Decomposition Solver for *hp* FEM

The solver will be discussed with respect to the model problem defined below:

Find $u \in \mathcal{V}$ such that $\mathcal{B}(u, v) = \mathcal{L}(v)$ $\forall v \in \mathcal{V}$

where $\mathcal{V} = \{v \in H^1(\Omega) : v = 0 \text{ on } \partial\Omega\}$, and $\mathcal{B}(u, v)$ is the bilinear form resulting from the weak formulation of a two-dimensional second order elliptic PDE with Dirichlet boundary conditions on the boundary $\partial\Omega$.

Define $\mathcal{V}^h \subset \mathcal{V}$ as a finite dimensional subspace constructed by a series of affine mappings of the following functions defined on a 'master element'.

Vertices (Nodes):

$$\widehat{\psi}_i(\xi, \eta) = \frac{1}{4}(1 \pm \xi)(1 \pm \eta) \qquad i = 1, 2, 3, 4$$

Edges:

$$\widehat{\zeta}_i^k(\xi, \eta) = \begin{cases} \dfrac{(-1)^i}{2}(1 \pm \eta)\rho_k(\xi) & i = 1, 3 \\[2mm] \dfrac{(-1)^i}{2}(1 \pm \xi)\rho_k(\eta) & i = 2, 4 \end{cases}$$

Interior (Bubble):

$$\widehat{b}_{ij}(\xi, \eta) = \rho_i(\xi)\rho_j(\eta) \qquad 2 \leq i, j \leq p$$

$$\rho_k(\xi) = \sqrt{\frac{2k-1}{2}} \int_{-1}^{\xi} P_{k-1}(s)\,ds$$

with P_{k-1} the Legendre polynomial of degree $k-1$

Assume domain Ω is decomposed into N_D sub-domains. Each sub-domain Ω_i is associated with a subspace $\mathcal{V}_i^h$. Then

$$\mathcal{V}^h = \sum_{i=1}^{N_D} \mathcal{V}_i^h, \qquad \mathcal{V}_i^h \subset \mathcal{V}^h \subset \mathcal{V} \quad \forall i$$

Each subspace is further decomposed into

$$\mathcal{V}_i^h = \mathcal{X}_i^N + \mathcal{X}_i^S + \mathcal{X}_i^V + \mathcal{X}_i^E + \mathcal{X}_i^B$$

where $\mathcal{X}_i^N$ and $\mathcal{X}_i^S$ are spaces spanned by vertex and edge functions on subdomain interfaces. $\mathcal{X}_i^V$, $\mathcal{X}_i^E$ and $\mathcal{X}_i^B$ are spaces spanned by vertex, edge and bubble functions in subdomain interiors. Then, the bilinear form can be written as:

$$\mathcal{B}(u_{hp}, u_{hp}) = \sum_{i=1}^{N_D} \mathcal{B}_i(u_{hp}^N + u_{hp}^S + u_{hp}^V + u_{hp}^E + u_{hp}^B, u_{hp}^N + u_{hp}^S + u_{hp}^V + u_{hp}^E + u_{hp}^B)$$

Now if the local trial functions are chosen to satisfy the orthogonality condition

(1) $$\mathcal{B}_{i,K}(\gamma_j, b_k) = 0 \quad \forall \gamma_j \in \mathcal{X}_i^V + \mathcal{X}_i^E, b_k \in \mathcal{X}_i^B$$

the element stiffness matrix reduces to the form

$$K_{elt} = \begin{bmatrix} \widetilde{VV} & \widetilde{VE} & 0 \\ \widetilde{EV} & \widetilde{EE} & 0 \\ 0 & 0 & BB \end{bmatrix}$$

where $\widetilde{VV}, \widetilde{EE}, \widetilde{VE}$ represent modified blocks of the original element stiffness matrix.

If in addition the trial functions satisfy the orthogonality condition

$$(2) \qquad \mathcal{B}_{i,K}(\tau_j, \varphi_k) = 0 \quad \forall \tau_j \in \mathcal{X}_i^N + \mathcal{X}_i^S, \varphi_k \in \mathcal{X}_i^V + \mathcal{X}_i^E + \mathcal{X}_i^B$$

the resulting subdomain stiffness matrices reduce to

$$K_i = \begin{bmatrix} \widetilde{NN} & \widetilde{NS} & 0 \\ \widetilde{SN} & \widetilde{SS} & 0 \\ 0 & 0 & II \end{bmatrix} \qquad II = \begin{bmatrix} \widetilde{II_{VV}} & \widetilde{II_{VE}} & 0 \\ II_{EV} & II_{EE} & 0 \\ 0 & 0 & BB \end{bmatrix}$$

where NN, NS, SN, SS   are shared degrees of freedom among subdomains. Note that the first condition causes the orthogonalization of the bubbles with respect to the edges and vertices while the second causes the orthogonalization of the interfaces with the interiors. The subdomain problems are now independent of the interface problem.

Remark 1:   Implementation of the first condition can be done at the element level and is thus completely parallelizable.

Remark 2:   The modifications of NN etc. to $\widetilde{NN}$ etc. are of the form

$$\widetilde{NN} = \sum_{i=1}^{N_D} (NN_i + \widehat{NN}_i)$$

Remark 3:   If an iterative solver (e.g. PCG,GMRES) is used, these modifications can then directly participate in the parallel matrix-vector product

$$K \cdot p = \sum_{i-1}^{N_D} (K_i + \widehat{K}_i) \cdot p$$

and there is no need for assembly of these components.

The parallel domain decomposition algorithm is summarized as follows:

Parallel Domain Decomposition Solution Algorithm

1. Partition the mesh into subdomains using any of the decomposition algorithms.
2. Create subdomain approximations transforming the algebraic system at the element level to satisfy orthogonality conditions (1) and (2).
3. Solve the reduced preconditioned system by an iterative method (e.g. PCG,GMRES) using coarser grid operator preconditioning.
4. Solve subdomain problems in parallel.

5. Transform the solution of the reduced system to the original system for condition
   (2) at subdomain level and condition (1) at element level.

4. Numerical Results And Conclusions

Poisson's equation in two dimensions is chosen as a test problem. In figure 2(a), both
iteration counts and condition number estimation are plotted against p (ranging
from 2 to 8). The condition number is controled under 16. Figure 2(b) shows
the residual and condition number estimation against iteration count for hp DD
and Conventional Jacobi. A variant of Lanczos connection is used to estimate
conditioning of the preconditioned operator[7].

In this study, several partitioning algorithms were developed and tested for adaptive hp meshes. Performance of RLBBO appears to superior. Two level orthogonalization of finite element basis functions produces good control of conditioning for
the algebraic system generated by hp finite element approximation. The resulting
domain decomposition solution algorithm is highly parallelizable.

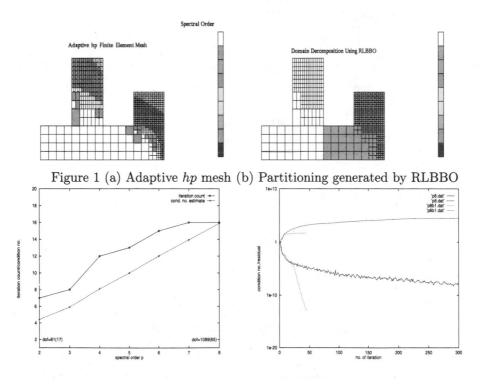

Figure 1 (a) Adaptive hp mesh (b) Partitioning generated by RLBBO

Figure 2.(a) hp Domain Decomposition Solver (4 sub-domain) (b) Comparison of
hp DD solver with conventional Jacobi Iterative Solver

References

1. L. Demkowicz, J. T. Oden, W. Rachowicz and O. Hardy "Toward A Universal hp Adaptive Finite Element Strategy, Part 1. Constrained Approximation and Data Structure" *Comput. Methods. Appl. Mech. and Engg.*, 77(1989), pp.79-112

2. J. T. Oden, Abani Patra and Y. S. Feng, " An hp Adaptive Strategy", *Adaptive, Multilevel and Hierarchical Computational Strategies*, A. K. Noor(ed)., AMD-Vol. 157, 1992, pp. 23-46.

3. G. L. Miller, S.Teng, W. Thurston, S. A. Vavasis, "Automatic Mesh Partitioning", CTC92TR112 , Advanced Computing Research Institute, Department of Computer Science, Cornell University, Ithaca, 1992.

4. E. A. Patrick, D. R. Anderson and F. K. Bechtel, "Mapping Multidimensional Space To One Dimension For Computer Output Display", *IEEE Transactions on Computers*, vol C-17, No. 10, October, 1968

5.J. Bramble, J. Pasciak, A. Schatz, "The Construction of Preconditioners for Elliptic Problems by Substructuring I", *Math. Comp.* 47, N 175 (1986), pp. 103-134

6. I. Babuska, A. Craig, J. Mandel, and J. Pitkaranta, "Efficient Preconditioning For The p version finite element method in Two Dimensions", *SIAM J. Numer. Anal*, Vol 28, No. 3, pp. 624-661, June 1991.

7. J. Tinsley Oden and Yusheng Feng, "A CG Extension for Nonsymmetric Linear Systems and Lanczos' Connection to PCG", Cornelius Lanczos International Centenary Conference, Raleigh, North Carolina , December 1993.

Texas Institute For Computational and Applied Mathematics,
The University Of Texas at Austin,
3500, West Balcones Ctr. Drive,
Austin, TX-78759
U.S.A

oden@ticom.ae.utexas.edu
abani@ticom.ae.utexas.edu
feng@ticom.ae.utexas.edu

Contemporary Mathematics
Volume **180**, 1994

Domain Decomposed Preconditioners
with Krylov Subspace Methods
as Subdomain Solvers

MICHAEL PERNICE

ABSTRACT. Domain decomposed preconditioners for nonsymmetric partial differential equations typically employ exact subdomain solves. Iterative methods for subdomain problems require less storage and allow flexibility in specifying accuracy on the subdomains. Savings in solution time is possible if the effectiveness of the domain decomposed preconditioner is not reduced by lower accuracy subdomain solutions. Numerical experiments compare the overall iteration count as the accuracy of the subdomain solutions is varied. The results demonstrate that the strategy is effective even for low accuracy subdomain solutions.

1. Introduction

Domain decomposition is a technique for constructing parallel algorithms for the solution of partial differential equations. Optimal and near-optimal methods have been developed, but many practical issues must be considered to obtain efficient, scalable implementations. This paper explores how the accuracy of subdomain solutions affects the performance of the overall method for nonsymmetric problems.

The solution of subdomain problems constitutes the largest single computational cost of a domain decomposition method and strongly influences the overall performance and cost of an implementation. Direct subdomain solves require storage for a banded matrix that represents the discrete subdomain operator. The small local memory of many multicomputers limits the size of problems that may be attempted. Also, the subdomain operator may not be explicitly available and may be too expensive to calculate. An obvious alternative is to solve the subdomain problems with an iterative method. This option leads to

---

1991 *Mathematics Subject Classification.* 65Y05, 65N55, 65F10.
This work was supported in part by a grant from the IBM Corporation.
This paper is in final form, and no version of it will be submitted for publication elsewhere.

other practical considerations, such as the accuracy of the subdomain solutions, possible degradation of the convergence rate of the overall procedure, and the choice of iterative method.

Some of these issues have been explored for symmetric problems [1, 6, 9]. Similar studies for nonsymmetric problems have not been pursued. This paper considers several transpose-free Krylov subspace methods to solve the subdomain problems: GMRES [8], CGS [10], TFQMR [4], and CGS with minimum residual smoothing (referred to as SCGS) [11].

2. The Domain Decomposition Framework

A nonoverlapping $p \times q$ partition for a total of $N_P = pq$ subdomains and a simplified version of the methods in [3, 2] are used. The discrete system

$$(1) \qquad\qquad\qquad Ax = b$$

is solved using preconditioned GMRES [8]. The preconditioner has the structure

$$M = \begin{pmatrix} \tilde{A}_\Omega & A_{\Omega\Gamma} & \\ & \tilde{A}_\Gamma & A_{\Gamma\chi} \\ & & \tilde{A}_\chi \end{pmatrix}$$

where $\tilde{A}_\Omega$ is a preconditioner for the subdomains, $\tilde{A}_\Gamma$ is a preconditioner for the interfaces, and $\tilde{A}_\chi$ is a preconditioner for the crosspoint problem [2]. $A_{\Omega\Gamma}$ and $A_{\Gamma\chi}$ couple subdomains, interfaces and crosspoints. The preconditioning step requires solution of a linear system involving M and is implemented as a block backsolve. A crosspoint system

$$\tilde{A}_\chi u_\chi = v_\chi$$

is solved first. The crosspoint system is duplicated on every processor using all-to-all communication and is solved using a banded factorization procedure. The operator $\tilde{A}_\chi$ is a coarse-grid analog of the discrete operator A.

The solution of the crosspoint system is used in problems on the interfaces

$$\tilde{A}_\Gamma u_\Gamma = v_\Gamma - A_{\Gamma\chi} u_\chi,$$

which is done in parallel. $\tilde{A}_\Gamma$ is constructed using IP(0) interface probing [3].

Finally the subdomain problems

$$(2) \qquad\qquad\qquad \tilde{A}_\Omega u_\Omega = v_\Omega - A_{\Omega\Gamma} u_\Gamma$$

are solved in parallel. Exact subdomain solves use $\tilde{A}_\Omega = A_\Omega$. Using an iterative method to solve (2) produces a variable preconditioner that cannot be accommodated by GMRES. A flexible variant [7] is used that allows variable preconditioning but doubles the required storage. Despite this a net savings in storage can be realized. Iterative methods on the subdomains are preconditioned by an MILU factorization [5] of the local discrete operator.

3. Numerical Results

Sharp estimates of convergence rates for the methods on the subdomains are not available, making it difficult to predict the amount of work needed to solve (2) and the resulting impact on the solution of (1). Consequently the advantages of using these methods are illustrated with numerical experiments.

3.1. Model Problem. The model problem is the convection-diffusion equation

$$
(3) \qquad c_1 u_x + c_2 u_y - \epsilon \Delta u = f \quad (x, y) \in \Omega.
$$

The equation is discretized using second-order centered differences for the diffusion term and first-order upwind differences for the convection terms.

Problem 1 solves (3) on $\Omega = [0,1] \times [0,1]$ with constant coefficients. Dirichlet boundary conditions and source term f are specified so that the exact solution is

$$
u(x, y) = \frac{e^{c_1 x/\epsilon} - 1}{e^{c_1/\epsilon} - 1} + \frac{e^{c_2 y/\epsilon} - 1}{e^{c_2/\epsilon} - 1}.
$$

All tests run for this problem were parameterized to produce a fixed $Re_c = 2$.

Problem 2 solves (3) on $\Omega = [-1,1] \times [0,1]$ with variable coefficients

$$
c_1(x, y) = 2y(1 - x^2) \qquad c_2(x, y) = -2x(1 - y^2)
$$

and mixed Dirichlet-Neumann boundary conditions:

$$
u(x, y) = \begin{cases} 0 & \text{if } x = -1, \text{ or } y = 1 \\ 0 & \text{if } y = 0 \text{ and } -1 \leq x < 0 \\ 100 & \text{if } x = 1 \end{cases},
$$

$$
u_n = 0 \text{ if } y = 0 \text{ and } 0 \leq x \leq 1
$$

and source term $f = 0$. All tests that were run for this problem were parameterized to produce $Re_c \leq 4$ throughout the domain.

3.2. Convergence behavior. Sample convergence histories of the tested methods are provided for reference. Results for problem 1 with a uniform mesh size of $h = 1/256$ and horizontal convection appear in Fig. 1. An initial approximation of $x_i = 1$ is used, and the iterations are halted when $\|r_n\| < 10^{-6}\|r_0\|$.

These histories indicate that relaxing the accuracy of the subdomain solutions will benefit GMRES most. They also indicate that CGS is likely to be the most economical method when accurate subdomain solutions are sought and that a fixed number of iterations may not be advisable for CGS and its smoothed variants. Convergence histories for other directions of convection and for problem 2 are similar.

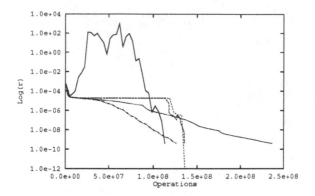

FIGURE 1. Convergence histories for problem 1. CGS: solid line; TFQMR: long dashed line; SCGS: short dashed line; GM-RES(20): dotted line; GMRES(5): dashed-dotted line.

3.3. Performance. This section compares the effectiveness of the various domain decomposed preconditioners as the accuracy of the subdomain solves is relaxed. In all cases $p \geq q$ was chosen to favor the computational and storage requirements of the direct methods on the subdomains with a natural ordering, irrespective of other considerations. The measurements were obtained on the nCUBE/2 at the San Diego Supercomputer Center. The time for constructing the preconditioner is included, since this is the dominant cost of using exact solves on the subdomains.

An initial approximation of $x_i = 0$ is used for the subdomain problems and the iterations on the subdomains are halted when $\|r_n\| < \eta\|r_0\|$ for various values of η. In the case of iterative solvers on the subdomains, the subdomain problems needed for interface probing were solved with the iterative method and a fixed tolerance of 10^{-6}, making $\tilde{A}_\Gamma$ independent of η. The overall FGMRES method uses an initial approximation of $x_i = 1$ and the iterations are halted when $\|r_n\| < 10^{-6}$.

Table 1 shows the results for problem 1. In most cases, decreasing η to 10^{-2} does not substantially increase the iteration counts. When it does, the work saved on the subdomains more than compensates for this. For $h = 1/256$ the memory per node was inadequate for a direct method on the subdomains when $N_P = 4$ or 8. For both problem sizes, reducing η to 10^{-1} greatly increases the overall iteration count but surprisingly is still beneficial for GMRES. Similar results were obtained with directions of convection.

Table 2 shows the results for problem 2. Smaller gridsizes were used for this problem because of memory constraints and a uniform meshsize. These results were quite similar to those for problem 1, except to note that for the small problem, the subdomain problems were more difficult for the iterative methods,

TABLE 1. Results for problem 1 with horizontal convection. Execution times are in seconds and iteration counts are in parentheses.

DIRECT METHOD ON SUBDOMAINS

$h = 1/128$			$h = 1/256$			
$p = 2$ $q = 2$	$p = 4$ $q = 2$	$p = 4$ $q = 4$	$p = 2$ $q = 2$	$p = 4$ $q = 2$	$p = 4$ $q = 4$	$p = 8$ $q = 4$
59.9 (8)	14.4 (12)	9.22 (18)	n/a	n/a	86.3 (19)	44.9 (59)

ITERATIVE METHODS ON SUBDOMAINS

$h = 1/128$	η	GMR(20)	GMR(5)	TFQMR	SCGS	CGS
$p = 2$	10^{-6}	53.1 (8)	40.4 (8)	29.9 (8)	30.8 (8)	26.2 (8)
$q = 2$	10^{-4}	39.7 (8)	31.2 (8)	26.0 (8)	26.7 (8)	23.0 (8)
	10^{-2}	23.6 (8)	20.5 (8)	21.2 (8)	22.3 (8)	19.2 (8)
	10^{-1}	18.3 (11)	16.0 (11)	17.2 (11)	18.2 (11)	19.1 (11)
$p = 4$	10^{-6}	22.9 (12)	21.6 (12)	16.5 (12)	17.0 (12)	14.9 (12)
$q = 2$	10^{-4}	17.6 (12)	16.3 (12)	14.1 (12)	14.3 (12)	12.6 (12)
	10^{-2}	10.8 (12)	10.8 (12)	12.5 (13)	12.7 (13)	10.4 (12)
	10^{-1}	8.55 (15)	8.52 (15)	11.1 (16)	11.4 (17)	10.7 (16)
$p = 4$	10^{-6}	17.2 (18)	16.2 (18)	12.8 (18)	13.0 (18)	11.4 (18)
$q = 4$	10^{-4}	13.3 (18)	12.2 (18)	10.7 (18)	10.9 (18)	9.50 (18)
	10^{-2}	8.87 (20)	8.79 (20)	9.97 (20)	9.52 (19)	8.44 (20)
	10^{-1}	8.18 (35)	8.06 (35)	14.5 (46)	20.1 (64)	13.2 (42)

$h = 1/256$	η	GMR(20)	GMR(5)	TFQMR	SCGS	CGS
$p = 2$	10^{-6}	308 (7)	221 (7)	174 (7)	180 (7)	153 (7)
$q = 2$	10^{-4}	239 (7)	170 (7)	154 (7)	161 (7)	132 (7)
	10^{-2}	149 (8)	114 (8)	136 (8)	137 (8)	110 (7)
	10^{-1}	121 (14)	99.4 (14)	127 (13)	132 (13)	115 (13)
$p - 4$	10^{-6}	156 (12)	120 (12)	92.8 (12)	95.1 (12)	81.8 (12)
$q = 2$	10^{-4}	118 (12)	91.5 (12)	80.9 (12)	82.8 (12)	72.0 (12)
	10^{-2}	66.7 (13)	59.7 (13)	69.5 (13)	74.9 (14)	65.3 (14)
	10^{-1}	51.4 (18)	47.5 (18)	65.3 (18)	71.0 (19)	69.9 (19)
$p = 4$	10^{-6}	123 (19)	99.9 (19)	75.1 (19)	76.7 (19)	65.5 (19)
$q = 4$	10^{-4}	92.7 (19)	74.0 (19)	63.4 (19)	65.1 (19)	55.8 (19)
	10^{-2}	50.3 (20)	45.6 (20)	59.0 (22)	56.1 (20)	48.8 (20)
	10^{-1}	43.3 (39)	41.2 (39)	259 (163)	420 (263)	347 (224)
$p = 8$	10^{-6}	106 (59)	99.8 (59)	77.8 (59)	79.5 (59)	68.8 (59)
$q = 4$	10^{-4}	78.9 (59)	71.6 (59)	64.5 (59)	65.7 (59)	57.7 (59)
	10^{-2}	44.2 (60)	44.6 (60)	53.1 (59)	53.7 (60)	47.6 (59)
	10^{-1}	46.3 (120)	46.2 (120)	95.8 (157)	260 (423)	203 (345)

MICHAEL PERNICE

TABLE 2. Results for problem 2. Execution times are in seconds and iteration counts are in parentheses.

DIRECT METHOD ON SUBDOMAINS

$h = 1/64$			$h = 1/128$			
$p = 2$ $q = 2$	$p = 4$ $q = 2$	$p = 4$ $q = 4$	$p = 2$ $q = 2$	$p = 4$ $q = 2$	$p = 4$ $q = 4$	$p = 8$ $q = 4$
26.3 (5)	6.15 (9)	4.64 (19)	n/a	64.4 (9)	42.3 (19)	19.5 (50)

ITERATIVE METHODS ON SUBDOMAINS

$h = 1/64$	η	GMR(20)	GMR(5)	TFQMR	SCGS	CGS
$p = 2$	10^{-6}	16.8 (5)	12.6 (5)	11.4 (5)	11.3 (5)	10.1 (5)
$q = 2$	10^{-4}	12.4 (5)	9.92 (5)	9.47 (5)	9.90 (5)	8.44 (5)
	10^{-2}	8.52 (5)	7.12 (5)	8.09 (5)	8.17 (5)	7.01 (5)
	10^{-1}	8.61 (9)	7.22 (9)	7.89 (7)	8.01 (7)	6.63 (6)
$p = 4$	10^{-6}	12.6 (9)	10.1 (9)	8.61 (9)	8.74 (9)	7.55 (9)
$q = 2$	10^{-4}	9.42 (9)	7.89 (9)	7.15 (9)	7.21 (9)	6.24 (9)
	10^{-2}	6.67 (10)	6.06 (10)	5.47 (10)	5.68 (9)	5.02 (9)
	10^{-1}	5.89 (14)	5.26 (14)	5.28 (11)	5.29 (11)	4.75 (10)
$p = 4$	10^{-6}	11.1 (19)	9.75 (19)	7.54 (19)	7.79 (19)	6.66 (19)
$q = 4$	10^{-4}	8.01 (19)	7.45 (19)	6.38 (19)	6.50 (19)	5.65 (19)
	10^{-2}	5.29 (19)	4.88 (19)	5.08 (19)	5.10 (19)	4.49 (19)
	10^{-1}	5.57 (35)	5.26 (35)	5.27 (24)	5.89 (28)	4.79 (24)

$h = 1/128$	η	GMR(20)	GMR(5)	TFQMR	SCGS	CGS
$p = 2$	10^{-6}	106 (5)	83.4 (5)	78.0 (5)	82.0 (5)	68.4 (5)
$q = 2$	10^{-4}	84.3 (5)	66.3 (5)	67.7 (5)	72.0 (5)	60.5 (5)
	10^{-2}	53.2 (5)	45.4 (5)	56.7 (5)	57.7 (5)	49.6 (5)
	10^{-1}	53.5 (9)	43.8 (9)	53.9 (7)	51.5 (6)	50.2 (7)
$p = 4$	10^{-6}	79.0 (9)	66.4 (9)	52.0 (9)	53.8 (9)	45.5 (9)
$q = 2$	10^{-4}	61.7 (9)	53.2 (9)	44.6 (9)	46.0 (9)	40.1 (9)
	10^{-2}	36.7 (9)	35.7 (9)	36.4 (9)	37.2 (9)	31.8 (9)
	10^{-1}	27.5 (13)	24.9 (13)	31.7 (10)	32.7 (10)	29.8 (10)
$p = 4$	10^{-6}	68.9 (19)	63.2 (19)	45.5 (19)	46.8 (19)	40.0 (19)
$q = 4$	10^{-4}	52.6 (19)	49.3 (19)	39.1 (19)	40.4 (19)	34.5 (19)
	10^{-2}	30.4 (19)	29.1 (19)	30.4 (19)	31.5 (19)	27.7 (19)
	10^{-1}	23.3 (35)	23.2 (36)	28.0 (23)	25.5 (20)	22.9 (20)
$p = 8$	10^{-6}	75.7 (53)	57.9 (53)	45.2 (50)	46.7 (50)	40.0 (50)
$q = 4$	10^{-4}	49.1 (53)	43.4 (53)	37.4 (50)	38.4 (50)	33.4 (50)
	10^{-2}	29.1 (55)	27.7 (55)	28.8 (51)	29.1 (51)	25.5 (50)
	10^{-1}	21.6 (80)	21.1 (80)	43.3 (96)	45.3 (102)	30.5 (74)

resulting in faster performance for the direct method on the subdomains with $N_P = 16$ and 32.

In general, using an iterative method to solve (2) is more effective for small values of N_P and all values of η. For larger values of N_P the strategy is not competetive unless large values of η are used.

4. Conclusions

The domain decomposed preconditioners remain effective until η is decreased to 10^{-1}. Performance improvement can be achieved by reducing the accuracy of the subdomain solutions. The lower memory requirements of the iterative methods allow larger problems to be attempted, which can be a critical factor when incorporating these techniques into actual applications.

Direct methods are preferred when the subdomain problems are small. For larger subdomain problems CGS and GMRES with a small restart parameter were the most effective of the iterative methods. The results for $\eta = 10^{-1}$ suggest that a fixed number of iterations should not be used with CGS and related methods. For the model problems that were evaluated, the additional cost of smoothing the CGS residuals does not appear to be justified. Future work will extend these ideas to overlapping domain partitions.

REFERENCES

1. C. Börgers, *The Neumann-Dirichlet domain decomposition method with inexact solvers on the subdomains*, Numer. Math. **55** (1989), 123–136.
2. X.-C. Cai, W. D. Gropp, and D. E. Keyes, *A comparison of some domain decomposition algorithms for nonsymmetric elliptic problems*, Fifth International Symposium on Domain Decomposition Methods for Partial Differential Equations (David E. Keyes, Tony F. Chan, Gérard Meurant, Jeffrey S. Scroggs, and Robert G. Voigt, eds.), 1992, pp. 224–235.
3. T. F. Chan and D. E. Keyes, *Interface preconditionings for domain-decomposed convection-diffusion operators*, Third International Symposium on Domain Decomposition Methods for Partial Differential Equations (T. F. Chan, R. Glowinski, J. Périaux, and O. B. Widlund, eds.), 1989, pp. 245–262.
4. R. W. Freund, *A transpose-free quasi-minimal residual algorithm for non-Hermitian linear systems*, SIAM J. Sci. Comput. (1993), 470–482.
5. I. Gustafsson, *A class of first order factorization methods*, BIT (1978), 142–156.
6. A. Meyer, *A parallel preconditioned conjugate gradient method using domain decomposition and inexact solvers on each domain*, Computing **45** (1990), 217–234.
7. Y. Saad, *A flexible inner-outer preconditioned GMRES algorithm*, SIAM J. Sci. Comput. **14** (1993), 401–469.
8. Y. Saad and M. H. Schultz, *GMRES: A generalized minimum residual algorithm for solving nonsymmetric linear systems*, SIAM J. Sci. Statist. Comput. **7** (1986), 856–869.
9. B. F. Smith, *A parallel implementation of an iterative substructuring algorithm for problems in three dimensions*, SIAM J. Sci. Comput. **14** (1993), 406–423.
10. P. Sonneveld, *CGS, a fast Lanczos-type solver for nonsymmetric linear systems*, SIAM J. Sci. Statist. Comput. **10** (1989), 36–52.
11. L. Zhou and H. F. Walker, *Residual smoothing techniques for iterative methods*, SIAM J. Sci. Comput. **15** (1994), 297–312.

UTAH SUPERCOMPUTING INSTITUTE, UNIVERSITY OF UTAH, SALT LAKE CITY, UTAH 84112
E-mail address: usimap@sneffels.usi.utah.edu

Contemporary Mathematics
Volume **180**, 1994

ELLIPTIC PRECONDITIONERS USING FAST SUMMATION TECHNIQUES

L. RIDGWAY SCOTT

ABSTRACT. We introduce the idea of using fast summation methods together with Green's function techniques to reduce the work in approximating the inverse matrix for the discretization of elliptic partial differential equations. The resulting method has a high degree of parallelism.

There are several reasons for studying the iterative methods introduced in this paper. For perspective, we list two. One arises in multigrid solvers, where there is a need to have a fast method on the coarsest grid. The coarse grid equation is solved repeatedly and may be of substantial size in engineering applications. In parallel multigrid solvers, the "coarsest" may need to be coarser than in the sequential case in order to keep good parallel efficiency. Another application of these methods arises in time-stepping methods, whether linear or nonlinear, where a particular system is inverted repeatedly in many algorithms [2]. If time integration is long, typically an implicit method would be employed which requires solving a linear system. In this case, the efficiency of the linear solver is critical. Such a system could be so small and irregular that a multi-level procedure would not be considered. We now elaborate these two examples.

Multigrid techniques for solving the linear systems arising from the discretization of elliptic partial differential equations involve a reduction in problem size to a coarse grid. This coarse grid may be relatively large due to the need to resolve the domain geometry (or the problem solution). Since the coarse-grid problem must be solved repeatedly, the limiting factor in the efficiency of the overall method may be the ability to solve the coarse-grid problem quickly. On

1991 *Mathematics Subject Classification.* Primary 65N22, 65N55, 65Y05; Secondary 65NF10, 65N06, 65N30.

We acknowledge generous support from the National Science Foundation, award number ASC-9217374 (which includes funds from DARPA), and award number DMS-9105437.

This paper is in final form and no version of it will be submitted for publication elsewhere.

the other hand, it is not necessary to solve the coarse-grid problem very accurately in the context of multigrid techniques. If a fast and sufficiently accurate preconditioner is available, then simple iteration can be used to produce a suitable approximation to the solution to the coarse-grid problem. We introduce and analyze a new technique for doing this using fast summation methods [6].

In practical problems, the coarsest mesh that can describe the geometry is quite large. Since the resulting linear system related to the coarse grid is unstructured, a popular technique for for solving it is direct factorization [7]. This has the advantage that the repeated coarse-grid solves require only back-solves, after the initial factorization. This can be quite efficient for two-dimensional problems if the unknowns are suitably ordered [7]. However, for three-dimensional problems even optimal orderings lead to work estimates that are less than ideal (cf. [3]). Moreover, the sequential nature of a back-solution makes it difficult to parallelize efficiently. The techniques presented here have the same level of efficiency independent of the domain dimension, and they are trivially parallel.

The basic idea of the technique is as follows. The solution to a linear system can be written as the product of the inverse matrix and the data vector. Moreover, this has a significant amount of parallelism in that each component of the solution can be computed independently of the others. The difficulty is that, for an $n \times n$ system, this requires $\mathcal{O}(n^2)$ work. However, the inverse matrix for the discretization of elliptic partial differential equations has significant structure that can be exploited to approximate its action with significantly less work. We explore this in detail in the context of finite element methods. We introduce the idea of using fast summation methods [6] together with Green's function techniques [9] to reduce the work to $\mathcal{O}(n \log n)$ in the sequential case. The key point is that this technique does not use any particular structure of the mesh, only structure of the differential equation being solved. Fast summation methods can be viewed as adaptive domain decomposition methods together with a projection on each subdomain.

1. A Model Problem

Let $\Omega \subseteq \mathbb{R}^d$ be a convex polygon and define

$$(1.1) \qquad\qquad a(u,v) = \int_\Omega \nabla u \cdot \nabla v \, dx.$$

We consider the variational formulation of the Dirichlet problem for Laplace's equation, as follows: find $u \in V := \mathring{H}^1(\Omega)$ such that

$$(1.2) \qquad\qquad a(u,v) = (f,v) \quad \forall \; v \in V$$

where $f \in L^2(\Omega)$.

Elliptic regularity (cf. Grisvard [8]) implies that $u \in H^2(\Omega) \cap \mathring{H}^1(\Omega)$. To approximate u, we consider a triangulation $\mathcal{T}$ of Ω, where h denotes the mesh size of $\mathcal{T}$. Let $V_\mathcal{T}$ denote C^0 piecewise polynomial functions of degree $\leq k$ with respect to $\mathcal{T}$ that vanish on $\partial\Omega$ [4]. The discretized problem is the following: find $u_h \in V_{\mathcal{T}}$ such that

$$(1.3) \qquad\qquad a(u_h, v) = (f, v) \quad \forall\ v \in V_{\mathcal{T}}.$$

It is well known [4] that

$$(1.4) \quad \|u - u_h\|_{H^s(\Omega)} \leq C\, h^{k+1-s}\, \|u\|_{H^{k+1}(\Omega)} \text{ for } s = 0, 1 \text{ and } 1 \leq m \leq k + 1.$$

Throughout this paper C, with or without subscripts, denotes a generic constant independent of h.

The Green's function, g^x, for (1.2) satisfies

$$(1.5) \qquad\qquad v(x) = a(g^x, v)$$

for suitably smooth functions v, say $v \in W_p^1(\Omega)$ for $p > d$. By Sobolev's inequality and [12] we see that

$$(1.6) \qquad\qquad \|g^x\|_{W_q^1(\Omega)} \leq C \sup_{0 \neq v \in \mathring{W}_p^1(\Omega)} \frac{a(g^x, v)}{\|v\|_{W_p^1(\Omega)}} \leq C$$

where $\frac{1}{p} + \frac{1}{q} = 1$.

Correspondingly, there is a discrete Green's function, $g_h^x \in V_{\mathcal{T}}$, which satisfies

$$(1.7) \qquad\qquad v(x) = a(g_h^x, v) \quad \forall\ v \in V_{\mathcal{T}}.$$

Using this discrete Green's function, we may write the solution to (1.3) via

$$u_h(x) = a(g_h^x, u_h) = (f, g_h^x) \quad \forall\ x \in \Omega.$$

If we let $\{x_i : 1 \leq i \leq \dim V_{\mathcal{T}}\}$ denote the interior vertices of $\mathcal{T}$, then the coefficients of u_h with respect to the standard Lagrange basis, $\{\psi_i : 1 \leq i \leq \dim V_{\mathcal{T}}\}$, for $V_{\mathcal{T}}$ can be determined via

$$(1.8) \qquad\qquad u_h(x_i) = (f, g_h^{x_i}) \quad \forall\ i = 1, \ldots, \dim V_{\mathcal{T}}.$$

Thus the $g_h^{x_i}$ are closely related to the rows of the inverse matrix for the problem (1.3).

2. Approximating the Green's function

The basic singularity of g^x is given by

$$(2.1) \qquad\qquad G^x(y) = \begin{cases} \log|x - y| & d = 2 \\ |x - y|^{2-d} & d \geq 3 \end{cases}$$

in the sense that there is a constant α such that

$$g^x - \alpha G^x \in \mathring{H}^1(\Omega).$$

See Figure 2.1 which depicts the discrete Green's function with singularity near one of the corners of a square domain. In fact, we can think of defining g^x by writing $g^x = w^x + \alpha G^x$ where $w^x = -\alpha G^x$ on $\partial\Omega$ and

$$a(w^x, v) = 0 \quad \forall \ v \in V.$$

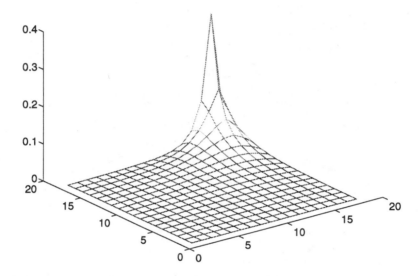

Figure 2.1. Plot of the discrete Green's function with singularity near one of the corners of a square domain with a regular mesh.

The integrals $(f, g_h^{x_i})$ can be approximated efficiently since the Green's function is quite smooth away from the singularity (see Figure 2.1). Near the singularity we compute the integral exactly, and far away we replace $g_h^{x_i}$ by appropriate averages. To define the process precisely, we fix $x = x_i \in \Omega$ for the moment. Let $\{S_j : j = 1, \ldots, J_h^x\}$ be a subdivision of Ω consisting of groups of triangles in $\mathcal{T}$ with the following property:

$$(2.2) \qquad \text{diam}\,(S_j) \leq \rho \inf\,\{|x - y| : y \in S_j\} \quad \forall \, j = 2, \ldots, J_h^x$$

for some constant $\rho < \infty$. Figure 2.2 depicts a subdivision with $\rho = 2$. Thus S_1 consists of the triangles close to the point x whereas the other members of the subdivision have a size comparable to the distance from the closest point to x.

We begin by describing a piecewise constant approximation. Let

$$(2.3) \qquad \gamma_j^x := \frac{1}{\text{meas}\,(S_j)} \int_{S_j} g_h^x(y)\,dy$$

and define $K^x : L^1(\Omega) \to \mathbb{R}$ by

(2.4) $$K^x f := \int_{S_1} g_h^x(y) f(y) \, dy + \sum_{j=2}^{J_h^x} \gamma_j^x \int_{S_j} f(y) \, dy.$$

Define $\mathcal{K} : L^1(\Omega) \to V_{\mathcal{T}}$ via

$$\mathcal{K} f := \sum_{i=1}^{\dim V_{\mathcal{T}}} (K^{x_i} f) \, \psi_i.$$

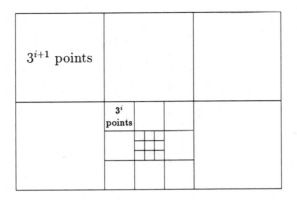

Figure 2.2. Mesh subdivision (aggregation of triangles in $\mathcal{T}$) where $\rho = 2$ in condition (2.2).

Define a function e^x on Ω by

$$e^x(y) = \begin{cases} 0 & \forall \, y \in S_1 \\ g_h^x(y) - \gamma_j^x & \forall \, y \in S_j, \; j \geq 2 \end{cases}$$

for all $x = x_i$. We have the following error representation:

(2.5) $$u_h(x) - K^x f = \int_{\Omega} e^x(y) f(y) \, dy.$$

We can define a linear operator $\mathcal{I}^x$ by

$$\mathcal{I}^x v(y) = \begin{cases} v(y) & \forall \, y \in S_1 \\ \frac{1}{\operatorname{meas}(S_j)} \int_{S_j} v(z) \, dz & \forall \, y \in S_j, \; j \geq 2 \end{cases}$$

for all $x = x_i$. Using this notation, (2.4) becomes

$$K^x f := \int_{\Omega} (\mathcal{I}^x g_h^x)(y) \, f(y) \, dy,$$

and (2.5) becomes

(2.6) $$(u_h - \mathcal{K}f)(x) = \int_{\Omega} (g_h^x - \mathcal{I}^x g_h^x)(y) f(y) \, dy.$$

More generally, we can imagine defining a family of linear operators $\mathcal{I}^x$ for $x \in \Omega$, each of which is the identity on (the respective) Ω_1 and such that $\mathcal{I}^x v$ is a polynomial of degree r in each S_j, $j \geq 2$, and that (2.6) holds. In the case that $\mathcal{I}^x v \in V_h$ for $v \in V_h$, we may write

$$(u_h - \mathcal{K}f)(x) = a(u_h, g_h^x - \mathcal{I}^x g_h^x).$$

We may formally think of $u_h = A_h^{-1} f$, where A_h denotes the operator associated with the variational problem (1.3). Thus, this becomes

$$\left(A_h^{-1} - \mathcal{K}\right) f(x) = a(u_h, g_h^x - \mathcal{I}^x g_h^x),$$

or, since $f = A_h u_h$, we may write

$$(\mathcal{I} - \mathcal{K}A_h) u_h(x) = a(u_h, g_h^x - \mathcal{I}^x g_h^x).$$

Let $y = x_j$ for some other nodal point, and consider the quantity

$$(\mathcal{I} - \mathcal{K}A_h) u_h(x) - (\mathcal{I} - \mathcal{K}A_h) u_h(y) = a(u_h, e^x - e^y)$$

where $e^x := g_h^x - \mathcal{I}^x g_h^x$. Dividing by $|x - y|$ and taking the maximum over all x and y, we find

$$\|(\mathcal{I} - \mathcal{K}A_h) u_h\|_{W_\infty^1(\Omega)} \leq C \|u_h\|_{W_\infty^1(\Omega)} \sup_{x,y \in \Omega} \frac{\|e^x - e^y\|_{W_1^1(\Omega)}}{|x - y|}.$$

Thus we have proved the following.

THEOREM 2.7. *There is a constant C depending only on the variational problem such that*

$$\|\mathcal{I} - \mathcal{K}A_h\|_{W_\infty^1(\Omega) \to W_\infty^1(\Omega)} \leq C \sup_{x,y \in \Omega} \frac{\|e^x - e^y\|_{W_1^1(\Omega)}}{|x - y|}$$

where $e^x := g_h^x - \mathcal{I}^x g_h^x$.

3. Error analysis

We are unable to give a complete error analysis of the term arising in Theorem 2.7, but we can give some partial results as an indicator as to what might be important factors effecting the method. We expand the expression $e^x - e^y$ as

$$e^x - e^y = g_h^x - \mathcal{I}^x g_h^x - g_h^y + \mathcal{I}^y g_h^y$$
$$= (\mathcal{I} - \mathcal{I}^x)(g_h^x - g_h^y) - (\mathcal{I}^x - \mathcal{I}^y) g_h^y.$$

We will show that we can write

$$(3.1) \qquad g_h^x - g_h^y = |x - y| \int_0^1 \tilde{g}_h^{y + t(x-y)} \, dt$$

where $\tilde{g}_h^z$ denotes the derivative Green's function [9] defined by

$$(3.2) \qquad a(\tilde{g}_h^z, v) = \frac{(x - y)}{|x - y|} \cdot \nabla v(z) \quad \forall \, v \in V_h.$$

This can be seen by using the fundamental theorem of calculus,

$$\int_0^1 \frac{(x-y)}{|x-y|} \cdot \nabla v(y + t(x-y))\, dt = v(x) - v(y),$$

and using the definition of the discrete Green's function. Note that the integral on the right-hand side of (3.1) defines a function in the finite dimensional space V_h, and

$$\int_0^1 a\left(\tilde{g}_h^{y+t(x-y)}, v\right) dt = a\left(\int_0^1 \tilde{g}_h^{y+t(x-y)}\, dt, v\right)$$

because the map $z \to \tilde{g}_h^z$ is a continuous map from $\Omega \to V_h$.

Applying (3.1), we find

$$|x-y|^{-1} \left\|(\mathcal{I} - \mathcal{I}^x)(g_h^x - g_h^y)\right\|_{W_1^1(\Omega)} \leq \max_{t \in [0,1]} \left\|(\mathcal{I} - \mathcal{I}^x)\tilde{g}_h^{y+t(x-y)}\right\|_{W_1^1(\Omega)}$$

From [9], we have

$$\left\|(\mathcal{I} - \mathcal{I}^x)\tilde{g}_h^z\right\|_{W_1^1(\Omega)} \leq C \left\|(\mathcal{I} - \mathcal{I}^x)\tilde{g}^z\right\|_{W_1^1(\Omega)} + \left\|(\mathcal{I} - \mathcal{I}^x)(\tilde{g}^z - \tilde{g}_h^z)\right\|_{W_1^1(\Omega)}$$

where $\tilde{g}^z$ solves (1.2) with f being a directional derivative of a mollified Dirac delta function [9]. Then from (2.2) we find

$$
\begin{aligned}
\left\|(\mathcal{I} - \mathcal{I}^x)\tilde{g}^z\right\|_{W_1^1(\Omega)} &\leq C \sum_{j=2}^{J_h^x} \operatorname{diam}(S_j)^r \int_{S_j} |z-y|^{-n-r}\, dy \\
&\leq C\rho^r \sum_{j=2}^{J_h^x} \int_{S_j} |z-y|^{-n}\, dy \\
&\leq C\rho^r \int_{\Omega \setminus S_1} |z-y|^{-n}\, dy \\
&\leq C\rho^r |\log(\operatorname{diam}(S_1))|.
\end{aligned}
$$

(3.3)

By choosing ρ sufficiently small (or r sufficiently large, if $\rho < 1$), we can make this term as small as we like.

Correspondingly from [11] and [9] we deduce that

$$
\begin{aligned}
\left\|(\mathcal{I} - \mathcal{I}^x)(\tilde{g}^z - \tilde{g}_h^z)\right\|_{W_1^1(\Omega)} &\leq C \left\|\tilde{g}^z - \tilde{g}_h^z\right\|_{W_1^1(\Omega \setminus S_1)} \\
&\leq C h^k \operatorname{diam}(S_1)^{-k}.
\end{aligned}
$$

(3.4)

Provided $h/\operatorname{diam}(S_1)$ is sufficiently small, this term will be small.

Estimates (3.3) and (3.4) are in competition to the extent that the former increases and the latter decreases as a function of $\operatorname{diam}(S_1)$. The latter requires $\operatorname{diam}(S_1)$ to be comparable to h and then the former requires $\rho^r |\log h|$ to be small. For fixed $\rho < 1$, this means that r must be chosen to grow as $\mathcal{O}(\log|\log h|)$ as h tends to zero.

To complete the analysis, we need to consider an estimate for

$$|x-y|^{-1} \left\|(\mathcal{I}^x - \mathcal{I}^y)g_h^y\right\|_{W_1^1(\Omega)}.$$

This appears to require some restrictions on the regularity of the interpolation process. However, we postpone further analysis to a later publication.

4. Complexity of initialization

The application of the operator $\mathcal{K}$ requires a great deal of initial computation. In particular, information related to the complete inverse of A_h is apparently necessary. We now address the question of how much work is involved and to what extent this work can be done in parallel. There are several contexts in which one might be able to assess (and amortize) the cost of initialization. In a time stepping scheme, this cost could be compared with the savings accrued over a large number of time steps. In a multigrid application, the cost of initialization for a coarse grid solver can be compared with the cost of a fine grid sweep. This allows a more precise comparison, so we carry this out in some detail in an example.

We consider an example based on regular meshes in d dimensions. Suppose the fine mesh consists of N^d points, and the coarse mesh consists of n^d points. We consider the construction of the operator $\mathcal{K}$ on the coarse mesh. We need to solve n^d equations of the form (1.3) for $f = \delta^x$. Each of these can be done in parallel, as they are completely independent. The complexity of solution of (1.3) depends on the structure of the problem, but we make the simplifying assumption that we can use multigrid and solve each one in an amount of work $\mathcal{O}(n^d)$.

Let us assume that the S_j's are based on the coarser meshes arising in a multigrid solution. The formation of the required averages (2.3) can be done as follows. We describe the process for a triangular mesh in two dimensions ($d = 2$). First, we form the averages on pairs of neighboring triangles which form a rectangle. This takes $\mathcal{O}(n^d)$ work. Then we average these in groups of four rectangles, forming averages over S_j's consisting of eight triangles. This takes again $\mathcal{O}(n^d)$ work, but produces only one quarter as many new averages. Grouping these averages again in groups of four rectangles and forming averages over such groups takes only one quarter as much work as before and produces only one quarter as many new averages. Since these are progressing geometrically, we see the overall work and storage is $\mathcal{O}(n^d)$. The number of terms arising in (2.4) is $\mathcal{O}(\log n)$, but the averages of f can similarly be computed in $\mathcal{O}(n^d)$ work. These averages must be computed for each of $\mathcal{O}(n^d)$ points, but they can be done completely in parallel.

Thus the total work for the initialization phase takes

$$\mathcal{O}\left(\frac{n^{2d}}{P}\right) \quad \text{work}$$

for P processors provided $1 \le P \le n^d$. This is to be compared with the cost of a fine grid sweep, which can be done (for certain smoothers [1]) in

$$\mathcal{O}\left(\frac{N^d}{P}\right) \quad \text{work}$$

for P processors provided $1 \leq P \ll N^d$. Here we are assuming N^d is large with respect to P in order to ignore the cost of communication in, say, a typical V cycle.

Thus the cost of initialization will be comparable to a fine grid sweep when

$$(4.1) \qquad\qquad n^2 = \mathcal{O}(N),$$

and this result is essentially independent of the number of processors, provided $1 \leq P \leq n^d = \mathcal{O}(N^{d/2})$, and independent of dimension. To quantify this, we note that (4.1) holds when $n = 10$ and $N = 100$; this corresponds to a typical three-dimensional problem [1] by today's standards and would allow for $P = 1000$. It also holds for $n = 30$ and $N = 1000$, and this coarse grid size corresponds to some of the two-dimensional examples presented subsequently. The natural amount of parallelism would be $P = \mathcal{O}(n^d) = \mathcal{O}(N^{d/2})$ in general, and thus it is scalable.

5. Complexity of the fast-summation algorithm

In the computation of (2.4) in parallel, the averages of f have to be accumulated at each processor. After a communication phase to be described subsequently, each processor does

$$(5.1) \qquad\qquad \mathcal{O}\left(\frac{n^d \log n}{P}\right)$$

floating point operations as depicted in (2.4) without further communication. We now compare the cost of communication with (5.1).

It will not be necessary to have all averages at each processor. For processors far away, only the coarser averages are needed. These can be accumulated as follows. For simplicity, we assume that data is distributed in a parallel computer having a network that has a d-dimensional mesh as a subnetwork. In particular, we assume that each processor can communicate with a neighbor in a d-dimensional mesh simultaneously, without contention. To keep the analysis simple, we ignore the effect of the degree of approximation r which would grow like $\mathcal{O}(\log |\log h|)$ as h tends to zero.

First, we form the averages on groups of rectangles residing completely within one processor; this takes $\mathcal{O}\left(\frac{n^d}{P}\right)$ floating point operations. For simplicity, let us assume that groups of rectangles of a particular size fit precisely in one of the processors, so that only averages over boxes (rectangles) of larger size require an exchange of data. More precisely, we assume the set of triangles assigned to a processor forms one of the groups S_j. In $\mathcal{O}(\log P)$ steps, these P values can be accumulated at each processor, and the remaining combinations can be done redundantly.

Let λ denote the number of floating point operations that can be done in the amount of time corresponding to the latency of communication, and let F denote the number of floating point operations that can be performed in the time that it takes to communicate one word (typical values for λ would range from ten to one hundred and for F would range from three to thirty.) Then the time for these combinations is comparable to

$$\mathcal{O}(\lambda \log(P) + FP)$$

floating point operations. The data to be reduced locally corresponds to a d-dimensional mesh with P cells. Doing the remaining combinations thus requires $\mathcal{O}(P)$ floating point operations. Thus the majority of time is spent in the communication of the initial averages (assuming $F \gg 1$).

In a fixed number of processors around each given processor, local information will need to be communicated. The number of such processors is bounded by $\mathcal{O}(\rho^d)$. This contributes an amount of time [5] comparable to

$$\mathcal{O}\left(2d \lceil \rho^d \rceil \left(\lambda + F \frac{n^d}{P}\right)\right)$$

floating point operations, where $\lceil x \rceil$ indicates the smallest integer not less than x.

If P is larger than $\frac{n^d}{P}$ $(P \geq n^{d/2})$, then the algorithm above for combining averages over larger boxes may not be optimal. Instead a series of local combinations can be done. At the first stage, there are P averages; neighbors exchange their averages (in $2d$ steps) and then compute averages over 2^d neighboring boxes. This produces $P/2^d$ new averages. One of the processors from each group of 2^d is chosen to exchange the resulting average with other groups of 2^d processors in $2d$ steps. These values are averaged and the value is passed back to all 2^d processors in each subset in $2d$ steps. This process is then repeated recursively, by subdividing the subsets into subsets. This clearly terminates after $\mathcal{O}(\log P)$ steps, since the box size is doubling at each step. The number of communication steps is $\mathcal{O}\left(2^d \log P\right)$, with a constant amount of data exchanged at each step. This corresponds to an amount of time equivalent to

$$\mathcal{O}\left(\lambda 2^d \log P\right)$$

floating point operations. With the assumption $P \leq n^d$, this corresponds to less than

$$\mathcal{O}\left(\lambda 2^d d \log n\right)$$

floating point operations.

Thus, the amount of time required to do the computation in (2.4) is of the order of the communication required (or greater). For a machine with large latency λ, it could be useful to attempt to refine further the exchanges and computation of the averages, if it is desired to have $P \approx n^d$.

6. Numerical experiments

We now describe some numerical experiments done to explore the viability of the approximate inverse described in section 2. We took the simplest possible situation, namely, a regular mesh in two dimensions. Thus A_h corresponds to the five-point difference stencil. Moreover, we took a quite simple approach to the averaging. In the first case, which we refer to as the "constant" case, we approximate g_h^x by piecewise constants, averaging over boxes of size 3^i mesh points as indicated in Figure 6.1 in the "constant-1" case and Figure 6.2 in the "constant-2" case. More precisely, in the " -1" aggregation scheme, there are annuli made up of eight boxes of size 3^i mesh points in a square annulus surrounding a similar arrangement for $i - 1$. This is depicted in Figure 2.2.

The " -2" aggregation scheme is more difficult to describe, but the boxes utilized are apparent from Figure 6.2.

Table 6.1 indicates the spectral radius for the various approximation schemes. The different aggregation schemes are indicated by the appendage " -1" for the coarser aggregation (see Figure 2.2) and "-2" for the finer aggregation (see Figure 6.2). In the "linear" cases, we instead interpolate g_h^x bilinearly, but not continuously. For the sake of reference, we give the spectral radius of the red/black SOR iterative method (with the optimal ω), as it is a well known iterative method with a comparable level of parallelism.

Mesh size	constant-1	constant-2	linear-1	linear-2	R/B SOR
8×8	0.9248	0.7031	0.3785	0.1604	0.4903
10×10	1.4372	1.0288	0.5075	0.2035	0.5604
12×12	1.9436	1.5778	0.6284	0.2345	0.6138
14×14	2.4531	2.0219	0.5765	0.2593	0.6558
16×16	3.0231	2.5137	0.8618	0.2796	0.6895
18×18	3.5189	2.9493	0.9377	0.2962	0.7173
20×20	3.9470	3.4080	0.9976	0.3096	0.7406
22×22	4.4126	3.8490	1.0642	0.3206	0.7603
24×24	4.8771	4.3090	1.2120	0.3296	0.7773
26×26	5.3262	4.8157	1.3855	0.3371	0.7920

Table 6.1. Spectral radius for various iterative schemes. The columns marked "constant" and "linear" contain the spectral radius of $\mathcal{I} - \mathcal{K}A_h$ for different mesh schemes indicated in Figures 6.2 (" -1") and 6.3 (" -2"), respectively. The column marked "R/B SOR" gives the spectral radius for the red/black SOR iterative method (with the optimal ω). The "Mesh size" indicates the number of mesh points in each direction of a regular two-dimensional mesh.

What we conclude from Table 6.1 is that the method certainly works, but not arbitrarily well. In particular, it may be necessary to take the estimates (3.3) and (3.4) seriously, increasing the order r of approximation and reducing the constant ρ in condition (2.2), which controls the grading of the mesh, appropriately.

We note that the matrix $\mathcal{K}$ as defined here is not symmetric, but it is very nearly symmetric. We computed the norm of $\mathcal{I} - \widetilde{\mathcal{K}} A_h$, where $\widetilde{\mathcal{K}} = \frac{1}{2}(\mathcal{K} + \mathcal{K}^t)$ is the symmetric part of $\mathcal{K}$, and the results agreed with those in the table to three significant digits.

References

[1] B. Bagheri, A. Ilin & L. R. Scott, Parallelizing UHBD, Research Report UH/MD 167, Dept. Math., Univ. Houston, 1993, and Parallel 3-d mosfet simulation, In *Proceedings of the 27th Annual Hawaii International Conference on System Sciences* vol. 1, T.N.Mudge and B.D. Shriver, ed's, IEEE Computer Soc. Press, 1994, pp. 46–54.

[2] B. Bagheri, S. Zhang & L. R. Scott, Implementing and using high–order finite element methods, *Comp. Meth. Appl. Mech. & Eng.*, to appear.

[3] R. E. Bank & L. R. Scott, On the conditioning of finite element equations with highly refined meshes, *SIAM J. Num. Anal.* **26**, (1989), 1383–1394.

[4] S. Brenner, L. R. Scott, *The Mathematical Theory of Finite Element Methods*, Springer-Verlag, 1994.

[5] T. W. Clark, R. v. Hanxleden, J. A. McCammon, and L. R. Scott, Parallelizing molecular dynamics using spatial decomposition, Proc. Scalable High-Performance Computing Conference, IEEE Computer Soc. Press, 1994, 95-102.

[6] C. I. Draghicescu, An efficient implementation of particle methods for the incompressible Euler Equations, *SIAM J. Numer. Anal.* **31** (1994), to appear.

[7] A. George, J. W. Liu, *Computer Solution of Large Sparse Positive Definite Systems*, Prentice-Hall, 1981.

[8] P. Grisvard, *Elliptic Problems in Nonsmooth Domains*, Pitman Advanced Publishing Program, Boston, 1985.

[9] R. Rannacher and L. R. Scott, Some optimal error estimates for piecewise linear finite element approximations, *Math. Comp.* **38** (1982), 437–445.

[10] L. R. Scott, Optimal L^∞ estimates for the finite element method on irregular meshes, *Math. Comp.* **30** (1976), 681–697.

[11] L. R. Scott and S. Zhang, Finite element interpolation of non–smooth functions satisfying boundary conditions, *Math. Comp.*, 54:483–493, 1990.

[12] Simader, C. G. (1972) *On Dirichlet's Boundary Value Problem*, Lecture Notes in Math. v. 268, Springer-Verlag, Berlin, 1972.

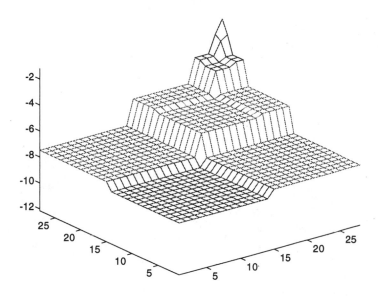

Figure 6.1. Plot of the logarithm of the piecewise constant approximation of the discrete Green's function. Coarse mesh aggregation where $\rho = 2$ in condition (2.2) as depicted in Figure 2.2.

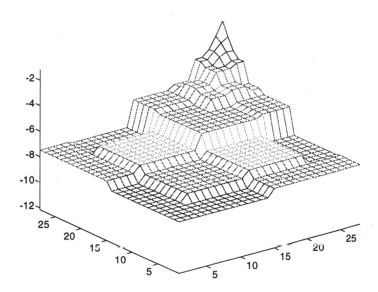

Figure 6.2. Plot of the logarithm of the piecewise constant approximation of the discrete Green's function. Finer mesh aggregation where $\rho = 2/3$ in condition (2.2).

DEPARTMENT OF MATHEMATICS, UNIVERSITY OF HOUSTON, HOUSTON, TX 77204-3476

E-mail address: scott@uh.edu

Contemporary Mathematics
Volume **180**, 1994

The Modified Vertex Space
Domain Decomposition Method
for Neumann Boundary Value Problems

JIAN PING SHAO

ABSTRACT. In this paper, we extend the vertex space domain decomposition method (VSDDM) [**18**] to solve singular systems arising from the discretization of partial differential equations with Neumann boundary conditions by finite differences or finite elements. We give a concrete discussion on how to deal with the null space in VSDDM so that it retains optimal condition number independent of sizes of coarse and fine grids. To reduce the complexity cost of VSDDM, we proposed several efficient variants VSDDM [**6**] based on Fourier approximation and a probing technique. Here, we further reduce the cost of the probing technique in VSDDM. Various numerical experiments have been conducted to test the efficiency of the modified VSDDM.

1. Introduction

The aim of this paper is to modify the vertex space domain decomposition method [**18**] so that this method can be applied to the symmetric positive semi-definite systems of linear algebraic equations arising from the discretization of elliptic systems with Neumann boundary condition by finite differences or finite elements. For these Neumann boundary value problems, the discrete stiffness matrix is singular and the solution is not unique. The VSDDM needs to be modified so that the results of preconditioning are orthogonal to the kernel space and the VSDDM retains an optimal convergence rate. Our motivation is to design

1991 *Mathematics Subject Classification.* Primary 65F10, 65N35, 65N55; Secondary 65Y05.

This work was supported in part by the Army Research Office under contract DAAL03-91-G-0150 and subcontract under DAAL03-91-C-0047, by the Office for Naval Research under contract ONR N00014-92-J-1890, by the National Science Foundation under grant ASC 92-01266 and ASC 93-10315.

This paper is in final form and no version of it will be submitted for publication elsewhere.

an efficient parallel algorithm for solving the Navier-Stokes equations. When our scheme is based on the velocity and stream function, we have to solve a symmetric semi-positive system, resulting from Laplace's equation with "Neumann like" boundary conditions, in each time step. We also focus on the improvement of the probing technique in the construction of edge and vertex approximations. Based on the Fourier approximation [9, 14, 2, 3] and the probing technique [7, 15, 16, 5], several efficient variants of the VSDDM have been proposed and tested [4, 6, 17]. Here, we further reduce the cost of the probing technique in the VSDDM for two dimensional problems. Only four instead of six probing vectors [6] are used to multiply by the Schur complement and to form the approximate edge and vertex matrices. Various numerical tests have been conducted to show that the modified VSDDM has an optimal convergence rate for singular problems with smoothly varying coefficients, and highly jumping coefficients.

In section 2, we discretize a singular problem, reduce the problem on the whole domain to the interface and form the Schur complement system on the interface. We discuss the properties of the Schur complement matrix. In section 3, we apply the VSDDM to singular problems with known kernel spaces and state a theorem on convergence rate. In section 4, we further improve the probing technique in the VSDDM. Finally, in section 5 we conduct the numerical experiments on this modified VSDDM with various approximate edge and vertex matrices for solving singular problems with highly varying or jumping coefficients.

2. Neumann Boundary Value Problems

Let Ω in $\mathbf{R}^d$ be a polygonal domain. In the Sobolev space $\mathbf{V} = (H^1(\Omega))^q$, we introduce a symmetric, bounded and semi-positive definite bilinear form $a(\cdot, \cdot)$: $\mathbf{V} \times \mathbf{V} \to \mathbf{R}$. Let $(\cdot, \cdot)$ be the inner product in $(L^2(\Omega))^q$: $(\mathbf{f}, \mathbf{v}) = \int_\Omega \mathbf{f} \cdot \mathbf{v} dx$. The kernel space is defined by

$$KerA = \{\mathbf{u} | \mathbf{u} \in \mathbf{V}, \quad a(\mathbf{u}, \mathbf{v}) = 0, \quad \forall \mathbf{v} \in \mathbf{V}\},$$

which is known for most Neumann boundary value problems.

Consider a general variational problem with a natural boundary condition in the space $\mathbf{V}$: Find $\mathbf{u} \in \mathbf{V}$, and $\mathbf{u} \perp KerA$ such that

$$(2.1) \qquad\qquad a(\mathbf{u}, \mathbf{v}) = (\mathbf{f}, \mathbf{v}), \qquad \forall \mathbf{v} \in \mathbf{V},$$

where f satisfies the compatibility condition: $(\mathbf{f}, \mathbf{v}) = 0, \qquad \forall \mathbf{v} \in KerA$.

As an example, we consider equation:

$$(2.2) \qquad Lu = f \quad \text{in } \Omega, \quad \text{and} \quad \frac{\partial u}{\partial N} = 0 \quad \text{on } \partial\Omega,$$

where $Lv = -\sum_{i,j} \dfrac{\partial}{\partial x_i}(\alpha_{ij} \dfrac{\partial v}{\partial x_j})$ and $\dfrac{\partial v}{\partial N} = \sum_{i,j} \alpha_{ij}(x) \dfrac{\partial v}{\partial x_j} \cos(\vec{n}, \vec{e}_i)$ with α_{ij} uniformly bounded, and positive definite. Then $\mathbf{V} = H^1(\Omega)$ and $KerA = \text{span}\{\mathbf{1}\}$.

We partition the domain Ω as the the union of disjoint regions Ω_k of diameter H,

$$\bar{\Omega} = \cup_k \bar{\Omega}_k \qquad \text{and} \qquad \Omega_i \cap \Omega_j = \emptyset \qquad \text{if } i \neq j.$$

These subdomains form the elements of a coarse triangulation. We denote the union of these subdomain boundaries as $\Gamma = \cup_k \partial\Omega_k$. Each subdomain is further divided into the elements of diameter $O(h)$. So we have a fine triangulation on the domain Ω. Assume that these triangulations are shape regular in the sense common to finite element theory, cf. Ciarlet [8].

By using finite element or finite difference method, we obtain stiffness matrices A_h and A_H on the fine and coarse triangulations respectively. These matrices are symmetric, positive semi-definite. Denote the kernel spaces as $Ker A_h = \{\mathbf{v}_h | A_h \mathbf{v}_h = 0\}$ and $Ker A_H = \{\mathbf{v}_H | A_H \mathbf{v}_H = 0\}$. Then problem (2.1) has the following discrete forms:

(2.3) $\qquad$ Find $\mathbf{u}_h \perp Ker A_h \qquad$ such that $\qquad A_h \mathbf{u}_h = \mathbf{f}_h$.

on the fine grid and

(2.4) $\qquad$ Find $\mathbf{u}_H \perp Ker A_H \qquad$ such that $\qquad A_H \mathbf{u}_H = \mathbf{f}_H$,

on the coarse grid. Each of these problems has an unique solution if its right-hand side is orthogonal to the kernel space.

Now we restrict the problem on Ω to the interface Γ. By grouping the unknowns in the interior of the subdomains in the vector $\mathbf{u}_I$ and those on the interface Γ in the vector $\mathbf{u}_B$, we can rewrite problem (2.3) as block form

$$\begin{pmatrix} A_{II} & A_{IB} \\ A_{BI} & A_{BB} \end{pmatrix} \begin{pmatrix} \mathbf{u}_I \\ \mathbf{u}_B \end{pmatrix} = \sum_k \begin{pmatrix} A_{II}^{(k)} & A_{IB}^{(k)} \\ A_{BI}^{(k)} & A_{BB}^{(k)} \end{pmatrix} \begin{pmatrix} \mathbf{u}_I^{(k)} \\ \mathbf{u}_B^{(k)} \end{pmatrix} = \begin{pmatrix} \mathbf{f}_I \\ \mathbf{f}_B \end{pmatrix}.$$

where $\mathbf{u}^{(k)} = ((\mathbf{u}_I^{(k)})^T, (\mathbf{u}_B^{(k)})^T)^T$ is a vector associated with $\bar{\Omega}_k$ and $\mathbf{u}_h = \sum_k \mathbf{u}^{(k)}$. Here $\mathbf{u}_I^{(k)}$ is a vector corresponding to the interior of subdomain Ω_k and $\mathbf{u}_B^{(k)}$ is associated with the nodal points on $\partial\Omega_k$. Note that each interior variable $\mathbf{u}_I^{(k)}$ is associated with only one of the substructures, it can be eliminated locally and simultaneously. The reduced global equation, called the Schur complement on Γ, can be written in assembled form,

(2.5) $\quad S\mathbf{u}_B = (A_{BB} - A_{BI} A_{II}^{-1} A_{IB}) \mathbf{u}_B = \sum_k S^{(k)} \mathbf{u}_B^{(k)} = \mathbf{g} = \sum_k \mathbf{g}^{(k)}$

where $S^{(k)} = A_{BB}^{(i)} - A_{BI}^{(k)} (A_{II}^{(k)})^{-1} A_{IB}^{(k)}$ corresponds to the contribution from Ω_k to the boundary $\partial\Omega_k \subset \Gamma$, and $\mathbf{g}^{(k)} = \mathbf{f}_B^{(k)} - A_{BI}^{(k)} (A_{II}^{(k)})^{-1} \mathbf{f}_I^{(k)}$, comes from the value of $\mathbf{f}$ on interior points Ω_k and boundary points $\partial\Omega_k$. The action of the inverse of $A_{II}^{(k)}$ is equivalent to solving a local problem on Ω_k with Dirichlet boundary condition. Note the reduced Schur complement is still singular i.e.

$$Ker S = \{\mathbf{u}_B | S\mathbf{u}_B = 0\} \neq \{0\}.$$

Hence, problem (2.5) has an unique solution $\mathbf{u}_B \perp KerS$ only if right-hand side $\mathbf{g}$ satisfies the compatibility condition: $\mathbf{g} \perp KerS$. For Schur complement system (2.5), we have the following properties, which can be easily proved [1, 17].

LEMMA 2.1. *If element* $u_h = (u_I^T, u_B^T)^T \in KerA_h$, *then* $u_B \in KerS$. *If* $u_B \in KerS$, *there exists* u_I *such that* $u_h = (u_I^T, u_B^T)^T \in KerA_h$.

LEMMA 2.2. *If* $f_h \perp KerA$, *then* $g = f_B - A_{BI}A_{II}^{-1}f_I$ *is orthogonal to* $KerS$.

LEMMA 2.3. *If* A_h *is a symmetric semi-positive matrix, then for any* v_B

$$v_B^T S v_B = \min_{v_I} v_h^T A_h v_h$$

where $v_h = (v_I^T, v_B^T)^T$. *Hence, the capacitance matrix* S *is also symmetric positive semi-definite.*

LEMMA 2.4. *For any* v_B, *let* $v_h = (v_I^T, v_B^T)^T$ *be discrete harmonic extension of* v_B, *i.e.* $(A_{II} \quad A_{IB})v_h = 0$. *Then* $v_B^T S v_B = v_h^T A_h v_h$.

From these lemmas, we can conclude that the direct restriction of $KerA_h$ to the pseudo-boundary Γ equals $KerS$. We can loosely state that $KerS = KerA_h$.

3. The Modified VSDD Algorithm

Let the interface Γ be partitioned as the union of faces $\Gamma_{F_{ij}}$, edges $\Gamma_{E_{ij}}$ and cross points $\mathbf{x}_i$: $\Gamma = (\cup_{ij} F_{ij}) \cup (\cup_{ij} E_{ij}) \cup (\cup_i \mathbf{x}_i)$, where the face F_{ij} is the interface of two neighbor substructures Ω_i and Ω_j, and the edge E_{ij} is the set with all the points on the substructure boundaries in the cylinder with radius $0(H)$ and the central line ending by adjacent cross points of $\mathbf{x}_i$ and $\mathbf{x}_j$. Let the vertex space X_i be the region consisting of a vertex $\mathbf{x}_i$ and an overlap of order H onto adjacent faces and edges. We restrict the overlapping so that no portion of Γ is covered more than p times. Here p is a small finite integer. If Ω is in 2 dimension space, the interface Γ is partitioned into overlapping regions: edges and vertex spaces.

For each subregion $\hat{\Gamma}$, we introduce $R_{\hat{\Gamma}}$ as the pointwise restriction operator which returns only those unknowns that are associated with $\hat{\Gamma}$. Denote as $\mathbf{V}^h(\Gamma)$ the space of all the grid functions defined on Γ, and as $\mathbf{V}^H$ the space of grid functions on the coarse mesh. Denote face, edge and vertex space submatrices by $S_{F_{ij}} = R_{F_{ij}}^T S R_{F_{ij}}$, $S_{E_{ij}} = R_{E_{ij}}^T S R_{E_{ij}}$, and $S_{X_i} = R_{X_i}^T S R_{X_i}$, respectively.

The criterion for choosing the restriction operator R_H is this: for any function $\mathbf{g} \in \mathbf{V}^h(\Gamma)$ with $\mathbf{g} \perp KerS$, $R_H\mathbf{g}$ shall be orthogonal to the space $KerA_H$. Let

R_H^T be a linear interpolation operator from V^H to $V^h(\Gamma)$. Then, R_H is the corresponding weighted restriction operator from the space $\mathbf{V}^h(\Gamma)$ to the space $\mathbf{V}^H$. For most Neumann boundary value problems, the corresponding kernel space $KerS$ consists of linear functions. Therefore, a straightforward computation gives that $R_H\mathbf{g}$ is orthogonal to the kernel space $KerA_H$ when $\mathbf{g}$ is orthogonal to the kernel space $KerS$. Thus, the coarse problem $A_H\mathbf{u}_H = R_H\mathbf{g}$ is well defined and has only one solution $\mathbf{u}_H \perp KerA_H$.

We solve the singular Schur complement system (2.5) by using a preconditioned conjugate gradient iterative method with VSDD as a preconditioner M. The action of the inverse of the preconditioner M involves following block calculations.

Calculate $\mathbf{u}_B = M^{-1}g$ where $g \perp KerS$:

The VSDD Preconditioner

(i) Solve subproblems on all faces $\Gamma^{F_{ij}}$, edges $\Gamma^{E_{ij}}$, and vertex spaces Γ^{X_i} :

$$S_{F_{ij}}\mathbf{u}_{F_{ij}} = R_{F_{ij}}\mathbf{g}; \qquad S_{E_{ij}}\mathbf{u}_{E_{ij}} = R_{E_{ij}}\mathbf{g}; \qquad S_{X_l}\mathbf{u}_{X_l} = R_{X_l}\mathbf{g};$$

(ii) Find $\mathbf{u}_H \perp KerA_H$ so that: $A_H\mathbf{u}_H = R_H\mathbf{g}$;
(iii) Calculate $\mathbf{w}_B = R_H^T\mathbf{u}_H + \sum_{ij} R_{F_{ij}}^T\mathbf{u}_{F_{ij}} + \sum_{ij} R_{E_{ij}}^T\mathbf{u}_{E_{ij}} + \sum_l R_{X_l}^T\mathbf{u}_{X_l}$;
(iv) Find $\bar{\mathbf{w}}_B \in KerS$ such that $\bar{\mathbf{w}}_B + \mathbf{w}_B$ is orthogonal to $KerS$. Then

$$M^{-1}\mathbf{g} = \bar{\mathbf{w}}_B + R_H^T\mathbf{u}_H + \sum_{ij} R_{F_{ij}}^T\mathbf{u}_{F_{ij}} + \sum_{ij} R_{E_{ij}}^T\mathbf{u}_{E_{ij}} + \sum_l R_{X_l}^T\mathbf{u}_{X_l}.$$

Remark: All the subproblems in step (i) and (ii) can be solved simultaneously. The whole preconditioning procedure can be rewritten in a short form:

$$(3.1) \qquad M^{-1}\mathbf{g} = R_H^T A_H^{-1} R_H\mathbf{g} + \bar{\mathbf{w}}_B + \sum_{ij} R_{F_{ij}}^T S_{F_{ij}}^{-1} R_{F_{ij}}\mathbf{g}$$

$$+ \sum_{ij} R_{E_{ij}}^T S_{E_{ij}}^{-1} R_{E_{ij}}\mathbf{g} + \sum_l R_{X_l}^T S_{X_l}^{-1} R_{X_l}\mathbf{g}$$

where $\bar{\mathbf{w}}_B \in KerS$ is determined by making $M^{-1}\mathbf{g}$ be orthogonal to the kernel space $KerS$. For problem (2.2), the vector $-\bar{\mathbf{w}}_B \in KerS$ in step (iv) is equal to the mean value of $\mathbf{w}_B$ times a one-vector defined on the interface Γ.

After obtaining the approximate solution $\mathbf{u}_B \perp KerS$ on the interface Γ through using the PCG iterative method, we can calculate the approximate solution of problem (2.3) on the whole domain by solving concurrently all the subproblems defined on the substructures Ω_k with Dirichlet boundary value $\mathbf{u}_B$ on $\partial\Omega_k \subset \Gamma$:

$$A_{II}\mathbf{u}_I = f_I - A_{IB}\mathbf{u}_B.$$

However, the extended solution $\mathbf{u}_h = (\mathbf{u}_I^T, \mathbf{u}_B^T)^T$, is not orthogonal to the kernel space $KerA_h$. Therefore, we have to find a function $\mathbf{w}_h \in KerA_h$ such that the approximate solution $\mathbf{u}_h + \mathbf{w}_h$ is orthogonal to $KerA_h$.

Using the general additive Schwarz framework [10, 11, 12, 13, 19], we can estimate the condition number κ of the modified VSDDM. The proof of the following theorem is similar to that in [11, 18, 19].

THEOREM 3.1. *Suppose the overlapping size is δ_i, then:*

$$\kappa(M^{-1}S) \leq \frac{\lambda_{\max}(M^{-1}S)}{\lambda_{\min}(M^{-1}S)} \leq C(1 + \max_i \frac{H}{\delta_i}),$$

where C is a constant independent of H and h.

4. The Variants of the Modified VSDDM

Assume that $\Omega \subset \mathbf{R}^2$. It is extremely expensive to form exact edge matrices $S_{E_{ij}}$ and vertex matrices S_{X_k} in the VSDDM. This expense in computation and storage can be significantly reduced when exact edge and vertex matrices are replaced by spectrally equivalent approximations [6] which are based on the Fourier approximation and the probing technique. In this section, we further reduce the cost of forming approximations by the probing technique. We use four instead of six probe vectors [6] in the modified VSDDM to form approximate edge matrices and vertex matrices. For simplicity, we first describe the procedure for constructing of these probing approximations on the common edge $\Gamma^{E_{ij}}$ of two adjacent rectangular substructures Ω_i and Ω_j. This technique can easily be extended to more general geometries. On the edge $\Gamma^{E_{ij}}$, we construct a symmetric tridiagonal matrix $\tilde{S}_{E_{ij}}$ to approximate the exact Schur complement $S_{E_{ij}}$ by using matrix vector products of $S_{E_{ij}}$ with two probing vectors. A heuristic motivation for using the tridiagonal approximations is that the entries of each $S_{E_{ij}}$ decay rapidly away from the diagonal : $(S_{E_{ij}})_{lm} = 0(\frac{1}{(l-m)^2})$ for l, m away from the diagonal [14].

Let us introduce two probing vectors:

$$p_1 = [1, 0, 1, 0, \cdots]^T \quad \text{and} \quad p_2 = [0, 1, 0, 1, \cdots]^T.$$

From the fact

$$\tilde{S}_{E_{ij}}[p_1, p_2] = \begin{bmatrix} a_1 & b_1 & & & \\ b_1 & a_2 & b_2 & & \\ & b_2 & a_3 & b_3 & \\ & & b_3 & a_4 & \ddots \\ & & & & \ddots \end{bmatrix} \begin{bmatrix} 1 & 0 \\ 0 & 1 \\ 1 & 0 \\ 0 & 1 \\ \vdots & \vdots \end{bmatrix} = \begin{bmatrix} a_1 & b_1 \\ b_1 + b_2 & a_2 \\ a_3 & b_2 + b_3 \\ b_3 + b_4 & a_4 \\ \vdots & \vdots \end{bmatrix},$$

we can construct a symmetric tridiagonal approximate matrix $\tilde{S}_{E_{ij}}$ by letting

$$[\tilde{S}_{E_{ij}} p_1, \tilde{S}_{E_{ij}} p_2] = [S_{E_{ij}} p_1, S_{E_{ij}} p_2].$$

Computing the matrix vector product $S_{E_{ij}} p_k$ requires solving one problem on each sub-domain Ω_i and Ω_j. Hence, an approximate submatrix $\tilde{S}_{E_{ij}}$ can be obtained from the matrix vector products $S_{E_{ij}} p_k$ by using the algorithm [15, 16]:

Symmetric Probe Algorithm

$$\text{For} \quad l = 1, \cdots, n_{ij}$$
$$a_l = \begin{cases} (S_{E_{ij}} p_1)_l & \text{if } l \text{ is odd} \\ (S_{E_{ij}} p_2)_l & \text{if } l \text{ is even} \end{cases}$$
$$b_1 = (S_{E_{ij}} p_2)_1$$
$$\text{For} \quad l = 2, \cdots, n_{ij} - 1$$
$$b_l = \begin{cases} (S_{E_{ij}} p_1)_l - b_{l-1} & \text{if } l \text{ is even} \\ (S_{E_{ij}} p_2)_l - b_{l-1} & \text{if } l \text{ is odd} . \end{cases}$$

THEOREM 4.1. *Assume that $S_{E_{ij}}$ is a symmetric diagonally dominant $n_{ij} \times n_{ij}$ M-matrix and*

$$|(S_{E_{ij}})_{l,l}| \geq |(S_{E_{ij}})_{l,l+1}| \geq \cdots \geq |(S_{E_{ij}})_{l,n_{ij}}| \qquad for\ l = 1, 2, \cdots, n_{ij}.$$

Then the symmetric approximate matrix $\tilde{S}_{F_{ij}}$ is also an M-matrix.

PROOF. Without loss of generality, we assume $n_{ij} = 2k + 1$. From the assumption, we have

$$(4.1) \qquad |(S_{E_{ij}})_{k,l}| \geq |(S_{E_{ij}})_{s,t}| \qquad \text{if } s \leq k, l \leq t,$$

and for $1 \leq l \leq n_{ij}$:

$$a_l = \begin{cases} (S_{E_{ij}})_{l,1} + (S_{E_{ij}})_{l,3} + \cdots + (S_{E_{ij}})_{l,2k+1} & \geq 0, & \text{if } l \text{ is odd} \\ (S_{E_{ij}})_{l,2} + (S_{E_{ij}})_{l,4} + \cdots + (S_{E_{ij}})_{l,2k} & \geq 0, & \text{if } l \text{ is even} . \end{cases}$$

$$b_1 = (S_{E_{ij}})_{1,2} + (S_{E_{ij}})_{1,4} + \cdots + (S_{E_{ij}})_{1,2k} \leq 0$$

follows from the property of M-matrix $S_{E_{ij}}$. Then,

$$|a_1| - |b_1| = (S_{E_{ij}})_{1,1} + (S_{E_{ij}})_{1,2} + (S_{E_{ij}})_{1,3} + \cdots + (S_{E_{ij}})_{1,n_{ij}} \geq 0.$$

The equation $\quad b_1 + b_2 = (S_{E_{ij}})_{2,1} + (S_{E_{ij}})_{2,3} + \cdots + (S_{E_{ij}})_{2,2k+1} \leq 0, \quad$ implies

$$b_2 = (S_{E_{ij}})_{2,1} + (S_{E_{ij}})_{2,3} + \cdots + (S_{E_{ij}})_{2,2k+1}$$
$$- (S_{E_{ij}})_{1,2} - (S_{E_{ij}})_{1,4} - \cdots - (S_{E_{ij}})_{1,2k} \qquad \leq 0,$$

by using inequality (4.1). Then,

$$\begin{aligned} |a_2| - |b_1| - |b_2| &= (S_{E_{ij}})_{2,2} + (S_{E_{ij}})_{2,4} + \cdots + (S_{E_{ij}})_{2,2k} \\ &+ (S_{E_{ij}})_{1,2} + (S_{E_{ij}})_{1,4} + \cdots + (S_{E_{ij}})_{1,2k} \\ &+ (S_{E_{ij}})_{2,1} + (S_{E_{ij}})_{2,3} + \cdots + (S_{E_{ij}})_{2,2k+1} \\ &- (S_{E_{ij}})_{1,2} - (S_{E_{ij}})_{1,4} - \cdots - (S_{E_{ij}})_{1,2k} \\ &= (S_{E_{ij}})_{2,1} + (S_{E_{ij}})_{2,2} + (S_{E_{ij}})_{2,3} + \cdots + (S_{E_{ij}})_{2,n_{ij}} \geq 0. \end{aligned}$$

By induction, for any $3 \leq l \leq n_{ij}$, we have

$$
\begin{aligned}
b_l &= (S_{E_{ij}})_{l,1} + (S_{E_{ij}})_{l,3} + \cdots + (S_{E_{ij}})_{l,2k+1} \\
&\quad - (S_{E_{ij}})_{l-1,2} - (S_{E_{ij}})_{l-1,4} - \cdots - (S_{E_{ij}})_{l-1,2k} \\
&\quad + b_{l-2} \qquad \text{for} \quad l \quad \text{even} ,
\end{aligned}
$$

and

$$
\begin{aligned}
b_l &= (S_{E_{ij}})_{l,2} + (S_{E_{ij}})_{l,4} + \cdots + (S_{E_{ij}})_{l,2k} \\
&\quad - (S_{E_{ij}})_{l-1,1} - (S_{E_{ij}})_{l-1,3} - \cdots - (S_{E_{ij}})_{l-1,2k+1} \\
&\quad + b_{l-2} \qquad \text{for} \quad l \quad \text{odd} .
\end{aligned}
$$

So, we can obtain $b_l \leq 0$ through using $b_{l-1} \leq 0$ and $b_{l-2} \leq 0$, and inequality (4.1). Hence,

$$
\begin{aligned}
|a_l| - |b_{l-1}| - |b_l| &= a_l + b_{l-1} + b_l \\
&= (S_{E_{ij}})_{l,1} + (S_{E_{ij}})_{l,2} + (S_{E_{ij}})_{l,3} + \cdots + (S_{E_{ij}})_{l,n_{ij}} \geq 0.
\end{aligned}
$$

Thus, the tridiagonal symmetric matrix $\tilde{S}_{E_{ij}}$ is positive definite. Note that above inequalities can be replaced by strict inequalities, if $S_{E_{ij}}$ is strictly diagonally dominant. $\square$

The assumption of this main theorem is always true according to our numerical tests. However, the theoretical proof of this assumption is still an open question.

To minimize the computational work and the approximate errors arising from boundary value on other edges in the constructing procedure of all approximate edge matrices, we will specify the same probe vectors p_k either on all horizontal edges simultaneously or on all vertical edges simultaneously. Let's define $\mathbf{p}_k$ for $k = 1, 2$:

$$
\mathbf{p}_k = \begin{cases} p_k & \text{on all horizontal edges} \\ 0 & \text{on all vertical edges} \end{cases}
$$

and

$$
\mathbf{p}_{k+2} = \begin{cases} 0 & \text{on all horizontal edges} \\ p_k & \text{on all vertical edges} . \end{cases}
$$

which are as drawn in Fig.1.

Analogously, these approximations $\tilde{S}_{E_{ij}}$ resulting from the simultaneous probe vectors $\mathbf{p}_k$ above preserve strict diagonally dominance and positive definiteness. Since the edge matrices $\tilde{S}_{E_{ij}}$ are tridiagonal, it is cheap and easy to calculate $\tilde{S}_{E_{ij}}^{-1} g_{E_{ij}}$.

THEOREM 4.2. *If the Schur complement S on Γ is a strictly diagonally dominant M-matrix, then all the edge approximations $\tilde{S}_{E_{ij}}$ obtained from above are strictly diagonally dominant and positive definite.*

PROOF. From the assumption that M-matrix S is strictly diagonally dominant, $\tilde{S}_{E_{ij}}$ can be proved to be strictly diagonally dominant in the same way as in the proof of Theorem 4.1. $\square$

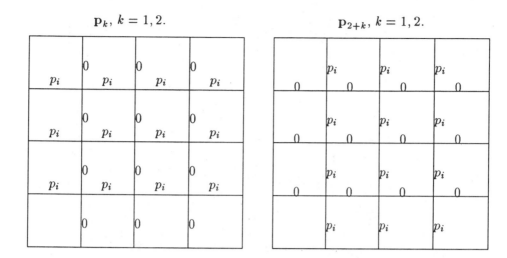

FIGURE 1. Probing Vectors in the Modified VSDDM

The probing approximate vertex matrices $\tilde{S}_{X_i}$ can be easily constructed from the results of probing matrix-vector products $\{S\mathbf{p}_k\}_{k=1}^4$. Details of this technique can be found in [6]. The same argument as in [6] can be used to show that the vertex approximations $\tilde{S}_{X_i}$ obtained from above probing procedure are diagonally dominant if Schur complement S is a diagonally dominant M-matrix.

5. Numerical Results

Now we present the numerical tests on the convergence rate of the modified VS methods with various edge and vertex approximations. The tests were performed for scalar elliptic problem (2.2) with Neumann boundary condition. The following three coefficient functions have been used in our tests:

(i) $\alpha(x, y) = 1$, the Laplace operator;
(ii) $\alpha(x, y) = e^{10xy}$, highly varying smooth coefficients;
(iii) Highly discontinuous coefficients defined as Fig. 2.

The square domain $[0, 1]^2$ is first divided into $1/H^2$ square sub-domains with uniform size H. Then each square sub-domain was triangulated into finite element with uniform mesh size h on the square domain. These problems are discretized by standard finite element method with five stencil.

u is a randomly generated solution of the scalar elliptic problem normalized so that the mean value of u is zero. The integer K is defined to be the number of iterations required to reduce the A-norm of the error $e_n = u - u_n$ by a factor 10^{-5}. We list the iteration number K and the estimated condition number κ for these discrete problems with various coarse mesh size H and fine mesh size h in the following tables. We fix the size of vertex space matrices as 5×5. So,

$\alpha = 300$	$\alpha = 10^{-4}$	$\alpha = 31400$	$\alpha = 5$
$\alpha = 0.05$	$\alpha = 6$	$\alpha = 0.07$	$\alpha = 2700$
$\alpha = 10^6$	$\alpha = 0.1$	$\alpha = 200$	$\alpha = 9$
$\alpha = 1$	$\alpha = 6000$	$\alpha = 4$	$\alpha = 140000$

FIGURE 2. Coefficient Function $\alpha(x, y)$

the overlapping size is $h/H = H^{-1}/h^{-1}$. FVS and PVS represent the modified VSDDM with the Fourier [6] and the probing approximations, respectively, on edges and vertex spaces. In our program, all sub-problems on coarse grid and on sub-domains are solved with high precision. The numerical results show that the modified VSDDM still has an optimal convergence rate for elliptic problems with Neumann boundary condition.

To compare six probing vectors with four probing vectors in probing VSDD method, we list the results in Table 2 for problem (2.2) with harmonic Dirichlet boundary condition and various coefficient defined above. Ovlp is denoted as the overlapping size. It can be observed from these numerical results that our probing technique improves efficiency as well as retains the optimal convergence.

Acknowledgments: The author is indebted to Tony F. Chan, Graeme Fairweather and Tarek P. Mathew for their advice and discussion on this project. The author would like to thank the referee for helpful corrections and comments.

REFERENCES

1. Chris Anderson, *Manipulating fast solvers - changing their boundary conditions and putting them on multiple processor computers*, Tech. Report CAM 88-37, Department of Mathematics, UCLA, 1988.
2. James H. Bramble, Joseph E. Pasciak, and Alfred H. Schatz, *The construction of preconditioners for elliptic problems by substructuring, I*, Math. Comp. **47** (1986), no. 175, 103–134.

TABLE 1. The Convergence of the Modified VSDDM

coeffi.	Laplace				e^{10xy}				discont.			
mesh	FVS		PVS		FVS		PVS		FVS		PVS	
H^{-1}/h^{-1}	κ	K	κ	K	κ	K	κ	K	κ	K	κ	K
4/64	7.3	14	5.1	11	10.8	15	7.1	12	27.2	18	13.8	13
8/64	5.5	13	3.7	10	6.9	14	4.7	11	19.1	16	9.8	12
16/64	4.4	12	3.0	9	5.3	12	3.4	10	11.4	14	6.3	11
32/64	5.7	13	5.5	13	5.5	13	5.9	13	6.4	13	5.5	13
4/128	9.5	15	6.5	12	14.8	17	7.9	12	35.9	20	8.5	13
8/128	7.4	14	4.9	11	8.1	14	6.0	11	26.6	19	13.3	13
16/128	5.5	13	3.7	10	6.3	14	4.2	10	18.8	16	9.8	12
32/128	4.4	11	2.9	9	4.8	12	3.3	9	12.1	14	6.6	11
64/128	5.7	13	5.4	13	5.7	13	5.6	13	6.3	13	5.5	13
4/256	11.5	16	11.0	15	16.9	19	13.3	16	45.7	21	27.4	16
8/256	9.6	15	6.2	12	11.1	16	7.5	12	36.1	20	18.3	15
16/256	7.3	14	4.6	11	7.5	14	4.9	11	24.6	18	13.1	13
32/256	5.6	13	3.6	10	5.7	13	3.7	10	18.9	16	9.6	12
64/256	4.4	12	2.9	9	4.5	12	3.1	9	5.0	12	5.8	10
128/256	5.7	13	5.3	13	5.7	13	5.4	13	6.1	13	5.4	13

TABLE 2. Comparison 6 with 4 Probing Vectors

coefficients		Laplace		$1 + 10(x^2 + y^2)$		e^{10xy}		discont.	
Probe Vec.		6	4	6	4	6	4	6	4
mesh	Ovlp	PVS		PVS		PVS		PVS	
H^{-1}/h^{-1}	h/H	K	K	K	K	K	K	K	K
4/64	1/16	9	10	9	10	9	10	11	12
8/64	1/8	9	9	9	9	8	9	10	10
16/64	1/4	8	9	8	9	8	8	9	10
32/64	1/2	13	13	13	13	13	13	13	13
4/128	1/32	10	10	10	10	10	11	11	12
8/128	1/16	9	10	9	10	9	10	11	10
16/128	1/8	9	9	8	9	8	9	9	10
32/128	1/4	8	9	8	9	8	8	9	10
64/128	1/2	13	13	13	13	13	13	13	13
4/256	1/64	13	13	13	13	13	11	13	13
8/256	1/32	10	11	10	11	10	11	11	11
16/256	1/16	9	10	9	10	9	10	11	11
32/256	1/8	9	9	9	9	8	9	10	9
64/256	1/4	8	9	8	9	8	8	9	10
128/256	1/2	13	13	13	13	13	13	13	13

3. Tony F. Chan, *Analysis of preconditioners for domain decomposition*, SIAM J. Numer. Anal. **24** (1987), no. 2, 382–390.

4. Tony F. Chan and Tarek P. Mathew, *An application of the probing technique to the vertex space method in domain decomposition*, Fourth International Domain Decomposition Methods for Partial Differential Equations (Philadelphia) (Roland Glowinski, Yu. A. Kuznetsov, G. Meurant, Jacques Périaux, and Olof Widlund, eds.), SIAM, 1991.

5. _____, *The interface probing technique in domain decomposition*, SIAM J. Matrix Analysis and Applications **13**(1) (1992), 212–238.

6. Tony F. Chan, Tarek P. Mathew, and Jian Ping Shao, *Efficient variants of the vertex space domain decomposition algorithm*, Tech. Report CAM 92-07, Department of Mathematics, UCLA, 1992, To appear in SIAM J. Sci. Comp.

7. Tony F. Chan and Diana C. Resasco, *A survey of preconditioners for domain decomposition*, Tech. Report /DCS/RR-414, Yale University, 1985.

8. Philippe G. Ciarlet, *The finite element method for elliptic problems*, North-Holland, New York, 1978.

9. Maksymilian Dryja, *A capacitance matrix method for Dirichlet problem on polygon region*, Numer. Math. **39** (1982), 51–64.

10. Maksymilian Dryja and Olof B. Widlund, *An additive variant of the Schwarz alternating method for the case of many subregions*, Tech. Report 339, also Ultracomputer Note 131, Department of Computer Science, Courant Institute, 1987.

11. _____, *Some domain decomposition algorithms for elliptic problems*, Proceedings of the Conference on Iterative Methods for Large Linear Systems held in Austin, Texas, October 1988, to celebrate the Sixty-fifth Birthday of David M. Young, Jr., Academic Press, Orlando, Florida, 1989.

12. _____, *Towards a unified theory of domain decomposition algorithms for elliptic problems*, Tech. Report 486, also Ultracomputer Note 167, Department of Computer Science, Courant Institute, 1989.

13. _____, *Additive Schwarz methods for elliptic finite element problems in three dimensions*, The Fifth International Symposium on Domain Decomposition Methods for Partial Diferential Equations (Philadelphia) (Tony Chan, D. E. Keyes, G. A. Meurant, J. S. Scroggs, and R. G. Voigt, eds.), SIAM, 1992.

14. Gene Golub and D. Mayers, *The use of preconditioning over irregular regions*, Computing Methods in Applied Sciences and Engineering, VI (Amsterdam, New York, Oxford) (R. Glowinski and J. L. Lions, eds.), North-Holland, 1984, (Proceedings of a conference held in Versailles, France, December 12-16,1983), pp. 3–14.

15. David E. Keyes and William D. Gropp, *A comparison of domain decomposition techniques for elliptic partial differential equations and their parallel implementation*, SIAM J. Sci. Stat. Comput. **8** (1987), no. 2, s166–s202.

16. _____, *Domain decomposition techniques for the parallel solution of nonsymmetric systems of elliptic BVPs*, Domain Decomposition Methods (Philadelphia) (Tony Chan, Roland Glowinski, Jacques Périaux, and Olof Widlund, eds.), SIAM, 1989.

17. Jian Ping Shao, *Domain decomposition algorithms*, Tech. Report CAM 93-38, Department of Mathematics, UCLA, Sept. 1993, Ph.D thesis.

18. Barry F. Smith, *An optimal domain decomposition preconditioner for the finite element solution of linear elasticity problems*, SIAM J. Sci. Stat. Comput. **13** (1992), no. 1, 364–378.

19. Olof B. Widlund, *Some Schwarz methods for symmetric and nosymmetric elliptic problems*, The Fifth International Symposium on Domain Decomposition Methods for Partial Diferential Equations (Philadelphia) (Tony Chan, D. E. Keyes, G. A. Meurant, J. S. Scroggs, and R. G. Voigt, eds.), SIAM, 1992.

CENTER FOR COMPUTATIONAL SCIENCES, UNIVERSITY OF KENTUCKY, LEXINGTON, KENTUCKY 40506, U.S.A.

E-mail address: shao@s.ms.uky.edu

Contemporary Mathematics
Volume **180**, 1994

A Multi-color Splitting Method and Convergence Analysis for Local Grid Refinement *

Tsi-min Shih, Chin-bo Liem, Tao Lu and Aihui Zhou

Abstract

This paper consists of two parts. In the first part, a multi-color splitting method is proposed for a multi-processor computer, which can be viewed as an algorithm combining successive subspace correction with parallel subspace correction. In the second part, error estimation and numerical test for a discrete Green's function are presented on local refinement grids.

1. Algorithm of Splitting by color

The features of a typical SIMD (Single Instruction Multiple Data) computer are:

- it contains many processors and hence it works highly in parallel;

- it works highly synchronously.

The method of splitting by color is specially designed for the SIMD architecture. We illustrate it as follows. Consider a second order elliptic equation

$$Lu \equiv - \sum_{i,j=1}^{d} \frac{\partial}{\partial x_i}(a_{ij} \frac{\partial}{\partial x_j}u) \;\; = \;\; f, \quad \text{in } \Omega \subseteq \mathcal{R}^d, \qquad (1.1)$$

$$u \;\; = \;\; 0, \quad \text{on } \partial\Omega.$$

The associated variational problem is to find $u \in H_0^1(\Omega)$ satisfying

---

*1991 Mathematics Subject Classification, 65N05, 65N30.

This work is supported by Hong Kong UPGC Earmarked Grant HKP 139/92E.

This paper is in final form and no version of it will be submitted for publication elsewhere.

$$a(u,v) \equiv \int_\Omega \sum_{i,j=1}^{d} a_{ij} \frac{\partial u}{\partial x_i} \frac{\partial v}{\partial x_j} dx = \int_\Omega fv dx, \quad \forall v \in H_0^1(\Omega). \qquad (1.2)$$

In order to solve (1.2) on a SIMD computer, subdivide Ω into nonoverlapping subregions: $\bar{\Omega} = \cup \bar{\Omega}_{ij}$, where Ω_{ij} denotes the jth element belonging to the ith color ($i = 1, 2, \cdots, N_c; j = 1, 2, \cdots, m$), and assume that for each color i, Ω_{ij} and Ω_{ik} are disjoint, i.e. $\bar{\Omega}_{ij} \cap \bar{\Omega}_{ik} = \emptyset$, $\forall j, k = 1, \cdots, m$ and $j \neq k$. Let H denote the grid size of the initial grid $\{\Omega_{ij}\}$, h denote that of the refined grid, and S_0^h denote the corresponding finite element space. If $\mathcal{N}(\Omega_{ij})$ denotes the nodal index set of $\bar{\Omega}_{ij}$, $V_{ij} = Span\{\varphi_k \in S_0^h; k \in \mathcal{N}(\Omega_{ij})\}$, and $V_i = \bigcup_{j=1}^{m} V_{ij}$, then

$$S_0^h = V_1 + V_2 + \cdots + V_{N_c}. \qquad (1.3)$$

We are now looking for $u_h \in S_0^h$ satisfying

$$a(u^h, v) = (f, v), \quad \forall v \in S_0^h. \qquad (1.4)$$

Define the projection operator $P_i : S_0^h \longrightarrow V_i$ satisfying

$$a(P_i u, v) = a(u, v), \quad \forall v \in V_i$$

and $P_{ij} : S_0^h \longrightarrow V_{ij}$ satisfying

$$a(P_{ij} u, v) = a(u, v), \quad \forall v \in V_{ij}.$$

Evidently, $P_i = \sum_{j=1}^{m} P_{ij}$ and $P_{ij} P_{ik} = 0$ $(j \neq k)$.

Algorithm 1 (Parallel Algorithm of Splitting by color)

Step 1. Choose an initial $u^0 \in S_0^h$ and relaxation factor $\omega \in (0, 2)$. Set $n := 0$, and $i := 0$.

Step 2. For $j = 1, 2, \cdots, m$, solve for $\delta_{ij} \in V_{ij}$ in parallel, according to

$$a(\delta_{ij}, v) = (f, v) - a(u^{n+i/N_c}, v), \quad \forall v \in V_{ij}.$$

Step 3. Set $u^{n+(i+1)/N_c} = u^{n+i/N_c} + \omega \sum_{j=1}^{m} \delta_{ij}$.

Step 4. If $i + 1 < N_c$, let $i := i + 1$ and goto Step 2;
If $i + 1 = N_c$, let $i := 0, n := n + 1$ and goto Step 2.

Denote the error by $e^{n+i/N_c} = u - u^{n+i/N_c}$, obviously

$$\begin{aligned}
e^{n+1} &= (I - \omega P_{N_c})(I - \omega P_{N_c-1}) \cdots (I - \omega P_1) e^n \\
&= \prod_{i=1}^{N_c} \prod_{j=1}^{m} (I - \omega P_{ij}) e^n. \qquad (1.5)
\end{aligned}$$

Hence Algorithm 1 mainly belongs to the framework of SSC(successive subspace correction), but Step 2 is a PSC(parallel subspace correction) algorithm [8].

In order to estimate the convergence rate of Algorithm 1, let $T = P_1 + \cdots + P_{N_c}$. Evidently,

$$\lambda_{max}(T) \leq \|T\| \leq N_c. \tag{1.6}$$

Proposition 1. *There exists a constant $C > 0$ independent of H and h such that*

$$\lambda_{min}(T) \geq C\frac{h^2}{H^2}. \tag{1.7}$$

Proof. Let $\hat{\Omega}_{ij} = \bigcup_{\varphi \in V_{ij}} Supp(\varphi)$, $\hat{\Omega}_{ij} \supset\supset \Omega_{ij}$, and $\{\hat{\Omega}_{ij}\}$ be an open covering of Ω. Construct the piecewise constant functions:

$$Q_i(x) = \begin{cases} 1, & x \in \bar{\Omega}_{ij}, \; j = 1, 2, \cdots, m, \\ 0, & \text{elsewhere}, \end{cases}$$

and

$$\varphi_i(x) = \frac{Q_i(x)}{\sum_{i=1}^{N_c} Q_i(x)},$$

then $\sum_{i=1}^{N_c} \varphi_i(x) \equiv 1$. Notice that $diam(\hat{\Omega}_{ij}) = O(H)$, and consider

$$u^h = I^h u^h = \sum_{i=1}^{N_c} I^h(\varphi_i u^h) = \sum_{i=1}^{N_c} u_i^h, \tag{1.8}$$

where $u_i^h = I^h(\varphi_i u^h) \in V_i$, and I^h is the interpolating operator on S_0^h. By the theory of inverse estimation, we have

$$a_\Omega(u_i^h, u_i^h) = \sum_{j=1}^{m} a_{\hat{\Omega}_{ij}}(u_i^h, u_i^h) \leq C_1 \sum_{j=1}^{m} \|u_i^h\|_{1,\hat{\Omega}_{ij}}^2$$

$$\leq C_2 \frac{H^2}{h^2} \sum_{j=1}^{m} \|u_i^h\|_{0,\hat{\Omega}_{ij}}^2. \tag{1.9}$$

Again from

$$\sum_{j=1}^{m} \|u_i^h\|_{0,\hat{\Omega}_{ij}}^2 = \sum_{j=1}^{m} \int_{\hat{\Omega}_{ij}} (I^h(\varphi_i u^h))^2 dx \leq C_3 \sum_{j=1}^{m} \int_{\hat{\Omega}_{ij}} (u^h)^2 dx,$$

we have

$$\sum_{i=1}^{N_c} a(u_i^h, u_i^h) \leq C_3 \sum_{i=1}^{N_c} \sum_{j=1}^{m} \int_{\tilde{\Omega}_{ij}} (u^h)^2 dx$$

$$\leq C_4 \frac{H^2}{h^2} \int_{\Omega} (u^h)^2 dx \leq C_5 \frac{H^2}{h^2} a(u^h, u^h). \tag{1.10}$$

Finally, (1.7) follows from Lions' Lemma [5].

Corollary 1 (cf.[8]). *The norm of $E = (I-\omega P_{N_c})\cdots(I-\omega P_1)$ can be estimated by*

$$\|E\|^2 \leq 1 - \frac{Ch^2(2-\omega)}{H^2(1+N_c)^2}. \tag{1.11}$$

2.   Estimation of the Convergence Rate of Local Grid Refinement

The error estimation of the composite grid method, in particular the error at the neighbourhood of singular points, is essential for engineering problems. Ewing, Lazarov and Vassilevski [2] discussed the error behavior of the finite difference scheme. For the finite element scheme, under the assumption that $u \in H^{1+\alpha}(\Omega)$ $(0 < \alpha < 1)$, Lin and Yan [4] proved that the superconvergence is of $O(H^{3/2} + h^\alpha)$ in the $H^1(\Omega)$-norm. From the engineering point of view, the accuracy of solving a problem with logarithmic singularity on a locally refined grid is very important. For example, consider the following elliptic equation

$$\begin{aligned} Lu &= \delta(x_1 - z_1)\delta(x_2 - z_2), &&\text{in } \Omega \subset \mathcal{R}^2, &&(2.1)\\ u &= 0, &&\text{on } \partial\Omega, \end{aligned}$$

where $\delta(x)$ is the Dirac-function and $z = (z_1, z_2) \in \Omega$ is a singular point. The solution u is the Green function $G_z(x_1, x_2)$ which has a logarithmic singularity at z. In fact, $G_z \notin H_0^1(\Omega)$, but $G_z \in \overset{0}{W_p^1}(\Omega), 1 \leq p < 2$.

If $z \in \Omega_1 \subset\subset \Omega$, let H be the grid size of the original coarse grid of Ω, h be the grid size of the refined grid of Ω_1, $S_0^H(\Omega) \subset H_0^1(\Omega), S_0^h \subset H_0^1(\Omega_1)$ be the linear finite element spaces on the original coarse grid and the refined grid respectively, and $H_c = S_0^H(\Omega) + S_0^h(\Omega_1)$ be the finite element space on the composite grid.

Denote the finite element approximation of G_z by G_z^H which satisfies

$$a(G_z^H, v) = v(z), \qquad\qquad \forall v \in H_c. \tag{2.2}$$

Let Ω_0 be a small subregion of Ω_1, and $z \in \Omega_0 \subset\subset \Omega_1 \subset\subset \Omega$. We have the following error estimation of G_z^H on Ω_0.

Proposition 2. *If (1) the coefficients $a_{ij}, b_{ij} \in W_1^\infty(\Omega)$, and (2) $[a_{ij}]$ is uniformly positive definite, then there is an $L_p(\Omega)$ estimation*

$$\|G_z - G_z^H\|_{0,p,\Omega_0} \leq C(hH^{2/p-1} + H^2)|\ln H|^2, \quad 1 \leq p < 2, \tag{2.3}$$

where $\|\cdot\|_{k,p,\Omega_0}$ denotes the $W_p^k(\Omega_0)$ norm, $C > 0$ is a constant that is independent of h and H, but dependent on Ω_0 and Ω_1.

Proof. Construct a function $\omega(x) \in C_0^\infty(\Omega_1)$ satisfying $\omega(x) \equiv 1, \forall x \in \Omega_0$. Also for all $\varphi \in L_q(\Omega), (\frac{1}{p} + \frac{1}{q} = 1)$, construct an auxiliary function $w(x)$ satisfying:

$$\begin{aligned} Lw &= \varphi, &&\text{in } \Omega, &&(2.4)\\ w &= 0, &&\text{on } \partial\Omega. \end{aligned}$$

Evidently, by the Sobolev Embedding Theorem,

$$c_0\|w\|_{1,\infty,\Omega} \le \|w\|_{2,q,\Omega} \le C\|\varphi\|_{0,q,\Omega}, \quad 2 < q < \infty. \tag{2.5}$$

Denote ωu by $\tilde{u}$. Since $G_z \in W_p^1(\Omega), 1 < p < 2$, from integration by parts, we have

$$\begin{aligned}
(\widetilde{G_z - G_z^H}, \varphi) &= a(\widetilde{G_z - G_z^H}, w) + I = a_{\Omega_1}(G_z - G_z^H, \tilde{w}) + I \\
&\le Ch\|G_z - G_z^H\|_{1,p,\Omega_1}\|w\|_{2,q,\Omega_1} + I \\
&\le Ch\|G_z - G_z^H\|_{1,p,\Omega}\|\varphi\|_{0,q,\Omega} + I, \tag{2.6}
\end{aligned}$$

where

$$\begin{aligned}
I &= \int_\Omega \{\sum_{i,j=1}^2 a_{ij}\frac{\partial \omega}{\partial x_i}\frac{\partial w}{\partial x_j} + \frac{\partial}{\partial x_i}(a_{ij}\frac{\partial \omega}{\partial x_j}w)\}(G_z - G_z^H)dx \\
&\le C\|G_z - G_z^H\|_{0,1,\Omega}\|w\|_{1,\infty,\Omega} \\
&\le C\|G_z - G_z^H\|_{0,1,\Omega}\|\varphi\|_{0,q,\Omega}. \tag{2.7}
\end{aligned}$$

Substituting (2.7) into (2.6),

$$|(\widetilde{G_z - G_z^H}, \varphi)| \le C(h\|G_z - G_z^H\|_{1,p,\Omega} + \|G_z - G_z^H\|_{0,1,\Omega})\|\varphi\|_{0,q,\Omega}. \tag{2.8}$$

Using the following two known results [9],

$$\|G_z - G_z^H\|_{1,p,\Omega} \le CH^{2/p-1}|\ln H|,$$

$$\|G_z - G_z^H\|_{0,1,\Omega} \le CH^2|\ln H|^2,$$

it follows that

$$\|G_z - G_z^H\|_{1,p,\Omega_0} \le \|\widetilde{G_z - G_z^H}\|_{0,p,\Omega} \le C(hH^{2/p-1} + H^2)|\ln H|^2.$$

With more detailed analysis [3], the following estimations can be obtained:

$$\|G_z - G_z^H\|_{0,p,\Omega} \le \begin{cases} C(h^{2/p} + H^2), & 1 < p < +\infty, \\ C(h^2|\ln h|^2 + H^2), & p = 1, \end{cases}$$

and

$$\|G_z - G_z^H\|_{1,p,\Omega_0} \le \begin{cases} C(h + H^2)h^{2/p-2}|\ln h|^{1/2}, & 1 < p < \infty, \\ C(h + H^2)|\ln h|, & p = 1. \end{cases}$$

Numerical Test. Consider $L = \triangle$, $\Omega = (0,2) \times (0,2)$ in equation (2.1) with $z = (1,1)$. Locally refine $(1/2, 3/2) \times (1/2, 3/2)$ by size $H/2$ and $(1/4, 5/4) \times (1/4, 5/4)$ by size $H/4$. Computed results using the Fast Adaptive Composite Grid Method (FAC)[6] for $G_z(599/600, 601/600)$ are as follows:

Uniform grid			Local refinement grid			
H	Error	CPU	H	Level No.	Error	CPU
1/600	1.30E-2	12.84		1	1.30E-2	12.84
1/1200	2.61E-3	1175.74	1/600	2	2.12E-3	243.97
				3	4.53E-4	452.96

References

[1] G.F. Carey (ed.), *Parallel Supercomputing :Methods, Algorithms and Applications*, John Wiley & Sons, 1989.

[2] R.E. Ewing, R.D. Lazarov & P.S. Vassilevski, *Local Refinement Techniques for Elliptic Problem on Cell-centered Grids: I. Error Analysis*, Math. Comp., 56(1991), 437–461.

[3] C. B. Liem, T.M. Shih, A.H. Zhou & T. Lu, *Error Analysis for Finite Element Approximations on Local Refinement Grids*, Research Report IMS-51, Institute of Mathematical Sciences, Academia Sinica, Chengdu, China, 1993.

[4] Q. Lin & N.N. Yan, *A Rectangle Test for Singular Solution with Irregular Meshes*, Proceedings of Systems Science & Systems Engineering, Great Wall Culture Publish Co., 1991, 236–237.

[5] P.L. Lions, *On the Schwarz Alternating Method I*, Proceedings of First International Symposium on Domain Decomposition Methods for Partial Differential Equations, Edited by R. Glowinski, G.H. Golub, G.A. Meurant and J. Periaux, SIAM, Philadelphia, 1988, 1–42.

[6] S.F. McCormick & J. Thomas, *The Fast Adaptive Composite Grid Method (FAC) for Elliptic Boundary Value Problems*, Math. Comp., 46(1986), 439–456.

[7] J. Nitsche & A.H. Schatz, *Interior Estimates for Ritz-Galerkin Methods*, Math. Comp., 28(1974), 937–955.

[8] J. Xu, *Iterative Methods by Space Decomposition and Subspace Correction*, SIAM Review, 34(1992), 581–613.

[9] Q.D. Zhu & Q. Lin, *Superconvergence Theory for Finite Elements*, Hunan Science Press, 1989 (in Chinese).

Authors' address:

T.M. Shih and C.B. Liem
Department of Applied Mathematics, Hong Kong Polytechnic, Hong Kong.

T. Lu
Institute of Chengdu Computer Applications, Academia Sinica, Chengdu, 610041, China.

A.H. Zhou
Institute of Systems Science, Academia Sinica, Beijing, 100080, China.

Contemporary Mathematics
Volume **180**, 1994

Boundary Elements in
Domain Decomposition Methods

OLAF STEINBACH

May 11, 1994

ABSTRACT. We give a brief survey of domain decomposition methods with boundary elements for the model problem of the potential equation, and we offer different parallel algorithms for solving these problems numerically.

1. Introduction

Let us consider the Dirichlet potential problem

(1.1)
$$\begin{aligned} -\operatorname{div} a(x)\nabla u(x) &= 0 && \text{for } x \in \Omega \subset I\!\!R^2 \,, \\ u(x) &= g(x) && \text{for } x \in \Gamma := \partial\Omega, \end{aligned}$$

where Ω is a plane, bounded domain with a piecewise Lipschitz continuous boundary Γ. Suppose

(1.2)
$$a(x) = a_i = constant \ \ \text{in } \Omega_i \,,$$

where the Ω_i are non–overlapping subdomains in a given domain decomposition

(1.3) $\overline{\Omega} = \displaystyle\bigcup_{i=1}^{p} \overline{\Omega}_i \,, \ \ \Gamma_i = \partial\Omega_i \,, \ \ \Omega_i \cap \Omega_j = \emptyset$ and $\Gamma_{ij} = \overline{\Omega}_i \cap \overline{\Omega}_j$ for $i \neq j$.

Now we introduce the skeleton Γ_S and the global coupling boundary Γ_C by

(1.4)
$$\Gamma_S = \bigcup_{i=1}^{p} \Gamma_i \,, \ \ \Gamma_C = \Gamma_S \backslash \Gamma \,,$$

and a q–dimensional set ω of all cross points.

1991 *Mathematics Subject Classification.* Primary 65N55, 65N38; Secondary 68Q22.

This work was partially supported by the Priority Research Program "Boundary Element Methods" of the German Research Foundation DFG under Grant Nb. We 659/8–5 .

This paper is in final form and no version of it will be submitted for publication elsewhere.

The variational formulation of the model problem (1.1) reads:

Find $u \in H^1(\Omega)$ with $u_{|\Gamma} = g(x)$ such that

$$(1.5) \qquad \int_\Omega a(x)\, \nabla u(x)\, \nabla v(x)\, dx \;=\; 0$$

for all test functions $v \in H_0^1(\Omega)$.

Before we apply the domain decomposition to (1.5) , we should introduce boundary integral equations corresponding to the local potential problems by use of the Calderon projector

$$(1.6) \quad \begin{pmatrix} u_{|\Gamma_i} \\ t_i \end{pmatrix} = \begin{pmatrix} \tfrac{1}{2}I - K_i & V_i \\ D_i & \tfrac{1}{2}I + K_i' \end{pmatrix} \begin{pmatrix} u_{|\Gamma_i} \\ t_i \end{pmatrix} = C_i \begin{pmatrix} u_{|\Gamma_i} \\ t_i \end{pmatrix},$$

where V_i is the simple layer potential, K_i the double layer potential, K_i' its adjoint, and D_i the hypersingular operator. Further, $t_i = \frac{\partial u}{\partial n}_{|\Gamma_i}$ denotes the normal derivative of the potential u with respect to Γ_i. From (1.6) we get two representations of the local Steklov–Poincaré operator [12]:

$$(1.7) \qquad \begin{aligned} t_i = S_i u_{|\Gamma_i} &= V_i^{-1}(\tfrac{1}{2}I + K_i)u_{|\Gamma_i} \\ &= \left[(\tfrac{1}{2}I + K_i')V_i^{-1}(\tfrac{1}{2}I + K_i) + D_i \right] u_{|\Gamma_i} \end{aligned}$$

Using the given domain decomposition (1.3) and Green's formula we may rewrite (1.5) into a variational formulation on the skeleton.

Find $u \in H^{1/2}(\Gamma_S)$ with $u_{|\Gamma} = g(x)$ such that

$$(1.8) \qquad \sum_{i=1}^{p} \int_{\Gamma_i} S_i u_{|\Gamma_i}\, v_{|\Gamma_i}\, ds \;=\; 0$$

for all test functions $v \in H_0^{1/2}(\Gamma_S)$.

2. Iterative methods with the "Symmetric Formulation"

One possibility for discretization of the Steklov–Poincaré operator in (1.8) is to replace the corresponding normal derivative t_i by the second equation of the Calderon projector. Furthermore, we need the first boundary integral equation to compute the local Cauchy datum [3].

Find $u \in H^{1/2}(\Gamma_S)$ with $u_{|\Gamma} = g$ and $t_i = \frac{\partial u}{\partial n}_{|\Gamma_i} \in H^{-1/2}(\Gamma_i)$ such that

$$(2.1) \qquad \begin{aligned} \sum_{i=1}^{p} a_i \left\{ \langle D_i u_{|\Gamma_i}, v_{|\Gamma_i} \rangle_{\Gamma_i} + \tfrac{1}{2}\langle t_i, v_{|\Gamma_i} \rangle_{\Gamma_i} + \langle t_i, K_i v_{|\Gamma_i} \rangle_{|\Gamma_i} \right\} &= 0 \\ a_i \left(\langle \tau_i, V_i t_i \rangle_{\Gamma_i} - \tfrac{1}{2}\langle \tau_i, u_{|\Gamma_i} \rangle_{|\Gamma_i} - \langle \tau_i, K_i u_{|\Gamma_i} \rangle_{\Gamma_i} \right) &= 0 \end{aligned}$$

for all test functions $v \in H_0^{1/2}(\Gamma_S)$ and $\tau_i \in H^{1/2}(\Gamma_i)$ and $i = 1, \ldots, p$.

The equivalent discrete system can be written in the block form

$$(2.2) \qquad \begin{pmatrix} V_h & -\frac{1}{2}M_h - K_h \\ \frac{1}{2}M_h^\top + K_h^\top & D_h \end{pmatrix} \begin{pmatrix} \underline{t} \\ \underline{u} \end{pmatrix} = \begin{pmatrix} \underline{f}_t \\ \underline{f}_u \end{pmatrix},$$

where $\underline{t}$ denotes the vector of the local coefficients of the t_i and $\underline{u}$ the coupling values of the potential on the skeleton. The system matrix is obviously block skew–symmetric, but positive definite. Note that the first block equation in (2.2) consists of local equations only; this means that we can compute

$$(2.3) \qquad \underline{t}_i = V_{h,i}^{-1} \left(\frac{1}{2}M_{h,i} + K_{h,i} \right) \underline{u}_{|\Gamma_i} + V_{h,i}^{-1} \underline{f}_{|\Gamma_i}$$

independently on the subdomains. Replacing the corresponding values in the second block equation, which describes the coupling across the boundaries, we get the BEM–Schur complement system

$$(2.4) \qquad \left[\left(\frac{1}{2}M_h^\top + K_h^\top \right) V_h^{-1} \left(\frac{1}{2}M_h + K_h \right) + D_h \right] \underline{u} = \underline{f}$$

with a symmetric and positive definite stiffness matrix S_h . Therefore we can use a preconditioned conjugate gradient method to solve (2.4) in parallel. We note that the action of S_h includes the inverse of the simple layer potential V_h. However, in (2.3) we need only the action of the local operators $V_{h,i}^{-1}$, this can be done by local preconditoned conjugate gradient methods.

Alternatively, for solving system (2.2) without these additional inner iterations we can use a result of [2] . Suppose, that we have given a symmetric and positive definite matrix C_V which is spectrally equivalent to V_h satisfying the inequalities

$$(2.5) \qquad \gamma_1 (C_V \underline{t}, \underline{t}) \leq (V_h \underline{t}, \underline{t}) \leq \gamma_2 (C_V \underline{t}, \underline{t})$$

for all vectors $\underline{t}$ of the structure according to (2.2). If $\gamma_1 > 1$, we can define a new inner product by

$$(2.6) \qquad \left[\begin{pmatrix} \underline{t} \\ \underline{u} \end{pmatrix}, \begin{pmatrix} \underline{\tau} \\ \underline{v} \end{pmatrix} \right] = ((V_h - C_V)\underline{t}, \underline{\tau}) + (\underline{u}, \underline{v})$$

and we have as a main result of [2] that the transformed matrix

$$(2.7)$$
$$A_h = \begin{pmatrix} C_V^{-1} V_h & -C_V^{-1}(\frac{1}{2}M_h + K_h) \\ (\frac{1}{2}M_h^\top + K_h^\top)C_V^{-1}(C_V - V_h) & D_h + (\frac{1}{2}M_h^\top + K_h^\top)C_V^{-1}(\frac{1}{2}M_h + K_h) \end{pmatrix}$$

now is self–adjoint and positive definite with respect to the newly defined product (2.6). Moreover, there hold spectral equivalence inequalities

$$(2.8)$$
$$\lambda_1 \left[R \begin{pmatrix} \underline{t} \\ \underline{u} \end{pmatrix}, \begin{pmatrix} \underline{t} \\ \underline{u} \end{pmatrix} \right] \leq \left[A_h \begin{pmatrix} \underline{t} \\ \underline{u} \end{pmatrix}, \begin{pmatrix} \underline{t} \\ \underline{u} \end{pmatrix} \right] \leq \lambda_2 \left[R \begin{pmatrix} \underline{t} \\ \underline{u} \end{pmatrix}, \begin{pmatrix} \underline{t} \\ \underline{u} \end{pmatrix} \right]$$

with

$$(2.9) \qquad R = \begin{pmatrix} I & 0 \\ 0 & D_h + (\frac{1}{2}M_h^\top + K_h^\top)V_h^{-1}(\frac{1}{2}M_h + K_h) \end{pmatrix} .$$

Therefore we can use a parallel preconditioned conjugate gradient method with respect to the transformed matrix (2.7) in the special inner product (2.6) to solve the original system (2.2) [10]. As preconditioner for the discrete Schur complement in R we use the BPS preconditioner, as proposed in [1].

3. Dirichlet Schur complement CG with hierarchical bases

In the methods described above it is necessary to discretize the local hypersingular operators D_i, which may be complicated. If we use only the first integral equation, we will get from (1.8) a new variational formulation:

Find $u \in H^{1/2}(\Gamma_S)$ and $t_i = \frac{\partial u}{\partial n}_{|\Gamma_i} \in H^{-1/2}(\Gamma_i)$ such that

$$(3.1) \qquad \begin{aligned} \sum_{i=1}^{p} a_i \langle t_i, v_{|\Gamma_i} \rangle_{\Gamma_i} &= 0 \\ \langle \tau_i, V_i t_i \rangle_{\Gamma_i} - \tfrac{1}{2}\langle \tau_i, u_{|\Gamma_i} \rangle_{\Gamma_i} - \langle \tau_i, K_i u_{|\Gamma_i} \rangle_{\Gamma_i} &= 0 \end{aligned}$$

for all test functions $v \in H_0^{1/2}(\Gamma_S)$ and $\tau_i \in H^{-1/2}(\Gamma_i)$ and $i = 1, \ldots, p$.

In analogy to (2.2) we can write the equivalent discrete system in the block form

$$(3.2) \qquad \begin{pmatrix} M_h^\top & 0 \\ V_h & \frac{1}{2}M_h + K_h \end{pmatrix} \begin{pmatrix} \underline{t} \\ \underline{u} \end{pmatrix} = \begin{pmatrix} 0 \\ \underline{f}_u \end{pmatrix} .$$

Using (2.3) we can replace $\underline{t}$ in the first block equation to get the discrete Dirichlet representation S_D of the Steklov–Poincaré operator

$$(3.3) \qquad M_h^\top V_h^{-1} \left(\frac{1}{2}M_h + K_h \right) \underline{u} = \underline{f} .$$

According to the properties of the symmetric representation of S_h in (2.4) we can symmetrize S_D. Therefore we can use a preconditioned conjugate gradient method to solve (3.3) in parallel.

If we use the BPS preconditioner, a transformation to a hierarchical basis is required to solve a coarse grid system to get the preconditioned values at the vertices. Here we use this idea already with respect to the ansatz functions. We introduce a coarse grid function

$$(3.4) \qquad u_H(x) = \sum_{k=1}^{q} u_k \varphi_k^H(x), \quad u_k = u(x_k) \text{ for all } x_k \in \omega,$$

to a given function $u(x)$. Then the resulting fine grid function $\tilde{u}(x) = u(x) - u_H(x)$ vanishes at all cross nodes $x_k \in \omega$. In general, this decomposition is not unique. Choosing discrete harmonic functions as basis functions in (3.4), the projection of a function $u(x)$ onto the coarse grid is unique. Such basis functions can be described as follows:

(i) $\varphi_k^H(x_l) = \delta_{kl}$ for all $x_l \in \omega$ and $k = 1, \ldots, q$,

(ii) $\varphi_k^H(x)$ are piecewise linear on all edges Γ_{ij},

(iii) $\Delta \varphi_k^H(x) = 0$ in all subdomains Ω_l, $l = 1, \ldots, p$.

Using the hierarchical bases constructed above the Galerkin discretization of (3.1) leads to the block system

$$(3.5) \qquad \begin{pmatrix} K_V & K_{VE} \\ K_{EV} & K_E \end{pmatrix} \begin{pmatrix} \underline{u}_V^H \\ \underline{u}_E^H \end{pmatrix} = \begin{pmatrix} \underline{f}_V \\ \underline{f}_E \end{pmatrix}$$

where every block matrix describes the action of the discrete Steklov–Poincaré operator (3.3) with respect to the corresponding basis functions. Finally we use a preconditioned conjugate gradient method to solve the Schur complement system to (3.5)

$$(3.6) \qquad \left[K_E - K_{EV} K_V^{-1} K_{VE} \right] \underline{u}_E^H = \underline{f},$$

which requires solving a coarse grid system in every iteration step.

4. Numerical results

As a simple example, we consider the problem (2.1) with a constant coefficient function $a(x) \equiv 1$ and a domain decomposition of the unit square into 16 subdomains. Method 1 is the symmetric Schur complement system (2.4). The conjugate gradient method with respect to the transformed matrix (2.7) is the method 2. Both cases result from the symmetric formulation including the full Calderon projector. In the other cases of the Dirichlet formulation it is necessary to solve local Dirichlet problems using the first boundary integral equation to realize the Steklov–Poincaré operator. Then we get method 3 with respect to the normal nodal bases and method 4 by using hierarchical bases.

All methods described above are iterative schemes which stop at a predefined bound with respect to a residual norm. For a relative error reduction of $\varepsilon = 10^{-6}$ we get the following results.

Table 1.

Nodes per	Symmetric Formulation		Dirichlet Formulation	
Subdomain	Method 1	Method 2	Method 3	Method 4
16	12	21	12	9
32	13	23	14	11
64	14	24	15	13
128	15	25	16	16
256	17	26	17	19

As preconditioners of the coupling nodes we use the BPS preconditioner. In all cases presented we need locally spectral equivalent matrices to the simple layer potentials $V_{h,i}$; here we use circulant matrices derived from the single layer potential with respect to a circle [11].

All the algorithms presented in this paper can be extended to more general problems, as e.g., to linear elasticity and, moreover, to three–dimensional problems.

All computations are executed on different parallel computer systems like the *Parsytec MC3*, the *Intel Paragon*, or at workstation clusters.

REFERENCES

1. J. H. Bramble, J. E. Pasciak and A. H. Schatz, *The construction of preconditioners for elliptic problems by substructuring* I, Math. Comput. **47** (1986), 103–134.
2. J. H. Bramble and J. E. Pasciak, *A preconditioning technique for indefinite systems resulting from mixed approximations of elliptic problems*, Math. Comput. **50** (1988), 1–17.
3. M. Costabel, *Symmetric methods for the coupling of finite elements and boundary elements*, Boundary Elements IX (C. A. Brebbia, G. Kuhn and W. L. Wendland eds.), Springer, Berlin, 1987, pp. 411–420.
4. M. Dryja, *A capacitance matrix method for Dirichlet problem on polygon region*, Numer. Math. **39** (1982), 51–64.
5. M. Dryja and O. B. Widlund, *Schwarz methods of Neumann–Neumann type for three–dimensional elliptic finite element problems*, Comm. Pure Appl. Math. (to appear).
6. R. Glowinski and M. F. Wheeler, *Domain decomposition and mixed finite element methods for elliptic problems*, First International Symposium on Domain Decomposition Methods for Partial Differential Equations (R. Glowinski et.al., eds.), SIAM, Philadelphia, 1988, pp. 144–172.
7. G. C. Hsiao and W. L. Wendland, *Domain decomposition in boundary element methods*, Fourth International Symposium on Domain Decomposition Methods for Partial Differential Equations (R. Glowinski et.al., eds.), SIAM, Philadelphia, 1991, pp. 41–49.
8. _____ *Domain decomposition via boundary element methods*, Numerical Methods in Engineering and Applied Sciences Part I (H. Alder et. al., eds.), CIMNE, Barcelona, 1992, pp. 198–207.
9. B. N. Khoromskij and W. L. Wendland, *Spectrally equivalent preconditioners for boundary equations in substructuring techniques*, East–West J. Numer. Math. **1** (1992), 1–25.
10. U. Langer, *Parallel iterative solution of symmetric coupled FE/BE–equations via domain decomposition*, Contemp. Math. **157** (1994), 335–344.
11. S. Rjasanow, *Effective iterative solution methods for boundary element equations*, Boundary Elements XIII (C. A. Brebbia and G. S. Gipson, eds.), CMP, Southampton, 1991, pp. 889–899.
12. A. H. Schatz, V. Thomée and W. L. Wendland, *Mathematical Theory of Finite and Boundary Element Methods*, Birkhäuser, Basel, 1990.
13. H. Yserentant, *On the multi–level splitting of finite element spaces*, Numer. Math. **49** (1986), 379–412.

DIPL.–MATH. O. STEINBACH, MATHEMATISCHES INSTITUT A, UNIVERSITÄT STUTTGART, PFAFFENWALDRING 57, 70569 STUTTGART, GERMANY
E-mail address: steinbach@mathematik.uni-stuttgart.de

Contemporary Mathematics
Volume 180, 1994

AN OVERDETERMINED SCHWARZ ALTERNATING METHOD

VICTOR H. SUN AND WEI-PAI TANG

ABSTRACT. In this paper, a new type of coupling on the artificial boundary layer is proposed for the classical SAM. The convergence rate is demonstrated for this algorithm. Numerical testing has been carried out for a variety of problems.

1. INTRODUCTION

It is well known that the rate of convergence of the Schwarz alternating method increases with the amount of overlap, yet large overlap is not desirable. One interesting problem is how to improve convergence speed without increasing overlap. Tang proposed the *generalized Schwarz Alternating method* (*GSAM*) a few years ago [7], which applies the Robin boundary condition, $\omega u + (1 - \omega)\frac{\partial u}{\partial n}$, on the artificial boundaries. A similar idea was presented by P. Lions [3]. Tang's work was generalized recently by Tan and Borsboom [6]. It was shown in [7] that rapid convergence can be achieved with the minimum overlap, if an optimal ω is chosen. The convergence rate is even faster than the classical SAM with a large overlap, but convergence can be a sensitive function of this parameter.

The *GSAM* applies the Robin condition only on the artificial boundaries. In this paper, we introduce an artificial boundary layer along the artificial boundary. The Robin condition is imposed on the entire layer. As we will demonstrate, this is beneficial to the convergence behavior of this approach.

In the next section, a new method, the *OSAM*, is introduced. The equivalent theorem and convergence behavior for the model problem is given in Section 3, and numerical results are shown in Section 4. Section 5 concludes the paper.

2. OVERDETERMINED SAM AND BOUNDARY LAYER RECONSTRUCTION

Consider the following boundary value problem:

$$(1) \qquad \begin{cases} L(u) &= f, \quad \Omega, \\ u &= g, \quad \partial\Omega, \end{cases}$$

1991 *Mathematics Subject Classification.* 65F10, 65N10.

This research was supported by the Natural Sciences and Engineering Research Council of Canada.

This paper is in final form and no version of it will be submitted for publication elsewhere.

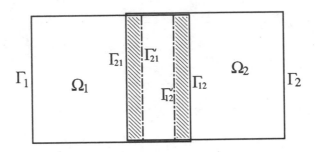

FIGURE 1. Artificial Boundary Layers

where L represents a general linear second order elliptic operator, and Ω is a bounded region in R^2. For simplicity, we consider only the two overlapping subdomain case (see Fig. 1). The generalization to an irregular solution domain or multi-subdomain problem is straightforward.

The rectangular solution region Ω is partitioned into two overlapping subdomains Ω_1 and Ω_2. Let

$$\Gamma_{21} = \partial\Omega_2 \cap \Omega_1, \ \Gamma_{12} = \partial\Omega_1 \cap \Omega_2, \ \Gamma_1 = \partial\Omega \cap \partial\Omega_1, \ \Gamma_2 = \partial\Omega \cap \partial\Omega_2.$$

We introduce two artificial boundary layers $\mathcal{L}_1$ and $\mathcal{L}_2$ (the shaded regions in Fig. 1), which are next to the corresponding artificial boundaries Γ_{12} and Γ_{21}. The thickness of these layers depends on the grid. When a uniform grid is employed, the corresponding thickness of the boundary layer will be the grid size h. For a general triangular mesh, the boundary layer will be the union of all triangles for which at least one of its edges or nodes is on the artificial boundary. The motivation for this choice is to allow the minimum overlap needed in the new algorithm. Let Γ'_{12} and Γ'_{21} denote the inner boundaries of the boundary layers $\mathcal{L}_1$ and $\mathcal{L}_2$, respectively.

The Robin condition is imposed on the entire boundary layer. Consequently, a new problem can be formulated as follows:

$$(2) \qquad \begin{cases} L(u_1) = f, \ \Omega_1, \\ u_1|_{\Gamma_1} = g, \\ b(u_1)|_{\mathcal{L}_1} = b(u_2)|_{\mathcal{L}_1}, \end{cases} \qquad , \qquad \begin{cases} L(u_2) = f, \ \Omega_2, \\ u_2|_{\Gamma_2} = g, \\ b(u_2)|_{\mathcal{L}_2} = b(u_1)|_{\mathcal{L}_2}, \end{cases}$$

where $b(u) = \omega u + (1-\omega)\frac{\partial u}{\partial n}$. There are many choices for the function ω. In this paper, we investigate only a very special case:

$$\omega = \begin{cases} 1, & (x,y) \in artificial\ boundary, \\ 0, & otherwise. \end{cases}$$

Problem (2) is overdetermined. In general, no solution may exist for an overdetermined problem, but in this case, the following result is trivially true:

Lemma 2.1. *If a solution of (1) exists and is unique, then a solution for problem (2) also exists and is unique.*

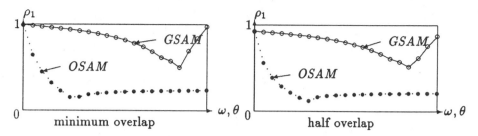

FIGURE 2. Maximum eigenvalues for 2-D seven subdomains

Actually, the solutions of (1) and (2) are identical. On the other hand, an iterative algorithm for solving problem (2) can easily be extended from the classical *SAM*.

To design a deterministic algorithm, reconstruction of the coupling on the boundary layer is necessary. When the problem is discretized, the constraint on the artificial boundary layer affects only the grid nodes on the boundary of the layer.

We observe that the solution on the inner side of the boundary layer satisfies both the difference operator and the derivative coupling between two solutions on the neighboring subdomain. Therefore, a natural approach is to apply a weighted combination of these two conditions. Namely, a linear combination of the difference operator and the Neumann condition is used to eliminate the extra constraint. In our numerical tests, the parameter $\theta(0 < \theta \leq 1)$ represents the weight for the difference operator part.

3. CONVERGENCE RESULT

Applying matrix analysis as in [7], the *OSAM* can be formulated as the solution to an enhanced matrix problem $\tilde{A}'x' = \tilde{b}'$ of $Ax = b$, which is the discretized form of (1)[5]. The convergence rate is then determined by the spectral radius of the Jacobian iterative matrix. For the model problem, we have the following results.

Theorem 3.1. *For the model problem, matrix A is equivalent (in the sense of [7]) to its enhanced OSAM matrix $\tilde{A}'$, for $0 \leq \theta \leq 1$.*

Figure 2 shows the maximum eigenvalues of the Jacobi iterative matrix as a function of the parameters. Detailed analysis can be found in [5]. It can be seen that the *OSAM* shows better convergence behavior than *GSAM* or *SAM* (which is the special case of the *GSAM* with $\omega = 1$). The sensitivity of the convergence rate for the *OSAM* to its parameter is much reduced. Moreover, *OSAM* with a minimum overlap is still much better than *SAM* with a half overlap.

4. NUMERICAL TESTS

Results for several test problems in the 2-D and 3-D cases are presented in this section. The differential equations are discretized by the standard central difference scheme. For each 2-D test problem, the solution domain is decomposed into a different number of subdomains. In all the cases, each subdomain contains a 20×20 grid and minimum overlap is considered. Domain decomposition is used as a preconditioner and Bi-CGSTAB is employed for the acceleration scheme. The convergence test is to require a residual reduction of 10^5.

In the tables of results, "SAM" represents the result for the traditional Schwarz alternating method. The "*" and "**" mean that the $OSAM$ does not converge within 100 and 200 iterations, respectively, and the iteration is stopped. "Iter" and "SubD" represent the number of linear iterations needed to reach the precision and the total number of subdomains, respectively. We also define the improvement factor to be $\tau = \frac{SAM - OSAM}{SAM}$. The notation τ_b represents the best improvement factor of the results.

4.1. Stress in helical spring. The first problem tested is [2]

$$\Phi_{xx} + \Phi_{yy} + \frac{3}{5 - y}\Phi_x = f, \ x \in \Omega,$$

where $\Omega = (-.5, .5) \times (-1, 1)$, and $\Phi = 0$ on the boundary. The problem has an exact solution [4]: $\Phi = (1 - y^2)(1 - 4x^2)(5 - y^3)(0.0004838y + 0.0010185)$.

The numerical results are shown in Table 1a. It can be seen that the $OSAM$ performs better than the SAM with only one exception of $\theta = 0.1$. For the optimal case, $\theta = 0.3$, $OSAM$ takes only half the number of iterations of SAM.

4.2. Discontinuous coefficient problem. In this test, the following equation was considered

$$(K_x u_x)_x + (K_y u_y)_y + u_x + u_y = \sin(\pi xy),$$

where $u = 0$ on the boundary of unit square, and

$$K_x = K_y = \begin{cases} 1, & [0, .5] \times [0, .5] \cup [.5, 1] \times [.5, 1], \\ 10^{+3}, & [.5, 1] \times [0, .5], \\ 10^{-3}, & [0, .5] \times [.5, 1]. \end{cases}$$

Stress in helical spring					*Discontinuous coefficients*			
SubD	8×12	6×8	3×4		SubD	10×10	8×8	4×4
θ	Iter	Iter	Iter		θ	Iter	Iter	Iter
0.1	68	42	18		0.1	*	*	*
0.2	17	12	7		0.2	87	63	*
0.3	14	11	6		0.3	16	15	10
0.4	16	13	6		0.4	16	12	8
0.5	17	14	6		0.5	13	11	7
0.6	17	13	6		0.6	16	11	7
0.7	18	14	7		0.7	17	12	7
0.8	20	15	7		0.8	19	11	9
0.9	20	15	7		0.9	16	11	9
1.0	19	14	9		1.0	17	12	10
SAM	29	20	10		*SAM*	31	22	13
τ_b	0.517	0.450	0.400		τ_b	0.548	0.500	0.462
a					*b*			

TABLE 1.

Results for indefinite problem

SubD	10×10		8×8		4×4	
θ	Iter		Iter		Iter	
	c=-20	c=-70	c=-20	c=-70	c=-20	c=-70
0.1	*	**	*	**	55	**
0.2	*	**	*	**	29	173
0.3	*	**	*	194	18	76
0.4	21	49	18	38	10	29
0.5	20	58	17	31	9	26
0.6	21	46	18	38	9	27
0.7	23	43	18	37	9	21
0.8	22	49	18	33	9	19
0.9	26	60	19	48	10	19
1.0	25	82	18	41	11	19
SAM	37	144	30	85	14	42
τ_b	0.459	0.713	0.433	0.665	0.357	0.548

TABLE 2.

Harmonic weighting is used at points of discontinuity in K_x, K_y. The results are shown in Table 1b. It can be seen that the improvement factor in this test is larger than that in the previous case. From the result of this test, it appears that $OSAM$ shows great potential for solving difficult problems. This will be further demonstrated by the following test.

4.3. Variable-coefficient, indefinite problem. The problem tested is

$$
\begin{aligned}
Lu = \quad & -[(1 + \frac{1}{2}\sin(50\pi x))u_x]_x - [(1 + \frac{1}{2}\sin(50\pi x)\sin(50\pi y))u_y]_y \\
& + 20\sin(10\pi x)\cos(10\pi y)u_x - 20\cos(10\pi x)\sin(10\pi y)u_y + cu
\end{aligned}
$$

(3)

where $u = \exp(xy)\sin(\pi x)\sin(\pi y)$ is defined on a unit square [1], for $c < 0$. In this work, the cases for $c = -20, -70$ were tested. We will see that the second case requires much more work than the first one for the same decomposition form.

The numerical results are reported in Table 2. From these results, the difficulty of this problem is obvious as compared with the corresponding iteration count of previous tests.

For $c = -20$, the problem is more weakly indefinite than for $c = -70$ with the same grid size. The results show that the work for $c = -20$ is only a little more than the corresponding situations in the last two tests. The improvement factor varies from one-third to one-half. However, when $c = -70$, $OSAM$ demonstrates its superiority over SAM. The improvement factors for all three decomposition cases are greater than one-half. This test further demonstrates that $OSAM$ performs better for difficult problems.

4.4. Helmholtz equation. Here we present the results for the Helmholtz equation $-\Delta u + u = f$ in 3-D case. The true solution is $u = \exp(xy)\sin(\pi z)$ in the cube. The solution domain is decomposed into 8 strips in z-direction. The total number of unknowns is 73^3.

Results for Helmholtz equation

θ	0.1	0.2	0.3	0.4	0.5	0.6	0.7	0.8	0.9	1.0	SAM
Iter	8	7	6	7	6	5	6	6	6	6	10

TABLE 3.

The numerical results are presented in Table 3. In the best case, $OSAM$ took only one half as many iterations as did SAM.

5. CONCLUSION

In this paper, a new extension of the classical SAM – $OSAM$ – is proposed. In this new approach, a stronger coupling is imposed on the artificial boundary layers. The superior convergence behavior is demonstrated for a variety of test problems. In particular, the weighted parameter θ does not have the sensitivity problem from which the $GSAM$ suffers. So far, our testing is restricted to a single level approach. A multilevel preconditioner approach is the natural future extension of this work.

REFERENCES

1. X.-C. Cai, W. Gropp, and D. Keyes, *A comparison of some domain decomposition algorithm for nonsymmetric elliptic problems*, Fifth International Symposium on Domain Decomposition Methods for Partial Differential Equations (Philadelphia, PA) (T. Chan, D. Keyes, G. Meurant, J. Scroggs, and R. Voigt, eds.), SIAM, 1992.
2. L. Collatz, *The numerical treatment of differential equations*, 3rd ed., Spring-Verlag, New York, 1966.
3. P. L. Lions, *On the Schwarz alternating method III: A variant for nonoverlapping subdomains*, Third International Symposium on Domain Decomposition Methods for Partial Differential Equations (Philadelphia, PA) (T. Chan, R. Glowinski, J. Periaux, and O. Widlund, eds.), SIAM, 1989, pp. 202–216.
4. J. Rice and R. Boisvert, *Solving elliptic problems using ELLPACK*, Spring-Verlag, New York, 1985.
5. Huosheng Sun and Wei-Pai Tang, *An Overdetermined Schwarz alternating method*, Tech. Report CS-93-53, University of Waterloo, Dept. of Computer Science, 1993.
6. K.H. Tan and M.J.A. Borsboom, *Problem-dependent optimization of flexible couplings in domain decomposition methods, with an application to advection-dominated problems*, Seventh International Symposium on Domain Decomposition Methods for Partial Differential Equations (University Park, PA) (D. Keyes and J. Xu, eds.), AMS, 1993.
7. W.-P. Tang, *Generalized Schwarz splitting*, SIAM J. Sci. Stat. Comp. 13 (1992), 573–595.

DEPT. OF APPL. MATH., UNIVERSITY OF WATERLOO, WATERLOO, ON, CANADA, N2L 3G1
E-mail address, V.H. Sun: hsun@yoho.uwaterloo.ca

DEPT. OF COMP. SCI., UNIVERSITY OF WATERLOO, WATERLOO, ON, CANADA, N2L 3G1
E-mail address, W.-P. Tang: wptang@lady.uwaterloo.ca

Contemporary Mathematics
Volume **180**, 1994

Domain Decomposition For Linear And Nonlinear Elliptic Problems Via Function Or Space Decomposition

XUE-CHENG TAI

ABSTRACT. In this article, we use a function decomposition method and a space decomposition method of [**5**] to derive some parallel overlapping and nonoverlapping domain decomposition methods for self–adjoint linear and nonlinear elliptic problems. The function decomposition method and the space decomposition method use different starting points in doing the domain decomposition.

1. The function decomposition and space decomposition methods

In [**5**], function decomposition and space decomposition methods were proposed for a general convex programming problem. Here, we shall briefly show how we can use the methods for overlapping and nonoverlapping domain decomposition methods for linear and nonlinear elliptic problems.

It was shown, see [**5**], that by suitably decomposing the energy function for an elliptic problem, we can derive the classical Alternating Direction methods, see [**1, 4**]. By using different function decompositions, we can also derive some nonoverlapping domain decomposition methods for these problems. This shows that the Alternating Direction methods and the domain decomposition methods are just different ways of decomposing a problem, or in other words that we can use the splitting methods to get domain decomposition methods for some linear and nonlinear elliptic and parabolic problems.

The concept of space decomposition was first introduced in a review paper [**8**]. There many multigrid and domain decomposition methods are presented and analyzed. It is known that the overlapping domain decomposition methods,

1991 *Mathematics Subject Classification.* 65K10, 65N55, 65Y05.
Key words and phrases. Parallel algorithm, domain decomposition, function decomposition, space decomposition, nonlinear problems.
This paper is in final form and no version of it will be submitted for publication elsewhere.
This work is supported by the University of Bergen, Norway.

the substructuring methods, [2], and the multilevel methods, [6, 7], give some nice ways to decompose the finite element spaces. In the published papers, their convergence behaviour has been carefully analyzed for linear problems. By using the space decomposition approach of [5], we try to show that if these methods can be used for linear problems, they can also be used for some nonlinear problems.

First, let us recall the results of [5]. We consider minimization

$$(1) \qquad \min_{v \in K} F(v), \qquad K \subset V.$$

In case of function decomposition, we need to assume:

(F1). The space V is a Hilbert space and there exist Hilbert spaces V_i, $i = 1, 2, \cdots, m$, such that $V = \cap_{i=1}^m V_i$.

(F2). The function $F : V \mapsto R$ is convex, lower–semicontinuous in V and there exist convex, lower–semicontinuous functions $F_i : V_i \mapsto R$ in V_i, $i = 1, 2, \cdots, m$, such that $F(v) = \sum_{i=1}^m F_i(v)$, $\forall v \in V$.

(F3). The subset K is closed and convex in the norm of V. There exist convex subsets $K_i \subset V_i$, $i = 1, 2, \cdots, m$, such that $K = \cap_{i=1}^m K_i$.

(F4). There exists a Hilbert space H such that $V \subset V_i \subset H$, $i = 1, 2, \cdots, m$.

In case of space decomposition, we need to assume:

(S1). The space V is a reflexive Banach and there exist reflexive Banach spaces V_i, $i = 1, 2, \cdots, m$, such $V = V_1 + V_2 + \cdots + V_m$.

(S2). The subset K is closed and convex in the norm of V. There exist closed and convex subsets K_i of V_i such that $K = K_1 + K_2 + \cdots + K_m$.

(S3). The function $F(v)$ is convex, lower-semicontinuous in the norm of V and satisfies $\lim_{\|v\|_V \to +\infty} \frac{F(v)}{\|v\|_V} = +\infty$.

(S4). There exist constants C_0, C_1 such that $C_0 \left\| \sum_{i=1}^m v_i \right\|_V^2 \leq \sum_{i=1}^m \|v_i\|_V^2$, $\forall v_i \in V_i$, $i = 1, 2, \cdots, m$, and

$$\begin{cases} \forall v \in V, \ \exists v_i \in V_i, i = 1, 2, \cdots, m, \ \text{such that} \\ \displaystyle\sum_{i=1}^m v_i = v \ \text{and} \ \sum_{i=1}^m \|v_i\|_V^2 \leq C_1 \|v\|_V^2 . \end{cases}$$

Under (F1)–(F4), we find that the minimization (1) is equivalent to

$$(2) \qquad \min_{\substack{(v_1, v_2, \cdots v_m) \in \prod_{i=1}^m K_i \\ v_1 = v_2 = \cdots = v_m}} \sum_{i=1}^m F_i(v_i).$$

This is a minimization of a separable structure under the extra constraint $v_1 = v_2 = \cdots = v_m$. In order to use a parallel method, we need to introduce a new variable v and realize the above constraint by enforcing $v_i = v, i = 1, 2, \cdots, m$. We will use augmented Lagrangian methods to deal with it. We define L_r on $H \times \prod_{i=1}^m V_i \times H^m$ by

$$L_r(v, v_i, \mu_i) = \sum_{i=1}^m F_i(v_i) + \frac{1}{m} \sum_{i=1}^m (\mu_i, v_i - v)_H + \frac{r}{2m} \sum_{i=1}^m \|v_i - v\|_H^2 .$$

We will seek a saddle point for L_r over $H \times \prod_{i=1}^m K_i \times H^m$. We say (u, u_i, λ_i) is a saddle point if

$$L_r(u, u_i, \mu_i) \le L_r(u, u_i, \lambda_i) \le L_r(v, v_i, \lambda_i), \quad \forall v \in H, v_i \in K_i, \mu_i \in H .$$

It is easy to prove that, if (u, u_i, λ_i) is a saddle point for L_r, then u is a minimizer for (1). Under (F1)–(F4), we get the following parallel algorithm for (1):

ALGORITHM 1.

Step 1. *Choose initial values $u_i^0 \in K_i$ and $\lambda_i^0 \in H$ $(i = 1, 2, \cdots m)$, and positive numbers $r > 0$, and $\rho \in (0, \frac{1+\sqrt{5}}{2})r$.*

Step 2. *For $n \ge 1$, set*

$$u^n = \frac{1}{m} \sum_{i=1}^m u_i^{n-1} + \frac{1}{rm} \sum_{i=1}^m \lambda_i^{n-1} .$$

Step 3. *Find $u_i^n \in K_i$, $i = 1, 2, \cdots, m$ in parallel such that*

(3)
$$F_i(u_i^n) + \frac{1}{m}(\lambda_i^{n-1}, u_i^n)_H + \frac{r}{2m}\|u_i^n - u^n\|_H^2$$
$$\le F_i(v_i) + \frac{1}{m}(\lambda_i^{n-1}, v_i)_H + \frac{r}{2m}\|v_i - u^n\|_H^2 , \quad \forall v_i \in K_i .$$

Step 4. *Update the multipliers and go to step 2: $\lambda_i^n = \lambda_i^{n-1} + \rho(u_i^n - u^n)$.*

The following theorem (see [5]) shows the convergence:

THEOREM 1. *Suppose L_r has a saddle point over $H \times \prod_{i=1}^m K_i \times H^m$. There exists a unique solution u_i^n such that (3) is satisfied. If conditions (F1)–(F4) are valid, and F_i is Gateaux differentiable with the inner product of H, then we have estimate:*

$$\sum_{i=1}^N \sum_{i=1}^m (F_i'(u_i^n) - F_i'(u), u_i^n - u)_H \le \frac{C_2 m}{2\rho} + \frac{C_3 r}{2m}, \quad \forall N > 0 .$$

The constants C_2 and C_3 depend only on the initial functions λ_i^0, u_i^0 and the solution u of (1).

Next, we discuss the space decomposition. Under conditions (S1)–(S2), we can see that, if $(u_1, u_2, \cdots, u_m)$ is a minimizer for

(4)
$$\min_{(v_1, v_2, \cdots v_m) \in \prod_{i=1}^m K_i} F(v_1 + v_2 \cdots + v_m) ,$$

then $\sum_{i=1}^m u_i$ is a minimizer for (1). We use Jacobi method to find a solution for the minimization (4):

ALGORITHM 2.

Step 1. *Choose $u_i^0 \in K_i$ and relaxation parameters $\alpha_i > 0$ such that $\sum_{i=1}^m \alpha_i \le 1$.*

Step 2. *For $n \ge 1$, find $u_i^{n+\frac{1}{2}} \in K_i$ in parallel for $i = 1, 2, \cdots, m$ such that*

$$F\left(\sum_{k=1, k \ne i}^m u_k^n + u_i^{n+\frac{1}{2}}\right) \le F\left(\sum_{k=1, k \ne i}^m u_k^n + v_i\right) , \quad \forall v_i \in K_i .$$

Step 3. *Set u_i^{n+1} as: $u_i^{n+1} = u_i^n + \alpha_i(u_i^{n+\frac{1}{2}} - u_i^n)$, and go to step 2.*

For this algorithm, we have the following convergence result (see [5]):

THEOREM 2. *Under conditions (S1)–(S4), we assume each K_i is a bounded subset in V or $K_i = V_i$, function F is Gateaux differentiable and locally uniformly convex over bounded subsets in V and F' is uniformly continuous over bounded subsets in V, then we have for Algorithm 2 the convergence*

$$u^{n+1} = \sum_{i=1}^{m} u_i^{n+1} \to u \text{ strongly in } V \text{ as } n \to \infty .$$

As was observed in [8], the Jacobi method may not converge for general space decomposition problems. Here, by using a suitable under relaxation, we get the sufficient condition of convergence even for general minimization problems. In case that F' is Lipschitz continuous and coercive, an error estimate in a weak form was proved in [5], which shows the dependence of the convergence on constant C_1/C_0.

2. Applications to domain decomposition

Let us consider the model problem:

$$(5) \qquad \min_{v \in W_0^{1,s}(\Omega)} \left(\int_{\Omega} \left(\frac{1}{s} |\nabla v|^s - fv \right) dx \right) .$$

We assume $s \geq 2$. If $s = 2$, it represents a typical self–adjoint linear elliptic equation; if $s \neq 2$, it is a nonlinear elliptic equation. We will restrict our consideration only to the discrete case. As in Glowinski and Marrocco [3], if we replace the Sobolev space $W_0^{1,s}$ by a finite element space and carry out the minimization of (5) over it, the finite element solution will converge to the minimizer of (5).

Assume Ω has been partitioned into finite elements $\mathcal{T}_h$ and the union of the finite elements form a discrete domain Ω_h. Let us define S_h as the nonconforming finite element space and V_h as the conforming finite element space of kth order polynomials, i.e.

$$S_h = \{v_h \mid \quad v_h \in P_k(e), \forall e \in \mathcal{T}_h, v_h = 0 \text{ on } \partial \Omega_h\} ,$$
$$V_h = \{v_h \mid \quad v_h \in C^0(\Omega), v_h \in P_k(e), \forall e \in \mathcal{T}_h, v_h = 0 \text{ on } \partial \Omega_h\} .$$

We define the inner product of S_h and V_h as $(u,v)_{S_h} = \sum_{e \in \mathcal{T}_h} (u,v)_{H^1(e)}$. The discrete version of (5) is:

$$(6) \qquad \min_{v_h \in V_h} \sum_{e \in \mathcal{T}_h} \int_e \left(\frac{1}{s} |\nabla v_h|^s - fv_h \right) dx .$$

We assume Ω_h has been partitioned into nonoverlapping subdomains Ω_1 and Ω_2, and each subdomain is the union of some elements of $\mathcal{T}_h$. This does not limit us to two parallel processors, because each Ω_i can again contain many disjoint subdomains. Here we will consider only the case that each Ω_i is a single connected subdomain. We will report elsewhere on the case that each Ω_i contains many disjoint subdomains. In order to use our algorithm, let us take $m = 2$, $V_1 = V_2 = V = S_h$, $H = S_h$, $K = V_h$, and

$$F_{h,i}(v_h) = \sum_{e \in T_h \cap \Omega_i} \int_e \left(\frac{1}{s} |\nabla v_h|^s - f v_h \right) dx, \forall v_h \in S_h, i = 1, 2.$$

In order to satisfy (F3), we extend each subdomain Ω_i to a larger subdomain O_i, $i = 1, 2$, and each O_i is the union of some elements of T_h. The subdomains Ω_i are nonoverlapping subdomains, while the subdomains O_i will overlap with each other. We define $K_i = \{v_h | \quad v_h \in P_k, \forall e \in T_h, v_h \in C^0(O_i)\}$. If O_1 and O_2 overlap suitably, we can have $K = K_1 \cap K_2$. Thus, the assumptions (F1)–(F4) are all satisfied and $F'_{h,i}$ satisfies [3]:

$$(F'_{h,i}(v_{h,1}) - F'_{h,i}(v_{h,2}), v_{h,1} - v_{h,2})_{S_h} \geq \alpha \int_{\Omega_i} |\nabla(v_{h,1} - v_{h,2})|^s dx, i = 1, 2.$$

Therefore, we get the following convergent algorithm for (6) from Algorithm 1.

ALGORITHM 3.

Step 1. *Choose initial values and constants* r, ρ. *For* $n \geq 1$, *set*

$$u_h^n = \frac{1}{2}(u_{h,1}^{n-1} + u_{h,2}^{n-1}) + \frac{1}{2r}(\lambda_{h,1}^{n-1} + \lambda_{h,2}^{n-1}) .$$

Step 2. *Solve* $u_{h,1}^n \in H^1(O_1), u_{h,2}^n \in H^1(O_2)$ *in parallel from:*

$$(|\nabla u_{h,1}^n|^{s-2}\nabla u_{h,1}^n, \nabla v_h)_{L^2(\Omega_1)} + \frac{1}{2} \sum_{e \in T_h \cap O_1} (\lambda_{h,1}^{n-1}, v_h)_{H^1(e)}$$

$$(7) \qquad + \frac{r}{2} \sum_{e \in T_h \cap O_1} (u_{h,1}^n - u_h^n, v_h)_{H^1(e)} = (f, v_h)_{L^2(\Omega_1)}, \quad \forall v_h \in V_h ,$$

$$(|\nabla u_{h,2}^n|^{s-2}\nabla u_{h,2}^n, \nabla v_h)_{L^2(\Omega_2)} + \frac{1}{2} \sum_{e \in T_h \cap O_2} (\lambda_{h,2}^{n-1}, v_h)_{H^1(e)}$$

$$(8) \qquad + \frac{r}{2} \sum_{e \in T_h \cap O_2} (u_{h,2}^n - u_h^n, v_h)_{H^1(e)} = (f, v_h)_{L^2(\Omega_2)}, \quad \forall v_h \in V_h .$$

We obtain the value of $u_{h,1}^n$ *in* $\Omega \backslash O_1$, *and the value of* $u_{h,2}^n$ *in* $\Omega \backslash O_2$ *through:*

$$u_{h,1}^n = u_h^n - \frac{1}{r}\lambda_{h,1}^{n-1} \text{ in } \Omega \backslash O_1, \qquad u_{h,2}^n = u_h^n - \frac{1}{r}\lambda_{h,2}^{n-1} \text{ in } \Omega \backslash O_2 .$$

Homogeneous Dirichlet boundary conditions should be enforced on $\partial\Omega$ *for (7) and (8).*

Step 3. *Update the multipliers as:* $\lambda_{h,i}^n = \lambda_{h,i}^{n-1} + \rho(u_{h,i}^n - u_h^n)$, *in* $\Omega, i = 1, 2$, *and go to step 2.*

In [5] several other algorithm were also obtained for (5) and other linear self-adjoint elliptic problems.

Next, we use overlapping domain decomposition for (6). As before, we assume we have partitioned Ω into finite elements. We then decompose Ω_h into overlapping subdomains and each subdomain is the union of some elements of

$\mathcal{T}_h$. We assume that the subdomains can be marked by m colors, so that the subdomains with the same color do not intersect with each other. We denote the union of the subdomains with the ith color as Ω_i. Let us take

$$V_{h,i} = \left\{ v_h \mid \quad v_h \in C^0(\Omega), v_h \in P_k, \forall e \in \mathcal{T}_h \cap \Omega_i, v_h = 0 \text{ in } \Omega_h \backslash \Omega_i \text{ and on } \partial\Omega_h \right\} .$$

If the subdomains overlaps suitably, we will have $V_h = \sum_{i=1}^m V_{h,i}$, and the constants C_0, C_1 can be explicitly estimated, see [2, 9]. For simplicity, we define for Algorithm 2:

$$u^n = \sum_{i=1}^m u_i^n, \quad w_i^{n+1} = \sum_{k=1, k \neq i}^m u_k^{n+1} + u_i^{n+\frac{1}{2}} = u^n - u_i^n + u_i^{n+\frac{1}{2}}, \forall i, n .$$

We get from Algorithm 2 the following overlapping algorithm for (6):

ALGORITHM 4.

Step 1. *Choose* $u_{h,i}^1 \in V_{h,i}$ *and constants* $\alpha_i > 0$ *such that* $\sum_{i=1}^m \alpha_i \leq 1$.

Step 2. *For* $n \geq 1$, *solve in parallel in each subdomain* Ω_i *the following problem:*

$$\begin{cases} (|\nabla w_{h,i}^{n+1}|^{s-2} \nabla w_{h,i}^{n+1}, \nabla v_h)_{L^2(\Omega_i)} = (f, v_h)_{L^2(\Omega_i)}, \qquad \forall v_h \in V_{h,i} , \\[2mm] w_{h,i}^{n+1} = 0 \text{ on } \partial\Omega_i \cap \partial\Omega_h , \quad w_{h,i}^{n+1} = u_h^n = \sum_{k=1, k \neq i}^m u_{h,k}^n \text{ on } \partial\Omega_i \backslash \partial\Omega_h . \end{cases}$$

Step 3. *Set* $u_{h,i}^{n+1}$ *as:* $u_{h,i}^{n+1} = u_{h,i}^n + \alpha_i(w_{h,i}^{n+1} - u_h^n)$ *in* $\Omega_i, i = 1, 2, \cdots m$, *and go to the next iteration.*

REFERENCES

1. J. Douglas and H. H. Rachford, *On the numerical solution of heat conduction problems in two and three space variables*, Trans. Amer. Math. Soc. . **82** (1956), 421–439.
2. M. Dryja and O. Widlund, *Multilevel additive methods for elliptic finite element problems*, Parallel Algorithms for Partial Differential Equations (W. Hackbush, ed.), Proceeding of the 6th GAMM seminar, Kiel, Jan. 19–21, 1990, Vieweg & Sons, Braunchweig, 1991, pp. 58–69.
3. R. Glowinski and A. Marrocco, *Sur l'approximation par éléments finis d'ordre un, et lan résolution par pénalisation-dualité, d'une classe de problémes de Dirichlet non linéaires*, Rev. Fr. Autom. Inf. Rech. Oper. Anal. Numér. R-2 (1975), 41–76.
4. D. H. Peaceman and H. H. Rachford, *The numerical solution of parabolic and elliptic differential equations*, SIAM J. Appl. Math. **3** (1955), 24–41.
5. X.-C. Tai, *Parallel function and space decomposition methods with applications to optimization, splitting and domain decomposition*, Preprint No. 231-1992, Institut für Mathematik, Technische Universität Graz (1992).
6. H. Yserentant, *On the multilevel splitting of finite element spaces*, Numer. Math. **49** (1986), 379-412.
7. J. C. Xu, *Theory of multilevel methods. Doctoral thesis. Cornell, Rep. AM-48, Penn. State U.* (1989).
8. _____, *Iteration methods by space decomposition and subspace correction jour SIAM Rev.* **34** (1992).
9. X. J. Zhang, *Multilevel Schwarz methods*, Numer. Math. **63** (1992), 521-539.

DEPARTMENT OF MATHEMATICS, UNIVERSITY OF BERGEN, ALLEGT 55, 5007, BERGEN, NORWAY.

E-mail address: Tai@mi.uib.no.

Contemporary Mathematics
Volume 180, 1994

ELLAM-Based Domain Decomposition and Local Refinement Techniques for Advection-Diffusion Equations with Interfaces

H. WANG, H.K. DAHLE, R.E. EWING, T. LIN, AND J.E. VÅG

ABSTRACT. We combine Eulerian-Lagrangian localized adjoint methods (EL-LAM) with domain decomposition and local refinement techniques to develop two nonoverlapping iterative schemes for advection-diffusion equations with various physical/numerical interfaces.

1. Introduction

Advection-diffusion equations arise from many important applications and often present serious numcrical difficulties. Most numerical methods exhibit some combination of excessive numerical dispersion or nonphysical oscillation. Moreover, practical advection-diffusion problems often have various interfaces that introduce extra difficulties. Physical interfaces arise from the modeling of transport processes in composite media and lead to advection-diffusion equations with discontinuous coefficients. Numerical interfaces arise from the application of domain decomposition and local refinement techniques. An identifying feature of groundwater contaminant transport and many other applications is the presence of large scale fluid flows coupled with transient transport of physical quantities such as pollutants, chemical species, radionuclides, and temperature, which are generally smooth outside some small regions and may have sharp fronts inside where important chemistry and physics take place. An extremely fine global mesh in both space and time is impossible due to the excessive computational

1991 *Mathematics Subject Classification.* 65M50, 65M55, 65M60.

Key words and phrases. Lagrangian methods, domain decomposition, grid refinement.

This research was supported in part by DOE, DE–AC05–840R21400, Martin Marietta, Subcontract, SK965C and SK966V, by NSF Grant No. DMS–8922865, by funding from the Institute for Scientific Computation at Texas A&M University, and by funding from the Norwegian Research Council, RV/412.92/003.

This paper is in final form and no version of it will be submitted for publication elsewhere.

cost. A feasible approach is to apply domain decomposition and local refinement techniques by partitioning the global domain into a number of subdomains and solving the problems with fine meshes in both space and time within the sharp front regions (subdomains) and coarse meshes outside (other subdomains). This way, both accuracy and efficiency can be guaranteed, but at a cost of introducing numerical interfaces between different subdomains.

Many domain decomposition and local refinement techniques have been developed for elliptic and parabolic equations, but it is more difficult to develop these techniques for advection-dominated equations. In this case, locally generated errors at the interfaces can be propagated into the domain so that the overall accuracy is decreased; improper treatment of the interfaces might destroy the stability of the numerical methods. Most existing methods for interface problems for advection-dominated equations employ the Eulerian approach and often yield numerical solutions with some combination of excessive numerical dispersion or oscillation. Extremely small time steps have to be used to maintain the accuracy and stability of these methods. While Eulerian-Lagrangian methods can overcome these problems to some extent, they cannot treat general boundary conditions and are therefore difficult to implement for interface problems for advection-dominated equations.

Eulerian-Lagrangian localized adjoint methods (ELLAM) [1, 2] (and the references cited there) have been successfully applied to solve advection-dominated equations and have yielded numerical solutions free of oscillations or numerical dispersion. ELLAM maintain mass conservation and treat boundary conditions systematically. In this paper we present two types of ELLAM-based decomposition and local refinement (in both space and time) techniques for the interface problems for the following one-dimensional model equation

$$(1) \qquad \mathcal{L}u \equiv u_t + (V(x,t)u - D(x,t)u_x)_x = f(x,t), \ x \in (a,b), \ t \in (0,T].$$

The boundary conditions at $x = a$ and $x = b$ can be Dirichlet, Neumann, or flux conditions, and different types of boundary conditions may be specified at $x = a$ and $x = b$. In addition, an initial condition is needed to close the system.

In a physical interface case, $V(x,t)$ and $D(x,t)$ are smooth except for the interface $x = d$ where either $V(x,t)$ or $D(x,t)$ or both have the first type of discontinuity with respect to x. Then, (1) is closed by the interface conditions:

$$(2) \qquad \begin{aligned} u(d-,t) &= u(d+,t), & t &\in [0,T], \\ (Vu - Du_x)(d-,t) &= (Vu - Du_x)(d+,t), & t &\in [0,T]. \end{aligned}$$

Numerical interfaces arise when different meshes are imposed over different subdomains. We may have both physical and numerical interfaces at the same locations. For simplicity, only one interface $x = d$ has been assumed to be present; generalization to several interfaces is straightforward. Also, $V(x,t)$ is assumed to be positive, so $x = a$ and $x = b$ are always the inflow and outflow boundaries, respectively.

2. An ELLAM Scheme

In this section we present an ELLAM scheme for equation (1) with smooth coefficients. Let I and N be two positive integers. We define the partitions of space and time as $x_i = a + i\Delta x$ $(i = 0, 1, \ldots, I)$ and $t^n = n\Delta t$ $(n = 0, 1, \ldots, N)$. In the numerical scheme, we use a time-marching algorithm to solve (1). We consider space-time test functions w that vanish outside of $[a, b] \times (t^n, t^{n+1}]$ and are discontinuous in time at each time level t^n. With these test functions, we can write the weak form for equation (1) as

$$
\begin{aligned}
(3) \quad & \int_a^b u(x, t^{n+1})w(x, t^{n+1})dx + \int_{t^n}^{t^{n+1}} \int_a^b Du_x w_x \, dx \, dt \\
& + \int_{t^n}^{t^{n+1}} (Vu - Du_x)w|_a^b \, dt - \int_{t^n}^{t^{n+1}} \int_a^b u(w_t + Vw_x)dx \, dt \\
& = \int_a^b u(x, t^n)w(x, t_+^n)dx + \int_{t^n}^{t^{n+1}} \int_a^b fw \, dx \, dt,
\end{aligned}
$$

where $w(x, t_+^n) = \lim_{t \to t_+^n} w(x, t)$.

Based on the ideas of the localized adjoint method and the Lagrangian nature of equation (1), we define the test functions w to be the standard hat functions at the time t^{n+1} (or at the outflow boundary) and to be constant along the approximate characteristics from t^{n+1} (or the outflow boundary) to the time t^n (or the inflow boundary), and to be discontinuous in time at time level t^n [1, 2].

Putting the test functions w above into the variational form (3), we can derive an ELLAM formulation. The first terms on both the left-hand and the right-hand sides of equation (3) are already defined at time levels t^{n+1} and t^n, respectively. The last term on the right-hand side of (3) is the source term that can be computed directly. The last term on the left-hand side of (3) measures the errors of the characteristic tracking and is negligible. (In fact, it vanishes when we track the characteristics exactly.) The inflow and outflow boundary conditions are naturally incorporated into (3) by the third term on the left-hand side of (3) when the test functions are not zero at the boundaries. Applying a one-point backward Euler quadrature to the second term on the left-hand side of (3) reduces this term to a term at the time t^{n+1} as well as terms at the inflow and outflow boundaries. Thus, with the known solution at time t^n as well as the inflow and outflow boundary conditions, our ELLAM scheme yields the numerical solution at time t^{n+1}, and the numerical solution at the outflow boundary for outflow Neumann/flux boundary conditions, or the total flux at the outflow boundary for the outflow Dirichlet boundary condition [1].

3. Generalized ELLAM Schemes for Interface Problems

3.1. Overview. In this section, we present two types of ELLAM-based domain decomposition and local refinement techniques to solve equation (1) with physical or numerical interfaces or both.

We first demonstrate the ideas by recalling the ELLAM-based scheme for the interface problems for first-order advection-reaction equations [2], which is an extreme case of equation (1) in that the diffusion coefficient $D(x,t)$ vanishes. Therefore, only an inflow Dirichlet boundary condition is needed at $x = a$, and no outflow boundary condition should be specified at $x = b$. In contrast to many existing methods that have difficulties in solving these problems, ELLAM can naturally be applied to do so. In fact, the ELLAM scheme presented in the last section is valid when $D(x,t)$ vanishes and yields the numerical solution at time t^{n+1} and at the outflow boundary $x = b$ with the given inflow boundary condition at $x = a$ and the known numerical solution at time t^n. The corresponding interface condition reduces to the second equation in (2), which imposes the continuity requirement on the advective flux across the interface $x = d$. Applying the ELLAM scheme to the advection-reaction equation on (a,d) yields the solution at time t^{n+1} and the left-limit of the solution at the interface $x = d$. Then, the interface condition gives the right-limit of the solution at the interface $x = d$. With this right-limit as the inflow boundary condition at $x = d$ and the known solution at time t^n, apply the ELLAM scheme to solve the advection-reaction equation on (d,b). Moreover, the spatial and temporal meshes used in $(a,d) \times [0,T]$ and $(d,b) \times [0,T]$ are independent of each other. Therefore, a noniterative and nonoverlapping ELLAM-based domain decomposition and local refinement technique is naturally derived, which treats various (physical/numerical) interface problems for advection-reaction equations in a universal way, and fully utilizes the intrinsic physics behind them.

3.2. A Dirichlet-Flux Algorithm. In this section we develop ELLAM-based domain decomposition and local refinement techniques for the interface problems for equation (1). Due to the effect of the diffusion term, the downstream values of the solutions also affect their upstream values. Thus, an iterative procedure should be used. Among other questions in the development of the schemes are the following: Which types of outflow and inflow boundary conditions should be imposed at the interface $x = d$ to close (1) over the subdomains (a,d) and (d,b)? What values should be chosen for these boundary conditions? Our studies [1] show the following observations. The numerical solutions with inflow flux/Dirichlet boundary conditions are very accurate, while an inflow Neumann condition is not physically reasonable and the corresponding solutions are not so accurate as those with flux/Dirichlet conditions. An outflow Neumann boundary condition is physically reasonable and the corresponding numerical solutions are accurate. For an outflow Dirichlet condition, the numerical solutions are accurate if a "right" value is specified, and may not be accurate otherwise because boundary layers will arise in this case. An outflow flux condition is not numerically stable and should be avoided.

Based on these observations, we propose a Dirichlet-flux iterative scheme for interface problems of equation (1). With the known solution $u(x,t^n)$ at time t^n as well as the inflow and outflow boundary conditions at $x = a$ and $x = b$, we compute the solution $u(x, t^{n+1})$ by the following procedure:

(i) Choose an initial guess $u(d_-, t) = u(x_d^*(t), t^n)$ for $t \in [t^n, t^{n+1}]$, where $x_d^*(t)$ is the foot of the (approximate) characteristics emanating backward from (d, t) at the interface.

(ii) With the given inflow boundary condition at $x = a$ and $u(d_-, t)$ as the outflow Dirichlet boundary condition at $x = d$, use the ELLAM scheme to solve (1) on (a, d) and obtain the solution $u(x, t^{n+1})$ at time t^{n+1} and the total flux $(Vu - Du_x)(d_-, t)$ for $t \in [t^n, t^{n+1}]$ at $x = d$.

(iii) The second equation in (2) gives $(Vu - Du_x)(d_+, t) = (Vu - Du_x)(d_-, t)$, $t \in [t^n, t^{n+1}]$.

(iv) With $(Vu - Du_x)(d_+, t)$ as the inflow flux condition at $x = d$ and the given outflow boundary condition at $x = b$, use the ELLAM scheme to solve (1) on (d, b) and obtain the solution $u(x, t^{n+1})$ at t^{n+1} as well as $u(b, t)$ or the total flux at $x = b$ depending on which type of boundary condition was specified at $x = b$.

(v) Compute $u(d_+, t)$ by projecting $u(x, t^{n+1})$ back along the (approximate) characteristics or by a linear interpolation of the solution u with its values at t^n and t^{n+1} along the (approximate) characteristics.

(vi) The first equation in (2) yields $u(d_-, t) = u(d_+, t)$. Go back to Step (ii) and repeat the process until the algorithm converges.

Since the exact solution of equation (1) is smooth along the characteristics, the initial guess chosen in Step (i) provides a "right" value for the Dirichlet condition at the interface $x = d$. Then, the ELLAM scheme yields the total flux $(Vu - Du_x)(d_-, t)$. Applying the second equation in (2) generates a most desirable inflow flux condition $(Vu - Du_x)(d_+, t)$ (Step (iii)) at $x = d$ for equation (1) on (d, b), which guarantees a full mass conservation when we move from (a, d) to (d, b). The continuity of the solutions is imposed in Step (vi) when we move back from (d, b) to (a, d).

3.3. A Neumann-Dirichlet Algorithm. A major concern for the Dirichlet-flux algorithm presented in the last part is that Step (ii) in the algorithm might introduce some potential error because we computed the flux out of a Dirichlet condition. In this section, we propose an alternative Neumann-Dirichlet iterative scheme. Since a Neumann condition is most appropriate at the outflow boundary, we impose an outflow Neumann condition at $x = d$ for equation (1) on (a, d). While an initial guess $u_x(d_-, t) = u_x(x_d^*(t), t^n)$ for $t \in [t^n, t^{n+1}]$ can be chosen, it is one-order less accurate than the solution itself due to the numerical differentiation involved. Some post-processing techniques can be used to enhance the accuracy. That is, choose $u_x(d_-, t) = (\mathcal{P}u)_x(x_d^*(t), t^n)$ for $t \in [t^n, t^{n+1}]$, where $\mathcal{P}u$ is a post-processed solution obtained from u. Our Neumann-Dirichlet algorithm can be presented as follows:

(i) Choose an initial guess $u_x(d_-, t) = (\mathcal{P}u)_x(x_d^*(t), t^n)$ for $t \in [t^n, t^{n+1}]$.

(ii) With the prescribed inflow boundary condition at $x = a$ and $u_x(d_-, t)$ as the outflow boundary condition at $x = d$, apply the ELLAM scheme to solve (1) on (a, d) and obtain the solution $u(x, t^{n+1})$ at t^{n+1} and $u(d_-, t)$ at the interface $x = d$.

(iii) The first equation in (2) gives $u(d_+, t) = u(d_-, t)$.

(iv) With $u(d_+, t)$ as the inflow Dirichlet boundary condition at $x = d$ and the prescribed outflow boundary condition at $x = b$, use the ELLAM to solve (1) on (d, b) and obtain the solution $u(x, t^{n+1})$ at t^{n+1} as well as the solution or the total flux at the outflow boundary depending on which type of outflow boundary condition was specified at $x = b$.

(v) Compute $u(d_+, t)$ and $(Vu - Du_x)(d_+, t)$ by projecting $u(x, t^{n+1})$ and $(Vu - D\mathcal{P}u_x)(x, t^{n+1})$ back along the approximate characteristics or by a linear interpolation at t^n and t^{n+1}, respectively.

(vi) Applying both equations in (2) yields $u(d_-, t) = u(d_+, t)$ and $(Vu - Du_x)(d_-, t) = (Vu - Du_x)(d_+, t)$. Then obtain the diffusive flux $-Du_x(d_-, t) = (Vu - Du_x)(d_-, t) - Vu(d_-, t)$. Go back to Step (ii) and repeat the process until the algorithm converges.

This algorithm uses an outflow Neumann condition at $x = d$ for equation (1) on (a, d) and avoids some potential numerical difficulties from a possibly improperly specified outflow Dirichlet condition at $x = d$ and from the numerical flux computed out of a Dirichlet condition. Then, imposing $u(d_+, t) = u(d_-, t)$ (the first equation in (2) yields a continuous numerical solution across the interface $x = d$. The continuity of mass is imposed when we move back from (d, b) to (a, d) to maintain mass conservation.

References

1. R.E. Ewing and H. Wang, *Eulerian-Lagrangian localized adjoint methods for variable-coefficient advective-diffusive-reactive equations in groundwater contaminant transport*, Proceedings of Sixth IIMAS-UNAM Workshop on Numerical Analysis and Optimization, Oaxaca, Mexico, 1992, (To appear).

2. H. Wang, T. Lin, and R.E. Ewing, *ELLAM with domain decomposition and local refinement techniques for advection-reaction problems with discontinuous coefficients*, Computational Methods in Water Resources IX. Vol. I: (Russell et al., eds.), Computational Mechanics Publications and Elsevier Applied Science, London and New York, 1992, 17–24.

DEPARTMENT OF MATHEMATICS, UNIVERSITY OF SOUTH CAROLINA, COLUMBIA, SC 29208
E-mail address: hwang@math.scarolina.edu

DEPARTMENT OF MATHEMATICS, UNIVERSITY OF BERGEN, BERGEN, NORWAY
E-mail address: Helge.Dahle@mi.uib.no, Jan-Einar.Vag@mi.uib.no

INSTITUTE FOR SCIENTIFIC COMPUTATION, TEXAS A&M UNIVERSITY, COLLEGE STATION, TX 77843–3404
E-mail address: ewing@ewing.tamu.edu

DEPARTMENT OF MATHEMATICS, VIRGINIA POLYTECHNIC INSTITUTE AND STATE UNIVERSITY, BLACKSBURG, VA 24061
E-mail address: tlin@mthiris.math.vt.edu

PART III

Parallelism

Contemporary Mathematics
Volume 180, 1994

Parallel Domain Decomposition Applied to Coupled Transport Equations

PETTER E. BJØRSTAD, W. M. COUGHRAN, JR. AND

ERIC GROSSE

ABSTRACT. Modeling semiconductor devices is an important technological problem. The traditional approach solves coupled advection-diffusion carrier-transport equations in two and three spatial dimensions on either high-end scientific workstations or traditional vector supercomputers. The equations need specialized discretizations as well as nonlinear and linear iterative methods. We will describe some of these techniques and our preliminary experience with coarse-grained domain-decomposition techniques applied on a collection of high-performance workstations connected at a 100Mb/s shared network.

1. Introduction

Simulation is widely used in the semiconductor industry to explore new manufacturing techniques, to characterize novel device technologies, and to validate and characterize circuit designs.

The transport physics of carriers in semiconductor devices has been studied for many years. Most of the existing model equations are derived from the Boltzmann transport equation (BTE) although the ideal set of equations would include Maxwell's and Schrödinger's equations [10]. Usually, Maxwell's equations are replaced by a Poisson equation via an assumption of locally constant dielectric behavior. Schrödinger's equation is obviated by making assumptions about the electronic states and band structure of the semiconductor. A simple model, called drift-diffusion, is effective for silicon devices down to $0.5\mu m$ or even

---

1991 *Mathematics Subject Classification*. Primary 65M55, 65N55; Secondary 68M10, 68N25.

This work was done while the first author was visiting the Department of Computer Science at Stanford Univ. and RIACS. Support from NFR is also acknowledged.

This paper is in final form and will not be published elsewhere.

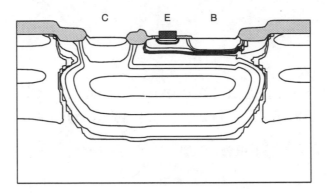

FIGURE 1. This is a two-dimensional cross-section of a three-dimensional 0.8μm BiCMOS bipolar transistor [13]. The contour lines represent impurity concentrations. The shaded gray areas are silicon oxide. The darker shaded region is a polysilicon wire. C, E, and B represent the collector, emitter, and base, respectively.

0.25μm feature sizes. It requires the solution of a Poisson equation coupled to two, formally parabolic, advection-diffusion equations

$$-\nabla \cdot (\epsilon \nabla \psi) = -q(n - p - N), \tag{1.1}$$

$$\frac{\partial n}{\partial t} = \nabla \cdot J_n - R, \tag{1.2}$$

$$\frac{\partial p}{\partial t} = -\nabla \cdot J_p - R, \tag{1.3}$$

where the current densities are

$$J_n = -\mu_n n \nabla \psi + D_n \nabla n, \tag{1.4}$$

$$J_p = -\mu_p p \nabla \psi - D_p \nabla p. \tag{1.5}$$

In silicon oxide (insulating) regions, these coupled equations are replaced by $-\nabla \cdot (\epsilon_0 \nabla \psi) = 0$. Here

- ϵ and ϵ_0 are dielectric constants;
- q is the magnitude of an elementary charge;
- $x \in \mathsf{R}^d$ and $t \in \mathsf{R}$ represent space and time, respectively;
- $N(x)$ is the impurity or doping concentration;
- μ_n, μ_p are carrier mobilities; D_n, D_p are diffusion coefficients;
- $R(n, p)$ represents electron-hole recombination-generation;
- $\psi(x, t)$ is the electrostatic potential;
- $n(x, t)$ and $p(x, t)$ are the electron and hole concentrations, respectively.

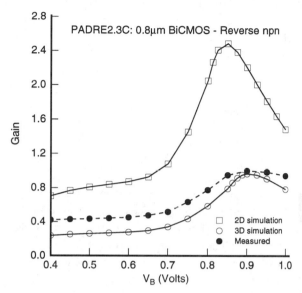

FIGURE 2. This represents the results of two- and three-dimensional simulations as well as measurements [13].

Geometry of the device structure and material interfaces are important; Figure 1 shows a cross-section of a bipolar transistor. Boundary conditions are commonly applied voltages at contacts along the perimeter of the device specified via ψ. Variables n and p range over several orders of magnitude while ψ varies less. The recombination-generation term, R, couples n and p. For power devices, it is sometimes necessary to add a coupled heat equation to model substrate heating.

Since the mid-1960s, numerous algorithms have been developed to solve the drift-diffusion equations. Specialized exponential upwinding schemes were invented for one-dimensional problems. These were generalized to tensor-product grids in two and three dimensions. The discretizations were also adapted to general finite volumes and elements. There are significant nonlinearities in this model so considerable research on Newton-like and nonlinear Gauss-Seidel methods has been done. Static grid adaption and continuation techniques have been used to explore devices with complex responses. Specialized sparse direct and iterative methods have been developed as well.

The state-of-the-art is such that complex two-dimensional problems are solved routinely. Sparse direct methods with triangular grids are effective and efficient for two-dimensional problems on conventional supercomputer architectures, like a Cray Y-MP. A snapshot of the state-of-the-art in 1990 is given by [11]. For drift-diffusion modeling, three-dimensional simulation is becoming more important (see Figure 2). However, grid generation and adaption as well as linear iterative methods are still inadequate for routine simulation.

There are more advanced partial-differential equation (PDE) models for semi-conductor devices. The so-called energy-balance or energy-transport equations are similar to the drift-diffusion equations in mathematical character. Additional equations for carrier energy variables have to be solved. The "hydrodynamic" model involves formally hyperbolic equations, which call for alternate algorithmic approaches. Our motivation comes from systems of a form analogous to the drift-diffusion or energy-transport models, rather than the hydrodynamic model.

Over the years, AT&T Bell Laboratories has relied on a series of Cray Research computers to do its large-scale simulation. We currently employ a Cray Y-MP M92/256 as a production workhorse. Our evaluations of parallel computer systems have been mixed, particularly since the specialized MIMD architectures carry a significant development and communication-infrastructure overhead. However, high-performance microprocessors in the form of workstations are changing things. We embarked on a prototype PDE solver designed to run on workstations connected via a "high-speed" network interface. Our hope is that such an approach will lead to cost-effective computing for two- and especially three-dimensional simulation. From our brief discussion of device modeling, it is apparent that the ability to solve general PDEs on complex geometries is crucial.

2. Domain decomposition approach

We are developing a collection of C codes to deal with general elliptic and advection-diffusions PDEs on composite grids made up of simplices, prisms, pyramids, and bricks. Each of the following has to be dealt with:
- geometry specification;
- grid generation;
- temporal discretizations;
- spatial discretizations, including a variety of upwinding techniques suitable for advection-diffusion equations;
- Newton-like and other iterative methods for dealing with the associated nonlinear systems;
- continuation methods for characterization studies;
- automatic differentiation techniques for linearization;
- sparse direct and iterative methods for the associated nonsymmetric linear systems of equations.

Each of these subproblems is significant by itself, but all of them must be dealt with to produce a generically useful simulation toolbox. Too often researchers allow themselves to use simplified approaches only suitable for model problems on trivial geometries.

For two-dimensional, single device, simulation problems with well-adapted grids (typically less than 10^4 unknowns), it is difficult to beat sparse direct factorization methods for solving the linear systems. However, numerous possibilities exist for iterative methods. Domain decomposition appears to us as a natural means to exploit coarse-grained parallelism.

"Domain decomposition" refers to any method that divides the original problem domain into pieces and solves locally on each subdomain [12]. The particular variant that we use, multiplicative overlapping Schwarz [4, 6], defines subdomains with overlap and solves local problems in succession, propagating data from the solution on the interior of one subdomain to the boundary values of the next subdomain. To allow information to propagate more rapidly, a coarse grid solution is also used. Finally, rather than cycling through all the subdomains repeatedly, as in a pure alternating Schwarz method, we use one sweep as a preconditioner for an iterative linear system solver.

To be more explicit, suppose we wish to solve $Lu = f$ on Ω, $u = b$ on $\partial\Omega$, which we discretize as $Au = Rf - Bb$ using a finite volume or "box" approach. Without loss of generality, we will call the modified right hand side in this system f and assume $b = 0$. The domain Ω is subdivided into k overlapping subdomains Ω_i with local operators that are restrictions of the global ones in the sense that at an interior point j of the global grid corresponding to interior point ι in subdomain i, we have $[f - Au]_j = [f_i - A_i u_i - B_i w_i]_\iota$. Here u_i means the local interior unknowns and w_i means the local boundary values, which are 0 or u values from a neighboring subdomain.

Define a preconditioner $v = M^{-1}g$ for solving $Av = g$ with a Krylov space method by the steps

(2.1)	$g_c = I_f^c g$	(restrict to coarse grid)
(2.2)	$A_c v_c = g_c$	on Ω
(2.3)	$v_f = I_c^f v_c$	(prolongate to fine grid)
(2.4)	from v_f send values into w_i	
(2.5)	for $1 \leq i \leq k$	
(2.6)	$A_k v_k = g_k - B_k w_k$	(local solve on Ω_k)
(2.7)	from v_k send values to w_j for neighboring Ω_j	

The final v is obtained by assembling values from the subdomains.

The multiplicative Schwarz method is attractive to us because we want a good serial code, and in our workstation cluster environment we seek only coarse parallelism. In contrast, the additive Schwarz variant provides more parallelism at the cost of roughly twice as many iterations [2, 3]. So, while other computer configurations would call for other methods, for the next several years we see workstation networks as the best price/performance choice.

Although in theory one must increase the number of rows of overlap between subdomains as the grid is refined in order to keep the same rate of convergence, previous experience has shown that this is balanced by the increased work [7, 14]. Following those earlier conclusions, we simply take one row of unknowns for the overlap.

3. Communication structure

As mentioned in the previous section, our interest in this project has been to explore whether an inexpensive network of workstations is capable of the VLSI simulation modeling that had previously required Cray supercomputer resources. Can one get by without shared memory or specialized communication fabric? The cautionary words of Baskett and Hennessy [13 Aug 1993, Science, p.868] make this seem unlikely.

> "One potentially attractive approach to building parallel processors is to use workstations connected on a local area network, often called a workstation cluster. This approach, however, has proved suitable only for applications where the parallel computations are so coarse-grained as to be essentially independent."

Surely the coupling between neighboring subdomains keeps the subproblems from being "essentially independent"?

Our laboratory was purchasing Silicon Graphics Indigo 50MHz MIPS R4000 workstations connected by a Fiber Distributed Data Interface (FDDI) network for a multimedia experiment; we decided to see if we could put them to use solving differential equations when they weren't otherwise needed. FDDI is nominally a 12 megabyte per second (MB/s) ring with a maximum length of 100km using a timed token-passing protocol, in which each packet goes all the way around the ring. The SGI implementation we use has been found to be reasonably easy to administer, and we consistently measure 7MB/s or better for large transfers (user space on one machine to user space on another). In contrived circumstances in which few pages of memory have to be touched, we measure as much as 11MB/s. The measured latency of about 1ms is far higher than the intrinsic hardware latency of 10us around the ring, but as we shall see is sufficient for domain decomposition.

We expect the FDDI network to be replaced in the next year or so by Asynchronous Transfer Mode (ATM), but are happy with "dull but deliverable" FDDI for now. The number of workstations we expect to be cooperatively solving one problem is limited more by sociological and economic constraints than by scaling limits of the network. Also, while the transfer rates needed by domain decomposition are high compared to the typical loaded campus Ethernet, the traffic is rather bursty and can be accommodated adequately on a shared medium. This is in contrast to the demands of the same set of workstations if they are transmitting video signals to each other.

One specific disappointment with the hardware as seen at the user level is that, although each packet transits each workstation, there appeared to be no feasible way to use this for data broadcast. We were using TCP as the transport protocol, not wishing to bear the burden of adding reliability onto UDP [5]. A sufficiently sophisticated communications package might do hardware broadcast, but as far as we could tell none of the popular implementations is yet up to this.

Certainly the one we tried, pvm [9], was far from using the full FDDI bandwidth even for direct process-to-process communication.

The communication needs of domain decomposition turn out to be so simple that a trivial communication package built directly on sockets and read/write is sufficiently expressive and efficient. One "coarse-grid process" formulates, computes, and distributes the coarse grid solution; multiple "fine-grid processes" talk to the coarse-grid process and to fine-grid processes for neighboring domains. All processes execute the same program and hold the coarse grid and subdomain topology, but only store and compute fine grid values for the subdomains "owned" by that process. This leads to send/receive pairs that are close together in the source code, with a static matching known at compile time, so correct sequencing and typing of messages is immediate.

There are three classes of communication: dot products for the outer Krylov iteration, neighbor communication between fine-grid processes, and communication between fine-grid processes and the coarse-grid process.

Our implementation of BICGSTAB [15] is shown in Figure 3. In order to bring to the surface the essential global synchronization points, we have written this in a rather unusual style. The function B takes as its first argument a string representing in readable form an operation that can be completed with, at most, one communication. Our C statement

```
    omega = B("dot{r*MAr}/dot{MAr*MAr}");
```
would ordinarily be written as

```
    omega = dot(r,MAr)/dot(MAr,MAr);
```
an unusually clever communications package might be able to batch the two sum operations together, but probably not without further hints. Alternatively, one might write the vector primitives to use lazy evaluation, but the reader has to trust that sophisticated operations behind the scenes are doing what is intended. Our coding shows the reader immediately that one communication is involved, without any need to understand and debug scheduling policies.

For neighbor communication, each process loops through all the subdomains in order. If a process "owns" a subdomain, then it performs a local solve, using y12m [16] for the experiments reported below, and sends boundary data to its neighbor processes; if instead its neighbor owns the subdomain, then the process waits to receive data from its neighbor; otherwise, the process examines the next subdomain. This arrangement is self-synchronizing, with no explicit task management. Subdomains are colored by a greedy algorithm and different color subdomains assigned statically to separate processes. If four colors are needed, four processes are assigned to each physical processor. Obviously, we attempt to place neighbors on the same processor when possible. Assuming the work can be predicted well enough, and partitioned evenly enough, this static assignment has given good overlap of communication and computation, as will be seen in the next section.

```
#include "subdom.h"
void bicgstab( void (*A)(int,int,void*), void *Auser,
void (*M)(int,int,void*), void *Muser,
int (*done)(int,double,Chk*),Chk *conuser){
  int n;
  double rho, rho_old, omega, alpha, beta, sigma;
  B("set userdata",Auser);
  A(Gu,Gw,Auser);
  B("w=f-w");
  M(Gw,Gr,Muser);
  B("r0=p=r");
  rho = B("dot{r0*r}");
  for(n = 1; n<1000; n++){
     if(n>1){
      rho_old = rho;
      rho = B("dot{r0*r}");
      beta = (rho/rho_old) * (alpha/omega);
      B("p=beta*(p-omega*MAp)+r",beta,omega);
     }
     A(Gp,Gw,Auser);   M(Gw,GMAp,Muser);
     alpha = rho/B("dot{r0*MAp}");
     B("r -= alpha*MAp",alpha);
     A(Gr,Gw,Auser);   M(Gw,GMAr,Muser);
     omega = B("dot{r*MAr}/dot{MAr*MAr}");
     sigma = B("u+=alpha*p+omega*r; r-=omega*MAr; norm2{r}",
               alpha,omega);
     if(done(Amults,sigma,conuser)) break;
  }
}
```

FIGURE 3. An implementation of BICGSTAB in C, using ad hoc
vector primitives ("BAHS") rather than BLAS in order to reveal
communication.

The amount of overlap is known at setup time, so there would be no trouble allocating fixed buffers. The local solves are expensive enough that one or two buffers provide sufficient queuing.

The neighbor communication, though it involves scatter/gather references to memory, deals with fairly small amounts of data. Communication between the coarse grid process and the fine grid processes for the exchange of coarse grid data in prolongation and restriction is just the opposite: bulky data in contiguous memory. In the experiments described below, we transmitted the entire coarse grid to each fine grid process, to avoid the complication of determining exactly which coarse grid points were really needed.

It may be worth mentioning the size of the communication code, to dispel the notion that dealing with bare sockets is an overwhelming burden. About 200 lines of C code are involved in providing network connection primitives; the domain decomposition program proper has about another 300 lines to set up the connections in the right order and to issue sends and receives. All this is dwarfed by the grid and discretization modules.

4. Results

Consider a model system of two equations in two space dimensions, derived from the linearized drift-diffusion equations with one carrier

$$(4.1) \qquad\qquad \Delta\psi + n = -r$$
$$(4.2) \qquad n_o\Delta\psi + \nabla\psi_0\nabla n - \Delta n = 0.$$

An appropriate upwinding scheme was used [1]. This captures some aspects of the coupled transport without the full complexity. For convenient benchmarking, we discretize on a square domain Ω divided into an 8 by 8 mesh of subdomains Ω_i. The typical subdomain has 7 by 7 interior grid points, though subdomains near $\partial\Omega$ have fewer grid points; we start with a uniform subdivision of Ω and define Ω_i by "growing" each cell. Neighbors overlap by one mesh line of unknowns. This yields a total of 4418 fine unknowns. Finally, we overlay a coarse grid with 7 by 7 interior points, for a total of 98 coarse grid unknowns.

Using adjoint as the restriction operator and BICGSTAB as the accelerator, we measured the wall clock time of program execution on a relatively unloaded network of uniprocessor workstations, taking the best of three runs. The timings here are for an experimental code compiled with -g, but we did not observe any dramatic changes with -O. All workstations had enough memory to avoid swapping.

A serial version of the program ran in 38.12 seconds elapsed, 37.7 cpu. With two workstations (and hence one coarse grid process and eight fine grid processes) our socket-based parallel code ran in 23.6 seconds elapsed time, for a speedup of 1.6. The corresponding time for a pvm-based version of the code was 34.8 seconds, for a speedup of 1.1. With 4 workstations, our code ran in 13.2 seconds, for a

TABLE 1. Convergence rate for model problem.

Amults	$\log_{10} \|u - u_*\|_2$	$\log_{10} \|M(f - Au)\|_2$
3	1.1	-0.4
5	-2.7	-2.9
7	-2.7	-2.9
9	-4.4	-4.4
11	-5.6	-5.6
13	-7.2	-7.2
15	-8.8	-8.9
17	-11.1	-11.1

speedup of 2.9. Because of a bug we were unable to isolate, our pvm-based version hung after initialization in the four workstation (17 processes) case.

Although we are using relatively small subdomains in this experiment, because we are solving a system of differential equations and use a complicated discretization, there is enough computation to mask most of the communication delay. For a simple Poisson equation that would ordinarily be used in primitive benchmarks like this, the situation is reversed. On the same grid described above but with one equation (49 coarse and 2209 fine unknowns) with the serial code we measure 4.5 seconds elapsed; on two workstations we measure 4.0 seconds (speedup 1.13) with the socket-based code, 8.2 seconds (speedup 0.6) with the pvm-based code; on four workstations, the socket-based code we measure 2.4 seconds (speedup 1.9).

Another way in which these timings are conservative is that for this model, the cost of matrix setup is rather low. In real codes, it may account for half of the total execution time, and fortunately scales linearly with processors.

Exactly the same computations are being done in the serial and parallel codes, and as Table 1 shows the program is converging satisfactorily. Total cpu times for the 4 workstations were only a few seconds more than the serial cpu time.

Because the domain decomposition preconditioner is so effective, the preconditioned residual $\|M(f - Au)\|_2$ is an excellent estimate for the true error $\|u - u_*\|_2$, as is apparent from Table 1.

For a more detailed look at the degree of overlap of communication and computation, we profiled the program by inserting `gettimeofday` calls before and after each read and write and recorded the size of the message. Elapsed time increased by factor 1.075. For a small run on two workstations lasting 6 seconds, there were 6112 messages transmitting 372 kilobytes, for an average of 7.6 doubles per message. A schematic drawing of when each process was busy confirmed our expectations of how processes would be naturally scheduled by the operating system, and we did not feel the need to tune this further by using threads instead of separate processes on each processor.

To help the reader avoid one blind alley, we give in Table 2 a cautionary set

TABLE 2. Number of iterations for different restriction operators and Krylov methods on a simple Poisson equation on a square.

BICGSTAB				TFQMR			
n_{fine}	fine	adjoint	interp	n_{fine}	fine	adjoint	interp
25	7	5	5	25	6	4	5
81	11	5	7	81	11	4	6
289	15	5	7	289	17	4	6
1089	29	3	7	1089	34	4	6
289	9	7	7	289	9	6	8
841	11	5	13	841	14	5	13
2401	23	5	29	2401	23	4	22
9409	37	5	39	9409	44	5	39
25	7	7	7	25	7	6	6
49	9	7	9	49	9	6	8
225	15	5	17	225	16	5	15
961	27	5	25	961	30	4	31

of intermediate results from our testing. Iterations to reduce the residual by a fixed factor are given for 1) only fine grid solves; 2) coarse and fine grid with bilinear interpolation as prolongation and the adjoint as restriction; 3) bilinear interpolation for both prolongation and restriction. As expected, the number of iterations increases with n for case 1) and is nearly constant for case 2). The variational theory depends on using the adjoint as the restriction operator, but experience from multigrid and elementary considerations of approximation on unequally spaced grids led us to guess that interpolation as in 3) would be a good choice. The numerical evidence dashed this hope.

The experiments were all done on uniform fine grids on a square, with three slightly different ways of defining subdomains and slightly different coarse grids. The experiments were also repeated with another well-respected Krylov method, TFQMR [8]. There were no significant differences for this example between BICGSTAB and TFQMR.

We are pleased with the speedup of 2.9 on four workstations, and look forward to expanding to a dozen workstations, larger subdomain solves, and additional work in matrix setup and solution rendering that is known to scale linearly. We are satisfied with these speedups, and have plenty of other work for any spare cycles left over because of load imbalance. Happily, domain decomposition appears to be one of those "essentially independent parallel computations" that works well for a modest number of workstations on modern networks.

Acknowledgements. We have benefited greatly from collaborations with Randy Bank, Mark Pinto, Don Rose, and Kent Smith on algorithms for advection-diffusion equations. Claude Pommerell introduced us to ways of structuring communication for parallel iterative methods. Dave Presotto and Phil Winterbottom were welcome sources of advice on TCP. Linda Kaufman and Margaret Wright improved the presentation.

REFERENCES

1. R. E. Bank, J. F. Bürgler, W. Fichtner, and R. K. Smith, *Some upwinding techniques for finite element approximations of convection-diffusion equations*, Numer. Math. **58** (1990), 185–202.
2. P. E. Bjørstad, R. Moe, and M. Skogen, *Parallel domain decomposition and iterative refinement algorithms*, (W. Hackbush, ed.), Vieweg Verlag, Braunschweig, 1991.
3. J. H. Bramble, J. E. Pasciak, J. Wang, and J. Xu, *Convergence estimates for product iterative methods with applications to domain decomposition*, Math. Comp. **57** (1991), 1–21.
4. X.-C. Cai and O. Widlund, *Multiplicative Schwarz algorithms for some nonsymmetric and indefinite problems*, SIAM J. Numer. Anal. **30** (1993), no. 4, 936–952.
5. D. E. Comer and D. L. Stevens, *Internetworking with TCP/IP, Vol. III: client-server programming and applications, BSD socket version*, Prentice-Hall, 1993.
6. M. Dryja, *An additive Schwarz algorithm for two- and three-dimensional finite element elliptic problems*, Domain Decomposition Methods (Tony Chan, Roland Glowinski, Jacques Périaux, and Olof Widlund, eds.), SIAM, 1989.
7. M. Dryja and O. B. Widlund, *Domain decomposition algorithms with small overlap*, SIAM J. Sci. Stat. Comput. **15** (1994).
8. R. W. Freund, *A transpose-free quasi-minimal residual algorithm for non-Hermitian linear systems*, SIAM J. Sci. Comput **14** (1993), 470–482.
9. A. Geist, A. Beguelin, J. Dongarra, W. Jiang, R. Manchek, and V. Sunderam, *PVM3 user's guide and reference manual*, Tech. Report ORNL/TM-12187, Oak Ridge National Laboratory, Oak Ridge TN 37831, 1993, *http://www.netlib.org/pvm3/ug.ps*.
10. K. Hess, *Advanced theory of semiconductor devices*, Prentice Hall, 1988.
11. K. Hess, J. P. Leburton, and U. Ravaioli, *Computational electronics: Semiconductor transport and device simulation*, Kluwer, 1991.
12. P. L. Lions, *On the Schwarz alternating method. I.*, First International Symposium on Domain Decomposition Methods for Partial Differential Equations (R. Glowinski, G. H. Golub, G. A. Meurant, and J. Périaux, eds.), SIAM, 1988.
13. M. R. Pinto, D. M. Boulin, C. S. Rafferty, R. K. Smith, W. M. Coughran, Jr., I. C. Kizilyalli, and M. J. Thoma, *Three-dimensional characterization of bipolar transistors in a submicron BiCMOS technology using integrated process and device simulation*, Int. Electr. Dev. Meeting Technical Digest '92, 1992, pp. 923–926.
14. M. D. Skogen, *Schwarz methods and parallelism*, Ph.D. thesis, Department of Informatics, University of Bergen, Norway, February 1992.
15. H. A. van der Vorst, *Bi-CGSTAB: a fast and smoothly converging variant of BI-CG for the solution of nonsymmetric linear systems*, SIAM J. on Scientific and Statistical Computing **13** (1992), 631–644.
16. Z. Zlatev, *Computational methods for general sparse matrices*, Kluwer, 1991.

INSTITUTT FOR INFORMATIKK, UNIVERSITETET I BERGEN, N-5020 BERGEN, NORWAY
E-mail address: Petter.Bjorstad@ii.uib.no

AT&T BELL LABORATORIES, MURRAY HILL, NEW JERSEY 07974
E-mail address: wmc@research.att.com, ehg@research.att.com

Contemporary Mathematics
Volume **180**, 1994

MENUS-PGG : A MAPPING ENVIRONMENT FOR UNSTRUCTURED AND STRUCTURED NUMERICAL PARALLEL GRID GENERATION

NIKOS CHRISOCHOIDES, GEOFFREY FOX, AND JOE THOMPSON

ABSTRACT. MENUS-PGG is a problem solving environment (PSE) for developing parallel algorithms that generate structured and unstructured static and adaptive grids (or meshes) required for the implementation of scalable parallel partial differential equation (PDE) solvers based on domain decomposition methods. Whereas the first generation PSEs for the numerical solution of PDEs on distributed memory multiprocessor systems are based on the data mapping of sequentially generated grids and support only the data parallel programming model, MENUS-PGG generates and maintains grids on the processors of parallel/distributed systems and combines the most valuable aspects of the data parallel programming model with the flexibility of the task parallel programming model. MENUS-PGG assumes a machine model that consists of homogeneous and heterogeneous clusters of processors operating in a distributed address space implemented on remote memory modules via message passing through a high-speed interconnection network. The major contribution of MENUS-PGG should be the reduction of the pre-processing overhead required by the data parallel PDE solvers and the efficient maintenance of the distributed data structures that support h, p, and hp-refinements. We present preliminary results indicating that the parallel grid generation results in a substantial reduction of the pre-processing overhead needed for the solution of the data mapping problem.

1. Introduction

Parallel computing and specifically high performance software for scientific computing will be made more attractive to scientists and engineers if these systems will provide support for all aspects of a simulation : grid generation, PDE solvers, adaptive refinement, and I/O. While a fair amount of work has been carried out for field solvers limited or none made towards parallel algorithms and software modules for numerical grid generation. Our aim in this paper is to describe a software system for the parallel numerical generation of structured and unstructured static and adaptive grids.

Besides the geometric modeling and grid visualization we identify and focus on four additional stages for parallel numerical grid generation. The first stage requires

The research of the first author was supported by the Alex G. Nason Prize Award.
AMS(MOS) key words : parallel, numerical, grid, software.
1991 *Mathematics Subject Classification.* Primary 35A40. 65N50, 68Q22
This paper is in final form and no version of it will be submitted for publication elsewhere.

the decomposition of the continuous domain into a set of non-overlapping subdomains with simpler shape (e.g. four- or six-sided polygons for 2- or 3-dimensional
spaces). The second stage requires the partitioning and placement of the subdomains on the processors of the target machine so that certain criteria are satisfied. The third stage requires independent grid generation on the subdomains
and sometimes maintenance of grid conformity and continuity on the interfaces
of the subdomains. Finally, the fourth stage requires the run-time migration of
grids among the processors so that processors workload (computation and communication) is balanced and local communication and synchronization among the
processors are minimized during the parallel execution of PDE computation. The
MENUS-PGG design is based on these stages. In the rest of the paper we describe
the MENUS-PGG software infrastructure and the individual software components
that correspond to the above stages.

2. SOFTWARE INFRASTRUCTURE

The design objective of MENUS-PGG is to provide a uniform environment and
the basic software components to implement, analyze and test scalable algorithms
for parallel grid generation. The software infrastructure of MENUS-PGG consists
of three major subsystems, namely, the *front-end subsystem*, corresponding to the
domain definition and the first two stages described above, the *grid generation
subsystem*, corresponding to the last two stages, and the *back-end subsystem*, corresponding to the visualization of grid and performance data. Next we describe the
software modules of the subsystems and their functionalities. Fig. 1 depicts the
software architecture of MENUS-PGG and the interaction among these modules.

Software Module	Computational Phase / Functionality
Front-end :	
Geometry Modeler	Specify and discretize the boundary curves or surfaces of the domain.
Domain and Subdomain Decomposer	Decompose the domain into four or six-sided subdomains and discretize internal subdomain boundaries.
Subdomain Mapper	Map subdomain to the processors of the target machine.
Grid Generation Subsystem :	
Static & Adaptive Grid Generator	Concurrently generate grids on the subdomains and maintain grid conformity - if it is necessary.
Scheduler	Maintain computation and communication load balance of the processors during the PDE computation.
Back-end :	
Visualization Tool	Visualize grids, solution and performance data.

The **geometry modeler** and the **3D visualization tool** are provided by the
National Grid Project which is under development at NSF Engineering Research
Center for Computational Field Simulation, ERC-CFS, at Mississippi State University [14]. The geometry modeler constructs the boundary surface (or curve) of
the region we want to discretize. The boundary representation of the region can
be received as patches from an external CAD system or it can be computed by an

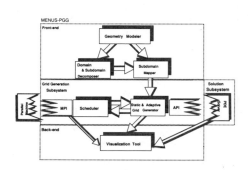

FIGURE 1. MENUS-PGG software architecture and NGP's Geometry Modeler.

internal available CAD system. An instance of the geometry modeler and the 3D visualization tool are depicted in Fig. 1. The **3D visualization tool** is a user friendly interface with functionalities allowing graphical movement of block structures (for the graphical construction of grid topology) as well as scaling, rotation, and transformation of surface segments.

The **domain and subdomain decomposer** is the module that decomposes the given domain into four or six-sided subdomains. Such decompositions can be achieved either interactively using graphical user interface tools [14] or automatically using computational geometry tools, like the Medial Axis Transformations [12]. The domain decomposer is essential for the conventional transfinite mapping techniques used for the generation of structured grids. The composite block structures generated by this module can also be used for parallel unstructured grids.

The **subdomain mapper** is based on a library of Mapping Templates for Load Balancing (MTLB) [4]. The MTLB library of templates provides algorithmic and software infrastructure for the mapping of the subdomains to the processors of the target parallel machine. Such mappings often involve the solution of an intractable combinatorial optimization problem that optimizes a number of criteria, like the minimization of the number of grid points that separate the subdomains residing on different processors, the proper distribution of grids so that the computation and communication work load of processors is balanced, or the appropriate placement of subgrids so that network contention is minimized. Although many heuristics for finding good suboptimal mapping solutions have been proposed in the literature (see in [6], [1], [7], [11], [8] and [2]) there is no general-purpose software that can be used independently of the specific characteristics and data structures of the application. The MTLB library aims to provide a common software framework for the implementation and evaluation of existing and new heuristics for the data mapping problem.

The **static and adaptive grid generator** will provide the algorithmic and software infrastructure for the concurrent grid generation of the subdomains and for the maintenance of grid continuity and conformity on the interfaces. Grids (or meshes) are classified into *structured grids*, formed by intersecting grid lines, and *unstructured grids or meshes*, formed by first creating all the node points and then connecting the nodes to form "best" possible triangles. Another classification of grids is based on their static or dynamic evolution during the PDE computation. Grids that remain the same throughout the PDE computations are called static and grids that evolve according to the behavior of predefined error estimators are called dynamic or adaptive. Fig. 2 depicts a taxonomy of the five different methods and types of grids that are under development within MENUS-PGG environment and they are : (1) parallel structured grids for 2 and 3-dimensional complex domains based on composite block structures [15] and (2) parallel unstructured grids using Delaunay triangulation, (3) parallel adaptive semi-structured grids based on a combination of nested structured grids and local equidistribution approaches (4) moving structured grids and (5) adaptive unstructured grids. A detailed discussion of the individual tools and issues related to parallel numerical grid generation is out of the scope of this paper and appears elsewhere [3], [4].

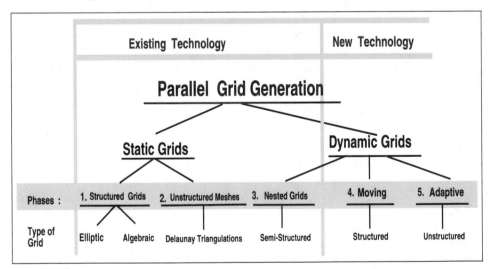

FIGURE 2. Taxonomy of grid types to be generated by MENUS-PGG.

The **scheduler** is a machine independent library of communication and synchronization routines required for the implementation of the computational engine on distributed memory MIMD/SIMD machines using the data and task-parallel message passing programming model. The portability of the computational engine will be guaranteed by developing the scheduler on the Message Passing Interface standard (MPI) defined recently [10]. A run-time support system suitable for parallel adaptive grid generation computations is the major software component of the scheduler. The run-time support system is limited to a very small set of parallel adaptive PDE computations. The system is organized in four layers (from the lower to higher layer) : (0) Network Interface Layer (interrupt and polling driven), (1) Multithread priority based scheduling layer (2) Thread Scheduling Layer (scheduling algorithms for different pairs of grid/architecture), and (3) Interface Layer to static and adaptive grid generator. The run-time support system will be available in the second version of the software. In the mean time the system will be based on existing incremental data mapping methods.

3. SUMMARY AND CONCLUSIONS

In this paper we discussed the MENUS-PGG software architecture and the modules that correspond to the six steps needed for the parallel numerical grid generation. Preliminary results indicate substantial reduction in the overhead introduced for the solution of the data mapping problem. Another advantage of the MENUS-PGG is that it generates and maintains grids on the processors of parallel/distributed systems and combines the most valuable aspects of the data parallel programming model with the flexibility of the task parallel programming model. With the use of composite block structures (CBS approach) as a tool for contracting the size of the problem not only we reduce the pre-processing overhead but we achieve optimality of the mapping.

Table 1 depicts performance data associated to the time (in sec) required to sequentially generate (Grid-Gen.), partition and store (Mapping) the subgrids onto the 64 nodes of the nCUBE 2, and the time (in sec) to sequentially generate and partition $C_f(\Omega)$ ($C_f(\Omega)$-Proc.) plus the time (in sec) to generate on a 64 node nCUBE 2 an algebraic grid using $C_f(\Omega)$. Columns fourth and seventh indicate the total sum of times (in sec) for the sequential grid generation and CBS approaches respectively. A SPARC workstation was used for the sequential generation and partitioning of grids and $C_f(\Omega)$.

TABLE 1. Performance data for sequentially generating, partitioning and storing grids on the 64 processors of nCUBE 2 and data for sequentially generating, partitioning and storing composite block structures together with the data for parallel grid generation on a nCUBE 2 with with 64 processors.

Grid Points	Grid-Gen.	Mapping	Total	$C_f(\Omega)$-Proc.	(//) Grid-Gen.	Total
2.5×10^3	0.38	17.81	18.19	3.44	0.08	3.52
10×10^3	1.61	46.45	48.06	4.38	0.35	4.73
22.5×10^3	3.61	100.15	103.76	8.48	0.71	9.19
40×10^3	6.45	195.13	201.58	16.06	1.25	17.31

ACKNOWLEDGMENTS

The authors are grateful to Wayne Mastin and Adam Gaither for their help with the NGP code. We also thank David Keyes for his detailed and constructive comments that improved the presentation of the paper.

REFERENCES

1. Charbel, F., A simple and efficient automatic FEM domain decomposer, *Computers and Structures*, Vol. 28, pp 579–602, 1988.
2. Chrisochoides, N., E. Houstis and J. Rice, Mapping Algorithms and Software Environment for Data Parallel PDE Iterative Solvers, *Special Issue of the Journal of Parallel and Distributed Computing on Data-Parallel Algorithms and Programming,* Vol. 21, No 1, pp 75–95, 1994.
3. Chrisochoides, N., An Alternative to Data Mapping for Parallel PDE Solvers : Parallel Grid Generation, *Proceedings of Scalable Parallel Libraries Conference*, pp 36–44, October, 1993.
4. Chrisochoides, N., Mapping Templates for Load Balancing. To be submitted to ACM Trans. of Mathematical Software. (In preparation)
5. Fox, G., M. Johnson, G. Lyzenga, S. Otto, J. Salmon and D. Walker *Solving Problems on Concurrent Processors.* Prentice Hall, New Jersey, 1988.
6. Fox, G., P. Messina R.Williams. Parallel Computing Works! Morgan Kaufmann, San Mateo, CA 1994
7. Hammond, W. S., Mapping Unstructured Grid Computations to Massively Parallel Computers, Ph.D Thesis, Rensselaer Polytechnic Institute, Troy, NY.
8. Mansour N., *Physical Optimization Algorithms for Mapping Data to Distributed-Memory Multiprocessors.* PhD thesis, Computer Science Department, Syracuse University, 1992.
9. Mansour N., R. Ponnusamy, A. Choudhary, and G. Fox. Graph Contraction for Physical Optimization Methods: A Quality-Cost Tradeoff for Mapping Data on Parallel Computers. *International Supercomputing Conference*, Japan, July 1993, ACM Press.
10. MPI Forum, Message-Passing Interface Standard, April 15, 1994.
11. Simon, H., Partitioning of Unstructured Problems for Parallel Processing, RNR-91-008, NASA Ames Research Center, Moffet Field, CA, 94035.
12. Tam T.K.H., Price M., Amstrong C., and McKeag. Computing the critical points of the medial axis of planar object using a Delaunay point triangulation algorithm. Submitted to IEEE, PAMI, 1993.
13. Thompson, J., Z. U. A. Warsi and C. Wayne Mastin, *Numerical Grid Generation.* North-Holland, New York, 1985.
14. Thompson, J., The National Grid Project, NSF Engineering Research Center for Computational Field Simulation, 1991.
15. Thompson, J., A survey of composite grid generation for general three-dimensional regions. *Numerical Methods for Engine-Airframe Integration,* S.N.B. Murthy and G. C. Paynter eds., 1984.

NORTHEAST PARALLEL ARCHITECTURES CENTER, SYRACUSE UNIVERSITY, 111 COLLEGE PLACE, SYRACUSE, NY, 13244-4100
E-mail address: nikos@npac.syr.edu

NORTHEAST PARALLEL ARCHITECTURES CENTER, SYRACUSE UNIVERSITY, 111 COLLEGE PLACE, SYRACUSE, NY, 13244-4100
E-mail address: gcf@npac.syr.edu

ENGINEERING RESEARCH CENTER FOR COMPUTATIONAL FIELD SIMULATION, MISSISSIPPI STATE UNIVERSITY, P.O. BOX DRAWER 6176, MISSISSIPPI STATE, MS 39762
E-mail address: joe@erc.msstate.edu

Contemporary Mathematics
Volume **180**, 1994

A comparison of three iterative algorithms based on domain decomposition methods

PATRICK CIARLET JR

ABSTRACT. In this paper we compare three domain decomposition preconditioners for the capacitance matrix and the conjugate gradient methods
to solve linear systems arising from the discretization of elliptic partial differential equations. The methods have been implemented on parallel and
superscalar architectures. Numerical experiments show that using these
preconditioners leads to competitive algorithms, both in terms of number
of iterations and parallelization rates.

1. Introduction

In this work, we compare the parallel and superscalar implementations of three
different iterative methods designed to solve the finite element discretization of
elliptic problems such as

$$-\operatorname{div}\left(\lambda \, \nabla u\right) = \alpha \text{ in } \Omega, \text{ with } \lambda(x,y) = \begin{pmatrix} a(x,y) & 0 \\ 0 & b(x,y) \end{pmatrix}$$

$$u = 0 \text{ on } \partial\Omega.$$

Here $\Omega =]0, 1[\times]0, 1[$ and a, b and α are given functions, a and b being nonnegative over the domain. We approximate these problems by using a P_1 finite
element method with right triangles, leading to a five-point centered scheme.
Note that by doing this, we are able to handle problems with discontinuous coefficients without any difficulty, as long as the jumps occur on the sides of the
triangles. This gives us a linear system $Ax = f$, with a n by n symmetric positive
definite matrix A.

We propose to solve this linear system by using either the Capacitance Matrix and Conjugate Gradient methods or the Preconditioned Conjugate Gradient
(**PCG**) method. Our goal is to obtain a good trade-off between the convergence

1991 *Mathematics Subject Classification.* 65F10, 65F50, 65N55.
This paper is in final form and no version of it will be submitted for publication elsewhere.

rate for the iterative methods and the Mflops rate on the computers. In order to define the preconditioners, we use domain decomposition methods based on a partition of the domain into boxes or strips. Moreover, to be able to use the iterative methods for solving large problems, we focus our interest on preconditioners requiring reasonable storage, i.e. "sparse" preconditioners.

In the next section, we briefly recall the definition of the Capacitance Matrix method. In section 3, we define the preconditioners. Finally, we compare our iterative methods in the last section.

2. The Capacitance Matrix Method

This method was first investigated by Buzbee and al in [1]. Let B be a nonsingular n by n matrix and S an extension matrix such that, using block notation, we have

$$A = \begin{pmatrix} A_{11} & A_{12} \\ A_{21} & A_{22} \end{pmatrix}, \ B = \begin{pmatrix} A_{11} & A_{12} \\ B_{21} & B_{22} \end{pmatrix} \text{ and } S = \begin{pmatrix} 0 \\ I_q \end{pmatrix}.$$

Here only the last q rows of A and B differ. The q by q matrix $C = S^T A B^{-1} S$ is called the capacitance matrix. One can easily prove that the capacitance matrix is nonsingular. Then the original problem $Ax = f$ can be replaced by

$$\begin{cases} Bv = h, \text{ with } h_1 = f_1, \ h_2 = 0, \\ C\omega = g, \text{ with } g = S^T (f - Av), \\ Bx = h + S\omega. \end{cases}$$

If C is small enough, then a direct solver can be used. Otherwise the linear system with C can be solved iteratively. In the following, we use either the Diagonally Preconditioned Conjugate Gradient (**DPCG**) method or a Conjugate Gradient-like method with matrix B as a "preconditioner", see [11] and [2]. Particularly, this means that one linear system in B is solved at each iteration for the second method.

3. The DD preconditioners

The construction of the preconditioners is based on partitioning the domain as indicated on Fig. 1, that is either into strips or boxes. We propose three preconditioners in the following. The first one is based on the partition into strips and the other two on the partition into boxes. Note that here we consider only nonoverlapping subdomains (strips or boxes).

The stripwise partition was introduced by Dryja and Proskurowski in [6]. Briefly, they defined a preconditioner of the original problem by keeping the same operator inside the strips and adding boundary conditions on the interfaces: Dirichlet boundary conditions for the white strips and Neumann boundary conditions for the black strips. Here, we replace the Neumann boundary conditions

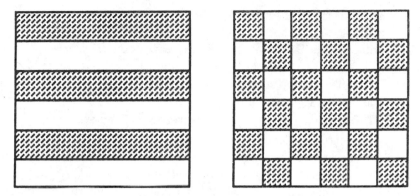

FIGURE 1 DECOMPOSITIONS OF THE DOMAIN

by mixed Neumann-Dirichlet boundary conditions for the black strips, i.e.

$$\mu u + \frac{\partial u}{\partial n} = 0, \quad \mu > 0.$$

Indeed, taking a prescribed nonnegative value for μ improves greatly the convergence rate of the method[1]. See [3] for numerical examples and details of implementation. Let B_1 be the discretized preconditioner. We have three types of unknowns: the nodes in the white strips (W), the nodes in the black strips (B) and the nodes on the interfaces called separators (S).

If we rewrite A and B_1 by block, it can easily be seen that A and B_1 differ only by their SS diagonal block. We therefore use the Capacitance Matrix and Conjugate Gradient methods to solve the problem in this case. Also, solving one linear problem with B_1 can be done in two steps:

    (1) one parallel solve on the extended black strips,

    (2) one parallel solve on the white strips.

Extended black strips include their surrounding separators. The resulting method is called **Cap1**.

The boxwise partitioning was introduced by Proskurowski and al in [7], [8] and [10]. The second preconditioner B_2 is similar to B_1, except that purely Neumann boundary conditions are considered for the black boxes. Indeed, for a variety of numerical examples ([3] or [4]), one can see that when the domain is divided into boxes, then the purely Neumann boundary condition is "optimal". Adding a fourth type of unknowns: the crosspoints (C), we use the same strategy as before to derive an iterative method also based on the Capacitance Matrix and Conjugate Gradient methods. Now, solving a linear problem with B_2 is done in two steps:

    (1) one solve on the extended black boxes and the crosspoints,

    (2) one parallel solve on the white boxes.

---

[1]This is also true for overlapping strips, as proved by Tang [12].

In short, to uncouple the extended black boxes (which include their surrounding separators), we introduce a problem defined only on the crosspoints. The corresponding matrix is sparse with at most seven nonzero entries per row (see [2]). The problem (1) which was originally defined on the extended black boxes and the crosspoints is replaced by two problems on the extended black boxes and one problem on the crosspoints:

(1.1) two parallel solves on the extended black boxes,

(1.2) one (parallel) solve on the crosspoints.

The problem defined on the crosspoints is solved by the DPCG method to get a parallel solver. The resulting method is called **Cap2**.

The third preconditioner, called M, is equal to one of the preconditioners studied by Ciarlet and Meurant. M corresponds to the second preconditioner in their terminology; see [5] for details. Solving a linear problem with M requires:

(1) two parallel solves on the (black and white) boxes,

(2) two parallel solves on the separators,

(3) one (parallel) solve on the crosspoints.

Step (2) is parallel because it can be shown that M_{SS} is block diagonal, each block corresponding to the nodes around a black box. For step (3), the DPCG method is used. As a matter of fact, the problem defined on the crosspoints for B_2 and M are identical. This last method, based on the PCG method, is called **MPCG**.

4. A few numerical examples

We solve the linear system $Ax = f$ on two computers. The first one is a Sequent Symmetry S81 with 20 Intel 80386 processors: a shared memory parallel architecture. The second computer is an IBM RS6000/560 with a RisC6000 processor: a superscalar architecture. The programming language is Fortran 77. We run the same code on both machines, with a preprocessing step using KAP [9], an automatic parallelizer, on the Sequent.

We will study the iterative methods in terms of

(1) the number of floating operations to solve the linear problem,

(2) the number of floating operations per second,

(3) the speed-up on the parallel architecture, from 1 to 16 processors,

(4) the CPU times.

Moreover, we will compare these results with those obtained for a well known iterative parallel method, the **DPCG** method.

The set of problems to be solved is

Problem #1

$a = 1$ in Ω

$b = 1$ in Ω

Problem #2

$a = \begin{cases} 10 & \text{if } 0 \le y < \frac{1}{2} \\ 1 & \text{elsewhere} \end{cases}$

$b = \begin{cases} 10 & \text{if } \frac{1}{2} < x \le 1 \\ 1 & \text{elsewhere} \end{cases}$

Problem #3

$a = \begin{cases} 10^2 & \text{if } \frac{1}{4} \le x \le \frac{3}{4} \\ 1 & \text{elsewhere} \end{cases}$

$b = 1$ in Ω

We set the number of unknowns, n, to 65025 ($= 255^2$). As we iteratively solve the problems, a stopping criterion is prescribed: it is reached when the norm of the residual has been reduced by a factor of 10^{-6}.

Then, we have to choose the number of strips (called n_0) and the number of boxes (called n_0^2). The bigger this number is, the greater the parallelism of the method, as $\frac{1}{2}n_0$ (resp., $\frac{1}{2}n_0^2$, n_0^2) subproblems are solved in parallel for **Cap1** (resp., **Cap2**, **MPCG**). However, note that by the way the domain is partitioned (Fig. 1), n_0 can, at most, be as large as $\frac{1}{4}\sqrt{n}$. Finally, recall that we want "sparse" preconditioners: here, this means that the number of nonzero entries of the preconditioners is proportional to n, as the storage of the original matrix and the vectors for the DPCG method is approximately $11n$. In [**3**], it is shown that this also leads to a value of n_0 proportional to $\sqrt{n}$. Therefore, potentially parallel preconditioners are "sparse" and *vice versa*. In the following, we choose $n_0 = 64$ for the strips and $n_0^2 = 32^2$ for the boxes, leading to the following storage for the original matrix, the preconditioner and the vectors for each iterative method: $20n$ for **Cap1**, $37n$ for **Cap2** and $43n$ for **MPCG**.

The parameter μ, arising in the mixed Neumann-Dirichlet boundary condition for **Cap1**, is set to n_0.

Table 1 compares the number of floating operations for Problem #1.

Table 1

Number of floating operations

Method	DPCG	Cap1	Cap2	MPCG
FLOP	928.10^6	343.10^6	244.10^6	312.10^6

The results are quite similar across all methods. The reference method gives by far the worst results. The numbers of iterations required to solve Problem #1 are the following: 645 for **DPCG**, 87 for **Cap1**, 12 for **Cap2** and 17 for **MPCG**. Unfortunately for the box-based methods, the average number of subiterations to solve the problem defined on the crosspoints is equal to 61 for the first method and 63 for the other.

Tables 2 and 3 give the average number (over the three problems) of floating point operations per second for each method. The results are in MFLOPs (millions of FLOPs). Table 2 gathers the results obtained on the RS6000. In Table 3, results are given for one and sixteen processors of the Sequent, and the corresponding speed-up is derived.

Table 2

Number of floating operations per second on the RS6000

Method	DPCG	Cap1	Cap2	MPCG
MFLOPs	25	24	20	16

Table 3

Number of floating operations per second and speed-up [bracketed] on the Sequent

Method	DPCG	Cap1	Cap2	MPCG
MFLOPs$_{\{1\}}$	0.31	0.41	0.32	0.38
MFLOPs$_{\{16\}}$	4.8 [15.7]	5.5 [13.6]	3.6 [11.3]	3.9 [10.5]

Surprisingly, the reference method is not the fastest one on the parallel architecture, though it is for this method that the highest speed-up is reached. In terms of parallel versus sequential parts of the execution of the code, using Amdahl's law, one finds that well over 99% of the reference method is executed in parallel, in comparison to a little less than 99% for the strip-based method and around 97% for the box-based methods. Note that these box-based implementations are actually less parallelized than the strip-based one, although they are potentially more parallelizable.

Finally, we show the CPU times needed to solve Problem #2 on the RS6000 in Table 4 and the CPU times on the 16-processor Sequent to solve Problem #3 in Table 5.

Table 4

CPU times on the RS6000

Method	DPCG	Cap1	Cap2	MPCG
time (s)	87	17	16	27

Table 5

CPU times on the 16-processor Sequent

Method	DPCG	Cap1	Cap2	MPCG
time (s)	990	47	161	177

The best methods are **Cap1** and **Cap2**, though both box-based methods are competitive versus the strip-based method. Moreover, a closer look at the CPU time shows that one third of the time is spent in the subroutines computing the matrix defined on the crosspoints for the box-based methods. Therefore, one way to reduce the CPU time for these methods is to find a better way to compute this matrix. Note that the strip-based method is particularly efficient for Problem #3, although the first coefficient of the operator λ is only piecewise constant inside each strip.

4. Conclusion

In this paper, we have presented three preconditioners for CG-like methods that seem well suited for parallel architectures. We have shown that the three of them compare very favorably to a fully parallel method. Nevertheless, the box-based methods have to be implemented on a massively parallel architecture to further enhance their merit, as they are potentially more parallelizable than a strip-based method. For the box-based methods, a solution has to be found to the very costly computation of the matrix defined on the crosspoints. Also, an efficient solver has to be derived for this matrix.

REFERENCES

1. B. Buzbee, F. Dorr, A. George and G. Golub, *The direct solution of the discrete Poisson equation in irregular regions*, SIAM J. Numer. Anal. **8** (1971), 722–736.
2. P. Ciarlet, Jr, *Méthodes itératives de résolution de problèmes elliptiques en 2D adaptées à des architectures (massivement) parallèles*, Technical Report, Note CEA 2688 (1992).
3. P. Ciarlet, Jr, *Etude de préconditionnements parallèles pour la résolution d'équations aux dérivées partielles elliptiques. Une décomposition de l'espace $L^2(\Omega)^3$*, Ph.D. thesis, Univ. Paris VI, France, 1992.
4. P. Ciarlet, Jr, *Implementation of a domain decomposition method well-suited for parallel architectures*, Int. J. of High Speed Computing (to appear).
5. P. Ciarlet, Jr and G. A. Meurant, *A class of domain decomposition preconditioners for massively parallel computers*, Domain Decomposition Methods in Science and Engineering (A. Quarteroni, J. Périaux, Y. A. Kuznetzov and O. B. Widlund, eds.), AMS, 1994, pp. 353–359.
6. M. Dryja and W. Proskurowski, *Capacitance matrix method using strips with alternating Neumann and Dirichlet boundary conditions*, Applied Numer. Math. **1** (1985), 285–298.
7. M. Dryja, W. Proskurowski and O. Widlund, *Numerical experiments and implementation of a domain decomposition method with cross points for the model problem*, Advances in Computer Methods for Partial Differential Equations VI (R. Vichnevetsky and R.S. Stepleman, eds.), IMACS, 1987, pp. 23–27.
8. M. Haghoo and W. Proskurowski, *Parallel implementation of a domain decomposition method*, Technical Report, CRI 88–06 (1988).
9. Kuck and Associates, Inc., *KAP/Sequent user's guide version 6*, 1988.
10. W. Proskurowski and S. Sha, *Performance of the Neumann-Dirichlet preconditioner for substructures with intersecting interfaces*, Domain Decomposition Methods for Partial Differential Equations (T.F. Chan, R. Glowinski, J. Périaux and O. Widlund, eds.), SIAM, 1990, pp. 322–337.
11. W. Proskurowski and O. Widlund, *A finite element-capacitance matrix method for the Neumann problem for Laplace's equation*, SIAM J. Sci. Stat. Comput. **1** (1980), 410–425.
12. W. P. Tang, *Generalized Schwarz splittings*, SIAM J. Sci. Stat. Comput. **13** (1992), 573–595.

CEA, CEL-V/ D.MA, 94195 Villeneuve St Georges Cedex, France

E-mail address: ciarlet@limeil.cea.fr

Contemporary Mathematics
Volume 180, 1994

IBLU Preconditioners for Massively Parallel Computers

E. DE STURLER

December 1, 1994

ABSTRACT. ILU and MILU are more or less the standard for preconditioning, but their inherent sequentiality precludes efficient parallel implementation. Blocked variants improve the parallelism, but generally they increase
the iteration count.

To improve this we adapt approaches from [2, 4], and we define parameterized block preconditioners based on subdomains with minimal overlap.
Although the choice of the parameters is an open problem, numerical experiments suggest that the iteration count can be kept almost constant going
from one to 400 subdomains. We also report on the performance on a 400-
processor Parsytec Supercluster at the Koninklijke/Shell-Laboratorium in
Amsterdam.

1. Introduction

For efficient iterative solvers on massively parallel computers, the iteration
must be efficient and the increase in the number of iterations must be low.
Here we focus on preconditioners that have a simple implementation to facilitate
their use in large existing programs ported to massively parallel computers. Incomplete Block LU (IBLU) preconditioners with the blocks corresponding to the
local equations of the subdomains are good candidates, but they tend to increase
the number of iterations significantly. We use adaptations to the approaches in
[2, 4] to overcome this problem. We take subdomains with minimal overlap,
and we modify the local equations on the artificial boundaries. The substitution
of the artificial boundary conditions into the matrix gives the so-called gener-

1991 *Mathematics Subject Classification*. Primary 65F10; Secondary 65Y05.

The author wishes to acknowledge Shell Research B.V. and STIPT for the financial support
of this research.

This paper is in final form and no version of it will be submitted for publication elsewhere.

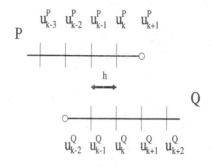 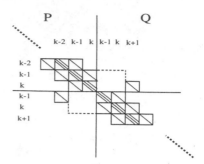

FIGURE 1.   Overlapping   FIGURE 2.   The SEM for the
subdomains P and Q.   overlap of P and Q.

alized Schwarz enhanced matrix (SEM) [4]. From this matrix we use the ILU
factorizations of the local equations to form the IBLU preconditioner.

2. Construction of the preconditioner

We describe the construction for a one-dimensional decomposition of a two-
dimensional domain and the extension to higher-dimensional decompositions and
more general meshes.

2.1. One-dimensional decomposition.
We assume a block tridiagonal ma-
trix which has been derived from the well-known 5-point discretization star. We
decompose the domain in the y-direction and we duplicate on each subdomain a
number of grid lines from the neighbouring subdomain. Figure 2 gives the matrix structure for duplicated grid lines corresponding to an overlap as indicated
in Figure 1, where u_k corresponds to the approximation on the k-th grid line in
the y-direction, and we have duplicated two grid lines, u_{k-1} and u_k. We will use
the artificial boundary conditions

$$(2.1) \qquad \beta_1 u_{k-1}^P + \alpha_1 u_k^P - u_{k+1}^P \;=\; \beta_1 u_{k-1}^Q + \alpha_1 u_k^Q - u_{k+1}^Q,$$

$$(2.2) \qquad \beta_2 u_k^P + \alpha_2 u_{k-1}^P - u_{k-2}^P \;=\; \beta_2 u_k^Q + \alpha_2 u_{k-1}^Q - u_{k-2}^Q,$$

to give equations for the exterior grid lines u_{k+1}^P and u_{k-2}^Q that are substituted
into the SEM to construct the generalized SEM. If we have only one overlapping
grid line we cannot use the parameter β. We can use very general boundary conditions, because the generalized SEM only defines the IBLU preconditioners, and
the only requirement for convergence is the nonsingularity of the preconditioner.

2.2. Higher dimensional decomposition and general meshes.
Higher
dimensional decompositions and more general meshes make the construction
of the generalized SEM directly from the matrix complicated. Therefore, we
adapt the discretization star on the boundaries of the subdomain instead; see
[1]. For example, a boundary condition like (2.1) leads to the discretization
star in Figure 3. This approach also permits artificial boundary conditions in

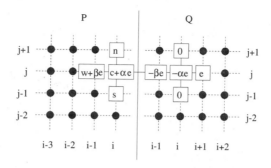

FIGURE 3. The adapted discretization star

directions non-orthogonal to the boundary; see [3].

2.3. Definition of the preconditioner. Let Ω denote the original domain, and let $\tilde{\Omega}$ be the decomposed domain with overlap. The operator $S : \Omega \mapsto \tilde{\Omega}$ assigns to a grid point in $\tilde{\Omega}$ the value of the corresponding grid point in Ω (also for duplicates). The operator $J_\omega : \tilde{\Omega} \mapsto \Omega$ is a projection using a weighted average on duplicate grid points. The operators S and J_ω satisfy the following relations. $J_\omega S : \Omega \mapsto \Omega = I$, and $S J_\omega : \tilde{\Omega} \mapsto \tilde{\Omega}$ computes a weighted average over duplicate grid points and is the identity operator on other grid points.

From the generalized SEM ($\tilde{\Omega} \mapsto \tilde{\Omega}$) we compute the (block) factors $\tilde{L}$, $\tilde{D}$ and $\tilde{U}$. We define the following operators over Ω, $J_\omega \tilde{L}^{-1} S$, $J_\omega \tilde{D} S$, and $J_\omega \tilde{U}^{-1} S$, and we define the preconditioner as $K : \Omega \mapsto \Omega = J_\omega \tilde{U}^{-1} S J_\omega \tilde{D} S J_\omega \tilde{L}^{-1} S$. We can reduce the communication cost by modifying $\tilde{D}$ such that $\tilde{D} S = S D$ for some $D : \Omega \mapsto \Omega$. This only requires averaging the entries of the diagonal matrix $\tilde{D}$ corresponding to duplicate grid points. Then, K is defined by

$$(2.3) \qquad K : \Omega \mapsto \Omega = J_\omega \tilde{U}^{-1} \tilde{D} S J_\omega \tilde{L}^{-1} S.$$

3. Experimental results

For the performance of the preconditioner, the iteration count is much more important than the extra computation for the overlaps and the communication with a few nearby processors. Therefore, we focus on the iteration count and discuss other overhead succinctly at the end of the section. We show only the potential of these preconditioners for reducing the number of iterations. For a discussion on the choice of parameters and the convergence see [1].

The first test problem is the Poisson equation on the unit square, $-\Delta u = 0$, with $u = 1$ for $y = 0$, $u = 0$ for $y = 1$, and $u' = 0$ for $x = 0$ and $x = 1$, where u' is the outward normal derivative on the boundary. We discretized this problem on a 100×100 grid using the finite volume method. The second test problem is given by $-(u_{xx} + u_{yy}) + b u_x + c u_y = 0$, on $[0,1] \times [0,4]$, where

$$b(x,y) = \begin{cases} 10 & \text{for } 0 \le y \le 1 \text{ and } 2 < y \le 3, \\ -10 & \text{for } 1 < y \le 2 \text{ and } 3 < y \le 4, \end{cases}$$

E. DE STURLER

TABLE 1. Problem 1 with local preconditioning.

decomposition	iteration count
1×1	$2 \times 50 + 17$
1×5	$3 \times 50 + 26$
1×10	$4 \times 50 + 2$
1×20	$4 \times 50 + 6$
20×20	$5 \times 50 + 46$

TABLE 2. Problem 2 with local preconditioning.

decomposition	iteration count
1×1	$11 \times 50 + 7$
1×5	$10 \times 50 + 35$
1×10	$11 \times 50 + 33$
1×20	$12 \times 50 + 10$
20×20	$19 \times 50 + 40$

and $c = 10$, with $u = 1$ on $y = 0$, $u = 0$ on $y = 4$, $u' = 0$ for $x = 0$ and $x = 1$. We discretized this problem on a 100×200 grid using the finite volume method. Using these problems we can compare the effects of the preconditioners for diffusion- and convection dominated flow.

We used GMRES(50) for both test problems to compare the convergence results. Tables 1 and 2 show the results for several decompositions without overlap. The one-dimensional decompositions are all in the y-direction. We give the iteration count as the number of complete cycles plus the number of iterations in the last cycle. We see that the effects of the decomposition are much larger for problem 1 than for problem 2.

Tables 3 and 4 show the iteration counts using overlapping grid lines and adapted artificial boundary conditions (2.1,2.2). We give the results without adaptation, $\alpha, \beta = 0$ and (if different) for the optimal α with $\beta = 0$ and for the 1×20 decomposition also with $\beta \neq 0$. Our experiments indicated that the optimal α increases if the number of subdomains increases, and that the convergence is not very sensitive to the choice of α. Although the iteration count as a function of the number of subdomains behaves differently for the two problems, the influence of the parameters α and β seems similar.

We will now discuss the convergence using preconditioners derived from two-dimensional decompositions on a 20×20 processor grid. For simplicity we have taken ω, the overlap size, and α and β (when appropriate) the same for all

TABLE 3. Problem 1 with adapted boundary conditions.

decomp.	ovl = 1		ovl = 2	
	α	it.count	α/β	it.count
1×5	0.0	$2 \times 50 + 28$	0.0/0.0	$2 \times 50 + 25$
	0.1	$2 \times 50 + 26$	0.4/0.0	$2 \times 50 + 24$
1×10	0.0	$2 \times 50 + 39$	0.0/0.0	$2 \times 50 + 38$
	0.4	$2 \times 50 + 37$		
1×20	0.0	$3 \times 50 + 17$	0.0/0.0	$3 \times 50 + 14$
	0.7	$3 \times 50 + 8$	0.6/0.0	$2 \times 50 + 37$
			0.7/0.4	$2 \times 50 + 17$

TABLE 4.   Problem 2 with adapted boundary conditions.

decomp.	ovl = 1		ovl = 2	
	α	it.count	α/β	it.count
1×5	0.0	$9 \times 50 + 15$	0.0/0.0	$10 \times 50 + 17$
			0.4/0.0	$9 \times 50 + 45$
1×10	0.0	$11 \times 50 + 44$	0.0/0.0	$11 \times 50 + 11$
	0.4	$10 \times 50 + 1$	0.3/0.0	$9 \times 50 + 41$
1×20	0.0	$10 \times 50 + 22$	0.0/0.0	$10 \times 50 + 28$
	0.5	$10 \times 50 + 8$	0.6/0.0	$10 \times 50 + 1$
			0.6/0.1	$9 \times 50 + 47$

TABLE 5.   Problem 1 with adapted boundary conditions on a 20×20 processor grid.

TABLE 6.   Problem 2 with adapted boundary conditions on a 20×20 processor grid.

α	β	iteration count
0.0	0.0	$7 \times 50 + 26$
0.5	0.0	$3 \times 50 + 15$

α	iteration count
0.0	$16 \times 50 + 10$
0.2	$13 \times 50 + 41$

directions.

Table 5 shows the results for problem 1 with two overlapping grid lines, and Table 6 shows the results for problem 2 with one overlapping grid line (best choices). The results for $\alpha, \beta = 0$ show that for two-dimensional decompositions without adapted boundary conditions the convergence may be poor. The other results are for the optimal values for α and β. Our experiments indicated that for two-dimensional decompositions the iteration count is much more sensitive to the choice of parameters than for one-dimensional decompositions. The results show that we can keep the increase in iterations small while running the algorithms on as much as 400 processors.

Finally, we discuss the performance for the tests with the minimal number of iterations (to illustrate the potential). We compared the measured runtimes to runtimes of a virtual preconditioned GMRES(50) that has the cycle time of GMRES(50) with a local preconditioner and the number of iterations of the sequential algorithm. This models a perfect speed-up for the preconditioner. We give the relative overhead of the preconditioners for the runtime of a cycle and for

TABLE 7.   Measured runtimes and relative performance for problem 1.

overlap (x/y)	measured runtimes			relative performance		
	cycle time (s)	iteration count	solution time (s)	cycle time	iteration count	solution time
0	1.77	$5 \times 50 + 46$	10.2	1.00	2.53	2.53
2	1.91	$3 \times 50 + 15$	6.90	1.08	1.41	1.52

TABLE 8.   Measured runtimes and relative performance for problem 2.

overlap (x/y)	measured runtimes			relative performance		
	cycle time (s)	iteration count	solution time (s)	cycle time	iteration count	solution time
0	2.21	$19 \times 50 + 40$	43.3	1.00	1.78	1.78
1	2.31	$13 \times 50 + 41$	31.8	1.05	1.24	1.30

the number of iterations. The product of these gives approximately the relative overhead for the total solution time. This can be considered as the inverse of the efficiency. The results are given in Tables 7 and 8. For comparison we also give the results for local preconditioning. The adapted preconditioners give a significantly better performance. Moreover, given the fact that we run on 400 processors in parallel, we can stay quite close to the optimal performance: about 65% for problem 1 and almost 80% for problem 2.

4. Conclusions

The IBLU preconditioners described have the potential to keep the iteration count almost constant going from one to 400 subdomains. The overhead within one cycle is only marginal, even on a large number of processors. The a priori choice of good parameters is an open problem, and future research in this direction is necessary. However, for block relaxations see [3, 4]. The generation of the preconditioner is straightforward and efficient; it does not introduce any significant parallel overhead.

REFERENCES

1. E. De Sturler. Incomplete Block LU preconditioners on slightly overlapping subdomains for a massively parallel computer. Technical Report CSCS-TR-94-03, Swiss Scientific Computing Center CSCS-ETHZ, 1994.
2. G. Radicati di Brozolo and Y. Robert. Parallel conjugate gradient-like algorithms for solving sparse nonsymmetric linear systems on a vector multiprocessor. *Parallel Comput.*, 11:223–239, 1989.
3. K.H. Tan and M.J.A. Borsboom. Problem-dependent optimization of flexible couplings in domain decomposition methods, with an application to advection-dominated problems. Technical Report 830, Mathematical Institute, University of Utrecht, 1993.
4. W.P. Tang. Generalized Schwarz splittings. *SIAM J. Sci. Statist. Comput.*, 13:573–595, 1992.

FACULTY OF TECHNICAL MATHEMATICS AND INFORMATICS, DELFT UNIVERSITY OF TECHNOLOGY, DELFT, THE NETHERLANDS
   Current address: Section of Research and Development, Swiss Scientific Computing Center CSCS-ETHZ, Via Cantonale, CH-6928 Manno, Switzerland
   E-mail address: sturler@cscs.ch

Contemporary Mathematics
Volume **180**, 1994

Tailoring Domain Decomposition Methods for Efficient Parallel Coarse Grid Solution and for Systems with Many Right Hand Sides

CHARBEL FARHAT

PO-SHU CHEN

ABSTRACT. We present and illustrate a methodology for extending the range of applications of scalable domain decomposition methods to problems with multiple and/or repeated right-hand sides, and for solving efficiently their coarse grid problems on massively parallel processors.

1. Introduction

The condition number of the interface problem associated with a *numerically scalable* domain decomposition (DD) method does not grow (or "grows weakly") asymptotically with the number of subdomains. One of the many reasons why numerical scalability is desirable is that increasing the number of subdomains is the simplest means for increasing the degree of parallelism of a DD based preconditioned conjugate gradient (PCG) algorithm. In other words, numerical scalability is critical for massively parallel processing. This optimal property is usually achieved via the introduction in a DD method of a coarse problem (or coarse grid, by analogy with multigrid methods) that relates to the original problem and that must be solved at each global CG iteration. Direct methods are often chosen for solving the coarse problem despite the fact that they are difficult to implement on a massively parallel processor and do not parallelize well. Therefore in many cases, a numerically scalable DD method loses its appeal because of its lack of *parallel scalability*. One way to restore parallel scalability is to solve iteratively the coarse problem, for example using a CG

1991 Mathematics Subject Classification Primary 65N20, 65N30, 65W05

Supported by RNR NAS at NASA Ames Research Center under Grant NAG 2-827, and by the National Science Foundation under Grant ASC-9217394.

This paper is in final form and will not be published elsewhere.

scheme. However, this approach raises the question of how to solve iteratively and efficiently a system with a constant matrix and repeated right-hand sides. Finding an answer to this question also extends the range of applications of DD methods to design problems, time dependent problems, eigenvalue problems, and several other applications where multiple and repeated right-hand sides always arise and challenge iterative solvers. The iterative solution of systems with multiple right-hand sides has been previously addressed in [1], and recently in [2,3]. Here, we present a CG based methodology for solving such problems that uses the same data structures as those employed in DD methods without a coarse grid. The numerical idea exposed in this paper is related to that analyzed in [1]. However, the specific algorithm proposed herein is different, simpler, and easier to parallelize than that described in [1], and faster but more memory consuming [4] than both schemes presented in [2].

2. Problem formulation and nomenclature

For the sake of clarity, we first discuss the problem and the proposed solution methodology in the absence of any DD method or coarse grid. In Section 4, we highlight the positive implications of domain decomposition on the advocated approach. We are interested in solving iteratively the following problems:

$$(1) \qquad\qquad A x_i \ = \ b_i \quad i \ = \ 1, \ ..., \ N_{rhs}$$

where A, $\{b_i\}_{i\,=\,1}^{i\,=\,N_{rhs}}$, and $\{x_i\}_{i\,=\,1}^{i\,=\,N_{rhs}}$ denote respectively a symmetric positive definite matrix, a set of N_{rhs} right-hand sides, and the corresponding set of N_{rhs} solution vectors. These problems can be transformed into the following minimization problems:

$$(2) \qquad\qquad \min_{x \in \mathcal{R}^{N_A}} \Phi_i(x) \ = \ \frac{1}{2}\, x^T A x \ - \ b_i^T x \quad i \ = \ 1, \ ..., \ N_{rhs}$$

where N_A is the dimension of matrix A, $\mathcal{R}$ is the set of real numbers, and T is the transpose superscript. If each minimization problem in (2) is solved with a PCG algorithm, the following Krylov subspaces are generated:

$$(3) \qquad \mathcal{S}_i \ = \ \{s_i^{(1)}, \ s_i^{(2)}, \ ..., \ s_i^{(k)}, \ ..., \ s_i^{(r_i)}\} \quad i \ = \ 1, \ ..., \ N_{rhs}$$

where $s_i^{(k)}$ and $r_i < N_A$ denote respectively the search direction vector at iteration k, and the number of iterations for convergence of the PCG algorithm applied to the minimization of $\Phi_i(x)$. Additionally, we introduce the following agglomerated subspaces:

$$(4) \qquad\qquad \overline{\mathcal{S}}_i \ = \ \bigcup_{j\,=\,1}^{j\,=\,i} \mathcal{S}_j \quad i \ = \ 1, \ ..., \ N_{rhs}$$

Let S_i denote the rectangular matrix associated with $\mathcal{S}_i$. From the orthogonality properties of the conjugate gradient method, it follows that:

$$(5) \qquad S_i^T A S_i \ = \ D_i \quad i \ = \ 1, \ ..., \ N_{rhs} \quad D_i \ = \ \begin{bmatrix} d_{i_1} & d_{i_2} & \cdots & d_{i_{r_i}} \end{bmatrix}$$

However, note that in general $\overline{S}_i^T A \overline{S}_i$ is not a diagonal matrix. Finally, we define $\overline{\overline{S}}_i$ as the matrix whose column vectors also span the subspace $\overline{S}_i$, but are orthogonalized with respect to matrix A. Hence, we have:
(6)

$$range\,(\overline{\overline{S}}_i) = \overline{S}_i \quad i = 1, ..., N_{rhs}, \quad \overline{\overline{S}}_i^T A \overline{\overline{S}}_i = \overline{\overline{D}}_i \quad i = 1, ..., N_{rhs}$$

$$\overline{\overline{D}}_i = \begin{bmatrix} \overline{\overline{d}}_{i_1} & \overline{\overline{d}}_{i_2} & \cdots & \overline{\overline{d}}_{i_{\overline{r}_i}} \end{bmatrix}, \qquad \overline{r}_i = \sum_{j=1}^{j=i} r_j$$

3. Projection and orthogonalization

Suppose that the first problem $Ax_1 = b_1$ has been solved in r_1 PCG iterations, and that the $N_A \times r_1$ matrix S_1 associated with the Krylov subspace S_1 is readily available. Solving the second problem $Ax_2 = b_2$ is equivalent to solving:

(7)
$$\min_{x \in \mathcal{R}^{N_A}} \Phi_2(x) = \frac{1}{2} x^T A x - b_2^T x$$

If $\mathcal{R}^{N_A}$ is decomposed as follows:

(8) $\quad \mathcal{R}^{N_A} = S_1 \oplus S_1^*, \quad dim\,(S_1^*) = N_A - r_1, \quad S_1$ and S_1^* are $A-$orthogonal

then the solution of problem (7) can be written as:

(9) $\quad x_2 = x_2^0 + z_2, \quad x_2^0 \in S_1, \; z_2 \in S_1^*, \;$ and $\; x_2^{0^T} A z_2 = z_2^T A x_2^0 = 0$

From Eqs. (7-9), it follows that x_2^0 is the solution of the minimization problem:

(10)
$$\min_{x \in S_1} \Phi_2(x) = \frac{1}{2} x^T A x - b_2^T x$$

and z_2 is the solution of the minimization problem:

(11)
$$\min_{z \in S_1^*} \Phi_2(z) = \frac{1}{2} z^T A z - b_2^T z$$

First, we consider the solution of problem (10). Since $x_2^0 \in S_1$, there exists a $y_2^0 \in \mathcal{R}^{r_1}$ such that:

(12)
$$x_2^0 = S_1 y_2^0$$

Substituting Eq. (12) into Eq. (10) leads to the following minimization problem:

(13)
$$\min_{y \in \mathcal{R}^{r_1}} \widetilde{\Phi}_2(y) = \frac{1}{2} y^T S_1^T A S_1 y - b_2^T S_1 y$$

whose solution y_2^0 is given by:

$$(14) \qquad S_1^T A S_1 \, y_2^0 \; = \; \tilde{b}_2, \quad \tilde{b}_2 \; = \; S_1^T b_2$$

From Eq. (5), it follows that the system of equations (14) is diagonal. Hence, the components $[y_2^0]_j$ of y_2^0 can be simply computed as follows:

$$(15) \qquad [y_2^0]_j \; = \; \frac{[\tilde{b}_2]_j}{d_{1j}} \qquad j \; = \; 1, \, ..., \, r_1$$

Next, we turn to the solution of problem (11) via a PCG algorithm. Since the decomposition (9) requires z_2 to be A-orthogonal to x_2^0, at each iteration k, the search directions $s_2^{(k)}$ must be explicitly A-orthogonalized to S_1. This entails the computation of modified search directions $\hat{s}_2^{(k)}$ as follows:

$$(16) \quad \hat{s}_2^{(k)} \; = \; s_2^{(k)} + \sum_{q=1}^{q=r_1} \alpha_q s_1^{(q)}, \quad \alpha_q \; = \; -\frac{s_1^{(q)^T} A s_2^{(k)}}{s_1^{(q)^T} A s_1^{(q)}} \; = \; -\frac{s_2^{(k)^T} A s_1^{(q)}}{s_1^{(q)^T} A s_1^{(q)}}$$

Except for the above modifications, the original PCG algorithm is unchanged. However, convergence is expected to be much faster for the second and subsequent problems than for the first one, because $\mathcal{S}_1^*$ and the subsequent supplementary spaces have smaller dimensions than $\mathcal{R}^{N_A}$, and a significant number of the solution components are included in the startup solutions of the form of x_2^0.

The generalization to the case of N_{rhs} right-hand sides of the two-step solution procedure described above is straightforward [3].

4. Application to coarse grids and DD interface problems

Despite its elegance and simplicity, the methodology described in Section 3 can be impractical when applied to the global solution of the problems $A x_i = b_i$, $i = 1, N_{rhs}$. Indeed, during the PCG solution of the first few problems — that is, before superconvergence can be reached — the cost of the orthogonalizations implied by Eq. (16) can offset the benefits of convergence acceleration via the optimal startup solution and the modified search directions $\hat{s}_i^{(k)}$. Moreover, storing every search direction $\hat{s}_i^{(k)}$ and the corresponding matrix-vector product $A\hat{s}_i^{(k)}$ can significantly increase the memory requirements of the basic PCG algorithm. However, the proposed methodology is computationally feasible in a domain decomposition context because it is applied only to the coarse grid and/or the interface problem.

5. Performance evaluation for coarse problems

First, we consider the static solution of a two-dimensional plane stress elasticity problem using the FETI domain decomposition method [5] on an iPSC-860 parallel processor. The FETI method is numerically scalable. For elasticity problems, the two-norm condition number of its preconditioned interface problem grows asymptotically as $\kappa_2 = O\left(1 + log^2 \frac{H}{h}\right)$, where H and h denote respectively the subdomain and mesh sizes [6]. At each k-th FETI global PCG iteration, the following coarse problem must be solved:

$$(17) \qquad\qquad (G_I{}^t G_I)\, x_k \;=\; b_k$$

where $G = [\, B_1 R_1 \quad \ldots \quad B_s R_s \quad \ldots \quad B_{N_s} R_{N_s}\,]$, B_s is a boolean matrix that extracts from a subdomain quantity its interface component, R_s spans the null space of the stiffness matrix associated with a singular subdomain s, N_s denotes the total number of subdomains, and b_k is related to the interface residual at the k-th global PCG iteration. Clearly, the systems in (17) have a constant matrix and repeated right-hand sides. The parallel solution of these problems via the CG scheme described in Section 3 is implemented using the existing subdomain based parallel data structures; it requires communication only between neighboring subdomains. Using 4-node plane stress elements, three finite element models corresponding to 4, 16, and 64 subdomains are constructed: each model has a different size but the ratio H/h is kept constant across all three finite element models. The performance results measured for all three problems are summarized in Table 1 where NEQ, N_p, N_{itr}, and T_{tot} denote respectively the number of equations generated by the finite element model, the number of processors, the number of FETI global PCG iterations, and the total CPU time for solving the plane stress problem.

TABLE 1
Two-dimensional elasticity problem:
performance results on an iPSC-860

h	H	NEQ	N_p	N_{itr}	T_{tot}
$\frac{1}{40}$	$\frac{1}{2}$	3,200	4	10	3.45 secs.
$\frac{1}{80}$	$\frac{1}{4}$	12,800	16	16	5.17 secs.
$\frac{1}{160}$	$\frac{1}{8}$	51,200	64	17	5.92 secs.

Clearly, the results summarized in Table 1 demonstrate the combined numerical and parallel scalability of the FETI method and its parallel implementation.

6. Performance evaluation for time dependent problems

Next, we apply the methodology described in this paper to the solution of repeated systems arising from the linear transient analysis using an implicit time-integration scheme of a three-dimensional line-pinched membrane with a circular hole. The structure is discretized in 5,680 4-node elements and 11,640 degrees of freedom. The finite element mesh is partitioned into 32 subdomains. The size of the interface problem is 1,892 — that is, 16.25% of the size of the global problem. The transient analysis is carried out on a 32 processor iPSC-860 system. After all of the usual finite element storage requirements are allocated, there is enough memory left to store a total number of 891 search directions. This number corresponds to 47% of the size of the interface problem. Using a transient version of the FETI method without a coarse grid [7], the system of equations arising at the first time step is solved in 322 iterations. After 3 time steps, 435 search directions are accumulated and only 20 iterations are needed for solving the fourth linear system of equations. After 16 time steps, the total number of accumulated search directions is only 536 — that is, only 28% of the size of the interface problem, and superconvergence is triggered: all subsequent time steps are solved in 1 or 2 iterations and in less than 1.0 second CPU (Fig. 1). When a parallel direct solver is applied to the above problem, at each time step, the pair of forward/backward substitutions consumes 15.0 seconds on the same 32 processor iPSC-860. Therefore, the proposed solution methodology is clearly an excellent alternative to repeated forward/backward substitutions on distributed memory parallel processors.

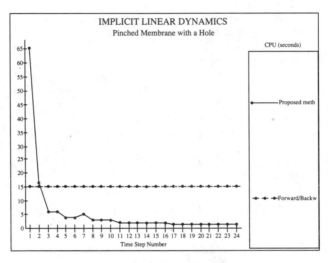

FIG. 1. *CPU history*

REFERENCES

1. Y. Saad, On the Lanczos Method for Solving Symmetric Linear Systems with Several Right-Hand Sides, Math. Comp., **48** No. 178 (1987), 651–662.

2. P. Fischer, Projection Techniques for Iterative Solution of Ax=b with Successive Right-Hand Sides, ICASE Rep. No. 93-90, NASA CR-191571.

3. C. Farhat, L. Crivelli and F. X. Roux, Extending Substructure Based Iterative Solvers to Multiple Load and Repeated Analyses, Comput. Meths. Appl. Mech. Engrg., (1994), in press.

4. P. Fischer, (1993), private communication.

5. C. Farhat and F. X. Roux, A method of finite element tearing and interconnecting and its parallel solution algorithm, Internat. J. Numer. Meths. Engrg., **32** (1991), 1205–1227.

6. C. Farhat, J. Mandel and F. X. Roux, Optimal Convergence Properties of the FETI Domain Decomposition Method, Comput. Meths. Appl. Mech. Engrg., (1994), in press.

7. L. Crivelli and C. Farhat, Implicit transient finite element structural computations on MIMD systems: FETI v.s. direct solvers, AIAA Paper 93-1310, AIAA 34th Structural Dynamics Meeting, La Jolla, California, April 19-21, 1993.

Department of Aerospace Engineering Sciences, Center for Aerospace Structures, University of Colorado, Boulder, Colorado 80309-0429

E-mail address: charbel@boulder.colorado.edu

Contemporary Mathematics
Volume **180**, 1994

Analysis and Implementation of DD Methods for Parallel FE Computations

HAI XIANG LIN

ABSTRACT. Applications of finite element methods (FEM) are known to be computation-intensive. Therefore, various forms of parallel numerical algorithms have been studied in order to speed up finite element computations. Many of these recently proposed parallelization methods use domain decomposition (DD) as a framework. In this paper, we analyze the load balance and communication overhead in relation to the number of subdomains and the number of processors. It is shown that for an efficient parallel computation the often used formulation of the partitioning problem needs to be reformulated. An efficient domain decomposer is presented which has been implemented as a parallelization preprocessor for the finite element software package DIANA.

1. Introduction

In [**3**], a methodology is presented for the parallelization of the FEM software packages (e.g. DIANA). Direct methods play a dominent role in commercial FEM software packages, so we first focus on the parallelization of the direct solution of a large sparse system of linear equations. Fig. 1 describes the major modules of the parallelization method. The proposed approach starts by extracting parallelism from the very beginning of a problem formulation: the description of the element model (i.e. element domain). This is accomplished by performing domain decomposition. The next module "ordering" determines a parallel elimination sequence.

Given an elimination sequence the non-zero structure of the matrix factor can be determined through the so called symbolic factorization. The matrix factor is stored as a doubly bordered block diagonal (DBBD) matrix (see Fig. 2c). All block submatrices in Region I are stored as skyline matrices, those in Region II are stored as a collection of sparse vectors, and those submatrices in Region III are stored as dense matrices. We define the operations (e.g. an LDL^T factorization, an update) with respect to an entire block submatrix as a single task. The task dependence graphs of the factorization and triangular solutions are then derived from the non-zero structure of the matrix factor. An efficient macro data flow execution scheme is used to implement the parallel execution of the task dependence graph.

1991 *Mathematics Subject Classification*. 65F05, 65Y05, 65F50.

DIANA is a registered tradmark of TNO Institute.

This paper is in final form and no version of it will be submitted for publication elsewhere.

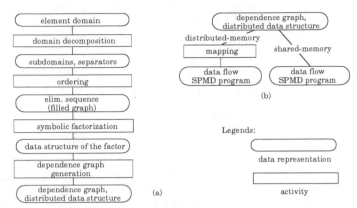

FIGURE 1. A strategy for the parallelization of direct solvers: the preparation phases; (b) Illustration of the PARASOL programs.

In this paper, we will concentrate on domain decomposition techniques. In order to handle regularly and irregularly structured domains with different types of elements, the domain is considered as a (general) connectivity graph. The problem of decomposing an element domain can now be considered as partitioning a graph. The so-called graph partitioning problem for the purpose of parallel computation is usually formulated as: *Divide a graph (element domain) into p (p is the number of processors) subgraphs separated from each other by the separators, such that the subgraphs are of the same size, i.e. consisting of the same number of elements or nodes, and that the number of separator nodes is minimal.*

It is known that smaller separators between the subgraphs imply fewer communication during a parallel computation. The requirement that all subgraphs having equal size are imposed to give a good load balance (we will show that more are required for a good load balance). Minimal separators and all subgraphs having equal sizes are two conflicting goals and the graph partitioning problem is known to be NP-complete. So, in practice heuristics have to be used.

2. A short review of previous work

The one-way and the nested dissection (ND) methods [2] were originally used to order the node elimination sequence for minimum fill-ins in sequential processing; recently they have been succefully applied as DD techniques for parallel computation. The dissection method starts by finding a "peripheral" node of the graph, then a subdivision is made through level-structuring. In case of one-way dissection, an s-partition can be obtained by selecting $(s-1)$ levels in the level structure as the sets of separators. In case of nested dissection, first a bisection is performed by chosing a "middle" level which separates the graph into two subgraphs of about the same size. Applying this bisection procedure on the resulting subgraphs repeatedly, a partition of $s = 2^t$ subgraphs is obtained after t iterations. Finding a good peripheral node usually requires a time of $O(n^2)$, so often a simpler approach is applied. The dissection methods are simple and fast (if a simple peripheral node is used), but do not give a very optimal separator length.

An alternative is to use a minimum-degree algorithm to perform a (sequential) ordering first. Then, a partitioning is performed on the elimination tree corresponding to the resulting ordering ([6], [1]). This method generally gives better results than the dissection methods in terms of separator length. However, the resulting

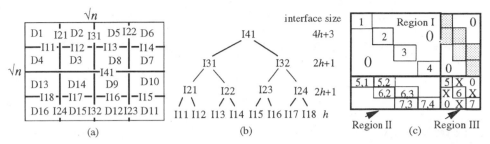

FIGURE 2. (a) A decomposition into 16 subdomains; (b) The elimination sequence of interfaces according to ND-ordering; (c) The DBBD matrix.

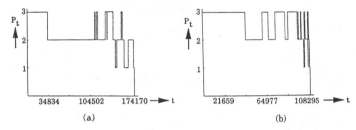

FIGURE 3. Execution profile of an example problem divided into (a) 3 subdomains; (b) 6 subdomains. $p = 3$. P_t is the number of tasks executed in parallel at a time.

subgraphs can vary considerably in size. Moreover, for a graph with n nodes a minimum-degree algorithm has a time complexity of $O(n^2)$ as compared to that of $O(n)$ to $O(n \cdot log(n))$ for the dissection methods. This means that the method using a minimum-degree algorithm has a higher time complexity than the solution of the sparse matrix system (e.g. $O(n^{3/2})$ for a $\sqrt{n} \times \sqrt{n}$ 2-D grid).

Another popular method is the spectral bisection (SB) method [7]. The SB method uses the second smallest eigenvector of the Laplacian matrix as the separator for a bisection of the graph. In terms of optimality the SB method is more robust than the dissection methods. However, it again has a time complexity of $O(n^2)$. So, speeding up the partitioning by means of parallel processing becomes an important issue. The parallel SB algorithm on the CM-5 is such an example.

3. Analysis of load balance in relation to the number of subdomains

In this section, we analyze the load balance and parallel efficiency as a function of subdomain's size and the number of processors. Because it is very hard to make a complexity analysis of a general element domain, we consider a $\sqrt{n} \times \sqrt{n}$ mesh and generalize the results to other problems. The $\sqrt{n} \times \sqrt{n}$ element mesh is subdivided into s subdomains each of size $\sqrt{n/s} \times \sqrt{n/s}$ as shown in Fig. 2a. We consider the factorization of $A = L \cdot D \cdot L^T$ with A being obtained through an ordering equivalent to the ND-ordering for the interfaces. We consider parallel sparse factorization of the type of medium to large granularity. Operations on the same block in the DBBD matrix form a single task and are scheduled on a single processor. In order to make the analysis simpler, we assume that the parallel factorization algorithm executes in a "lock-step" fashion. That is, all processors start with executing a group of tasks at the same time, they will wait for all the processors have completed their current group of tasks before entering a communication phase. After a communication phase, all processors will start to execute the next group of tasks.

The procedure of computing the LDL^T factor of the DBBD-matrix according to the ND-ordering can be described as follows.

1. For each $A_{i,i}$ in Region I, compute the LDL^T factor: $A_{i,i} \longrightarrow L_{i,i} \cdot D_{i,i} \cdot L_{i,i}^T$, for $i = 1, ..., s$. All the submatrices in Region I can be factorized independently.
2. Compute the factors in Region II: $L_{i,j} = A_{i,j} \cdot L_{i,i}^{-T} \cdot D_{i,i}^{-1}$ for $j = 1, ..., s$, and (interface segment) $i \in$ Adj{subdomain j}, followed by the updates in Region III (Schur-complement): $A_{k1,k2} = A_{k1,k2} - L_{k1,k} \cdot D_{k,k} \cdot L_{k2,k}^T$ for $k = 1, ..., s$ and interface segments $k2, k \in$Adj{subdomain $k1$}.
3. The interface segments induced by the last (recursive) bisection are eliminated first. Level $d = log(s)$ comprises the interfaces of the last bisections and level 1 the interface of the first bisection. The elimination sequence of the interfaces corresponds to an elimination tree in Fig. 2b. At each level l, the elimination can be performed simultaneously, followed by the updates: $A_{k1,k2} = A_{k1,k2} - L_{k1,k} \cdot D_{k,k} \cdot L_{k2,k}^T$ for $k1$ being an interface segment at level l and interface segments $k2, k \in$Adj{interface segment $k1$}.

Let each subdomain consisting of h^2 interior nodes. For $p = s = h^2$ (hence $n = s \cdot h^2 = h^4$), it has been derived in [3] that the parallel efficiency is

$$(3.1) \qquad \eta(p = s) \approx \frac{829/42}{8 \cdot n^{1/2}/p + 26 + (log(p) - 44)/2 \cdot p^{1/2} - 70/p} \approx 0.6$$

When $p = s/2 = h^2/2$, the parallel efficiency becomes:

$$(3.2) \qquad \eta(p = s/2) \approx \frac{829/42}{29 + 14 \cdot \sqrt{2} \cdot p^{-1/2}} \approx 0.68$$

In the calculation of the above efficiencies, the sequential time of the (complete) nested dissection ordering is used, which is known to be optimal in terms of the order of the number of floating point operations [2]. It has been shown in [3] that for $s \gg h \geq 5$ (i.e. n is very large), the parallel efficiencies are $\eta \approx 0.76$ for $p = s$ and $\eta \approx 0.92$ for $p = s/2$.

The results shown in (3.1) and (3.2) do not include the communication overhead, so they can be interpreted as the degree of load balance. It can be concluded that the load balance is significantly improved by partitioning a domain into $2 \cdot p$ instead of p subdomains. This is also illustrated in Fig. 3. Fig. 3a and 3b show the execution profiles of a problem partitioned into 3 and 6 subdomains respectively. The number of processors is 3. The execution profile shows the number of parallel tasks as a function of the time. The work load among the 3 processors in Fig. 3b is clearly beter balanced than in Fig. 3a.

When the parallel factorization is performed on a distributed memory system, there will be communication overhead. The load balance is increased by increasing the number of subdomains, however, the communication overhead will increase as well. For example, on a 2-D grid of $\sqrt{p} \times \sqrt{p}$ processors, the communication time is $T_{comm} = O(c_1 \cdot p \cdot h^2 + c_2 \cdot \sqrt{s} \cdot h^2)$. For a constant p, when s increases from p to $2 \cdot p$, the increase in communication time is bounded by a factor of $\sqrt{2}$. Thus, if T_{comm} is about 10% of the total execution time for $s = p$, then T_{comm} is less than 14% for $s = 2 \cdot p$. So, for a non-communication-dominant computation the improvement

Initialization: Mark all elements in G as unpartitioned;

While (not all elements in G are partitioned) do
    Select an unpartitioned element el with the minimum degree;
    Determine a set S of elements by a level-structuring algorithm using el as starting point; and which holds that all elements in S have not been partitioned previously and $min\_size \leq |S| \leq max\_size$;
    Determine the set of newly introduced separators I;
    Improve I by means of the bipartite matching technique and adjust the S accordingly;
    Add the set S as the new subdomain;
    Mark all elements in S in the graph as partitioned;

od;

Define the interface segments in the resulting partition of the graph;
Improve the subdomain connectivity of the resulting partition through adjusting the interface segments;

FIGURE 4. A sketch of the graph partition algorithm GP.

in load balance (between 8% and 16%) outweighs the increase in communication overhead. We conclude that it is generally better to partition a domain into $k \cdot p$ subdomains with k is a small integer larger than 1 (say 2 or 3).

4. A generalized graph partitioning algorithm

From the analysis in Section 3, we reformulate the partitioning problem as follows: *Divide a graph into $k \cdot p$ subgraphs, such that the size of the subgraphs is between* **min_size** *and* **max_size** *and that the number of separator nodes is minimal.*

The relaxation of the requirement of all subdomains having equal size creates the possibility for reducing the length of the separators. Load balance can still be achieved through combining smaller and larger subdomains together to a processor during the task assignment. Experiments show that choosing the values of min_size and max_size to differ about 10% from the mean value n/p gives good results.

Fig. 4 depicts a scheme of the graph partitioning algorithm. The algorithm starts with determining an initial partition of a subdomain by means of a variant of the dissection methods (similar to the Cuthill-McKee ordering). Then, in order to shorten the length of the (newly) introduced separators between the current subdomain and the unpartitioned parts, a bipartite maximum matching algorithm [6] is applied. This procedure repeats until the entire graph is partitioned. Finally, the separators are improved for a smaller subdomain interconnectivity [3].

5. Results

Fig. 5 shows two examples of domain decomposition produced by the algorithm GP. The time complexity of the algorithm GP is $O(n \cdot log(n))$, so the time required to partition a domain is typically less than 10% of the factorization time. Experiments on a Convex shared-memory vector computer with 4 processors show that typically efficiencies between 75% and 95% are obtained [4]. In Table 1 some simulation results of the parallel solver on a distributed memory computer are shown. The four problems are as follows. (1) A plate decomposed into 3 subdomains and the execution is simulated for 3 processors; (2) Mesh A, decomposed into 64 subdomains and simulated for 32 processors; (3) Mesh B, as in (2), but now simulated for 64 processors; (4) A stopcock, decomposed into 9 subdomains and simulated for 9 processors. T is the duration time of a parallel factorization (incl. the communication time) simulated using the timing characteristics of the Intel iPSC/2.

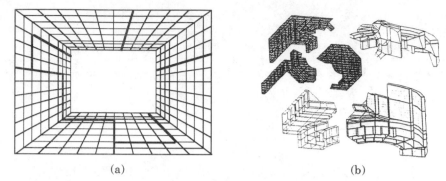

(a)   (b)

FIGURE 5. (a) A masonry wall partitioned into 5 subdomains; (b) A stop-cock partitioned into 6 subdomains.

Problem	Plate	Mesh A	Mesh B	Stopcock
α	64.8	73.8	57.0	79.7
β	14.3	7.7	15.2	11.8
T	133706	7572911	4840176	226948925

TABLE 1. Some simulation results of the parallel execution of the task graphs resulted from DD. T is in time unit equivalent to a floating point operation.

α is the ratio between the average busy time per processor and T. β is the ratio between the average communication time per processor and T. α and β indicate the load-balance and communication overhead respectively.

Conclusions We have analyzed the DD technique for parallel direct solution of FE systems. It is shown that a better load-balance is achieved if the number of subdomains is k times $(k > 1)$ the number of processors. We have reformulated the graph partitioning problem to allow a better resolution of the trade-offs between balanced subdomains and minimal separators. The results obtained from the parallelization of DIANA show that it is an efficient method for parallel FE computations.

REFERENCES

1. C.C. Ashcraft, *The domain/segment partition for the factorization of sparse symmetric positive definite matrices*, Eng. Comp. & Anal. TR ECA-TR-148, Boeing Compter Service, Nov. 1990.
2. A. George and J.W.H. Liu, *Computer solution of large sparse positive difinite systems*, Prentice-Hall, 1981.
3. H.X. Lin, *A methodology for parallel direct solution of finite element systems*, Ph.D. thesis, Delft University of Technology, 1993.
4. H.X. Lin and H.J. Sips, *Parallel direct solution of large sparse systems in finite element computations*, Proc. 1993 Int. Conf. on Supercomupting, Tokyo, July, 1993, pp. 261-270.
5. J.W.H. Liu, *A graph matching algorithm by node separators*, ACM Trans. Math. Softw. 15 (1989), 198 - 219.
6. J.W.H. Liu, *The role of elimination trees in sparse factorization*, SIAM J. Matrix Anal. Appl. 11 (1990), 134-172.
7. A. Pothen, H.D. Simon and K-P. Liou, *Partitioning sparse matrices with eigenvectors of graphs*, SIAM J. Matrix Anal. Appl. 11 (1990), 430-452.

DEPARTMENT OF APPLIED MATHEMATICS AND INFORMATICS, DELFT UNIVERSITY OF TECH-NOLOGY, MEKELWEG 4, NL-2628 CD, DELFT, THE NETHERLANDS
E-mail address: lin@pa.twi.tudelft.nl

Contemporary Mathematics
Volume **180**, 1994

Finite-Element/Newton Method for Solution of Nonlinear Problems in Transport Processes Using Domain Decomposition and Nested Dissection on MIMD Parallel Computers

M. REZA MEHRABI AND ROBERT A. BROWN

ABSTRACT. Finite-element discretizations of nonlinear steady-state problems describing transport processes are solved by Newton's method using a multilevel algorithm for direct factorization of the Jacobian matrix. The factorization algorithm is based on incomplete nested dissection and domain decomposition to distribute the equations to processors of a MIMD parallel computer, where the Jacobian matrix is concurrently factorized. Sample calculations for two flow problems show reasonable computation speeds and speedups on an Intel iPSC/860 hypercube. Moreover, the execution times for these codes compare favorably with the performances for conventional finite element software on a serial, vector supercomputer and for a highly parallel commercial program for solving incompressible flow problems on a MIMD parallel computer.

1. Introduction

The development of efficient and robust algorithms on parallel computer architectures is an outstanding problem for solution of the wide variety of problems in fluid flow and heat and mass transfer that are of interest in modeling materials processing and manufacturing. The mathematical models for this class of problems are typically composed of a group of differential equations, algebraic equations and integral constraints. For steady-state (time independent) models, finite element or finite difference discretization of these models leads to large sets of nonlinear differential equations with highly structured coupling between the

1991 *Mathematics Subject Classification.* Primary 65N55, 65Y05; Secondary 65N30, 65Y20.

This research was supported by the Advanced Research Projects Agency and the Microgravity Sciences and Applications Program of the National Aeronautics and Space Administration of the United States Government.

The final version of this paper will be submitted for publication elsewhere.

variables. These dependencies are naturally asymmetric: the linearization of the nonlinear equation sets leads to asymmetric Jacobian matrices. For serial, vector computers, Newton's method, coupled with direct LU-decomposition, has proven to be an efficient method for solution of these nonlinear equation sets, at least for the discretization of problems in two space dimensions, where the structure of the nonzero entries in the Jacobian matrix makes direct LU-decomposition reasonably efficient in terms of operation count and memory usage. Indeed, finite-element/Newton algorithms have been used to solve an enormous variety of transport problems and are the basis of several commercial computer codes used for this purpose. The popularity of this algorithm comes from the robustness of Newton's method for converging to the solution of extremely nonlinear problems and also from the availability of continuation methods, based on the computation of the factors of the Jacobian matrix [27, 17], for mapping out multiple solutions.

The goal of the development of the algorithm described in this paper is to extend finite-element/Newton algorithms for the solution of transport problems to MIMD parallel computers like the Intel iPSC/860 hypercube available to our research group. The major effort associated with this endeavor has been the development of a an LU-factorization algorithm for the solution of large, sparse, asymmetric linear equation sets based on domain decomposition and nested dissection for partitioning the equations to the processors of the MIMD computer. The algorithm used here and described in detail by Mehrabi and Brown [22] is based on dividing the geometrical domain recursively into subdomains and separators using nested dissection [8, 9] and assigning subdomains and associated separator data to each processor. The algorithm described in [22] extends the work of Lucas et al. [19] for symmetric, positive-definite matrices to avoid duplicate storage, to allow asymmetric matrices and to allow partial pivoting during factorization. The algorithm stores the lower and upper triangular forms and performs forward and backward solutions separately.

The LU-factorization algorithm is described in more detail in Section 2. Two test problems that arise in the solution of isothermal and nonisothermal incompressible flows are described in Section 3; these are the two-dimensional flow of an incompressible fluid in a lid-driven cavity and the natural convection motion of a Boussinesq fluid in a two-dimensional cavity with differentially heated lateral walls. Both problems have been widely used as test problems for incompressible flow calculations and become computationally difficult as the intensity of the convection is increased. This occurs with increasing the Reynolds number for the lid-driven cavity problem [24, 21] and with increasing the Grashof number for the thermal convection problem [6]. Because the finite element discretizations used to create the discrete equation sets are standard, we do not focus on the accuracy of the computations, but rather on their efficiency compared to implementation of the finite-element/Newton method on a serial, vector supercomputer and to a highly parallelizable spectral element code based on

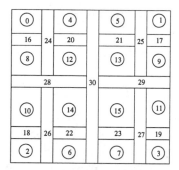

FIGURE 1. Nested dissection ordering of a rectangular two-dimensional domain. A binary reflected Gray code is used for numbering the subdomains to reduce hypercube communication [4].

pseudo-time-stepping to compute the steady-state solution.

2. Concurrent LU-Factorization and Storage: CFS

For simplicity, we consider a mathematical model for transport processes described in a two-dimensional geometrical region represented by the rectangular domain shown in Fig. 1. This region is partitioned into P quadrilateral subdomains, each of which is assigned to a processor of the MIMD parallel computer. The equations that correspond to each subdomain are ordered using incomplete nested dissection [9]. Then, each point on the elimination tree for the nested dissection ordering corresponds to a supernode or a group of equations [22].

In general, the mathematical model may be composed of differential and algebraic equations, integral constraints and boundary conditions. Finite element methods are used to define local approximations to the field variables and to discretize the partial differential equations and boundary conditions. The details of these discretizations are not important for the discussion here, as long as the approximations to the field variables have compact support within the elements. The discretized problem is a large set of nonlinear algebraic equations that is represented as

$$(2.1) \qquad\qquad \mathbf{R}(\mathbf{x}) = \mathbf{0}$$

where $\mathbf{R} \in \mathbf{R}^N$ and $\mathbf{x} \in \mathbf{R}^N$, where N is the total dimension of the discrete field variables. The variables and equations are associated with specific processors according to $\mathbf{x} = (\mathbf{x}^{(1)}, \mathbf{x}^{(2)}, \mathbf{x}^{(3)}, ..., \mathbf{x}^{(P)})$ and $\mathbf{R} = (\mathbf{R}^{(1)}, \mathbf{R}^{(2)}, \mathbf{R}^{(3)}, ..., \mathbf{R}^{(P)})$, where the dimensions of the variables $\{\mathbf{x}^{(i)}\}$ and equations $\{\mathbf{R}^{(i)}\}$ associated with each processor depend on the number of finite elements allocated to it.

We solve eqs. (2.1) by Newton's method using direct LU-decomposition of the Jacobian matrix at each Newton iteration. The Jacobian matrix is written as a partitioned matrix among the P-processors as

$$(2.2) \qquad\qquad \mathbf{J} = \left[\begin{array}{cc} \mathbf{J}^{(ii)} & \mathbf{J}^{(ie)} \\ \mathbf{J}^{(ei)} & \mathbf{J}^{(ee)} \end{array} \right]$$

where the Jacobian matrix is organized into data that is totally interior to each subdomain – the interior matrix $\{\mathbf{J}^{(ii)}, \mathbf{J}^{(ei)}, \mathbf{J}^{(ie)}\}$ – and data that corresponds to the borders shared by two or more subdomains, which are the separators in Fig. 1 – the exterior matrix $\mathbf{J}^{(ee)}$. The exterior matrix is partitioned among adjacent processors to minimize communication and memory requirements. The data structures used for storage of the interior and exterior matrices are described in [22]. Matrix data corresponding to each supernode is stored in block from to facilitate vectorization and pivoting during the LU-factorization.

The algorithm for the LU-decomposition of the matrix, eq. (2.2), described by Mehrabi and Brown [22] is operationally equivalent to direct computation and factorization of the Schur complement. The formation of the Schur complement corresponds to the LU-decomposition of the interior matrices of each processor $\{\mathbf{J}^{(ii)}\}$ and to the update of the exterior matrix $\mathbf{J}^{(ee)}$ by those factors. Data is communicated from each processor to form the exterior matrix using the *fan-in* method of Ashcraft et al. [1]. A hybrid *fan-in/fan-out* algorithm is used for updating the exterior matrices during factorization and subsequent updating. As discussed in [22], the fan-in and the hybrid fan-in/fan-out methods of update reduce the number of messages and the total message volume for LU-decomposition by approximately factors of 2 and 3 over the algorithm of Mu and Rice [23] developed using a grid-based, subtree-subcube assignment of exterior matrix elements to the processors.

The computer code was written in standard Fortran and C with only machine specific statements for send, receive and synchronization. The code was implemented on a 32-node Intel iPSC/860 hypercube with 8 Mbytes of memory per processor and compiled with Intel's *if77* and *icc* compilers. Hand-coded Level-1 BLAS by Kuck and Associates [18] were used to increase the speed. Each i860 node executes the LINPACK benchmark at 9.5 MFLOPS and the machine has a ratio of speed for communication to computation of 1:100, which lowers the parallel speedup when communication is large. Here the speedup is defined as the speed of a calculation on P-processors divided by the speed on a single processor of the same machine, where the single processor solves a problem of similar size to the problem of each of the P-processors. Below we report speedups for the LU-factorization, for forward and backward elimination steps of the solution of linear equation sets and for the formation of the matrix and forcing vector at each Newton iteration.

3. Test Problems and Results

The two test problems are shown schematically in Fig. 2. Each of these problems is described below.

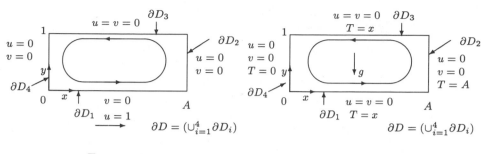

FIGURE 2a   FIGURE 2b

FIGURE 2. Schematic diagrams of the flow geometries for the problems of flow in a (a) lid-driven cavity and (b) thermal convection in heated side-wall cavity. The boundary conditions for each flow problem also are shown.

3.1. Lid-Driven Cavity. Consider the steady-state incompressible flow of a Newtonian liquid in a two-dimensional cavity with the bottom surface moving, as shown in Fig. 2a. The Navier-Stokes equations are made dimensionless with the speed of the bottom surface and the height of the cavity to give the dimensionless equations

$$(3.1) \qquad\qquad \text{Re } \mathbf{v} \cdot \nabla \mathbf{v} = \nabla^2 \mathbf{v} - \nabla p$$

$$(3.2) \qquad\qquad \nabla \cdot \mathbf{v} = 0$$

where the boundary conditions are shown in Fig. 2a. In addition to setting both components of velocity on each surface, the pressure is specified at a point to set a datum level. The problem is specified in terms of two dimensionless parameters: the Reynolds number Re, which scales the importance of fluid inertia, and the aspect ratio of the cavity A, which scales the cavity width to its height. Including a nonzero value of Re makes the problem nonlinear and the Jacobian matrix asymmetric. The calculations presented here are meant only as a demonstration of the parallelization of the algorithm; they were performed with parameter values of $A=4$ and Re=1. Calculations with large Re are feasible with the finite element meshes used here.

The eqs. (3.1)-(3.2) are discretized using a standard mixed finite element method for the Stokes problem [3]. The velocity components are interpolated using Lagrangian biquadratic polynomials defined on each element and the pressure by bilinear polynomials. The weak forms of the eqs. (3.1)-(3.2) are written using the isoparametric mapping to a unit element and the two-dimensional integrals are computed numerically using 9-point Gaussian quadrature. The Newton's iteration was started with the initial guess of no flow in the cavity and converged to an absolute tolerance of 10^{-12} in 4 iterations. A sample solution for a mesh of 36 elements in the vertical direction and 76 elements in the horizontal

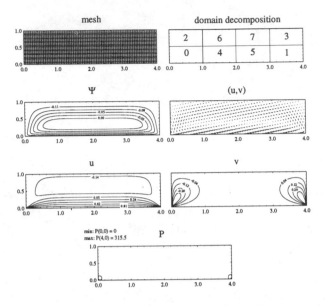

FIGURE 3. Sample mesh, subdomains, velocity components and pressure field for the flow in a lid-driven cavity. Calculations are for a 36 × 76 element mesh with N=25,000 degrees-of-freedom (DOF's) on 8 processors.

direction is shown in Fig. 3. Calculations with varying mesh size show a 5 percent oscillation in the maximum magnitude of the velocity due to singularities in the pressure field and its gradient at the corners where the moving and stationary walls meet. It is possible to circumvent the convergence problem caused by these singularities by changing the boundary conditions in the corners; however, we did not attempt to do so because we are only interested in the performance of the algorithm at this point, not in the details of the flow field.

The computation times for formulation of the Jacobian matrix and forcing vector and for LU-decomposition are shown in Fig. 4 for calculations with between 1 and 32 processors (P) and discretizations resulting in 4,000 to 80,000 degrees-of-freedom (DOF's). Two features are important. First, the time for calculation decreases proportionally to the number of processors, showing good parallel scaling for both operations. For the largest problem (N=80,000), the time for formulation (8.5 s) is a fraction of the time needed for LU-decomposition (28 s). The total time for a Newton iteration is the summation of these times with the time for forward and backward solution of the triangular equation sets. For the calculation with N=80,000, this is

$$\frac{\text{Total Time}}{\text{Newton Iteration}} = 8.5 \text{ s (Formulation)} + 28 \text{ s } (LU\text{-Decomposition})$$
$$+ \quad 1.1 \text{ s (Forward Elimination)}$$
$$+ \quad 0.6 \text{ s (Backward Elimination)}$$
$$= \quad 38.2 \text{ s}$$

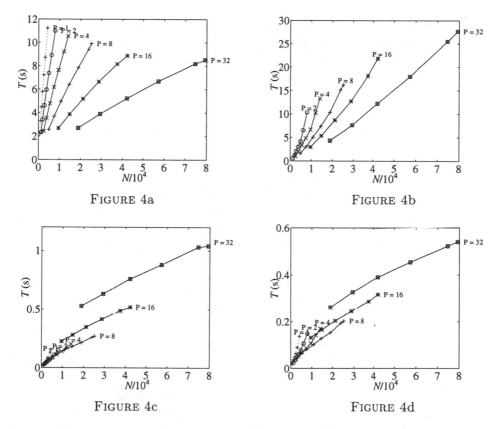

FIGURE 4a

FIGURE 4b

FIGURE 4c

FIGURE 4d

FIGURE 4. Execution times for (a) formulation, (b) LU-decomposition, (c) forward elimination and (d) backward elimination for solution of the lid-driven cavity problem as a function of problem size (N) and number of processors (P).

where formulation represents 22 percent and LU-decomposition 73 percent; the other two steps represent 3 and 2 percent of the total computation time, respectively.

The low percentage time spent performing the forward and backward solution algorithms offsets the poor parallelization of these two algorithms, as is well known; e.g. see Lucas et al. [19]. As seen in Fig. 4, the performance of both algorithms degrades with increasing the number of processors for a constant problem size (N) because of the increased communication needed in these steps of the algorithm.

The lid-driven cavity problem was used as a benchmark to establish the performance of the finite-element/Newton method using the CFS algorithm for solution of linear equation sets relative to the finite-element/Newton method implemented using a frontal solution method [16, 14, 15] on a vector super-computer. Two sets of calculations are reported. In the first, both the CFS and

TABLE 1. Comparison of finite-element/Newton method using the CFS LU-decomposition software and the serial frontal solution algorithm. The domain is dissected 6 times by CFS. Both programs are run on a single, dedicated processor of a Cray X-MP.

Quantity	Serial Frontal Code	CFS Code
Time for Newton Iteration (s)	162	144
Time for Solution (s)	900	780
Speed for LU-Decomposition (MFLOPS)	110	45

frontal algorithms are run on a single processor of the Cray X-MP computer at the MIT Supercomputer Facility to establish the advantages of the incomplete nested dissection algorithm relative to the frontal technique outside of the context of the parallel computer. In the second calculation, the CFS algorithm run on the Intel iPSC/860 is compared to the frontal algorithm running on the Cray X-MP.

The comparison between the frontal version of the finite-element/Newton method and the CFS implementation on the Cray X-MP is summarized in Table 1 for a calculation with a 68×68 mesh, giving $N=42,000$; the speeds for the two codes are included there. The speed for the frontal code is taken from diagnostic software supplied by Cray. The speed for the CFS code is calculated directly by counting the number of operations and dividing it by time. The frontal code uses Level-1 BLAS routines in the innermost loops of the LU-decomposition routine and runs at 110 MFLOPS on the Cray. By comparison, no explicit attempt has been made to vectorize the CFS code on the Cray; as a result it executes only at 45 MFLOPS. The test run by CFS is dissected 6 times, i.e. into 2^6 sections. Further dissection degrades this speed more because of smaller length of vectors and the larger overhead of integer addressing. Even though the CFS routine is not as fast as the frontal solver, it still executes in a slightly lower time, because of the lower operation count associated with the incomplete nested dissection ordering of the matrix [11, 10].

The comparison between the frontal version of the finite-element/Newton method running on the Cray X-MP and the CFS version running on the 32-processor Intel iPSC/860 for a discretization with $N=80,000$ (64×136 elements) is summarized in Table 2. Most significantly, the CFS version is almost 10 times faster for performing the same calculation. The Intel machine is performing at half of the maximum speed of $32 \times 9.5=304$ MFLOPS estimated from the LINPACK benchmark run on a single processor.

TABLE 2. Comparison of finite-element/Newton method using the CFS *LU*-decomposition software on a MIMD computer, where each subdomain is dissected 6 times, and the frontal implementation of *LU*-decomposition on a serial, vector supercomputer.

Quantity	Serial Frontal Code on a single processor of Cray X-MP	CFS Code on 32-Node Intel iPSC/860
Time for Newton Iteration (s)	273	38
Time for Solution (s)	1800	180

3.2. Thermal Convection in a Cavity with Heated Sidewalls. The geometry for this problem is shown schematically in Fig. 2b. The equations are made dimensionless with the height of the cavity, the temperature difference between the hot and cold walls and the buoyant velocity to yield the dimensionless momentum, continuity, and energy equations

$$(3.3) \qquad \sqrt{\text{Gr}}\, \mathbf{v} \cdot \nabla \mathbf{v} \;=\; \nabla^2 \mathbf{v} - \nabla p + \sqrt{\text{Gr}}\, \mathbf{e}_y\, T$$

$$(3.4) \qquad \nabla \cdot \mathbf{v} \;=\; 0$$

$$(3.5) \qquad \sqrt{\text{Gr}}\, \mathbf{v} \cdot \nabla T \;=\; \frac{1}{\text{Pr}} \nabla^2 T$$

which are expressed in terms of the Grashof number (Gr), the Prandtl number (Pr) and the aspect ratio A of the cavity. In eq. (3.3), $\mathbf{e}_y$ is the unit vector in the y-direction. The boundary conditions for the velocity and temperature field are shown in Fig. 2b; here the temperature field along the horizontal surfaces has been specified as a linear function of position to match the temperatures at the two vertical surfaces; these conditions are the same as those used by others [6] in the specification of this problem as a test for algorithms for computation of natural convection.

Equations (3.3)-(3.5) are discretized using a standard mixed finite element method for the natural convection problem [12]. The velocity components and the temperature are interpolated using Lagrangian biquadratic polynomials defined on each element and the pressure by bilinear polynomials. The weak form of the eqs. (3.3)-(3.5) is constructed in a similar way to what was described for the lid-driven cavity problem.

Calculations were carried out using the parameter values of $A=4$, $Pr=0.015$ and $Gr=20,000$ to demonstrate the robustness of the algorithm for computing a highly nonlinear steady-state. Setting the Prandtl number on this low value makes the time scales for heat and momentum diffusion very different and makes this computation particularly hard for any algorithm based on a pseudo-time-stepping procedure for computing the steady-state solution. The Newton's iterations converge in 8 iterations starting with the converged solution for $A=4$,

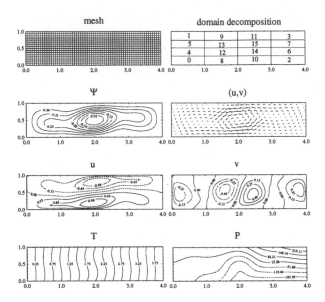

FIGURE 5. Sample mesh, subdomains, velocity components, temperature field and pressure field for the thermal convection in a cavity with heated sidewalls. Calculations are for a 16×64 element mesh with $N=14,000$ degrees-of-freedom (DOF's).

Pr=0.015, and Gr=5,000 as the initial approximation, which in turn converges in 5 iterations from the no flow state as the initial approximation. The solution contours computed with a mesh of 16×64 elements are shown in Fig. 5 and compare well with solutions reported by others [5, 2] for this same flow state.

The speedups associated with the calculations on the Intel iPSC/860 are shown in Fig. 6 for the formulation and LU-decomposition as a function of the DOF's and the number of processors P. The efficiency is approximately 0.85 for the formulation and 0.65 for LU-decomposition; both values are essentially independent of the problem size.

The computational efficiency of the finite-element/Newton method for solution of this thermal convection problem was compared directly to a highly optimized commercial incompressible flow code, Nekton, which uses spectral element discretizations coupled with a semi-implicit temporal operator splitting to compute steady-state solutions as the asymptotic limit of transients [20, 25, 13, 26, 7]. The operator splitting method requires only the solution of positive-definite linear equation sets at each time step. These systems are solved using a conjugate gradient algorithm with a two-level preconditioner. The results of the comparison tests run on the Intel iPSC/860 are summarized in Table 3. The CFS calculation is nearly 7 times faster. The comparison has been based on the solution of discretizations of similar size (N). For this discretization, the spectral element representation (11th-order polynomials for velocity and temperature

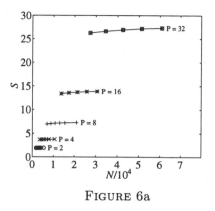

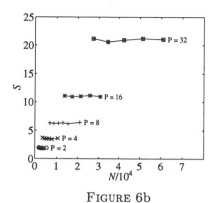

FIGURE 6a   FIGURE 6b

FIGURE 6. Speedups for (a) formulation, and (b) *LU*-decomposition for solution of the thermal convection problem as a function of problem size (N) and number of processors (P).

and 9th-order for pressure) is much more accurate than the finite element approximation. However, the CFS algorithm also is applicable to spectral element discretizations and will yield improved computation rates because of the reduced size of the exterior matrix relative to the interior matrix. A direct comparison on spectral element calculations will be reported later. The two test calculations reported here make a reasonably fair comparison.

4. Discussion

The calculations presented here demonstrate that the finite-element/Newton method coupled with the CFS algorithm for *LU*-factorization is a robust and efficient method for the solution of two-dimensional nonlinear transport problems using MIMD parallel computers. We believe that the utility of the method will be even greater for problems with more complicated physicochemical processes, such as complicated kinetics, radiative heat transfer and nonlinear constitutive behavior, that are present in many materials processing problems of interest. The results show that the speedup associated with the *LU*-decomposition is not particularly high; it is shown to be 0.65 for 32-processors and estimated to be as low as 0.5 for 512-processors. However, the robustness of the *LU*-decomposition for solution of the linear equation set and the rapid convergence of Newton's iterations for nonlinear problems makes the finite-element/Newton method a viable algorithm for solving any problem that can be fit into memory of the MIMD computer. Current machine sizes limit this approach to large two-dimensional and small three-dimensional problems.

TABLE 3. Comparison of finite-element/Newton method based on CFS algorithm with Nekton, a spectral element code based on semi-implicit, pseudo-time-stepping method with preconditioned conjugate gradient method. Both calculations were run on Intel iPSC/860. Each subdomain is dissected 6 times by CFS. For this calculation Gr=1, Pr=0.01 and A=4.

Quantity	Nekton	Finite-Element/Newton CFS method
Steps/Iterations	100 time steps	3 Newton iterations
Mesh	8 × 16 spectral elements	48 × 96 finite elements
Order of Approximations	Velocity: 11 Temperature: 11 Pressure: 9	Velocity: 2 Temperature: 2 Pressure: 1
DOF's	58,000	61,000
Number of Processors	P=8*	P=32
Solution Time for P=32	800	120

* Solution time on P=32 estimated by dividing time on P=8 by 4.

Acknowledgements

The authors would like to thank Paul Fischer for providing the results by Nekton, and Todd Salamon and David Bornside for providing the results by the serial frontal algorithm.

REFERENCES

1. C. Ashcraft, S. Eisenstat, and J. Liu, *A fan-in algorithm for distributed sparse numerical factorization*, SIAM J. Sci. Statist. Comput. **11** (1990), 593–599.
2. M. Behnia and G. de Vahl Davis, *Fine mesh solutions using stream function-vorticity formulation*, Numerical Simulation of Oscillatory Convection in Low-Pr Fluids (Braunschweig, Germany) (B. Roux, ed.), Friedr. Vieweg & Sohn Verlagsgesellschaft mbH, 1990, pp. 11–18.
3. G. F. Carey and J. T. Oden, *Finite elements*, vol. 6, Prentice Hall, Englewood Cliffs, NJ, 1986.
4. T. Chan and Y. Saad, *Multigrid algorithms on the hypercube multiprocessor*, IEEE Trans. on Computers **C-35** (1986), 969–977.
5. O. Daube and S. Rida, *Contribution to the GAMM workshop*, Numerical Simulation of Oscillatory Convection in Low-Pr Fluids (Braunschweig, Germany) (B. Roux, ed.), Friedr. Vieweg & Sohn Verlagsgesellschaft mbH, 1990.
6. B. Roux (ed.), *Numerical simulation of oscillatory convection in low-Pr fluids*, Notes on Numerical Fluid Mechanics, vol. 27, Friedr. Vieweg & Sohn Verlagsgesellschaft mbH, Braunschweig, Germany, 1990.
7. P. F. Fischer, *Spectral element solution of the navier-stokes equations on high performance distributed-memory parallel processors*, Ph.D. thesis, MIT, Cambridge, MA, June 1989.
8. A. George, *Nested dissection of a regular finite element mesh*, SIAM J. Numer. Anal. **10** (1973), 345–367.
9. A. George and J. W. H. Liu, *Computer solution of large sparse positive definite systems*, Prentice Hall, Englewood Cliffs, NJ, 1981.
10. A. George and E. Ng, *On the complexity of sparse QR and LU factorization of finite-*

element matrices, SIAM J. Sci. Statist. Comput. **9** (1988), 849–861.

11. A. George and H. Rashwan, *Auxiliary storage methods for solving finite element systems*, SIAM J. Sci. Statist. Comput. **6** (1985), 882–910.
12. P. M. Gresho, R. L Lee, S. T. Chan, and J. M. Leone, *A new finite element for Boussinesq fluids*, Proc. 3rd Intl. Conf. Finite Elements in Flow Problems (Alberta, Canada) (D. H. Norrie, ed.), Banff, 1980, pp. 204–215.
13. L. W. Ho and A. T. Patera, *A Legendre spectral element method for simulation of unsteady incompressible viscous free-surface flows*, Computer Methods in Applied Mechanics and Engineering **80** (1990), 355–366.
14. P. Hood, *Frontal solution program for unsymmetric matrices*, J. Num. Meth. Engng. **10** (1976), 379–399.
15. _____, *Note on frontal solution program for unsymmetric matrices*, J. Num. Meth. Engng. **11** (1977), 1055.
16. B. M. Irons, *A frontal solution program for finite element analysis*, Intl. J. Num. Meth. Engng. **2** (1970), 5–32.
17. H. B. Keller, *Numerical solution of bifurcation and nonlinear eigenvalue problems*, Applications of Bifurcation Theory (New York, San Francisco, London) (P. H. Rabinowitz, ed.), Academic Press, Inc., 1977, pp. 359–384.
18. Kuck and Associates, Inc., Champaign, IL, *CLASSPACK basic math library*, February 1992.
19. R. F. Lucas, T. Blank, and J. J. Tiemann, *A parallel solution method for large sparse systems of equations*, IEEE Trans. Computer-Aided Design **CAD-6** (1987), 981–991.
20. Y. Maday and A. T. Patera, *Spectral element methods for the navier-stokes equations*, State-of-the-Art Surveys in Computational Mechanics (New York, NY) (A. K. Noor and J. T. Oden, eds.), American Society of Mechanical Engineers, 1989.
21. A. Mahallati and J. Militzer, *Application of the piecewise parabolic finite analytic method to the three-dimensional cavity flow*, Numerical Heat Transfer, Part B **24** (1993), 337–351.
22. M. R. Mehrabi and R. A. Brown, *An incomplete nested dissection algorithm for parallel solution of finite element discretizations of partial differential equations.*, J. Sci. Computing **8** (1993), 373–387.
23. Mo Mu and J. R. Rice, *A grid-based subtree-subcube assignment strategy for solving partial differential equations on hypercubes*, SIAM J. Sci. Stat. Comput. **13** (1992), 826–839.
24. H. Nishida and N. Satofuka, *Higher-order solutions of square driven cavity flow using a variable-order multi-grid method*, Intl. J. Numer. Meths. Engng. **34** (1992), 637–653.
25. A. T. Patera, *A spectral method for fluid dynamics, laminar flow in a channel expansion*, Journal of Computational Physics **54** (1984), 468–488.
26. E. M. Ronquist, *Optimal spectral element methods for the navier-stokes equations*, Ph.D. thesis, MIT, Cambridge, MA, 1988.
27. Y. Yamaguchi, C. J. Chang, and R. A. Brown, *Multiple buoyancy-driven flows in a vertical cylinder heated from below*, Phil. Trans. R. Soc. Lond. **A312** (1984), 519–552.

DEPARTMENT OF CHEMICAL ENGINEERING, MASSACHUSETTS INSTITUTE OF TECHNOLOGY, CAMBRIDGE, MASSACHUSETTS 02139
E-mail address: reza@mit.edu and rab@mit.edu

Contemporary Mathematics
Volume **180**, 1994

Modeling with Collaborating PDE Solvers: Theory and Practice

MO MU AND JOHN R. RICE

ABSTRACT. We consider the problem of modeling very complex physical systems by a network of collaborating PDE solvers. Various aspects of this problem are examined from the points of view of real applications, modern computer science technologies, and their impact on numerical methods. The related methodologies include *network of collaborating software modules*, *object-oriented programming* and *domain decomposition*. We present a domain decomposition approach of collaborating PDE solvers based on interface relaxation. The mathematical properties and application examples are discussed. A software system RELAX is described which is implemented as a platform to test various relaxers and to solve complex problems using this approach. Both theory and practice show that this is a very promising approach for efficiently solving complicated problems on modern computer environments.

1. Introduction

Modeling physical phenomena with scientific computing is an interdisciplinary effort involving engineers, mathematicians and computer scientists. Practical physical systems are often mathematically modeled by complicated partial differential equations (*PDEs*). Their numerical solution requires high performance computers, large software systems and efficient algorithms. The design of numerical PDE algorithms must balance many factors. From the numerical analysis point of view, one often focuses on good approximation, fast convergence, low arithmetic expense, and other mathematical properties. In practical applications, one should be able to handle the complexity and generality of PDE problems. Among the major concerns for software development are software

1991 *Mathematics Subject Classification*. Primary 65N55, 65F10; Secondary 65Y05, 65c20.
The first author was supported in part by the National Science Foundation grant CCR-8619817 and the Hong Kong RGC DAG93/94.SC10. The second author was supported in part by the Air Force Office of Scientific Research grants, 88-0243, F49620-92-J-0069 and the Strategic Defense Initiative through Army Research Office contract DAAL03-86-K-0106.
The final version of this paper will be submitted for publication elsewhere.

productivity, complexity, reusability, maintenance, portability, and other quality issues. Modern software technologies and concepts are needed. In addition, the use of parallel computing leads to issues, such as parallel algorithms, communication cost, and scalability. Obviously, many of these objectives conflict with each other. The principal trade off is programming effort versus execution time efficiency. We examine various aspects of the simulation problem from the practical point of view. These considerations lead to the domain decomposition approach which we call *collaborating PDE solvers*. It aims to solve complex physical problems with the use of modern computer science technologies. It is based on the classical relaxation idea by iteratively solving local problems and adjusting interface conditions. A software system RELAX has been implemented as a platform to support this approach. One can use this system to model complex physical objects, specify mathematical problems and test various interface relaxation schemes. It is shown that this approach is promising in both theory and practice.

2. Collaborating PDE solvers

A physical system in the real world normally consists of a large number of components. They have different shapes, obey different physical laws, and collaborate with each other by adjusting interface conditions. An automobile engine system is such a typical example. Mathematically, it corresponds to a very complicated PDE problem with various formulations for the geometry, PDE, and interface/boundary condition in many different regions. Interface locations and conditions may also vary, as in systems with moving interfaces. One can imagine the great difficulty in creating a software system to model such a complicated real problem. Therefore, one needs an effective software development mechanism which first, is applicable to a wide variety of practical problems, second, allows for the use of advanced software technologies in order to achieve high productivity and quality, and finally, is suitable for some reasonably fast numerical methods.

Notice that most of the physical systems in practical applications can be modeled as a mathematical *network*. Here a network is a directed graph consisting of a set of nodes and edges. If we represent each physical component in a system by a node, then a pair of neighboring components are linked by an edge in the graph, with the edge directions used to indicate the necessary information transmission for the interface adjustment. Each node in the network is then assigned a key to represent the local physical law for the corresponding component. For numerical relaxation one may also assign certain weights to each edge in order to provide detailed control for the interface adjustments, such as for boundary values and their jumps across the interfaces. Moving interfaces are allowed in this network specification.

Usually, individual components are simple enough so that each node corresponds to a simple PDE problem with a single PDE defined on a regular geometry. There exist many standard PDE solvers that are well developed and can be applied to these local node problems. To solve the global problem, we let these local solvers collaborate with each other by invoking an interface controller. It collects boundary values from neighboring subdomains and adjusts interface conditions according to the network specifications. Therefore, the network abstraction of a physical system allows us to build a software system which is a network of collaborating PDE solvers. These networks can be very big for major applications. There are normally about 5 interfaces per subdomain. For a highly accurate weather prediction, for example, one needs 3 billion variables in a simulation with continuous input at 50 million places. This assumes a 3-D adaptive grid, otherwise the computation is much larger. Very optimistically, if one needs a new forecast every 2-3 hours, the answer is 100 gigabytes in size and requires 80 mega-giga FLOPs to compute. Such a network roughly consists of 3,000 subdomains and 15,000 interfaces. An "answer" is a data set that allows the accurate approximate solution to be displayed at any place. It is much smaller than the numerical solution from which it is derived. Another example to consider is a realistic vehicle simulation, where there are perhaps 100 million variables and many different time scales. This problem has very complex geometry and is very non-homogeneous. The answer is 20 gigabytes in size and requires about 10 tera FLOPs to compute. The network has 10,000 subdomains and 35,000 interfaces.

A software network of this type is a natural mapping of a physical system. It simulates how the real world evolves and thus normally produces a reasonable solution. It allows various advanced software technologies to be applied to create a high quality system in a very productive way. For instance, one can apply the networking technology to efficiently integrate a collection of software components into an entire system and to implement a neat and flexible system architecture for the model and its interface connections. This implies the use of the software parts technology that is the natural evolution of the software library idea with the addition of software standards. It allows software reuse for easy software update and evolution, which are extremely important in practice. The real world is so complicated and diverse that we believe there are no monolithic, universal solvers. Without software reuse, it is impractical for anyone to create on his own a large software system for a reasonably complicated application. For example, automobile manufacturers frequently change automobile models. Each change normally results in a new software system. Recreating such a system could easily take several months or years. In contrast, the execution time to perform the required computation might only be a few hours. Notice that such a physical change usually corresponds to replacing, adding, or deleting a few nodes in the network with a corresponding change in interface conditions. These can be modular manipulations on a network that do not affect the majority of the system components.

Object-oriented programming is a powerful software development methodology well suited to such a software system. In this methodology each physical component can be viewed both as a physical object and as a software object. Actions and interactions of objects are clearly defined by the network. Two basic principles of object-oriented programming are data structure abstraction and information hiding for each object. These principles are expressed here by the local solvers and the interface conditions. In addition, this network approach is naturally suitable for parallel computing as it exploits the potential parallelism in physical systems. One can handle issues like data partition, assignment, and load balancing on the physics level by the structure of a given physical system. Synchronization and communication are controlled by the network specification and restricted to interfaces of subdomains, which results in a coarse-grained computational problem. This is suitable for today's most advanced parallel supercomputers, such as the Intel's PARAGON system and the Thinking Machine's CM-5. The network approach also allows high scalability. Finally, this network approach naturally fits into the mathematical domain-decomposition framework with the overall geometry being viewed as automatically partitioned into a collection of subdomains. Note that subdomains and interfaces simply correspond to the network nodes and edges, respectively.

There have been many types of domain decomposition methods proposed over the past decade. Specific references to the literature are given in [2] and in other volumes of these proceedings. However, not all of them are suitable for, or directly applicable to, this network framework due to considerations from the practical and software points of view. First, the artificial subdomain overlapping introduced for mathematical convergence purposes obviously violates the basic principles of object-orientation. Each software object should correspond to a natural physical component without knowing part of data structures of other objects or exposing its local data structures to others. Second, it is not proper to apply the algebraic type of domain decomposition methods that first discretize a PDE problem on an entire domain and then partition the discrete system according to the geometric decomposition. In fact, the network framework implies the problem partition on the continuous problem level so that PDE solution techniques in different regions may be totally independent depending on local properties. One may use finite differences for one subdomain, and finite elements or even an analytic solution for another. In addition, the subdomain PDE operators are not necessarily extensible to interfaces so that global discretization is not always applicable. More importantly, the success of most of these methods relies on finding a good preconditioner for the interface Schur complement matrix, which is very difficult to do in practice for a complicated physical system. Another well-known class of methods are motivated by observing that the global solution of a Poisson equation on an entire domain is continuous on interfaces up to its first derivative, i.e., U and $\partial U/\partial n$. In order to match the continuity for both U and $\partial U/\partial n$, one starts with an initial guess for U and $\partial U/\partial n$ on interfaces, takes them as boundary data to solve a Dirichlet or Neumann boundary value problem on subdomains, then updates interface values using the new solution data, and iterates until convergence. A common approach is the alternating Dirichlet-Neumann algorithm. This is a non-overlapping method. Generally speaking, it converges slowly and may diverge, although preconditioning techniques may be applied to improve convergence. Theoretically, it is rather difficult to understand the convergence mechanism, especially when the so-called cross points are present on interfaces. In addition, interface conditions in practical applications usually appear in more complicated forms. Nevertheless, it is important that this subdomain-iteration based approach best fits into the network framework and is thus promising from the practical point of view. The challenge is then to extend it to general interface conditions and to guarantee its fast convergence.

3. Interface Relaxation

We now present a general subdomain-iteration approach based on the classical relaxation idea.

Let Γ_{ij} be a typical interface, that is, the common boundary piece of two neighboring subdomains Ω_i and Ω_j, i.e., $\Gamma_{ij} = \partial\Omega_i \cap \partial\Omega_j$. Each subdomain obeys a physical law locally. Namely, there is a PDE L_l and function U_l defined in each Ω_l so that

$$(3.1) \qquad L_l U_l = f_l \ in \ \Omega_l \ for \ l = i, j.$$

Notice that the interface condition on Γ_{ij} can usually be specified in the form

$$(3.2) \qquad g_{ij}\left(U_i, U_j, \frac{\partial U_i}{\partial n}, \frac{\partial U_j}{\partial n}\right) = 0.$$

In general, the left-hand side of (3.2) may also involve higher order derivatives. Without loss of generality, we only consider first order derivatives. For example, for the continuity conditions of the global solution and its normal derivative, (3.2) takes the form

$$(3.3) \qquad (U_i - U_j)^2 + \left(\frac{\partial U_i}{\partial n} - \frac{\partial U_j}{\partial n}\right)^2 = 0.$$

For some physical phenomena we might have different conditions to be satisfied on opposite sides of the interface so that the interface conditions need not be symmetric, i.e., we can have $g_{ij} \neq g_{ji}$. Denote by $BV(U_i, U_j) \equiv \{U_i, U_j, \frac{\partial U_i}{\partial n}, \frac{\partial U_j}{\partial n}\}|_{\Gamma_{ij}}$ the data set of boundary and derivative values of local solutions U_i and U_j on Γ_{ij}. Equation (3.2) can then be viewed as a constraint on $BV(U_i, U_j)$.

We now describe a general relaxation procedure as follows. Suppose that we have an initial guess for BV, denoted by BV^{old}, which satisfies the constraint

(3.2) for all interfaces. For each subdomain, we solve the boundary value problems with the corresponding PDEs in (3.1) and by using part of BV^{old} as the boundary data. With the newly computed local solutions, denoted by U_l^{new} for Ω_l, we evaluate boundary values to get $BV(U_i^{new}, U_j^{new})$ for all Γ_{ij}, which is denoted by BV' for brevity. In general, BV' does not satisfy the constraint (3.2) although part of BV^{old} may be preserved in BV' as the boundary data used in the local solve. The relaxation idea is to further change, i.e., to *relax*, certain components in BV' to obtain a new data set BV^{new} that (better) satisfies the constraint (3.2). This leads to solving equation (3.2) for the corresponding boundary components as the unknowns. The above two-phase procedure, consisting of local PDE solve and constraint relaxation, defines a mapping from BV^{old} to BV^{new}. Iterating this procedure until convergence, we obtain the global solution that satisfies both the local PDEs and interface constraints.

It is easy to create an object-oriented implementation of this relaxation procedure. The actions defined on a subdomain object are: (a) solving a PDE boundary value problem with the provided boundary data in BV from interfaces, and (b) evaluating boundary values of the resulting local solution. The actions defined on an interface object are: (a) collecting boundary values from neighboring subdomains, (b) checking for convergence by examining the interface constraints, (c) relaxing the constraint to update BV, and (d) invoking local solvers for neighboring subdomains.

There are various possible choices for the relaxation, depending on the boundary condition type for each subdomain solve and the way of relaxing the interface constraint. The alternating Dirichlet-Neumann approach is an example. An alternative is to apply a smoothing procedure along an interface, which blends the neighboring solutions to better satisfy the interface constraints along the interface. It is also possible to apply least squares to perform an overdetermined interface constraint relaxation rather than an exact relaxation. As usual, one may derive a multi-step type of relaxer by introducing certain relaxation parameters and taking a weighted average of previous and updated iterates.

We consider the following class of relaxers. First, we consider only *stationary relaxers*, those that use the same relaxation and PDE solution techniques at every iteration. There are non-stationary relaxers of serious interest, such as those that alternate between satisfying Neumann and Dirichlet conditions. Second, we consider only relaxers that use values and derivatives of PDE solutions along interfaces. That is, at each iteration a PDE is solved for U_l in Ω_l and the boundary values of U_l and its derivatives are the input to the relaxers. Discrete versions of the relaxers may involve such values on or near interfaces.

We define this class of relaxers precisely as follows. Let $I(l)$ be the indices of those subdomains that are neighbors of subdomain l. Let the PDE problem that is solved on Ω_l be

$$L_l U_l^{new} \;=\; f_l \; in \; \Omega_l,$$

(3.4) $$B_{lj} U_l^{new} \;=\; b_{lj} \; on \; \Gamma_{lj} \; for \; j \in I(l),$$

U_l^{new} satisfies the global boundary conditions on $\partial\Omega$,

where B_{lj} is a usual boundary condition operator and b_{lj} is defined as part of the relaxer as follows. Let $\vec{X}_{lj}^{old}$ be a vector of values which approximate U_l and its derivatives on Γ_{lj} for $j \in I(l)$. The length of the vector $\vec{X}$ is the number of derivatives of U_l used, it is normally 2 (using values and normal derivatives of U_l). Then a *relaxer* is a procedure that maps U_l^{old}, $\vec{X}_{lj}^{old}$ for $j \in I(l)$, $\vec{X}_{jl}^{old}$ for $j \in I(l)$ into b_{lj}.

Note that this definition of relaxers makes them domain-based and not interface-based. That is, the process of obtaining U_l^{new} is not easily interpreted as applying some independent set of procedures along the interfaces Γ_{lj} for $j \in I(l)$. Indeed, there might be interaction between different b_{lj} where, for example, two interfaces meet.

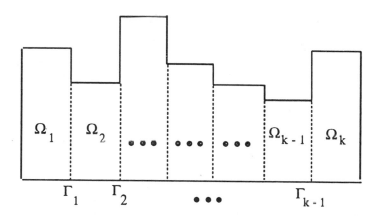

FIGURE 1. A "one dimensional" composite domain Ω.

To be more specific and for the sake of simplicity, let us assume that Ω is as in Fig. 1 and denote $\Gamma_{i,i+1}$ simply by Γ_i. Furthermore, without loss of generality, we assume that the global solution vanishes on $\partial\Omega$, and the interior interface condition is (3.3). Suppose that we impose Dirichlet condition on each Γ_{lj} for $j \in I(l)$ in (3.4). In our simplified notation, the solutions on both sides of any interface Γ_i have the same boundary values on Γ_i, denoted by X_i, at each iteration. Equation (3.3) is then reduced to

$$(3.5) \qquad \frac{\partial U_i}{\partial n} = \frac{\partial U_{i+1}}{\partial n} \ on \ \Gamma_i \ for \ i = 1, 2, \ldots, k-1.$$

In principle, one can apply any numerical method, such as finite differences, finite elements, or collocation to solve the local PDE problem (3.4). The corresponding discrete systems can be generally written as

$$(3.6) \qquad \begin{cases} A_l U_l^{new} = f_l + P_{\Omega_l, \Gamma_{l-1}} X_{l-1}^{old} + P_{\Omega_l, \Gamma_l} X_l^{old} \ for \ l = 1, 2, \ldots, k \\ \\ X_0^{old} \equiv X_k^{old} \equiv 0 \end{cases}$$

where the matrices A_l, $P_{\Omega_l, \Gamma_{l-1}}$ and P_{Ω_l, Γ_l} correspond to the discretization of the PDE operator L_l in the interior and next to the boundary pieces Γ_{l-1} and Γ_l of the subdomain Ω_l. We do not distinguish the notations for a continuous function and the vector of its discrete values.

After solving (3.4) to obtain $\{U_l^{new}\}_{l=1}^{k}$, we want to relax the interface conditions by adjusting the solution values on interfaces to better satisfy (3.5). Let $\{X_i^{new}\}_{i=1}^{k-1}$ denote the new interface values. The normal derivatives of the relaxed solutions on both sides of Γ_i can be approximated by the finite differences involving values on Γ_i and the discretization lines next to Γ_i. Denote these neighboring grid lines by $\Gamma_i^{\pm}$. A discrete approximation to (3.5) is then

$$(3.7) \qquad \frac{X_i^{new} - U_i^{new}|_{\Gamma_i^-}}{h_i^-} = \frac{U_{i+1}^{new}|_{\Gamma_i^+} - X_i^{new}}{h_i^+} \ for \ i = 1, 2, \ldots, k-1,$$

where $h_i^{\pm}$ denote the corresponding spacings between Γ_i and $\Gamma_i^{\pm}$. Solving (3.7) for X_i^{new} we obtain

$$(3.8) \qquad X_i^{new} = \alpha_i^- U_i^{new}|_{\Gamma_i^-} + \alpha_i^+ U_{i+1}^{new}|_{\Gamma_i^+} \ for \ i = 1, 2, \ldots, k-1,$$

with $\alpha_i^- = \frac{h_i^+}{h_i^- + h_i^+}$ and $\alpha_i^+ = \frac{h_i^-}{h_i^- + h_i^+}$. As in general relaxation methods, one can further introduce some relaxation parameters or make use of U_i^{new} and U_{i+1}^{new} values on other grid lines nearby Γ_i or use previous values X_i^{old}, U_i^{old}, etc., in order to accelerate the overall convergence. For example, one can define X_i^{new} by

$$(3.9) \qquad X_i^{new} = \omega X_i^{old} + (1 - \omega)(\alpha_i^- U_i^{new}|_{\Gamma_i^-} + \alpha_i^+ U_{i+1}^{new}|_{\Gamma_i^+}),$$

where ω is a relaxation parameter. Recall that $X_i^{old} \equiv U_i^{new}|_{\Gamma_i} \equiv U_{i+1}^{new}|_{\Gamma_i}$ in the present case. In general, we see that a *linear relaxer* can be expressed as

(3.10) $$X_i^{new} = \varphi_i(U_i^{new}, U_{i+1}^{new}) \ for \ i = 1, 2, \ldots, k-1,$$

where φ_i is a linear combination of U_i^{new} or U_{i+1}^{new} restricted to grid lines near to the interface Γ_i with certain weights. The choice of φ_i depends on the interface condition (3.2), the approximation accuracy of the finite differences to the normal derivatives, the relaxation techniques, and so on.

We may combine solving (3.4) for $\{U_l^{new}\}_{l=1}^k$ with (3.10) to obtain the matrix representation of $\{X_i^{new}\}$ in terms of $\{X_i^{old}\}$. The convergence analysis of the relaxation process is then reduced to the standard spectral analysis of the corresponding iteration matrix. We show in [2] the convergence of this iteration for the class of relaxers as described above for general elliptic PDE problems and general domain decomposition with cross interface points. Furthermore, under certain model problem assumptions for a rectangular domain as decomposed in Fig. 1, an explicit expression is obtained for the spectrum of the iteration matrix so that the convergence mechanism is fully understood. In addition, the optimal relaxation parameters are also determined. Extensive numerical experiments are reported in [2] to support the theoretical convergence analysis.

4. RELAX problem solving environment

In this section, we describe a problem solving environment RELAX [1] that is implemented as a platform to support the collaborating PDE solvers approach. RELAX provides both a computational and user interface environment. The computational environment coordinates teams of single-domain PDE solvers, which collaboratively solve mathematical systems called composite PDEs that model complex physical systems. The user interface environment coordinates multiple interactive user interface components, called editors, which display or alter any feature of a composite PDE problem. Editors may be both text-oriented (e.g., equation editors) and graphics-oriented (e.g., solution plotters).

RELAX is implemented using object-oriented programming technology. The system architecture is based upon a set of inter-communication software components. Editors and single-domain PDE solvers are examples of RELAX components – these particular ones are externally supplied (perhaps from libraries or other software systems). RELAX provides a message-passing mechanism for supporting the inter-component communication. It is capable of integrating existing scientific software for PDEs into a broader problem solving environment. It also has the capability of using pre-existing display and interaction components to form a flexible, dynamic user interface.

We briefly outline the function of each type of component of the RELAX architecture and refer to [1] for more details:

- *Primitive Objects*: These are externally supplied components which model and solve primitive PDE problems. Primitive objects are responsible for all aspects of solving a single-domain PDE problem, including the generation of numerical meshes, discretization of the PDE, and solving systems of equations.

- *Editors*: These are externally supplied components which provide an interface between the user and some feature of the system. The editor component is responsible for the complete presentation of the user interface, including all communication with the window system and/or graphics package.

- *Message Dispatcher*: This is a system supplied component which handles all transmission of messages within the system. The message dispatcher also registers targets and can assist editors in locating targets.

- *Composite Problem Platform*: This is a system component which maintains the data structures defining a composite PDE problem. For example, the composite problem platform stores topological information about which primitive objects share geometric interfaces, as well as equations defining the interface conditions along those interfaces. Additionally, the composite problem platform maintains data structures defining a global solution iteration, and is capable of executing such iterations. Finally, the composite problem platform is capable of defining composite PDE problems hierarchically.

- *Object Support Platform*: This is a system component responsible for integrating primitive objects into the system. The object support platform provides the attachment point for primitive objects – recall that they are external components and must be dynamically attached to the running system. The object support platform relays messages between primitive objects and the message dispatcher. Another feature of the object support platform is a *virtual object* mechanism which allows various primitive objects to filter the messages intended for other primitive objects.

- *Editor Support Platform*: This is a system component which provides an attachment point and communication interface for editors. The editor support platform relays messages to and from editors, and is also capable of parameterizing and controlling the message flow, for example, by copying and buffering messages.

With this environment, one can easily describe primitive PDE problems and interface conditions to compose a complex mathematical system, specify local solvers and relaxers to define an iterative procedure, and display the computed solution in various ways. As an application example, we use the RELAX system to solve a physical heat flow problem as shown in Fig. 2. The complex object consists of seven simple subdomains with nine interfaces. The radiation conditions allow heat to leave on part of the boundaries while the temperature U is zero on all the other boundaries. The mounting regions have heat dissipated.

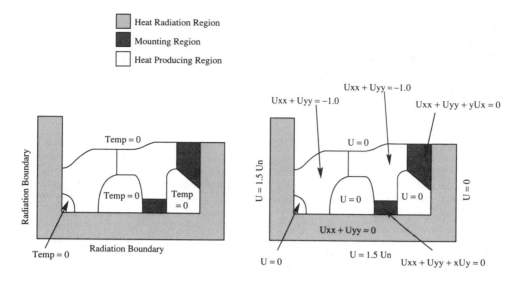

FIGURE 2. A heat flow problem for a complex domain along with the physical and mathematical descriptions.

The interface conditions are continuity of temperature U and its derivative. An approximate solution of 3-digit accuracy is obtained after 15 iterations, where the initial guess is zero and the relaxer used is as described in Section 3 with the relaxation parameter $\omega = 0$.

5. Conclusions

We examine in this paper various aspects of the real world simulation with the emphasis on software productivity and quality. Application of modern software technologies and the impact on numerical PDE methods are considered. We present a general approach for modeling complex physical systems by a network of collaborating PDE solvers. The related methodologies include *networks of collaborating software modules*, *object-oriented programming* and *domain decomposition*. They lead to a suitable subdomain-iteration based procedure with interface relaxation. Various types of relaxers are discussed. A software system RELAX is described which is implemented as a platform to test various relaxers and to solve complex problems using the network approach. Both theory and practice show that this is a promising approach for solving complicated problems on modern computer environments.

Acknowledgement. The RELAX environment is implemented by Dr. Scott McFaddin. We would like to thank him very much for providing the implementation details. We also thank Professor David Keyes for his valuable comments and suggestions.

REFERENCES

1. S. McFaddin and J. R. Rice, *Architecture of the RELAX problem solving environment*, CSD-TR-92-081 and CER-92-37, Department of Computer Sciences, Purdue University, West Lafayette, IN47907, October, 1992.
2. M. Mu and J. R. Rice, *Collaborating PDE solvers with interface relaxation*, CSD-TR-93-024 and CER-93-13, Department of Computer Sciences, Purdue University, West Lafayette, IN47907, April, 1993.

DEPARTMENT OF MATHEMATICS, THE HONG KONG UNIVERSITY OF SCIENCE & TECHNOLOGY, CLEAR WATER BAY, KOWLOON, HONG KONG
  E-mail address: mamu@usthk.bitnet

COMPUTER SCIENCES DEPARTMENT, PURDUE UNIVERSITY, WEST LAFAYETTE, IN 47907 U.S.A.
  E-mail address: rice@cs.purdue.edu

Contemporary Mathematics
Volume **180**, 1994

Parallelization of a Multigrid Solver via a Domain Decomposition Method

FRANCOIS-XAVIER ROUX AND DAMIEN TROMEUR-DERVOUT

July 8, 1994

ABSTRACT. This paper reports some experiments with the implementation of a multigrid solver on a distributed memory parallel system. The use of a domain decomposition method for the coarse grid problem is shown to improve the method from both numerical and parallel efficiency points of view. The domain decomposition method is accelerated via a restarting procedure using the direction vectors computed at the previous solutions of the coarse grid problem.

1. Introduction

Multigrid solvers are used in a wide range of applications. The principle of these algorithms is to perform a few iterations of an iterative solver at each grid level, in order to smooth out the components of the residual related to the eigenvectors associated with the highest eigenvalues of the discrete operator and then project it on the coarser grid level. At the coarsest level, the problem is completely solved. Then the error is interpolated back recursively on the finer grid levels. At each level, a few smoothing iterations are performed again. This set of operations defines a cycle of the multigrid algorithm. The complete method consists in iterating cycles to get a good approximation of the solution at the finest grid level. The simplest implementation of this method is made with two grid levels, with V-cycles.

One of the main problems with the parallel implementation of multigrid solvers lies in the fact that the coarse grid problem needs to be solved with a good relative precision. One strategy consists in solving the coarse grid problems with a direct method. But direct methods are difficult to implement efficiently

---

1991 *Mathematics Subject Classification.* Primary 65M55, 68Q22, 65M06.
This paper is in final form and will not be published elsewhere.

on distributed memory systems, especially for sparse matrices with small bandwidth. Iterative methods are easier to parallelize. But the granularity of the tasks depends upon the number of nodes of the problem, that tends to be small for the coarsest grid levels.

An alternative strategy consists in using a domain decomposition solver for the coarse grid problem, in order to increase the granularity of the tasks. In this paper we present some results with the implementation of a two-grid solver, using a domain decomposition method at the coarse grid level, on a distributed memory machine.

The paper is organized as follows : Section 2 presents some results with the implementation on a 128-processor iPSC-860 computer of a multigrid solver, using a classical iterative procedure at the coarse grid level. Section 3 recalls the principle of the dual Schur complement method. Section 4 presents a restarting technique for the domain decomposition method in order to reduce the number of iterations to be performed for each solution of the coarse grid problem. Section 5 is devoted to a discussion of the intrinsic properties of the dual Schur complement method that makes it a very convenient solver to be used within a multigrid approach, and the conclusions are derived in the last section.

2. Parallel implementation of a two-grid solver

In this section we present the performance of a two-grid method, in order to highlight the difficulties encountered with the parallel efficiency of iterative solvers for the coarse grid problem.

The test problem is a Poisson equation arising from the discretisation by finite difference methods of 3D incompressible Navier-Stokes equations with a velocity-vorticity formulation. The multigrid solver is used for solving the velocity problem at each time step. The fine, respectively coarse, mesh is a 128^3, respectively 64^3, node regular grid. The complete domain is divided into 128 cubic subdomains, and so both grids are decomposed into 128 non-overlapping subgrids. The iterative solver used for the smoothing iterations at the fine grid level and for the solution of the coarse grid problem is Gauss-Seidel for the diagonal blocks associated with the inner nodes of each subdomain, and Jacobi for the off-diagonal blocks associated with the interaction between subdomains.

This solver is naturally parallelizable on a distributed memory machine, each subdomain, with its two-grid, being treated by one processor. At each iteration, only interface nodes are involved in data transfers.

Table 1 presents the results with the implementation of a V-cycle, with 10 smoothing iterations at the fine grid level, and either 200 or 400 iterations for solving the coarse grid problem, on a 128-processor iPSC-860 machine.   These results show the degradation of the parallel efficiency due to the decrease of granularity at the coarsest grid level. Of course, the degradation is very sharp here, because the iPSC-860 is a very unbalanced machine: the ratio of the computing speed over the communication speed is greater than 10 for the present case, according to the time measurements. But on any machine, the time for data transfers should become dominant for some grid levels: if a subdomain consists of a n^3 nodes regular grid, the number of interface nodes is $6n^2$. As the amount of arithmetic operations depends linearly upon the number of nodes within the subdomain, and the amount of data transfers depends upon the number of interface nodes, the ratio of the communication cost over the computational cost would become large on any distributed memory machine for n small enough, which means for grids coarse enough.

Table 1 : Efficiency of multigrid with Gauss-Seidel solver

number of iterations for the coarse grid	efficiency on the fine grid	efficiency on the coarse grid	global efficiency
200	44%	16%	34%
400	44%	16%	27%

3. Solution of the coarse grid problem with the dual Schur complement

The dual Schur complement method for the Poisson equation consists in introducing λ, the flux along the interface between subdomains, and solving iteratively the associated condensed interface problem [1]. The solution of this problem is the λ for which the solution fields of the local Neumann problems are continuous along the interfaces between subdomains. Then, the global field whose restriction to each subdomain is defined as the solution of the local Neumann problem is continuous, and also its normal derivative. It is therefore the solution of the global problem.

In the case of a decomposition in two subdomains, Ω_1 and Ω_2, if A_1 and A_2 are the discretized operators on Ω_1 and Ω_2 with Neumann boundary conditions on the interface Γ_3, and B_1 and B_2 the trace operators over Γ_3 of functions defined upon Ω_1 and Ω_2, the hybrid formulation of the Poisson problem on the domain Ω can be written as follows :

(3.1) $$\begin{cases} A_1 u_1 + B_1^t \lambda & = & f_1 \\ A_2 u_2 - B_2^t \lambda & = & f_2 \\ B_1 u_1 - B_2 u_2 & - & 0 \end{cases}$$

By substitution of u_1 and u_2 given by the two first equations in the third one, the condensed interface problem for λ is :

(3.2) $$(B_1 A_1^{-1} B_1^t + B_2 A_2^{-1} B_2^t)\lambda = B_1 A_1^{-1} f_1 - B_2 A_2^{-1} f_2$$

This problem can be solved by the conjugate gradient algorithm. At each iteration of the condensed interface problem, the local Neumann problems are solved by a direct method, and then the discontinuity of the local solutions along the interfaces are computed. So, the communication costs per iteration are the same as for the Gauss-Seidel method, because only interface unknowns are to be transferred, but the granularity of the tasks is much larger, because the local task consists of a forward/backward substitution using the dense skyline factorization instead of the sparse initial matrix. Table 2 presents the results with the implementation of a V-cycle, with 10 smoothing iterations at the fine grid level, and either 200 Gauss-Seidel iterations or 20 dual Schur complement iterations for solving the coarse grid problem, on a 128-processor iPSC-860 machine.

Table 2 : Efficiency with Gauss-Seidel or dual Schur complement

coarse grid solver	efficiency on the fine grid	efficiency on the coarse grid	global efficiency
Gauss-Seidel	44%	16%	34%
dual Schur complement	44%	63%	53%

4. Acceleration of the dual Schur complement method

For solving the linear system of equations $Ax = b$, the conjugate gradient algorithm consists in computing a set of direction vectors that are conjugated for the dot product associated with the matrix A, and that generate the Krylov space $\mathrm{Span}\{g_0, Ag_0, \ldots, A^{p-1}g_0\}$, where g_0 is the initial residual $Ax_0 - b$. The approximate solution x_p at iteration p, minimizes the A-norm of the error, $(Ax_p - b, x_p - x)$, over the space $x_0 + \mathrm{Span}\{g_0, Ag_0, \ldots, A^{p-1}g_0\}$.

If a set of conjugate directions (w_i), $1 \le i \le p$, is given, then the element x_p of $x_0 + \mathrm{Span}\{w_1, w_2, \ldots, w_p\}$ that minimizes the residual can be easily computed:

$$(4.1) \qquad x_p = x_0 - \sum_{i=1}^{p} \frac{(g_0, w_i)}{(Aw_i, w_i)} w_i$$

Then, the iterations of the conjugate gradient algorithm can start from the optimal starting point x_p. But the application of the standard conjugate gradient algorithm does not make sure that the new direction vectors are conjugated to the vectors w_i, $i = 1, \ldots, p$. To enforce these conjugacy relations, the new direction vector d_j at iteration number j must be reconjugated to the vectors w_i through the following procedure :

$$(4.2) \qquad d_j = Mg_j - \frac{(Mg_j, Ad_{j-1})}{(Ad_{j-1}, d_{j-1})} d_{j-1} - \sum_{i=1}^{p} \frac{(Mg_j, Aw_i)}{(Aw_i, w_i)} w_i$$

where M is the preconditioner of the matrix A, and g_j is the gradient vector at iteration j, $g_j = Ax_j - b$.

This is equivalent to performing the new iterations of the conjugate gradient algorithm in the subspace A-conjugate to $\mathrm{Span}\{w_1, w_2, \ldots, w_p\}$. Then the algorithm is optimal in that sense that the actual dimension of the problem to be solved by the conjugate gradient algorithm is now equal to the dimension of A minus p.

In practice, the set of conjugate vectors (w_i) is built by accumulating the direction vectors computed for the solution of all coarse grid problems. In order to make the method more robust, in the face of finite precision arithmetic, a complete reconjugation procedure is performed :

$$(4.3) \qquad d_j = Mg_j - \sum_{k=0}^{j-1} \frac{(Mg_j, Ad_k)}{(Ad_k, d_k)} d_k - \sum_{i=1}^{p} \frac{(Mg_j, Aw_i)}{(Aw_i, w_i)} w_i$$

This acceleration procedure is very efficient: for 20 V-cycles of the test problem introduced in section 2, it reduces the global number of conjugate gradient iterations for the dual Schur complement method at the coarse grid level from 450 to 150.

Of course, this acceleration technique could be applied to any problem with multiple right hand sides solved by the conjugate gradient algorithm. But for sparse linear systems of equations, the procedure should be too expensive in computing time and memory requirement. Within a domain decomposition method, the procedure is relatively inexpensive, as the reconjugation procedure involves only interface unknows, when each iteration for the condensed interface problem requires the solution of all local problems. This is all the less expensive relatively when the domain decomposition method is used only for solving the coarsest grid problem within a multigrid approach.

Table 3 details the elapsed times in the various sections of the multigrid procedure whith the solution of the coarse grid problem by the Gauss-Seidel algorithm or the accelerated dual Schur complement method. The test problem is the same as in section 2, the total number of V-cycles is 20. The global error indicated in the last column is the ratio of the L_2 norm of the error over the L_2 norm of the exact solution on the complete domain.

Table 3 : Comparison of elapsed times

coarse grid solver	computation	communication	total	global error
Gauss-Seidel	39	91	140	$7 * 10^{-4}$
dual Schur complement	35	40	75	$8 * 10^{-5}$

This table shows that the multigrid method with the accelerated Schur complement method at the coarse grid level should be faster, even on a sequential machine, than the multigrid method with Gauss-Seidel iterations at the coarse grid level. Furthermore, the first method is clearly better suited than the second one for parallel computing.

5. Numerical efficiency of multigrid with solution of the coarse grid problem by the dual Schur complement method

Table 3 exhibits different values for the global error of the solution of the problem after 20 V-cycles, depending upon the solution method for the coarse grid problem. For both cases, the number of smoothing iterations at the fine grid level, and the stopping criterion for the residual of the coarse grid problem are the same. However, for all the test cases we studied, the results were identical: with the same stopping criterion at the coarse grid level, the multigrid algorithm converges faster when the coarse grid problems are solved by the dual Schur complement method.

A tentative explanation of this phenomenon relies upon the spectral properties of the dual Schur complement operator [2]. The largest eigenvalues of the dual Schur complement operator are related to the low frequencies of the primal problem. This implies that the components that are captured first by the conjugate gradient iterations for problem 3.2 are the low frequencies of the coarse grid. That is exactly what the coarse grid solver is supposed to do in order to make the multigrid method efficient.

Furthermore, these spectral properties of the dual Schur complement method allow a very fast convergence if the stopping criterion is not too small. Within the multigrid solver, the coarse grid problem needs to be solved only to within a relative residual slightly smaller than the global residual at the current cycle. Therefore, the dual Schur complement method turns out to be a very well suited iterative solver for the coarse grid level.

6. Conclusions

The results presented here show that the dual Schur complement method for solving the coarse grid problems in a multigrid algorithm has very good features from both numerical and parallel implementation points of view. Of course, although the test problem was a two-grid method, it can be used with any multigrid strategy. The acceleration technique introduced in section 4 can also be easily extended to non symmetric problems with another Krylov space method like the generalized conjugate residual. So, the coupling between multigrid and domain decomposition techniques looks very promising for the parallel solution of large scale numerical engineering problems.

REFERENCES

1. C. Farhat, F.-X. Roux, *An unconventional domain decomposition method for an efficient parallel solution of large-scale finite element systems*, Siam J. Sci. Stat. Comput., vol 13, pp. 379–396, 1992.
2. F.-X. Roux, *Dual and spectral properties of Schur and saddle point domain decomposition methods* , T. Chan, D. Keyes, G. Meurant, J. Scroggs, R. Voigt eds., Proceedings of the 5th International Symposium on Domain Decomposition Methods, SIAM, Philadelphia (1992), pp. 73–90 .

ONERA, PARALLEL COMPUTING DIVISION, 29 AVENUE DE LA DIVISION LECLERC, 92320 CHATILLON, FRANCE
E-mail address: roux@onera.fr

Contemporary Mathematics
Volume **180**, 1994

A Compiler for Parallel Finite Element Methods with Domain-Decomposed Unstructured Meshes

JONATHAN RICHARD SHEWCHUK AND OMAR GHATTAS

ABSTRACT. Archimedes is an automated system for finite element methods on unstructured meshes using distributed memory supercomputers. Its components include a mesh generator, a mesh partitioner, and a data-parallel compiler whose input is C augmented with machine-independent operations for finite element computations, and whose output is parallel code for a particular multicomputer. We describe an elegant implementation of domain decomposition and give preliminary performance results.

1. Introduction

Data-parallel languages such as High Performance Fortran make it possible to quickly and portably program multiprocessors. However, most current compilers are not satisfactory for programming finite element simulations, because they cannot support complicated parallel data structures.

There are several reasons why effective parallel compilers for finite elements are difficult to construct. If unstructured meshes are desired, the finite element code must use indirect addressing to process elements and to form stiffness matrices; but parallel indirect addressing is difficult. Communication costs will be high unless data structures are intelligently divided among processors. Furthermore, few data-parallel compilers provide explicit support for performing operations on a processor-by-processor basis; this makes it impossible to use domain decomposition methods to explicitly manage parallelism.

To address these problems, we are developing Archimedes, a system that generates finite element code for distributed memory supercomputers. The structure

---

1991 *Mathematics Subject Classification*. Primary 65Y05, 65M55; Secondary 68N20, 65F10.

The first author was supported in part by the Natural Sciences and Engineering Research Council of Canada. Both authors were supported in part by the National Science Foundation under Grant ASC-9318163.

This paper is in final form and no version of it will be submitted for publication elsewhere.

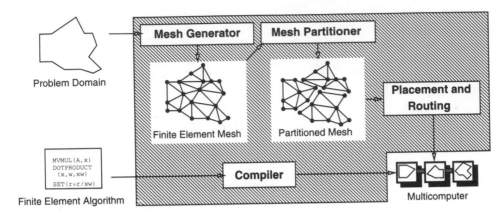

FIGURE 1.   Structure of Archimedes.

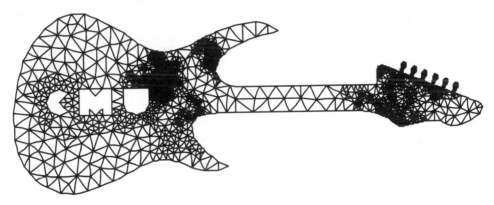

FIGURE 2.   Refined mesh of an electric guitar.

of Archimedes is diagrammed in Fig. 1. Its components include a mesh generator, a mesh partitioner, placement and routing heuristics, and a compiler.

The mesh generator uses an algorithm due to Ruppert [4] to create quality two-dimensional meshes on complex straight-line domains, and can also refine meshes based on *a posteriori* error estimates. An example of a mesh generated and refined this way is illustrated in Fig. 2.

Meshes are partitioned by a geometric algorithm due to Miller, Teng, Thurston, and Vavasis [3]. The partitioner serves three purposes. It divides a mesh into subdomains, to be mapped to separate processors (Fig. 3). It generates a nested dissection ordering on each subdomain, and thereby improves the performance of domain decomposition methods. Finally, a lesser known fact is that one can form a nested dissection ordering that improves memory cache performance because physically adjacent nodes tend to be grouped together in memory.

The communication graph for the partition of Fig. 3 is illustrated in Fig. 4. The nodes of this graph represent processors, and edges are drawn between any two processors having adjacent subdomains. On most multicomputers, com-

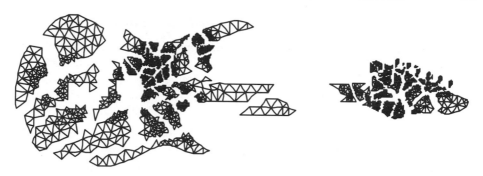

FIGURE 3.   Partitioned electric guitar mesh.

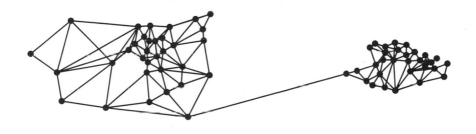

FIGURE 4.   Communication graph of partitioned mesh.

munication is faster if adjacent subdomains are mapped to nearby processors. Hence, we use placement heuristics to find such a mapping. Some multiprocessors can be sped up by explicitly choosing communication routes between processors; routing heuristics are provided for these systems. The placement and routing heuristics are described in detail by Feldmann, Stricker, and Warfel [2].

Archimedes' compiler takes as input C code with special machine-independent operations for finite element computations, and outputs parallel code for a particular multicomputer. Users write parallel code without knowing the underlying communication mechanisms of the parallel architecture. This simplifies the task of writing parallel finite element code, or experimenting with iterative linear solvers. The remainder of this paper describes the parallel operations provided by the compiler.

2. Parallel operations for domain decomposition

2.1. Data distribution and communicating operations. The data distribution of the stiffness matrix K is the key to our implementation. Let P be the number of processors. Each processor p holds a *processor stiffness matrix* K^p, which is a portion of the global stiffness matrix. Effectively, K^p contains zero rows and columns for each node not mapped to processor p; of course, these zeroes are not actually stored in memory. The value of the global stiffness matrix is $K = \sum_{i=1}^{P} K^i$.

Archimedes' partitioner maps each mesh element to only one processor, and K^p is defined by the set of elements mapped to processor p. Element stiffness matrices are assembled into processor stiffness matrices in parallel without communication, but the global stiffness matrix K is never actually formed. We say that K is *partially assembled*, because it is not assembled across processor boundaries.

Here, our methodology is at variance with traditional data-parallel compilers. Nodes and edges on subdomain boundaries are shared by multiple processors. Accordingly, a distributed stiffness matrix may have storage allocated for an edge on several processors. ("Edge" here should be read to include self-edges, i.e., diagonal entries of the stiffness matrix.) Each processor stores the nonzero portion of its processor stiffness matrix in Compressed Sparse Row format.

Distributed vectors may have storage for a node allocated on several processors. Ordinarily, they are stored so that the duplicated nodes have duplicated values. In other words, a vector x is distributed so that each processor knows the elements of x corresponding to the nodes mapped to that processor. For reasons that will become clear in the next paragraph, we say that x is *fully assembled*.

Performing a distributed matrix-vector product of the form $y = Kx$ is a two-step process. In the first step, each processor p takes the product $y^p = K^p x$. This step uses a standard sequential sparse matrix-vector product, and requires no communication. At this point, we say that y, like K, is partially assembled, because the true value of y is $y = \sum_{i=1}^{P} y^i$. The second step is to *fully assemble y*. To accomplish this, each processor communicates with its neighbors (along the routes of the communication graph in Fig. 4) and sums each processor's value for each shared node. For example, if processors p and q share node j, then both processors will take the sum $y_j^p + y_j^q$ as the value of y_j. We call this step a *communicating sum*.

Many iterative methods for solving $Kx = y$ can be implemented with only two communication operations: communicating sums, and parallel reductions (such as dot product). Several local operations are also required, such as sparse matrix-vector multiply and elementwise vector operations. If the stiffness matrix is unsymmetric, our data distribution makes it trivial to obtain the transpose of the global stiffness matrix without communicating; hence, iterative methods such as biconjugate gradients (which requires the product $K^T x$) are easy to implement.

We can also use the communicating sum to form a diagonal preconditioner. Each processor extracts the diagonal of its processor stiffness matrix, and a communicating sum is used to find the diagonal of the global stiffness matrix. Thereafter, the diagonal can be used as a preconditioner without further communication.

2.2. Domain decomposition. We present a domain decomposition method appropriate for a sequence of linear problems having the same global stiffness matrix, as arise in time-dependent problems.

Order the variables so that those interior to subdomain 1 come first, followed by subdomain 2, etc. Last comes the set $\mathbb{I}$ of variables corresponding to interface nodes (each shared by two or more subdomains). The system $Kx = y$ has the form

$$\begin{bmatrix} K_{11} & 0 & \cdots & 0 & K_{1\mathbb{I}} \\ 0 & K_{22} & & 0 & K_{2\mathbb{I}} \\ \vdots & & \ddots & & \vdots \\ 0 & 0 & & K_{PP} & K_{P\mathbb{I}} \\ K_{\mathbb{I}1} & K_{\mathbb{I}2} & \cdots & K_{\mathbb{I}P} & K_{\mathbb{I}\mathbb{I}} \end{bmatrix} \begin{bmatrix} x_1 \\ x_2 \\ \vdots \\ x_P \\ x_{\mathbb{I}} \end{bmatrix} = \begin{bmatrix} y_1 \\ y_2 \\ \vdots \\ x_P \\ y_{\mathbb{I}} \end{bmatrix}.$$

A standard nonoverlapping domain decomposition technique is to use block elimination of $x_1, x_2, \ldots, x_P$ to yield the Schur complement system $\tilde{K}_{\mathbb{I}\mathbb{I}} x_{\mathbb{I}} = \tilde{y}_{\mathbb{I}}$, where $\tilde{K}_{\mathbb{I}\mathbb{I}} = K_{\mathbb{I}\mathbb{I}} - \sum_{i=1}^{P} K_{\mathbb{I}i} K_{ii}^{-1} K_{i\mathbb{I}}$ and $\tilde{y}_{\mathbb{I}} = y_{\mathbb{I}} - \sum_{i=1}^{P} K_{\mathbb{I}i} K_{ii}^{-1} y_i$. This system is then solved by an iterative Krylov subspace method. Contrary to standard practice, we explicitly form the Schur complement matrix $\tilde{K}_{\mathbb{I}\mathbb{I}}$.

If we ignore zero rows and columns, each processor stiffness matrix is of the form $K^p = \begin{bmatrix} K_{pp} & K_{p\mathbb{I}} \\ K_{\mathbb{I}p} & K_{\mathbb{I}\mathbb{I}}^p \end{bmatrix}$, where $K_{\mathbb{I}\mathbb{I}} = \sum_{i=1}^{P} K_{\mathbb{I}\mathbb{I}}^i$. By factoring K_{pp} (using a nested dissection ordering), each processor p forms (without communicating) a *processor Schur complement* $\tilde{K}_{\mathbb{I}\mathbb{I}}^p = K_{\mathbb{I}\mathbb{I}}^p - K_{\mathbb{I}p} K_{pp}^{-1} K_{p\mathbb{I}}$. Afterward, the Schur complement is a partially assembled matrix, just like the stiffness matrix K — in other words, it has the property that $\tilde{K}_{\mathbb{I}\mathbb{I}} = \sum_{i=1}^{P} \tilde{K}_{\mathbb{I}\mathbb{I}}^i$. Hence, we can use $\tilde{K}_{\mathbb{I}\mathbb{I}}$ in Krylov methods, with the same data distribution and communicating sum used for K.

Each processor Schur complement is a dense matrix coupling the boundary nodal unknowns of that processor's subdomain. This density is advantageous, because most modern microprocessors perform dense matrix-vector multiplication at two to ten times the speed of sparse matrix-vector multiplication, and because we can easily apply a Neumann-Neumann preconditioner, as described by Bourgat et al. [1]. By applying a communicating sum to $\tilde{K}_{\mathbb{I}\mathbb{I}}^p$, each processor forms a fully assembled *processor preconditioner* $M_{\mathbb{I}\mathbb{I}}^p$. $M_{\mathbb{I}\mathbb{I}}^p$ is the submatrix of $\tilde{K}_{\mathbb{I}\mathbb{I}}$ that represents only the boundary of subdomain p. Although $\tilde{K}_{\mathbb{I}\mathbb{I}}^p$ may be singular, $M_{\mathbb{I}\mathbb{I}}^p$ generally is not, and each processor can easily factor or invert $M_{\mathbb{I}\mathbb{I}}^p$ (which is dense). Our inverse preconditioner (which approximates $\tilde{K}_{\mathbb{I}\mathbb{I}}^{-1}$) is $M^{-1} = \sum_{i=1}^{P} (M_{\mathbb{I}\mathbb{I}}^i)^{-1}$. (For simplicity, we are abusing notation: each inverse is taken by ignoring zero rows and columns, which represent nodes not in the subdomain; but the summation takes these zero rows and columns into account. To be pedantically correct, the above sum should read $\sum_{i=1}^{P} (R^i)^T (R^i M_{\mathrm{III}}^i (R^i)^T)^{-1} R^i$, where R^p is a global-to-processor restriction matrix.) M^{-1} is a partially assembled matrix, and may be manipulated in the same fashion as K and $\widetilde{K}_{\mathrm{II}}$ (although the inverted matrices that compose M^{-1} need not be explicitly formed).

Here we summarize our domain decomposition algorithm. We have obtained good performance on a 64-processor iWarp; timings are omitted due to space.

(i) Each processor assembles its processor stiffness matrix K^p.

(ii) Each processor forms its (dense) processor Schur complement $\widetilde{K}_{\mathrm{III}}^p$.

(iii) With a communicating sum, each processor forms its (dense) processor preconditioner M_{III}^p, which is then factored or inverted.

(iv) For each time step (or right-hand side):

   (a) Each processor assembles, element-by-element, its partially assembled force vector y^p.

   (b) From y^p and the factors of K^p, each processor forms its partially assembled reduced force vector $\widetilde{y}_{\mathrm{I}}^p$. A communicating sum is used to fully assemble the reduced force vector $\widetilde{y}_{\mathrm{I}}$.

   (c) The Schur complement system $\widetilde{K}_{\mathrm{II}} x_{\mathrm{I}} = \widetilde{y}_{\mathrm{I}}$ is solved iteratively. The values on the interface nodes (x_{I}) are thus found.

   (d) Using triangular backsubstitution with the factors of K^p, each processor finds from x_{I} the values on its interior nodes (x_p).

We recommend this approach because of its simplicity. By writing a communicating sum and parallel dot product, and using standard sparse matrix libraries, one can quickly implement an efficient domain decomposition solver.

References

1. J.-F. Bourgat, R. Glowinski, P. Le Tallec, and M. Vidrascu, *Variational formulation and algorithm for trace operator in domain decomposition calculations*, Second Int. Conf. Domain Decomposition Methods (T. Chan, R. Glowinski, J. Périaux, and O. Widlund, eds.), SIAM, 1989.

2. A. Feldmann, T.M. Stricker, and T.E. Warfel, *Supporting sets of arbitrary connections on iWarp through communication context switches*, Proc. 5th Annual ACM Symp. Parallel Algorithms and Architectures, 1993, pp. 203–212.

3. G.L. Miller, S.-H. Teng, W. Thurston, and S.A. Vavasis, *Automatic mesh partitioning*, Graph Theory and Sparse Matrix Computation (A. George, J.R. Gilbert, and J.W.H. Liu, eds.), Springer-Verlag, 1993.

4. J.M. Ruppert, *A Delaunay refinement algorithm for quality 2-dimensional mesh generation*, To appear in J. Algorithms, 1994.

School of Computer Science, Carnegie Mellon University, 5000 Forbes Avenue, Pittsburgh, Pennsylvania 15213-3891
   E-mail address: jrs@cs.cmu.edu

Department of Civil Engineering, Carnegie Mellon University, 5000 Forbes Avenue, Pittsburgh, Pennsylvania 15213-3891
   E-mail address: oghattas@cs.cmu.edu

PART IV

Applications

Contemporary Mathematics
Volume **180**, 1994

A χ-Formulation of the Viscous-Inviscid Domain Decomposition for the Euler/Navier-Stokes Equations

RENZO ARINA AND CLAUDIO CANUTO

ABSTRACT. In this paper we present an application of the χ-formulation for the solution of the Navier-Stokes equations. Subsonic laminar and transonic turbulent flows are calculated.

1. Introduction

Many flow phenomena, such as turbulent flows, involve a wide range of length scales, for this reason their numerical simulation is a challenging problem. In many situations, high accuracy is necessary only in limited parts of the domain, one possibility is to resolve the physics on a global uniform grid with the smallest desired mesh size. However this direct approach is far from being efficient. For these reasons several physically motivated domain decomposition methods have been proposed. In this paper, we present a domain decomposition method termed χ-formulation [**3**]. The key idea is to locally evaluate the magnitude of the diffusive part of the Navier-Stokes equations, and by this inspection to detect the smallest scales involved into the phenomenon under investigation. In this way, we obtain an automatic detection of the shear layers (such as boundary layers, wakes, etc...) where an highly accurate simulation is required, and a natural splitting of the domain can be performed.

The χ-formulation has already been successfully applied to a scalar model problem [**2**], showing its ability in optimizing the interface position. In the present paper, we present an application of this approach to the solution of the compressible Navier-Stokes equations for two-dimensional subsonic and transonic flows, in laminar and turbulent regimes.

1991 *Mathematics Subject Classification*. Primary 76N10,65M55; Secondary 35Q30.
The first author was supported in part by BRITE-ETMA CEE Contract.
This paper is in final form and no version of it will be submitted for publication elsewhere.

2. The χ-Formulation for the Navier-Stokes Equations

Consider the incompletely parabolic problem

(2.1)
$$\begin{cases} u_t + Bu_x + Cu = Au_{xx} \, , & x \in (0,L) \, , \ t > 0 \, , \\ + \text{ initial and boundary conditions} \, , \end{cases}$$

where $u = \binom{u_1}{u_2}$, $A = \left(\begin{smallmatrix} 0 & 0 \\ 0 & \nu \end{smallmatrix}\right)$ with $\nu > 0$, and B a symmetric matrix. The χ-formulation replaces problem (2.1) by a modified problem, in which the diffusive term is deleted when it is negligible. To this end, let us choose a cut-off parameter $\delta > 0$, and a further parameter $\sigma > 0$ such that $\sigma \ll \delta$, and let us introduce the monotone function $\chi = \mathbb{R} \to \mathbb{R}$ defined as:

(2.2)
$$\begin{cases} \chi(s) = \begin{cases} s & \text{if } s > \delta \, , \\ (s - \delta + \sigma)(\delta/\sigma) & \text{if } \delta - \sigma \leq s \leq \delta \, , \\ 0 & \text{if } 0 \leq s < \delta - \sigma \, , \end{cases} \\ \chi(s) = -\chi(-s) \, , \quad s < 0 \, . \end{cases}$$

We define the χ-formulation of problem (2.1), as the following modified problem

(2.3)
$$\begin{cases} u_t + Bu_x + Cu = A\chi(u_{xx}) \, , & x \in (0,L) \, , \ t > 0 \, , \\ + \text{ initial and boundary conditions} \, , \end{cases}$$

where the viscous term $Au_{xx} = (0, \nu u_{2,xx})^T$ is replaced by the modified term $(0, \nu \chi(u_{2,xx}))^T$, denoted by $A\chi(u_{xx})$. The linear ramp of size σ in the function χ yields a continuous transition between the state $\chi = 0$ (the inviscid state) and states $|\chi| \geq \delta$ (the fully viscous region). This formulation leads to a smooth behavior of the solution u of (2.3) at the viscous/inviscid interface.

In [3], it has been proved that the χ-formulation leads to a well posed problem for the scalar advection-diffusion problem, and that the maximun deviation of u from the solution of (2.1) is proportional to $\delta\nu$, providing an estimate for the choice of the parameter δ as a function of the diffusion parameter ν. Moreover it has been shown that u is continuously differentiable all over the domain, in particular across the interface. This property is peculiar to the χ-formulation. Indeed any a priori choice of the viscous/inviscid interface leads only to C^0 continuity. Further in [4] the same results have been proved for more general boundary conditions, and in [5] the convergence of the iterative self-adaptive domain decomposition has been investigated.

The χ-formulation introduces a nonlinearity into the diffusive part of the equations. However, if the original problem is already nonlinear, this added nonlinearity can be treated explicitly as the other nonlinearities. In the case of the viscous Burgers equation, it has been shown [2] that the χ-Burgers equation can be solved with a computational effort which is comparable with the cost of solving the Burgers equation on the same domain. In [1], Achdou and Pironneau have extended the χ-formulation to the incompressible Navier-Stokes equation, showing how the self-adaptive detection of the viscous subdomain greatly improve the efficiency of the simulations.

In this work we present an extension of [3] to the compressible Navier-Stokes equations, which for two-dimensional compressible flows, can be written in the nondimensional form

$$(2.4) \qquad \frac{\partial Q}{\partial t} + \frac{\partial F}{\partial x} + \frac{\partial G}{\partial y} = \frac{1}{Re}\left(\frac{\partial F_v}{\partial x} + \frac{\partial G_v}{\partial y}\right) \ ,$$

where $Q = (\rho, \rho u, \rho v, E)^T$,

$$F = \begin{pmatrix} \rho u \\ \rho u^2 + p \\ \rho u v \\ (E+p)u \end{pmatrix} \ , \quad G = \begin{pmatrix} \rho v \\ \rho u v \\ \rho v^2 + p \\ (E+p)v \end{pmatrix} \ ,$$

are the convective fluxes, and the viscous fluxes are given by

$$F_v = \begin{pmatrix} 0 \\ \tau_{xx} \\ \tau_{xy} \\ u\tau_{xx} + v\tau_{xy} + \frac{\mu}{(\gamma-1)Pr}\frac{\partial a^2}{\partial x} \end{pmatrix} \ , \quad G_v = \begin{pmatrix} 0 \\ \tau_{xy} \\ \tau_{yy} \\ u\tau_{xy} + v\tau_{yy} + \frac{\mu}{(\gamma-1)Pr}\frac{\partial a^2}{\partial y} \end{pmatrix} .$$

Denoting by $\underline{\tau}$ the viscous stress tensor, the viscous terms in the momentum equations can be written in the form $div\underline{\tau}$. One way of extending the χ-formulation to the Navier-Stokes is to define a monotonic function of the viscous terms as follows

$$(2.5) \qquad \chi_M(div\underline{\tau}) = \alpha_M(\|div\underline{\tau}\|) \cdot div\underline{\tau} \ ,$$

where, for a given $\delta > 0$ and $\sigma > 0$,

$$(2.6) \qquad \alpha_M(\|s\|) = \begin{cases} 1 & \text{if } \|s\| \geq \delta \ , \\ f(\|s\|) & \text{if } \delta - \sigma < \|s\| < \delta \ , \\ 0 & \text{if } \|s\| \leq \delta - \sigma \ , \end{cases}$$

with $f(\|s\|)$ any smooth monotonic function, with values between 0 and 1.

In the total energy equation, the scalar diffusive term can be written in the form

$$(2.7) \qquad -divq + div(\underline{\tau} \cdot V) = -divq + gradV \cdot \underline{\tau} + V \cdot div\underline{\tau} \ ,$$

with q the heat conduction vector and V the velocity vector. Two distinct dissipative mechanisms are present: the heat diffusion $divq$, and the part due to the viscous stress tensor $\underline{\tau}$. The quantity $V \cdot div\underline{\tau}$ is negligible if the viscous terms of the momentum equation $div\underline{\tau}$ are negligible. Therefore, the same function χ_M given in (2.5) can be applied to these terms. In the present application we are interested in subsonic and transonic adiabatic flows along walls. For such flows, the dissipation function $gradV \cdot \underline{\tau}$ is always small with respect to the term

$V \cdot div\underline{\tau}$. Moreover, in the case of air, the ratio between the thickness of the thermal boundary layer and the thickness of the velocity boundary layer is of order one. In this case, the function χ_M is also a good indicator of the magnitude of the term $divq$. From these considerations we argue that for the applications in which we are interested, the domain decompositon criterion can be based on the χ-function χ_M, and the χ-formulation of problem (2.4) reads

$$(2.8) \qquad \frac{\partial Q}{\partial t} + \frac{\partial F}{\partial x} + \frac{\partial G}{\partial y} = \frac{\alpha_M}{Re}\left(\frac{\partial F_v}{\partial x} + \frac{\partial G_v}{\partial y}\right) \quad .$$

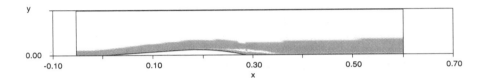

FIGURE 1. Laminar subsonic flow, $Re = 5.\,10^5$, $\frac{p_{exit}}{p_{tot}} = 0.95$.

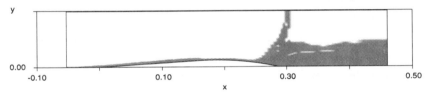

FIGURE 2. Turbulent transonic flow, $Re = 2.3\,10^7$, $\frac{p_{exit}}{p_{tot}} = 0.64$.

In order to prove the previous statement, we have solved the Navier Stokes equations (2.4), in the case of a channel flow. And then we have made an *a posteriori* evaluation of the function α_M. In Figure 1, we have the behavior of the function α_M in the case of a laminar subsonic flow in a channel, with $Re = 5.\,10^5$, and inflow $Mach = 0.25$. The gray region is formed by the points where $\alpha_M = 1$. Similarly in Figure 2, for the turbulent transonic flow ($Re = 2.3\,10^7$, and inflow $Mach = 0.65$). In both cases, $\delta = 1$ and $\sigma = \delta/100$. From the figures it may be observed that the boundary layer region is well detected, as well as the laminar separation bubble developing behind the bump. In the transonic case, we have a very thin turbulent boundary layer, and a complex shock wave structure, creating downstream a region with vorticity. All these features are detected by the χ-function.

3. Domain Decomposition via the χ Formulation

The previous calculation can be performed by splitting the domain into two parts. A viscous region along the wall, where problem (2.8) is solved, and an inviscid region, formed by the rest of the domain, from the interface to the upper boundary, along the channel symmetry line. In the inviscid region we solve the Euler equations, that is system (2.4) with $F_v = G_v = 0$.

The interface between the two regions is placed in such a way that the resulting viscous region contains the points where the function $\alpha_M \neq 0$, in addition $\alpha_M = 0$ on the grid points belonging to the interface itself. This last requirement along the interface, is very important in order to have a coupling between the Euler equations and the χ-Navier-Stokes equations, of inviscid type (that is Euler-Euler coupling). In this way it is possible to specify boundary conditions consistent with the hyperbolic character of the equations on both sides. By a one-dimensional analysis along the normal to the interface, it is possible to see that for the Euler equations we have to impose the velocity component along the normal and, if there is an incoming flow with respect to the inviscid region, the total enthalpy and the entropy. For the χ-Navier-Stokes equations, we have to specify the pressure and, if the flow is incoming with respect to the viscous subdomain, the total enthalpy and the entropy. We can see that the external inviscid flow is driving the development of the viscous region, by imposing the pressure distribution. And the viscous region interacts with the external inviscid flow by introducing displacement effects. This kind of interaction is well known from the classical boundary layer theory. However, in the present approach we do not introduce any kind of approximation, and the interaction between the two fields is much more general.

The Euler equations, and the χ-Navier-Stokes equations, are solved by a finite-volume method in curvilinear coordinates. The convective terms of the Euler equations are discretized in space by a second order flux vector splitting technique, while for the χ-Navier-Stokes equations a centred scheme is applied. Upwind methods, such as the flux-vector splitting, are very well suited for representing shock waves, but they are excessively dissipative inside the boundary layers. On the contrary, centred schemes give better results inside the viscous layers, but they present problems when capturing shock waves. With the present approach we try to combine the good properties of the two schemes, avoiding their drawbacks. The governing equations are integrated in time by an implicit technique.

After a first calculation of the Navier-Stokes equations, on the whole domain with a coarse grid, we detect the viscous region and we locate the interface following the requirements explained above. Introducing a fine grid in the viscous subdomain, the Euler equations and the χ-Navier-Stokes equations are alternatively integrated in time in each subdomain. After a step in one subdomain, the appropriate characteristic boundary conditions are imposed at the interface, and the other subdomain is calculated. During the coupled calculation, the interface is displaced if the above requirements are not fullfilled, that is if the size of the viscous region was underestimated. Similarly if the viscous region is overestimated, the interface is displaced, in order to optimize the calculation effort.

Complete details on the numerical algorithm, its efficiency, as well as further results, will be presented in a forthcoming paper.

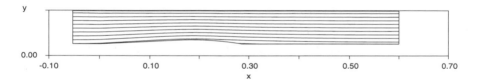

FIGURE 3. Laminar subsonic flow, $Re = 5.10^5$, $\frac{p_{exit}}{p_{tot}} = 0.95$, Inviscid subdomain.

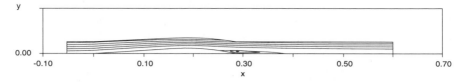

FIGURE 4. Laminar subsonic flow, $Re = 5.10^5$, $\frac{p_{exit}}{p_{tot}} = 0.95$, Viscous subdomain.

In Figure 3 and 4, we report the calculation of the laminar subsonic channel flow. Constant mass-flow lines are reported for the inviscid part of the domain (Figure 3) and the viscous subdomain (Figure 4). The interface was placed along a curvilinear coordinate line, and the mesh size was 11×41 points in the inviscid part, and 31×81 in the viscous subdomain. The accuracy of the solution oobtained by domain decomposition, is comparable with a solution of the Navier-Stokes equations, with a global grid of size 101×81.

REFERENCES

1. Y. Achdou, and O. Pironneau, *The χ-method for the Navier-Stokes equations*, C.R. Acad. Sci. Paris **168** (1991).
2. R. Arina, and C. Canuto, *A self adaptive domain decomposition for the viscous/inviscid coupling. I. Burgers equation*, J. Comput. Phys., **105** (1993), 290-300.
3. F. Brezzi, C. Canuto, and A. Russo, *A self-adaptive formulation for the Euler/Navier-Stokes coupling*, Comput. Methods Appl. Mech. and Eng., **73** (1989), 317-330.
4. C. Canuto, and A. Russo, *A viscous-inviscid coupling under mixed boundary conditions*, Math. Models Meth. Appl. Sci. **2**,4 (1992), 461-482.
5. _____ , *On the elliptic-hyperbolic coupling. I: the advection-diffusion equation via the χ-formulation*, I.A.N-C.N.R Report 841 (1992).

DEPARTMENT OF AEROSPACE ENG., POLITECNICO DI TORINO, TORINO, ITALY
E-mail address: arina@athena.polito.it

DEPARTMENT OF MATHEMATICS, POLITECNICO DI TORINO, TORINO, ITALY
E-mail address: ccanuto@polito.it

Contemporary Mathematics
Volume 180, 1994

DOMAIN DECOMPOSITION AND COMPUTATION OF TWO DIMENSIONAL DETONATION WAVES

WEI CAI

ABSTRACT. We present a hybrid domain decomposition numerical method for the computation of 2-D detonation waves. The numerical algorithm uses a multi-domain approach so that different numerical techniques and mesh resolution can be applied for different components of detonation waves. The detonation waves are assumed to undergo an irreversible, unimolecular reaction $A \rightarrow B$.

1. Introduction

In this paper, we present a hybrid numerical methods with domain decomposition technique for the computation of 2-D detonation waves. Previous work can be found in [8], [6] and [1] and the references therein. The motivation of our using domain decomposition is based on the special physical characteristics of detonation waves [5]. They include (1) a strong precursor detonation front; (2) Mach stem configuration of the "triple points" and transverse wave structures; (3) stiff chemical reactions. Thus, the flow field can be divided into regions of highly irregular and steep gradients near the detonation front and regions of strong but smooth pressure waves. Near the shock fronts, strong vorticity fields are expected from the roll-up of slip lines. The temporal changes of thermodynamic and chemical compositions also vary dramatically from region to

---

1991 *Mathematics Subject Classification.* Primary 65M99, 76Q05; Secondary 76V05.

The author was supported by NSF grant ASC-9113895 and AFOSR grant F49620-94-1-0317 and is partially supported by the National Aeronautics and Space Administration under NASA contract NAS1-19480 while the author was in residence at the Institute for Computer Applications in Science and Engineering (ICASE), NASA Langley Research Center, Hampton, VA 23681-0001. The computing facility was provided by a supercomputing grant from the North Carolina Supercomputer Center.

This paper is in final form and no version of it will be submitted for publication elsewhere.

region. We construct our numerical schemes according to these characteristics of multi-dimensional detonation waves.

The reaction rate depends on the flow temperature exponentially through the Arrhenius relation. Accurate computations of the flow field are extremely important in producing the correct chemical reactions and thus the correct cellular structures. Traditional shock capturing schemes, designed to smooth shock and contact discontinuities, introduce a considerable amount of numerical viscosity near those discontinuities. They have been shown to have a tendency to distort the real chemical reaction processes. In [3], the widely used P.P.M. high-order Godunov scheme was found to produce nonphysical weak detonations. Also in [4], the ENO finite difference scheme was shown to yield wrong detonation speeds in one dimensional ZND simulations. All these facts point out the importance of designing numerical methods without excessive numerical viscosity.

2. Governing Equations

Consider two dimensional overdriven detonation waves in a channel moving from left to right into unreacted gas mixtures. The channel is denoted by Ω (Figure 2), $\Omega = (-\infty, \infty) \times [-\frac{W}{2}, \frac{W}{2}]$ where W is the channel width.

The governing equations for reacting detonation waves with one step $A \longrightarrow B$ reaction are the following nondimensionalized Euler equations,

$$(2.1) \qquad \frac{\partial \vec{u}}{\partial T} + \frac{\partial f(\vec{u})}{\partial x} + \frac{\partial g(\vec{u})}{\partial y} = \Phi(\vec{u}),$$

where

$$\vec{u} = (\rho, \rho u, \rho v, \rho e, \rho_A)^\top,$$

$$f(\vec{u}) = \left(\rho u, \rho u^2 + p, \rho u v, \rho u(e + \frac{p}{\rho}), \rho_1 \right)^\top,$$

$$g(\vec{u}) = \left(\rho v, \rho v u, \rho v^2 + p, \rho v(e + \frac{p}{\rho}), \rho_A v \right)^\top,$$

$$\Phi(\vec{u}) = (0, 0, 0, 0, \omega)^\top,$$

and $(u, v), \rho, p, e$ are velocity vector, density, pressure, and total specific internal energy. ρ_A is the mass density of the reactant A. The source term is $\omega = -K\rho\lambda \exp(-\frac{E^+}{T})$ where $\lambda = \frac{\rho_A}{\rho}$ and E^+ is the activation energy. If we assume an exothermic reaction, the specific internal energy is

$$(2.2) \qquad e = \frac{p}{(\gamma - 1)\rho} + \frac{u^2 + v^2}{2} + \lambda Q$$

where Q is the specific heat formation and $\gamma = 1.2$ is the ratio of specific heats. All quantities above have been non-dimensionalized by the initial states in the unreacted gas mixture in front of the detonation fronts.

3. Domain Decomposition and Hybrid Algorithm

The computational region is composed of the detonation front moving to the right and the rear piston boundary and upper and lower solid walls. In the computation presented in this paper, we take this region to be $[-150\ell^*, 0] \times [-\frac{W}{2}, \frac{W}{2}]$ where ℓ^* is the half reaction distance. $x = 0$ is the position of initial plane detonation front at $t = 0$ which will be curved as time progresses. The computational region is decomposed with the detonation front as the right most boundary. Method of lines will be used to discretize the Euler equations (2.1). Third order Runge-Kutta is used as time integrator while a combination of three spatial discretization techniques is applied in different parts of computational region. They are

- Shock tracking algorithm for the detonation front;
- High order ENO finite difference scheme in the subdomain which contains the reflected shocks and contact discontinuities along the detonation front;
- Chebyshev collocation method for the strong vorticity and pressure fields from the interaction of transverse waves along the detonation front.

We refer the reader to [2],[7] for the descriptions of those numerical techniques and will only describe here the implementation of interface condition between adjacent subdomains.

4. Interface Conditions between Subdomains

The idea underlying the treatment of the interface and the boundary conditions is the propagation of information along characteristics of the hyperbolic systems. On a typical interface between two subdomains, say Γ between Ω_l and Ω_r, $\vec{v} = (P, u, v, \rho, \lambda)^\top$ is the primitive flow variable, and $\vec{v}^l, \vec{v}^r$ denote the solutions computed for the time step t^{n+1} in Ω_l, Ω_r respectively. $S_{\Gamma, I}$ denotes the normal speed of the interface at point I with the normal direction $n = (n_x, n_y)$. The characteristic variables are

$$(4.1) \qquad w_1 = p - \bar{a}\bar{\rho}u_n$$

$$(4.2) \qquad w_2 = u_t$$

$$(4.3) \qquad w_3 = p - \bar{a}^2\rho$$

$$(4.4) \qquad w_4 = \lambda$$

$$(4.5) \qquad w_5 = p + \bar{a}\bar{\rho}u_n$$

where the overbar denotes an average state between $\vec{v}^l$ and $\vec{v}^r$.

In order to update $\vec{v}$ at point I for the time step t^{n+1}, we make the following correction on $w_i, 1 \leq i \leq 5$ based on the sign of the difference between the eigenvalues and the normal speed of the interface $S_{\Gamma, I}$, i.e.

$$(4.6) \qquad w_i^{corrected} = \begin{cases} w_i^r & \text{if } \lambda_i - S_{\Gamma, I} < 0 \\ w_i^l & \text{if } \lambda_i - S_{\Gamma, I} \geq 0. \end{cases}$$

Finally, we set $\vec{v}^r = \vec{v}^l = \vec{v}^{corrected} = T^{-1}w^{corrected}$.

5. Numerical Results

We present one set of numerical results of 2-D detonation waves using the hybrid scheme. Consider a detonation wave with parameters $Q = 50, E^+ = 50, f = 1.6$, and channel width $W = 20\ell^*$, again ℓ^* is the half reaction distance. The domain of computation is chosen to be $[-150\ell^*, 0] \times [-10\ell^*, 10\ell^*]$ in (x, y) coordinate. Mesh convergence test [2] has shown that 13-15 points per half reaction distance is needed in the reaction zone, a uniform mesh based on this resolution will result in a mesh size $= 260 \times 1950 = 500,000$ points. By using domain decomposition, we can distribute most of the mesh points in the reaction region (in the right most subdomain) and the computational mesh consists of $\sum_{i=1}^{9}(n, m) = (50, 250) + (34, 70) + (20, 40) + (20, 30) + (20, 30) + (10, 20) + (10, 10) + (10, 10) + (10, 10) = 17,380$ points, which is a saving of 20 times over the uniform mesh. The number of 'Marker Points' on the shock front is 300. Figure 1 shows a two cell pattern produced by the trajectories of four triple points (A, B, C and D) which is obtained by recording the pressure distribution along the detonation front at different times. There is a larger cell with width approximately $10\ell^*$ (half the channel width) and a smaller one with width $5\ell^*$. Figure 2 contains six snapshots of the detonation at time $t = 20.25t^*, 21.5t^*, 22t^*, 22.5t^*, 23t^*, 23.5t^*$, t^* is the half reaction time. A random perturbation with magnitude $\epsilon = 0.3$ is used to perturb the shock front at $t = 0$.

6. Conclusion

The multi-domain approach allows different numerical techniques and mesh resolutions to be applied for different components of detonation waves. The propagation of waves across the interfaces of subdomains is smooth and the order of accuracy of the whole numerical scheme is only limited by the accuracy of the time integrator. Tracking of the detonation front prevents differences across the detonation front, thus avoiding excessive numerical viscosity in shock capturing schemes. The adaptivity in numerical schemes makes accurate simulation of 2-D detonation waves possible while adaptivity in mesh resolution yields big saving in CPU time.

REFERENCES

1. A. Bourlioux and A. J. Majda *Theoretical and Numerical Structure for Unstable Two-dimensional Detonations,* Combustion and Flam **90**: 211-229, 1992.
2. W. Cai, *High Order Numerical Simulations of Two Dimensional Detonation Waves,* ICASE Report 93-47, 1993.
3. P. Colella, A. Majda, and V. Roytburd, "Theoretical and Numerical Structure for Reacting Shock Waves," *SIAM J. Sci. Stat. Computing* 4 (1986) 1059–1080.
4. B. E. Engquist and B. Sjogreen, "Robust Difference Approximations of Stiff Inviscid Detonation Waves," *UCLA CAM REPORT* 91-03 (1991).
5. W. Fickett and W. C. Davis, *Detonation,* Univ. of California Press, 1979.

6. E. S Oran, J. P. Boris, M. Flanigan, T. Burks, and M Picone, *Numerical Simulations of Detonations in Hydrogen-Air and Methane-Air Mixtures",* Eighteenth Symposium (International) on Combustion, The Combustion Institute, Pittsburgh, 1641–1649, 1981.

7. C. W. Shu and S. Osher, *Efficient Implementation of Essentially Nonoscillatory Shock Capturing Schemes,* J. Compt. Phys., **77** 439-471, 1988.

8. Taki, S. and Fujiwara, T., "Numerical Analysis of Two Dimensional Nonsteady Detonations," <u>AIAA Journal</u>, Vol. 16, 1978, pp. 73-77.

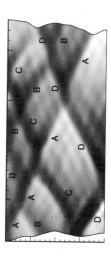

FIGURE 1. Cell pattern of detonation waves

DEPARTMENT OF MATHEMATICS, UNIVERSITY OF CALIFORNIA, SANTA BARBARA, CA 93106

E-mail address: wcai@math.ucsb.edu

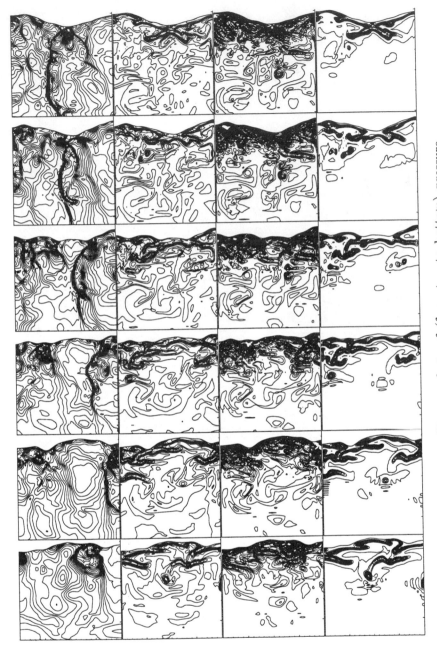

FIGURE 2. Six snapshots of (from top to bottom) pressure, temperature, vorticity, and mass fraction of reactant at $t = 20.25t^*, 21.5t^*, 22t^*, 22.5t^*, 23t^*, 23.5t^*$.

Contemporary Mathematics
Volume **180**, 1994

PARALLEL IMPLICIT METHODS FOR AERODYNAMICS

X.-C. CAI, W. D. GROPP, D. E. KEYES, AND M. D. TIDRIRI

ABSTRACT. Domain decomposition (Krylov-Schwarz) iterative methods are natural for the parallel implicit solution of multidimensional systems of boundary value problems that arise in aerodynamics. They provide good data locality so that even a high-latency workstation network can be employed as a parallel machine. Matrix-free (Newton-Krylov) methods are natural when it is unreasonable to compute or store a true Jacobian. We call their combination Newton-Krylov-Schwarz and report experimental progress on two algorithmic aspects: the use of a coarse grid in additive Schwarz preconditioning and the use of mixed discretization schemes in the (implicitly defined) Jacobian and its preconditioner. Two model problems in two-dimensional compressible flow are considered: the full potential equation and the Euler equations.

1. KRYLOV-SCHWARZ ALGORITHMS

Fully implicit linear solvers in aerodynamics allow more rapid asymptotic approach to steady states than time-explicit, approximate factorization, or relaxation solvers that hold the outer nonlinear iteration to small time steps. Nevertheless, the all-to-all data dependencies between the unknown fields in a fully implicit method have led to a resurgence of interest in less rapidly convergent methods in high-latency parallel environments. Resisting, we briefly overview two related efforts that lie along the route to parallel implicit computational aerodynamics. Though the governing equation formulations are mathematically very different – elliptic subsonic full potential and hyperbolic transonic Euler – a common implicit software core allows them to be treated together. Our ultimate interest is in applying Schwarzian domain decomposition techniques to industrial computations still being carried out in these (physically) primitive potential and Euler formulations, and in extending them to Navier-Stokes, for which implicit solvers are even more important. For a variety of reasons, industrial CFD groups are inclining towards the distributed network computing environment characterized by coarse to medium

1991 *Mathematics Subject Classification.* Primary 65N55, 65N22, 76G25.

The work was supported in part by the National Science Foundation and the Kentucky EPSCoR Program under grant STI-9108764 (XCC); by the Office of Scientific Computing, U.S. Department of Energy, under Contract W-31-109-Eng-38 (WDG); by the National Science Foundation under contract number ECS-8957475, the State of Connecticut and the United Technologies Research Center (DEK); and by the National Aeronautics and Space Administration under NASA contract NAS1-19480 while three of the authors (XCC,DEK,MDT) were in residence at the Institute for Computer Applications in Science and Engineering.

This paper is in final form. A more comprehensive version will appear elsewhere.

granularity, large memory per node, and very high latency, which creates a niche for domain decomposition methods.

Schwarz-preconditioned Krylov solvers for nonsingular linear systems, $Ax = b$, find the best approximation of the solution x in a small-dimensional subspace that is built up from successive powers of the preconditioned matrix on the initial residual. Such systems, in which A is a Jacobian matrix and b is the nonlinear residual, arise from discretization and linearization of the governing PDEs. A variety of parallel preconditioners, whose inverse action we denote by B^{-1}, can be induced by decomposing the domain of the underlying PDE, finding an approximate representation of A on each subdomain, inverting locally, and combining the results. Generically, we seek to approximate the inverse of A by a sum of local inverses:

$$B^{-1} = \sum_k R_k^T A_k^{-1} R_k,$$

where R_k is a restriction operator that takes vectors spanning the entire space into the smaller dimensional subspace in which A_k is defined.

The simplest domain decomposition preconditioner is block Jacobi, which can be regarded as a zero-overlap form of additive Schwarz [2]. The convergence rate of block Jacobi can be improved, at the price of a higher cost per iteration, with subdomain overlap and (for many problems) by solving an additional judiciously chosen coarse grid system.

2. NEWTON-KRYLOV METHODS

Evaluation of the discrete residuals of d-dimensional compressible flow formulations requires a large number of arithmetic operations. (For instance, a $(d + 2)$-dimensional eigendecomposition may be required at each grid point.) Their Jacobians, though block-sparse, have dense blocks and are usually an order of magnitude even more complex to evaluate, whether by analytical or numerical means. Hence, matrix-free Newton-Krylov methods, in which the action of the Jacobian is required only on a set of given vectors, are natural in this context. To solve the nonlinear system $f(u) = 0$, given u^0, let $u^{l+1} = u^l + \lambda^l \delta u^l$, for $l = 0, 1, \ldots$, until the residual is sufficiently small, where δu^l approximately solves the Newton correction equation $J(u^l)\delta u^l = -f(u^l)$, and λ^l is a damping parameter. The action of Jacobian J on an arbitrary Krylov vector w can be approximated by

$$J(u^l)w \approx \frac{1}{\epsilon} \left[f(u^l + \epsilon w) - f(u^l) \right].$$

Finite-differencing with ϵ makes matrix-free methods potentially much more susceptible to finite word-length effects than ordinary Krylov methods [5]. Steady aerodynamics applications require the solution of linear systems that lack strong diagonal dominance, so a secondary goal of our investigation is to verify that properly-scaled matrix-free methods can be employed in this context. For brevity, details are deferred to a more comprehensive paper. We simply note here that GMRES may have an advantage over other Krylov methods in the matrix-free context in that the vectors w that arise in GMRES have unit two-norm, but may have widely varying scale in other Krylov methods.

An approximation to the Jacobian can be used to precondition the Krylov process. Natural examples are: (1) the Jacobian of a related discretization that allows economical analytical evaluation of elements, (2) the Jacobian of a lower-order discretization, (3) a finite-differenced Jacobian computed with lagged values for expensive terms, and (4) domain-parallel preconditioners of the form

$$B^{-1} = R_0^T J_{0,u^l}^{-1} R_0 + \sum_{k=1}^{K} R_k^T J_{k,u^l}^{-1} R_k \ , \ \text{where } J_{k,u^l} = \left\{ \frac{\partial f_i(u^l)}{\partial u_j} \right\}$$

is the Jacobian of $f(u)$ for i and j in subdomain k, and subscript "0" corresponds to a possible coarse grid. The Newton-Krylov-Schwarz method (case 4) can be combined with any other split-discretization technique (cases 1–3), in principle. Right preconditioning of the Jacobian with an operator B^{-1} can be accommodated via

$$J(u^l)B^{-1}w \approx \frac{1}{\epsilon} \left[f((u^l + \epsilon B^{-1}w)) - f(u^l) \right].$$

3. MODEL PROBLEMS

For density ρ, velocity $\mathbf{v}$, specific internal energy e, and pressure p, the steady Euler equations of inviscid compressible flow are

$$\nabla \cdot (\rho \mathbf{v}) = 0, \quad \nabla \cdot (\rho \mathbf{v} \mathbf{v} + pI) = 0, \text{ and } \nabla \cdot ((\rho e + p)\mathbf{v}) = 0.$$

The full potential equation for velocity potential Φ,

$$\nabla \cdot (\rho(||\nabla \Phi||)\nabla \Phi) = 0,$$

follows from the additional assumptions of irrotationality, $\mathbf{v} \equiv \nabla \Phi$, and isentropy, $\nabla(p/\rho^\gamma) \equiv 0$. The density is given in terms of the potential by

$$\rho = \rho_\infty \left(1 + \frac{\gamma-1}{2} M_\infty^2 (1 - \frac{q^2}{q_\infty^2}) \right)^{\frac{1}{(\gamma-1)}},$$

where $q = ||\nabla \Phi||$ and $M_\infty = q_\infty/a_\infty$. Here, a is the sound speed, q the flow speed, and ∞ refers to the freestream. When the flow is everywhere subsonic the full potential formulation fits within the monotone nonlinear elliptic framework of additive Schwarz methods [1]. For a simple non-lifting model problem of an airfoil lying along the symmetry axis $y = 0$, we choose boundary conditions as follows:

- Upstream and Freestream: $\Phi = q_\infty x$ (zero angle of attack),
- Downstream: $\Phi_{,n} = q_\infty$,
- Symmetry: $\Phi_{,n} = 0$,
- On the parameterized airfoil with shape $y = f(x)$: $\Phi_{,n} = -q_\infty f'(x)$.

The farfield boundary conditions lead to inaccuracies if applied too near the airfoil, but our interest is in algebraic convergence rates.

Coarse Grid	0×0	4×5	8×9	12×13	16×17	20×21
Analytical	177	35	28	27	24	21
Matrix-free	183	41	28	27	25	23

TABLE 1. Average number of GMRES steps per Newton step for full potential Newton-Krylov-Schwarz solver with varying coarse grid size.

Table 1 shows convergence performance for a fixed-size problem of 128×128 uniform cells with a fixed number of subdomains in an 8×8 array as the density of the unnested uniform coarse grid varies. Bilinear rectilinear elements are used for both coarse and fine grids, and bilinear interpolation for intergrid transfers. An overlap $2h$ is employed on each of the subdomains, which are solved exactly. M_∞ is 0.1 and the airfoil is the scaled upper surface of a NACA0012. Nonlinear convergence is declared following a 10^{-3} relative reduction in the steady-state residual, which requires only three Newton steps independent of inner linear method. Inner iteration convergence is a relative residual reduction of 10^{-4}. The Krylov solver used throughout this paper is GMRES [7], because of previous comparisons [3] with other modern Krylov solvers on the same problem class that showed CPU cost differences to be small and unsystematic when well-enough preconditioned that any of the methods were practical. Here, we restart GMRES every 20 iterations and precondition on the right, in order to keep the preconditioner out of the residual norm estimates used in the convergence test. Key observations from this example are: (1) even a modest coarse grid makes a significant improvement in an additive Schwarz preconditioner; (2) a law of diminishing returns sets in at roughly one point per subdomain; and (3) matrix-free "matvecs" degrade convergence as much as 15-20% in the less well-conditioned cases.

| Precond. | Block Jacobi | | Add. Schwarz | | Mult. Schwarz | |
CFL No.	1	10^2	1	10^2	1	10^2
1×1	1	1	1	1	1	1
2×2	4	14	7	14	2	7
4×4	4	18	7	17	3	8
8×8	5	28	10	23	3	8

TABLE 2. Iteration counts for transonic flow Jacobians at local CFL numbers of 1 and 10^2, for various preconditioners and decomposition into 4, 16, or 64 subdomains.

Our Euler example is a two-dimensional transonic airfoil flow modeled using an EAGLE-derivative code [6] that employs a finite volume discretization over a body-fitted coordinate grid. Only C-grids of 128×16 or 128×32 cells (from [3]) around a NACA0012 airfoil at an angle of attack of $1.25°$ and an M_∞ of 0.8 are considered herein. To obtain a representative matrix/RHS pair on which to test the behavior of Euler Jacobians under Krylov-Schwarz, we first ran a demonstration case from [6] partway to convergence and linearized about the resulting flow state. Following the defect correction practice of [8], a flux vector split scheme is employed for the implicit operators, and $f(u)$ itself is discretized by a flux difference split scheme. Characteristic variable boundary conditions are employed at farfield boundaries using an explicit, first-order accurate formulation. For a given granularity of decomposition, curvilinear "box" decompositions are generally better than curvilinear "strip" decompositions for this problem. Table 2 shows that the zero-overlap results are only slightly less convergent than the corresponding h-overlapped additive Schwarz results at high Courant-Friedrichs-Lewy (CFL) number, and that h-overlapped multiplicative Schwarz is significantly better, though the latter is a much less parallel algorithm. Though we have not yet experimented with a coarse grid in the Euler context, [9] shows that even a piecewise constant coarse grid operator substantially improves Krylov-Schwarz convergence rates in unstructured problems.

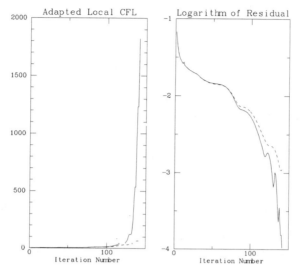

FIGURE 1. CFL and steady-state residual versus iteration count for defect correction and Newton-Krylov solvers.

To test the nonlinear matrix-free approach in a situation with four differently scaled components per gridpoint, we started over and approached the steady solution via a pseudo-transient continuation with an adaptively chosen local CFL number, as described in [3]. Use of the baseline approximate factorization defect correction algorithm produces the dashed curves in Fig. 1. To obtain the solid curves, the explicitly available (Van Leer) flux vector split Jacobian (J_{VL}) is used to precondition the implicitly defined (Roe) flux difference split Jacobian (J_R) at each implicit time step. In matrix terms, the corrections u are obtained as the approximate solutions of, respectively,

$$J_{VL}u = -f_R \quad \text{and} \quad (J_{VL})^{-1}J_R u = -(J_{VL})^{-1}f_R.$$

Unfortunately, in the retrofit of the existing code, transition to a full Newton method (CFL number approaching infinity) is precluded by explicit boundary conditions, but CFL number can be advanced, as shown in the figure, to $\mathcal{O}(10^3)$ with advantage.

Though space does not permit a meaningful discussion, we mention that both codes have been executed on an ethernet network of workstations using a package of distributed sparse linear system routines developed at Argonne National Laboratory by Gropp and Smith [4], with p4 as the data exchange layer. When exact solvers are used on each subdomain, speedups on a per iteration basis are seen on up to 16 processors, but exact solvers are an extreme case. As a serial preconditioner, global incomplete LU is superior to a Schwarz method using exact subdomain solvers.

4. Conclusions

By concentrating data dependencies locally, domain decomposition preconditioners exploit the two-level memory hierarchy of high-latency distributed-memory architectures. Low-communication zero or small overlaps between the preconditioner blocks are feasible with small convergence rate penalty, at least for intermediate granularities. The addition of a coarse grid has been shown to lead to major iteration count improvements, for a fixed problem size and algebraic residual reduction. Demonstrating the applicability of elliptic-based domain decomposition preconditioners to full potential and Euler problems is only a beginning. Further research will explore the limits of inconsistent preconditioners in matrix-free contexts, the cost versus benefits of the coarse grid solve in parallel contexts, and the relative tuning of inner and outer iteration convergence tolerances.

References

1. X.-C. Cai and M. Dryja, *Domain Decomposition Methods for Monotone Nonlinear Elliptic Problems*, these proceedings, 1994.
2. M. Dryja and O. B. Widlund, *An Additive Variant of the Alternating Method for the Case of Many Subregions*, Courant Institute, NYU, TR 339, 1987.
3. W. D. Gropp, D. E. Keyes and J. S. Mounts, *Implicit Domain Decomposition Algorithms for Steady, Compressible Aerodynamics*, in A. Quarteroni et al., Sixth International Symposium on Domain Decomposition Methods for Partial Differential Equations (Como, Italy), AMS, Providence, 1994, pp. 203–213.
4. W. D. Gropp and B. F. Smith, *Simplified Linear Equations Solvers Users Manual*, ANL-93/8, Argonne National Laboratory, 1993.
5. P. R. McHugh and D. A. Knoll, *Inexact Newton's Method Solutions to the Incompressible Navier-Stokes and Energy Equations Using Standard and Matrix-Free Implementations*, AIAA Paper, 1993.
6. J. S. Mounts, D. M. Belk and D. L. Whitfield, *Program EAGLE User's Manual, Vol. IV – Multiblock Implicit, Steady-state Euler Code*, Air Force Armament Laboratory TR-88-117, Vol. IV, September 1988.
7. Y. Saad and M. H. Schultz, *GMRES: A Generalized Minimal Residual Algorithm for Solving Nonsymmetric Linear Systems*, SIAM J. Sci. Stat. Comp. **7**(1986), 865–869.
8. J. L. Steger and R. F. Warming, *Flux Vector Splitting of the Inviscid Gasdynamics Equations with Applications to Finite-Difference Methods*, J. Comp. Phys. **40**(1981), 263–293.
9. V. Venkatakrishnan, *Parallel Implicit Unstructured Grid Euler Solvers*, AIAA Paper 94-0759, Reno, Nevada, January 1994.

COMPUTER SCIENCE DEPARTMENT, UNIVERSITY OF COLORADO, BOULDER, CO, 80309.
E-mail address: cai@cs.colorado.edu

ARGONNE NATIONAL LABORATORY, 9700 SOUTH CASS AVE., ARGONNE, IL 60439-4844.
E-mail address: gropp@mcs.anl.gov

COMPUTER SCIENCE DEPARTMENT, OLD DOMINION UNIVERSITY, NORFOLK, VA 23529-0162 AND ICASE, MS 132C, NASA-LaRC, HAMPTON, VA 23681-0001
E-mail address: keyes@icase.edu

ICASE, MS 132C, NASA-LaRC, HAMPTON, VA 23681-0001
E-mail address: tidriri@icase.edu

Contemporary Mathematics
Volume **180**, 1994

Parallel Domain-Decomposed Preconditioners in Finite Element Shallow Water Flow Modeling

Y. CAI AND I.M. NAVON

ABSTRACT. This paper is a highly condensed account of applying a domain-decomposed (DD) preconditioner approach to the parallel numerical solution of the shallow water equations by the finite element method. Three types of DD preconditioners are employed to accelerate, with right preconditioning, the convergence of several competitive non-symmetric linear iterative solvers of current interest. Analysis and comparisons are provided in this contribution. The resulting algorithms are parallelized at both the subroutine level to accommodate the subdomain by subdomain computation and at the loop level to exploit the parallelism of the finite element discretization. The implementations were carried out on the parallel vector supercomputer CRAY Y-MP/432.

1. Introduction

Although domain decomposition ideas are traceable to the work of Schwarz [17] in 1869 and that of engineers beginning from the 1960s [8,14,15], an efficient way to handle the coupling between artificially divided non-overlapping substructures was first proposed in [7]. In essence, this is a divide-and-feedback process which continues until a prescribed convergence criterion on the interfaces is satisfied.

However, the method mentioned above can be expensive in the absence of fast subdomain solvers. To remedy this disadvantage and, at the same time, retain the nice parallelization property, at least two other approaches have been

1991 *Mathematics Subject Classification.* Primary 65M55, 65Y05; Secondary 65N30, 65Y20.

The first author was supported in part by US DOE Grant #120054323 and DE-FC05-85ER250000, through the Supercomputer Computations Research Institute of the Florida State University. The authors are thankful to David Keyes for his valuable suggestions on the first draft of the paper.

The final version of this paper will be submitted for publication elsewhere.

proposed. One is the domain-decomposed preconditioner approach (DDPA) (also called full matrix domain decomposition in [**11**]) advocated in [**3**], the other being the recently proposed modified interface matrix approach (MIMA) (see [**13**]). Both approaches abandon the idea of the Schur complement matrix approach that decoupled subdomain problems are independently solved *only after* the solution on the interfaces is obtained.

The DDPA approach consists of the construction of a preconditioner in a domain-decomposed way such that approximate solutions in the subdomains and on the interfaces can be simultaneously updated at the cost of only inexact subdomain solvers. On the other hand, the MIMA approach successively improves the subdomain and interface approximate solutions with improved initial solutions. Thus it mitigates the disadvantage due to the absence of fast subdomain solvers.

In this paper, we shall concentrate our attention only on the application of DDPA to finite element shallow water flow simulation. A detailed comparison between DDPA and MIMA for the problem at hand will be addressed elsewhere in a separate research work.

We test all the proposed algorithms, to be presented shortly, on a shallow water equations model using the Grammeltvedt initial conditions [**9**] on a limited-area rectangular region — a channel on the rotating earth. The model is essentially the same as the one earlier used by Houghton, et al. [**10**].

Non-symmetric linear iterative solvers are important kernels to the current domain decomposition approach. Among many available algorithms, we are especially interested in three, namely, GMRES [**16**], CGS [**18**] and Bi-CGSTAB [**19**].

In section 2, we give three types of DD preconditioners under consideration. Carefully selected numerical results are given in section 3 and main conclusions are drawn.

2. Domain-decomposed preconditioners

The DD preconditioner approach is presented in this section using the notation adopted in [**13**]. We consider solving linear systems resulting from finite element discretization in space with an implicit temporal difference scheme [**4**].

Since we are interested in implementing the algorithms on parallel computers with powerful vector processors, the original domain is divided into strips along the west-east directions in order to yield longer vectors (see [**12**]). Thus, cross points are eliminated from consideration.

In order to explore the inherent parallelism at the subdomain level, following [**14**], we number the global nodes in a substructured way, such that the coefficient matrices of geopotential and velocities at each time step present the following block-bordered structures

$$(2.1) \qquad\qquad A = \begin{bmatrix} A_{dd} & A_{ds} \\ A_{sd} & A_{ss} \end{bmatrix}$$

The meaning of each element in the matrix is explained in [13]

Three types of DD preconditioners employed for our problem assume, respectively, the structurally symmetric form, the block upper and the lower triangular forms

(1)

(2.2)
$$B = \begin{bmatrix} B_{dd} & A_{ds} \\ A_{sd} & B_{ss} \end{bmatrix}$$

(2)

(2.3)
$$B = \begin{bmatrix} B_{dd} & A_{ds} \\ 0 & G \end{bmatrix}$$

(3)

(2.4)
$$B = \begin{bmatrix} B_{dd} & 0 \\ A_{sd} & G \end{bmatrix}$$

where B_{dd} approximates A_{dd} in some sense. For example, B_{ii} might be the relaxed incomplete LU factorization (RILU) [2] of A_{ii}. B_{ss} is given, for the first type of DD preconditioners, by

(2.5)
$$B_{ss} = G + \sum_{i=1}^{n} A_{si} B_{ii}^{-1} A_{is}$$

For all three cases, the matrix G is to be determined.

Let

(2.6)
$$AB^{-1} = \begin{bmatrix} P_{11} & P_{12} \\ P_{21} & P_{22} \end{bmatrix}$$

Then we may show that (P_{11}, P_{12} and P_{21} are given in [4])

(2.7)
$$P_{22} = \begin{cases} (A_{ss} - A_{sd}B_{dd}^{-1}A_{ds})G^{-1} & \text{for (2.2)} \\ (A_{ss} - A_{sd}B_{dd}^{-1}A_{ds})G^{-1} & \text{for (2.3)} \\ A_{ss}G^{-1} & \text{for (2.4)} \end{cases}$$

The matrix G is chosen such that P_{22} is an identity matrix. Thus we have that $G = A_{ss}$ for the third type of DD preconditioners and

(2.8)
$$G = A_{ss} - A_{sd}B_{dd}^{-1}A_{ds} = A_{ss} - \sum_{i=1}^{n} A_{si}B_{ii}^{-1}A_{is}.$$

for the first two types. For the latter case, nn_s number of inexact subdomain solves must be carried out for the construction of G such that P_{22} is an identity matrix, where n and n_s are, respectively, the number of subdomains and number of nodes on the interfaces. After one has constructed the preconditioner B, the major computational work required for solving the preconditioning linear system

TABLE 1. Amount of work for solving $Bp = q$

First type	$2n$ inexact subdomain solves
Second type	n inexact subdomain solves
Third type	n inexact subdomain solves

of the form $Bp = q$ at each time step corresponding to each of the three types of DD preconditioners is summarized in Table 1.

Similar to many other applications, it turns out that action of the matrix operator G is, in fact, predominantly local and the matrix G given by (2.8), although dense, allows itself to be reasonably approximated by a very low-bandwidth sparse matrix. The modified rowsum preserving interface probing ideas [4] (see also [5,6]) are employed for the approximate formation of G, for which only one inexact solve is needed in each subdomain.

With this particular interface probing construction of G, it is readily observed that the number of subdomain solves involved in using each of these preconditioners for the solution of a linear system at each time step is $(2k_1 + 3)n$, $(k_2 + 2)n$ and $(k_3 + 1)n$, respectively, where k_1, k_2 and k_3 are numbers of iterations required to satisfy a convergence criterion for an iterative method (where only one matrix-vector multiplication is needed in each iteration, like GMRES) using the first, second and third types of DD preconditioners. Note that the subdomain solves required for recovering the final solution $x = B^{-1}\tilde{x}$ have been included.

3. Numerical results and main conclusions

In this section, we present selected numerical results along with some discussions. All numerical experiments are carried out on the four-processor CRAY Y-MP/432 vector-parallel supercomputer. The numerical values of some parameters associated with the original PDEs are summarized in [13]. The problem is non-dimensionalized by choosing $\varphi_0 = 10^2 \ m^2/s^2$ [4].

In this set of numerical experiments, we test the relative efficiencies of the three types of DD preconditioners presented in section 2. As inexact solvers in the subdomains, we use RILU (relaxed incomplete LU) factorization along with forward and back substitutions. On the interfaces, the MIP(0) interface probe construction is used for the approximation of G given in (2.11) for the first two types of DD preconditioners. Table 2 gives some numerical results.

We see that GMRES, CGS and Bi-CGSTAB are very competitive with each other. GMRES requires the largest number of iterations to attain prescribed convergence. However we note that only one matrix-vector multiplication per iteration will drive GMRES to convergence, compared with two such multiplications required for CGS or Bi-CGSTAB at each iteration. The non-smooth convergence behavior (not shown here) of the CGS algorithm is observed in the current set of experiments. Bi-CGSTAB and GMRES methods generate smoother convergence history.

TABLE 2. A comparison of CPU time (number of iterations)

Preconditioner types		First type	Second type	Third type
40 by 35	GMRES	0.178 (12)	0.102 (12)	0.088 (11)
mesh	CGS	0.199 (7)	0.099 (6)	0.106 (7)
resolution	Bi-CGSTAB	0.219 (7)	0.099 (6)	0.092 (6)
80 by 75	GMRES	0.89 (14)	0.53 (15)	0.47 (14)
mesh	CGS	1.08 (9)	0.54 (8)	0.50 (8)
resolution	Bi-CGSTAB	0.97 (8)	0.54 (8)	0.50 (8)

Using the idea of m-step preconditioning [1], we can test the sensitivities of these preconditioners to inexact subdomain solvers. We observe that preconditioners of the third type can accelerate the convergence of three iterative methods at about the same rate as the other two types of preconditioners, except for cases when subdomain solvers may be considered to be exact or nearly exact.

In practice, the inexact subdomain solvers are far from exact, we consider preconditioners of the third type to be the best, although the second type of DD preconditioners is very competitive. Even if exact or nearly exact subdomain solvers are used for the first two types of DD preconditioners, the gain in the number of iterations is far from offsetting the additional computational cost required for constructing these solvers.

For parallelization, subdomain by subdomain computations are carried out at the subroutine level, and loop level parallelism is exploited for calculations involved in finite element discretization using a multicoloring scheme [4].

We integrated the finite element model of the shallow water equations for four different mesh resolutions for a period of five hours with corresponding time step sizes of $\Delta t = 1800$ s, 1000 s, 600 s and 400 s, respectively, using Bi-CGSTAB preconditioned by the third type of DD preconditioners. The speed-up results are summarized in Table 3. It should be pointed out that the automatic do-loop level parallelization as detected and exploited by the autotasking preprocessor does not yield a speed-up larger than two. The reason is that autotasking is unable to detect parallelism across subroutine boundaries.

TABLE 3. Parallel performance results on the CRAY

Mesh resolutions	19×15	34×27	49×43	64×55
Serial seconds	1.03	6.26	35.19	108.07
Parallel seconds	0.38	2.03	10.29	29.77
Speed-up ratios	2.7	3.1	3.4	3.6

The main conclusions of this research are

(1) Three types of DD preconditioners were found to work reasonably well

with GMRES, CGS and Bi-CGSTAB, with the third type being computationally the least expensive and the first type most expensive.

(2) For all cases, GMRES requires roughly twice as many iterations as required by the CGS or Bi-CGSATB. However, these three algorithms were
found to be approximately equally efficient in terms of CPU time.

(3) To achieve better speed-up results, it is important to remove the critical
regions in the assembly process by using multicoloring of the elements.

References

1. L. Adams, *m-Step preconditioned conjugate gradient methods*, SIAM J. Sci. Stat. Comput.
   6(2) (1985), 452–463.
2. O. Axelsson and G. Lindskog, *On the eigenvalue distribution of a class of preconditioning
   methods*, Numer. Math. **48** (1986), 479–498.
3. J.H. Bramble, J.E. Pasciak and A.H. Schatz, *The construction of preconditioners for elliptic problems by substructures*, Math. Comp. **47** (1986), 103–134.
4. Y. Cai and I.M. Navon, *Parallel block preconditioning techniques for the numerical simulation of the shallow water flow using finite element methods*, J. Comput. Phys. (to
   appear).
5. T.F. Chan and T.P. Mathew, *The interface probing technique in domain decomposition*,
   SIAM J. Matrix Anal. Appl. **13(1)** (1992), 212–238.
6. T.F. Chan and D. Resasco, *A survey of preconditioners for domain decomposition*, Technical Report 414, Computer Science Dept., Yale Univ., 1985.
7. M. Dryja, *A capacitance matrix method for Dirichlet problem on polygon regions*, Numer.
   Math. **39** (1982), 51–64.
8. T. Furnike, *Computerized multiple level substructuring analysis*, Computers and Structures
   2 (1972), 1063–1073.
9. A. Grammeltvedt, *A survey of finite difference scheme for the primitive equations for a
   barotropic fluid*, Mon. Wea. Rev. **97(5)** (1969), 384–404.
10. D. Houghton, A. Kasahara and W. Washington, *Long-term integration of the barotropic
    equations by the lax-wendroff method*, Mon. Wea. Rev. **94(3)** (1966), 141–150.
11. D.E. Keyes and W.D. Gropp, *A comparison of domain decomposition techniques for elliptic
    partial differential equations and their parallel implementation*, SIAM J. Sci. Stat. Comput.
    8(2) (1987), s166–s202.
12. G. Meurant, *Domain decomposition methods for partial differential equations on parallel
    computers*, Int. J. Supercomputer Appl. **2(4)** (1988), 5–12.
13. I.M. Navon and Y. Cai, *Domain decomposition and parallel processing of a finite element
    model of the shallow water equations*, Comput. Methods Appl. Mech. & Eng. **106(1-2)**
    (1993), 179–212.
14. J.S. Przemieniecki, *Matrix structural analysis of substructures*, AIAA J. **1** (1963), 138–147.
15. M.F. Rubinstein, *Combined analysis by substructures and recursion*, ASCE J. of the Struc-
    tural Division **93(ST2)** (1967), 231–235.
16. Y. Saad and M.H. Schultz, *GMRES: A generalized minimal residual algorithm for solving
    nonsymmetric linear systems*, SIAM J. Sci. Stat. Comput. **7(3)** (1986), 856–869.
17. H.A. Schwarz, *Über einige Abbildungsaufgaben*, Ges. Math. Abh. **11** (1869), 65–83.
18. P. Sonneveld, *CGS, A fast Lanczos-solver for nonsymmetric linear systems*, SIAM J. Sci.
    Stat. Comput. **10(1)** (1989), 36–52.
19. H.A. van der Vorst, *Bi-CGSTAB: a more smoothly converging variant of CG-S for the
    solution of nonsymmetric linear systems*, SIAM J. Sci. Stat. Comput. **13(3)** (1992), 631–
    644.

DEPARTMENT OF MATHEMATICS AND SUPERCOMPUTER COMPUTATIONS RESEARCH INSTI-
TUTE, FLORIDA STATE UNIVERSITY, TALLAHASSEE, FLORIDA 32306

E-mail address: cai@scri.fsu.edu

Contemporary Mathematics
Volume **180**, 1994

A Domain Decomposition Method
for Bellman Equations

F. CAMILLI, M. FALCONE, P. LANUCARA AND A. SEGHINI

ABSTRACT. We apply a domain decomposition technique with subdomains without overlapping to construct an approximation scheme for Bellman equations in $\mathbb{R}^n$. The algorithm is presented for a 2-domain decomposition where the original problem is split into two problems with state constraints plus a linking condition on the interface. We establish the convergence to the viscosity solution and show the results of a numerical experiment.

1. Introduction.

We deal with the numerical solution of the Bellman equation related to the infinite horizon problem with state constraints in an open bounded convex subset Ω of $\mathbb{R}^n$, namely

$$(B) \qquad \lambda u(x) + \max_{a \in A} \left[-b(x,a) \cdot \nabla u(x) - f(x,a) \right] = 0 \quad , \qquad x \in \Omega,$$

where λ is a positive real parameter and A is a compact subset of $\mathbb{R}^m$ representing the set of admissible controls. It is known (see Soner [8], Capuzzo Dolcetta-Lions [3]) that, under rather general assumptions, the value function of the problem is the unique constrained viscosity solution of (B). We mention that the numerical solution of (B) gives complete information about the control problem since it provides approximate optimal controls in feedback form and the corresponding approximate optimal trajectories. However, this solution requires to solve a partial differential equation in Ω and this can be unaffordable when the number of state variables is large (f.e. in many economic problems $\Omega \subset \mathbb{R}^n$ and $n >> 10$). This is the main obstacle that has limited the application of the dynamic programming approach to the solution of real problems. The application of a

---

1991 *Mathematics Subject Classification*. Primary 65N55, 49L25; Secondary 49L20.

This paper is in final form and no version of it will be submitted for publication elsewhere.
The third author gratefully acknoweledges the support of the CASPUR (Consorzio per le applicazioni di Supercalcolo per l'Università e Ricerca) during the period that this research was carried out.

domain decomposition strategy seems to be an answer to this problem since it permits a huge problem (in Ω) to be split into a number of problems (in Ω_r) of managable size. This strategy can be directly implemented on parallel machines enlarging the possibilities of the dynamic programming approach.

We refer to [6] for a first step in this direction, where we studied a splitting algorithm for (B) based on a domain decomposition *with overlapping* between the sets Ω_r of the decomposition, $\Omega = \bigcup_r \Omega_r$, $r = 1, \ldots, d$. Here we extend our result to the situation in which we do *not* have overlapping using a variable step technique. For simplicity we present our algorithm in the case of a 2-domains decomposition but the extension to an m-domains decomposition requires only technical adaptations.

Our approach is based on recent results in the numerical approximation of the infinite horizon problem with state constraints. We refer to [2] for an a priori estimate of the fully discrete scheme with fixed time-step and to the references therein for other numerical methods for Hamilton–Jacobi–Bellman equations. It is important to notice that the basic ideas of the method presented here are general enough to be applied to other first order Hamilton–Jacobi–Bellman equations. Finally, we should also mention the work [9] on the numerical solution of the Bellman equation related to an exit time problem for diffusion processes (i.e. for second order elliptic problems) wherein a different algorithm is considered.

2. The infinite horizon problem with state constraints

Let Ω be an open bounded convex subset of $\mathbb{R}^n$ with regular boundary ($\nu(x)$ being its outward normal at $x \in \partial\Omega$). We will make the following assumptions on $b : \mathbb{R}^n \times A \to \mathbb{R}^n$ and $f : \mathbb{R}^n \times A \to \mathbb{R}$:

$(A1)$ $\qquad\qquad\qquad$ b and f are continuous in $\overline{\Omega} \times A$

$(A2)$ $\quad |g(x_1, a) - g(x_2, a)| \leq L_g |x_1 - x_2| \qquad$ for $g = b, f$ and $a \in A$

Soner [8] has extended the notion of "viscosity solution" in order to characterize the value function for the constrained problem. To this end, he also proved that the value function is continuous provided there exists a positive constant c such that

$(A3) \qquad \forall x \in \partial\Omega \qquad \exists a \in A$ such that $< b(x, a), \nu(x) > \leq -c < 0,$

(see also [3] for further theoretical results on constrained problems). In [5] and [2] the following fully discrete scheme has been studied

$(B_h^k) \quad u(x_i) = \inf_{a \in A_h(x_i)} \{(1 - \lambda h) u(x_i + h b(x_i, a)) + h f(x_i, a)\}, \qquad i = 1, \ldots, N,$

where x_i, is a node of a regular triangulation of Ω, h is a positive parameter to be interpreted as a (fixed) time-step and for any $i \in I \equiv \{1, \ldots, N\}$ the set

$A_h(x_i)$ is defined as

(1)
$$A_h(x_i) = \{a \in A : x_i + hb(x_i, a) \in \Omega\}, \quad i \in I.$$

Notice that, due to (A1) and to the boundary condition (A3), there exists a $\overline{h} > 0$ such that for any $h \leq \overline{h}$

(2)
$$A_h(x_i) \neq \emptyset, \quad i \in I.$$

Working in the space of piecewise linear finite elements, (B_h^k) is reduced to a finite dimensional fixed point problem which admits a unique solution V^* by the contraction mapping theorem.

3. A convergence result for the domain decomposition method

In this section we modify the previous approach by considering a variable time step η in order to deal with a decomposition *without overlapping*.

Let Ω be partitioned into two open subdomains Ω_1 and Ω_2, such that $\Omega = \Omega_1 \cup \Omega_2$, and let Γ be the interface, i.e. $\Gamma \equiv \partial\Omega_1 \cap \partial\Omega_2$. Given the above decomposition and a positive parameter h, we define the variable step $\eta : \overline{\Omega} \times A \to \mathbb{R}_+$ as follows

$$\eta(x, a) \equiv \begin{cases} \eta^r(x, a) & (x, a) \subset \overline{\Omega}_r \setminus \Gamma \times A_h(x), \\ h & (x, a) \in \Gamma \times A_h(x) \end{cases}$$

where

$$\eta^r(x, a) \equiv \min\{\inf[t \in \mathbb{R}_+ : x + tb(x, a) \in \mathbb{R}^n \setminus \overline{\Omega}_r], h\}, \quad r = 1, 2.$$

We define the following operator $D_h : L^\infty(\Omega) \to L^\infty(\Omega)$,

$$[D_h u](x) \equiv \begin{cases} \displaystyle\inf_{a \in A_h(x)} \{\beta(x, a)u(x + \eta(x, a)b(x, a)) + \eta(x, a)f(x, a)\} & x \in \overline{\Omega} \setminus \Gamma, \\ \displaystyle\inf_{a \in A_h(x)} \{(1 - \lambda h)u(x + hb(x, a)) + hf(x, a)\} & x \in \Gamma \end{cases}$$

where $\beta(x, a) \equiv (1 - \lambda\eta(x, a))$. In order to establish the convergence of the domain decomposition algorithm, we study the properties of D_h.

THEOREM 3.1. *If (A1), (A3) are verified and $h \in (0, \frac{1}{\lambda}]$, then there exists a unique solution $u_h \in L^\infty(\Omega)$ of*

(3)
$$u(x) = D_h u(x) \qquad x \in \overline{\Omega}$$

PROOF. *Uniqueness.* Let us suppose that there exist two solutions $u, v \in L^\infty(\Omega)$ of (3). Let $x \in \Gamma$, then there exists $a \in A_h(x)$ such that

$$u(x) - v(x) \leq (1 - \lambda h)[u(x + hb(x, a)) - v(x + hb(x, a))]$$

and, by symmetry, we get

(4)
$$|u(x) - v(x)| \leq (1 - \lambda h)\|u - v\|_{L^\infty(\Omega)}$$

For $x \in \overline{\Omega} \setminus \Gamma$ we have

$$u(x) - v(x) \leq \beta(x, a)[u(x + \eta(x, a)b(x, a)) - v(x + \eta(x, a)b(x, a))].$$

If $\eta(x, a) = h$ we conclude as in (4). When $\eta(x, a) < h$ we have $x + \eta(x, a)b(x, a) \in \Gamma$ so that by (4), we obtain

$$u(x) - v(x) \leq \beta(x, a)(1 - \lambda h)\|u - v\|_{L^\infty(\Omega)}.$$

In conclusion, (4) is verified for any $x \in \overline{\Omega}$.

Existence.   Let $\|f\|_\infty \leq M_f$, $u_0 \equiv -M_f/\lambda$ and $u^0 \equiv M_f/\lambda$. It is simple to check that u_0 and u^0 are respectively a sub-solution and a super-solution of (3). Let $\mathcal{K}$ be the convex, closed subset of $L^\infty(\Omega)$ defined by $\mathcal{K} \equiv \{u \in L^\infty(\Omega) : u_0 \leq u \leq u^0\}$. The operator D_h is compact and $D_h(\mathcal{K}) \subset \mathcal{K}$, so Schauder's fixed point theorem (see e.g. [7]) implies that there exists a solution of (3). $\square$

Remark. In order to give a constructive method to compute the fixed point u_h, we use the following strategy. We define $\eta_\varepsilon(x, a) = \eta(x, a) \vee \varepsilon$, for $\varepsilon > 0$ and introduce a new operator $D_{h\varepsilon}$ which is obtained replacing $\eta(x, a)$ with $\eta_\varepsilon(x, a)$ in the definition of D_h. $D_{h\varepsilon}$ is a contraction map in $L^\infty(\Omega)$ and we will denote by $u_{h\varepsilon}$ the corresponding fixed point. Then we can prove that

$$\|u_h - u_{h\varepsilon}\|_{L^\infty} \leq \lambda(\varepsilon \vee \eta - \varepsilon)\|u_h\|_{L^\infty} + (1 - \lambda(\varepsilon \vee \eta))\|u_h - u_{h\varepsilon}\|_{L^\infty}$$

which implies

$$\|u_h - u_{h\varepsilon}\|_{L^\infty} \leq (1 - \frac{\eta}{\varepsilon \vee \eta})\|u_h\|_{L^\infty}.$$

Therefore the sequence of fixed points $u_{h\varepsilon}$ converges to u_h as $\varepsilon \longrightarrow 0^+$.

THEOREM 3.2. *Let (A1) and (A3) be verified. Then $\{u_h\}$ converges uniformly in $\overline{\Omega}$ to the unique constrained viscosity solution u of (B).*

PROOF. Let us define

$$\overline{u} = \limsup_{\substack{h \to 0^+ \\ y \to x}} u_h(y), \qquad \underline{u} = \liminf_{\substack{h \to 0^+ \\ y \to x}} u_h(y)$$

Then $\underline{u} \leq \overline{u}$. If we prove that $\underline{u}$ is a viscosity sub-solution in Ω of (B) and $\overline{u}$ is a viscosity super-solution in $\overline{\Omega}$, by the comparison theorem in [1] we can conclude that $u = \overline{u} = \underline{u}$ in $\overline{\Omega}$ and $u_h \longrightarrow u$ uniformly. The proof in $\overline{\Omega} \setminus \Gamma$ can be obtained by a straightforward modification of Theorem 2.7 in [2]. Therefore we limit ourselves to the case $x \in \Gamma$.

Given $\phi \in C^\infty(\overline{\Omega})$, let $x_0 \in \Gamma \cap \Omega$ be a maximum point for $\overline{u} - \phi$. Then, by definition of $\overline{u}$, there exist two sequences $\{h_n\}$ and $\{y_n\}$ such that $h_n \longrightarrow 0^+$, $y_n \longrightarrow x$

$$\overline{u}(x_0) = \lim_{n \to +\infty} u_{h_n}(y_n)$$

and y_n is a maximum point for $u_{h_n} - \phi$. Since $x_0 \in \Omega$, $A_{h_n}(y_n) = A$, for h sufficiently small. Since y_n is a maximum point for $u_{h_n} - \phi$, for any fixed $a \in A$, we have

$$\lambda \eta(y_n, a) u_{h_n}(y_n) - \beta(y_n, a) \left[u_{h_n}(y_n + \eta(y_n, a)b(y_n, a)) - u_{h_n}(y_n) \right] +$$
$$- \eta(y_n, a) f(y_n, a) \leq \lambda \eta(y_n, a) u_{h_n}(y_n) - \beta(y_n, a) \left[\phi(y_n + \eta(y_n, a)b(y_n, a)) + \right.$$
$$\left. - \phi(y_n) \right] + \eta(y_n, a) f(y_n, a)$$

By definition, $\eta(y_n, a) \longrightarrow 0^+$ for $n \to +\infty$ and $\eta(x, a) > 0$ for any $(x, a) \in \overline{\Omega} \times A_h(x)$, so dividing the above inequality by $\eta(y_n, a)$ and passing to the limit for n tending to $+\infty$ we obtain

$$\lambda \overline{u}(x_0) - b(x, a) \cdot \nabla \phi(x_0) - f(x_0, a) \leq 0.$$

Since a is arbitrary, this implies that $\overline{u}$ is a subsolution of (B) at x_0.
Let $x_0 \in \Gamma \cap \overline{\Omega}$ be a minimum point for $\underline{u} - \phi$. Then we can define a sequence $\{y_n\}$ of minimum points for $u_{h_n} - \phi$ such that

$$\underline{u}(x_0) = \lim_{n \to +\infty} u_{h_n}(y_n).$$

Let $\overline{a}_n$ be a control such that the infimum in (D_{h_n}) is obtained. Since $\{\overline{a}_n\}$ is contained in the compact set A, there exists a subsequence, still denoted by $\overline{a}_n$ and $\overline{a} \in A$ such that $\lim_{n \to +\infty} \overline{a}_n = \overline{a}$. Then a straightforward computation gives

$$0 \leq \lambda \eta(y_n, \overline{a}_n) u_{h_n}(y_n) - \beta(y_n, \overline{a}_n) \left[\phi(y_n + \eta(y_n, \overline{a}_n)b(y_n, \overline{a}_n)) - \phi(y_n) \right] +$$
$$- \eta(y_n, \overline{a}_n) f(y_n, \overline{a}_n)$$

Now, dividing the above inequality by $\eta(y_n, \overline{a}_n)$, we get for $n \to +\infty$

$$0 \leq \lambda \underline{u}(x_0) - b(x, \overline{a}) \cdot \nabla \phi(x_0) - f(x_0, \overline{a})$$

which implies that $\underline{u}$ is a supersolution of (B) at x_0. $\square$

In order to define the numerical algorithm we introduce the following notations,

(5) $A_h^r(x) \equiv \{a \in A : x + \eta(x, a)b(x, a) \in \overline{\Omega}_r\}, \quad x \in \overline{\Omega}_r, \quad r = 1, 2.$

The definition of $\eta(x, a)$ implies that

(6) $A_h(x) = A_h^r(x) \qquad \forall x \in \overline{\Omega}_r \setminus \Gamma \quad r = 1, 2$

(7) $A_h(x) = A_h^1(x) \cup A_h^2(x) \qquad \forall x \in \Gamma$

We shall always assume that the triangulation of Ω is such that

$(A4)$ no simplex crosses Γ.

We will divide the nodes x_i, $i \in I \equiv \{1, \ldots, N\}$, into three classes depending on the region to which they belong, defining

$$I_0 = \{i : x_i \in \Gamma\} \text{ and } I_r = \{i : x_i \in \overline{\Omega}_r \setminus \Gamma\}, \quad r = 1, 2.$$

Let N_r, $r = 1, 2$, be the number of nodes in $\overline{\Omega}_r$. We define the "discrete" restriction operator

$$(8) \qquad R_r : \mathbb{R}^N \to \mathbb{R}^{N_r}, \qquad R_r(U) = \{U_i\}_{i \in I_r \cup I_0}, \qquad r = 1, 2.$$

Since the numerical solution of (3) on the triangulation requires to compute the value of u in points which are not nodes, we use a linear interpolation defining

$$u(x_i + \eta(x_i, a)b(x_i, a)) \equiv \sum_{j=1}^{N} \lambda_{ij}(a)u(x_j),$$

where $\lambda_{ij}(a)$ are the baricentric coordinates of the point $x_i + \eta(x_i, a)b(x_i, a)$ with respect to the vertices of the simplex which contains it. The decomposition of Ω corresponds to split the $(N \times N)$–matrix $\Lambda(a)$ into two submatrices $\Lambda^{(1)}(a) = \{\lambda_{ij}^{(1)}(a)\}_{i,j \in I_1 \cup I_0}$ and $\Lambda^{(2)}(a) = \{\lambda_{ij}^{(2)}(a)\}_{i,j \in I_2 \cup I_0}$.

We introduce the two discrete operators D_1 and D_2, $D_r : \mathbb{R}^{N_r} \to \mathbb{R}^{N_r}$, $r = 1, 2$, related to the subdomains Ω_1 and Ω_2

$$[D_r(U)]_i \equiv \min_{a \in A_h^r(x_i)} \left\{ \beta_i(a) \sum_{j \in I_r \cup I_0} \lambda_{ij}^{(r)}(a)U_j + \eta_i(a)F_i(a) \right\}, \qquad i \in I_r \cup I_0,$$

where $\eta_i(a) = \eta(x_i, a)$, $\beta_i(a) = \beta(x_i, a)$ and $F_i(a) = f(x_i, a)$.

Finally, by D_1 and D_2 we define the operator $D : \mathbb{R}^N \to \mathbb{R}^N$ related to the domain Ω as follows:

$$[D(U)]_i = \begin{cases} [D_r(U^r)]_i & i \in I_r, r = 1, 2, \\ \min\{[D_1(U^1)]_i , [D_2(U^2)]_i\} & i \in I_0 \end{cases}$$

where $U^r = R_r(U), r = 1, 2$.

Remark. The discrete map D corresponding to the operator D_h and the discrete map corresponding to the operator $D_{h\varepsilon}$ coincide for $\varepsilon = \min\{\eta(x_i, a), i \in I, a \in A_h(x_i)\}$. Therefore the convergence of the fully discrete operator is guaranteed by the contraction mapping theorem.

Numerical experiment.

Let us set $\Omega \equiv (-2, 2)^2$, $\Omega_1 \equiv (-2, 0) \times (-2, 2)$, $\Omega_2 \equiv (0, 2) \times (-2, 2)$

$$A \equiv [0, 1], \quad \lambda = 1, \quad b(x, y, a) = (ay, 0) \text{ and } f(x, y, a) = (|x| - 1)^2.$$

The exact solution is known. Numerical results in double precision FORTRAN where obtained on an IBM 3090 using only 1 CPU. The computation of the solution in Figure 1 required 136 iterations for a total of 85 seconds of CPU time giving an L^∞ error of 0.025.

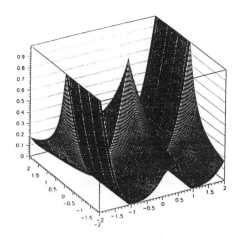 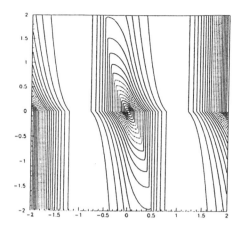

FIGURE 1. The approximate solution and its level curves ($h = 0.05$, $k = 0.025$).

REFERENCES

[1] G. Barles and B. Perthame, *Comparison principle for Dirichlet-type H.J.equations and singular perturbations of degenerate elliptic equations*, Appl. Math. Optim. **21** (1990), 21 44.

[2] F. Camilli and M. Falcone, *Approximation of optimal control problems with state constraints: estimates and applications* to appear on Proceedings of the IMA Workshop on Nonsmooth Analysis and Geometric Methods in Deterministic Optimal Control (B.S. Mordukhovich and H.J. Sussmann,eds.), IMA Volumes in Mathematics and its Aplications, Springer-Verlag, New York (1994).

[3] I. Capuzzo Dolcetta and P. L. Lions, *Viscosity solutions of Hamilton–Jacobi equations and state constraints*, Trans. Amer. Math. Soc. **318** (1990), 643–683.

[4] M.G. Crandall, H.Ishii and P.L. Lions, *User's guide to viscosity solutions*, Bull. Amer. Math. Soc. **27** (1992), 1–67.

[5] M. Falcone and A. Digrisolo, *An approximation scheme for optimal control problems with state constraints*, preprint (1992).

[6] M. Falcone, P. Lanucara and A. Seghini, *A splitting algorithm for Hamilton–Jacobi–Bellman equations*, to appear in Applied Numerical Mathematics.

[7] J. Smoller, *Shock waves and reaction-diffusion equations*, Springer-Verlag, 1983.

[8] H.M. Soner, *Optimal control problems with state-space constraints*, SIAM J. Control and Optimization **24** (1986), 552–562.

[9] M. Sun, *Domain decomposition algorithms for solving Hamilton–Jacobi–Bellman equations*, Numerical Functional Analysis and Optimization **14** (1993), 145–166.

DIPARTIMENTO DI MATEMATICA, UNIVERSITÀ DI ROMA "LA SAPIENZA", P.LE ALDO MORO, 2, 00185 ROMA, ITALY

E-mail address: falcone@sci.uniroma1.it

Contemporary Mathematics
Volume **180**, 1994

DOMAIN DECOMPOSITION FOR THE SHALLOW WATER EQUATIONS

J. G. CHEFTER, C. K. CHU, AND D. E. KEYES

ABSTRACT. Domain decomposition (Krylov-Schwarz) iterative methods are proposed for the parallel implicit solution of the unsteady geopotential equation that arises when semi-implicit, semi-Lagrangian (SISL) methods are employed in the long-time integration of the shallow water equations. SISL methods permit timesteps on the scale of Rossby wave dynamics, in contrast to the small timesteps required to resolve gravity waves, and thus satisfy the stability bound of an explicit method. The price of the semi-implicitness is a global elliptic problem on a multiply-connected semi-periodic domain with variable coefficients that become singular at the poles of latitude-longitude coordinate systems. Elliptic solvers based on domain decomposition offer flexibility in discretization and good algebraic convergence properties. They also provide good data locality with a view towards high-latency coarse- to medium-grained parallelism.

1. THE SHALLOW WATER EQUATIONS

The system of shallow water equations (SWEs) is a hyperbolic problem at the core of many models for the dynamics of the oceans or the atmosphere. In this paper, the SWEs are reformulated for large timestep by putting the computational burden on a scalar elliptic equation addressable through Schwarz-type domain decomposition methods. The three-dimensional Euler equations for a shallow layer of inviscid fluid on a sphere of radius a rotating with angular velocity Ω reduce under the assumptions of constant density, hydrostatic balance in the radial direction, and the Taylor-Proudman hypothesis to:

$$(1) \qquad \frac{du}{dt} = (2\Omega \sin\lambda + \frac{u}{a}\tan\lambda)v - \frac{1}{a\cos\lambda}\frac{\partial\Phi}{\partial\varphi},$$

$$(2) \qquad \frac{dv}{dt} = -(2\Omega \sin\lambda + \frac{u}{a}\tan\lambda)u - \frac{1}{a}\frac{\partial\Phi}{\partial\lambda},$$

$$(3) \qquad \frac{\partial\Phi}{\partial t} = -\frac{1}{a\cos\lambda}(\frac{\partial}{\partial\varphi}[(\Phi - \Phi_b)u] + \frac{\partial}{\partial\lambda}[(\Phi - \Phi_b)v\cos\lambda]).$$

1991 *Mathematics Subject Classification*. Primary 65M55, 65Y20, 76C20.

The work was supported in part by Schlumberger-Doll Research (JGC), by National Aeronautics and Space Administration under NASA cooperative grant NCC 5-34 (CKC), by the National Science Foundation under contract number ECS-8957475 (DEK), and by the National Aeronautics and Space Administration under NASA contract NAS1-19480 while the third author was in residence at the Institute for Computer Applications in Science and Engineering.

This paper is in final form and no version of it will be submitted elsewhere.

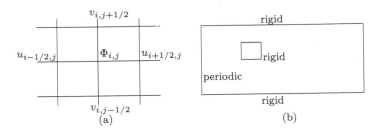

FIGURE 1. (a) Regular staggered grid employed for (u, v, Φ). (b) Computational domain.

The unknown fields are u and v, the vertically averaged velocities in the longitudinal (φ) and latitudinal (λ) directions, and the free-surface geopotential, Φ. Potentials $\Phi \equiv gh(\varphi, \lambda, t)$ and $\Phi_b \equiv gh_b(\varphi, \lambda)$, are defined in terms of the height of the top and bottom surfaces, and $\frac{d}{dt} \equiv \frac{\partial}{\partial t} + \frac{u}{a \cos \lambda} \frac{\partial}{\partial \varphi} + \frac{v}{a} \frac{\partial}{\partial \lambda}$ is the horizontal material derivative.

To integrate these equations, we use the semi-Langrangian scheme first proposed in [7, 8]. The integral of either of the momentum equations (1,2) over one timestep can be approximated to first-order accuracy in time as $\Xi^{k+1} - \Xi^k_* = \xi^{k+1}\tau$, where Ξ^{k+1} is the value of Ξ at the arrival point (φ, λ) of some streamline at time $(k+1)\tau$, Ξ^k_* is the value of Ξ at the departure point (φ_*, λ_*) of the same streamline at time $k\tau$, and $\xi^{k+1}\tau$ is the implicit approximation of the RHS. To find the departure point (φ_*, λ_*), one needs to integrate backwards along the characteristics, which can be approximated [4] by a sum over N equal intervals $\tau_1 = \frac{\tau}{N}$.

Implicit approximation of the integrals leads to better stability properties of the scheme in comparison with explicit time integration. In fact, explicit schemes for shallow water equations must respect the CFL condition [10]: $\tau < \delta/(|u| + |v| + \sqrt{gh})$, where δ is a cell diameter. The speed of the gravity wave $\sqrt{gh}$ usually poses the most restrictive condition on allowed timestep. For high latitudes this may lead to timesteps of the order of minutes, which are unreasonably small for desired simulation periods. Semi-implicit schemes remove the gravity wave speed from the CFL condition. Used with semi-Lagrangian integration of the material derivative, they often result in unconditionally stable schemes.

Applying semi-Lagrangian integration for momentum equations (1,2) we get a system of two linear equations. Solving it for u^{k+1} and v^{k+1}, substituting them into the implicitly integrated continuity equation (3), and cancelling cross-derivative terms, a self-adjoint elliptic equation for Φ^{k+1} results:

$$(4) \qquad \Phi^{k+1} \cos \lambda - \eta^2 \cos \lambda \frac{\partial}{\partial \varphi} \left(W^k \frac{\partial \Phi^{k+1}}{\partial \varphi} \right) - \zeta^2 \frac{\partial}{\partial \lambda} \left(W^k \cos \lambda \frac{\partial \Phi^{k+1}}{\partial \lambda} \right)$$

$$= \Phi^k \cos \lambda - \eta \cos \lambda \frac{\partial}{\partial \varphi} \left(W^k (u^k_* + F^k v^k_*) \right) - \zeta \frac{\partial}{\partial \lambda} \left(W^k \cos \lambda (v^k_* - F^k u^k_*) \right),$$

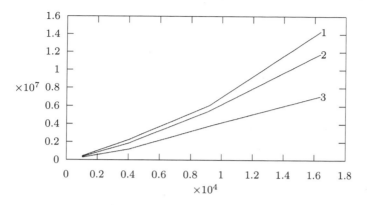

FIGURE 2. FLOPs versus number of unknowns for $||r||/||r_0|| <$ 10^{-8}. 1–ILU(4), 2–ASM, 3–MSM. Both Schwarz methods are based on four longitudinal strips.

and simple formulae for u^{k+1} and v^{k+1} in terms of Φ^{k+1}:

$$(5) \qquad u^{k+1} = \frac{1}{1+F^{k2}}\left(u_*^k - \eta\frac{\partial\Phi^{k+1}}{\partial\varphi} + F^k(v_*^k - \zeta\frac{\partial\Phi^{k+1}}{\partial\lambda})\right)$$

$$(6) \qquad v^{k+1} = \frac{1}{1+F^{k2}}\left(v_*^k - \zeta\frac{\partial\Phi^{k+1}}{\partial\lambda} - F^k(u_*^k - \eta\frac{\partial\Phi^{k+1}}{\partial\varphi})\right),$$

where $F^k = (2\Omega\sin\lambda + \frac{u^k}{a}\tan\lambda)\tau$, $\eta = \frac{\tau}{a\cos\lambda}$, $\zeta = \frac{\tau}{a}$, and $W^k = \frac{\Phi^k - \Phi_b}{1+F^{k2}}$.

Our staggered mesh is shown in Fig. 1(a). Using centered five-point space differencing, we get a first-order in time, second-order in space numerical scheme. In matrix form, for naturally ordered $\Phi_{i,j}^{k+1}$,

$$(7) \qquad\qquad\qquad A\Phi = g,$$

where A is a symmetric positive definite block-tridiagonal matrix. Linear stability analysis shows that the numerical scheme is unconditionably stable.

The computational domain with one island is shown in Fig. 1(b). For the simplest case of boundaries parallel to coordinate lines, the momentum equations at the boundaries are reduced to the following mixed-derivative implicit conditions for (4), for boundaries of constant λ and constant φ, respectively:

$$(8) \qquad \zeta\frac{\partial\Phi^{k+1}}{\partial\lambda} = -F^k(u_*^k - \eta\frac{\partial\Phi^{k+1}}{\partial\varphi})\ , \quad \eta\frac{\partial\Phi^{k+1}}{\partial\varphi} = F^k(v_*^k - \zeta\frac{\partial\Phi^{k+1}}{\partial\lambda}).$$

By treating the surface height gradients in the momentum equations explicitly we could get simple Neumann boundary conditions, but in our tests these conditions led to instabilities on the boundaries. Boundary conditions (8) are stable, but they make the system (7) slightly nonsymmetric.

Equation (4) with Dirichlet or Neumann boundary conditions is of a form for which overlapping domain decomposed preconditioners exist such that the convergence rate is independent of the resolution and the number of subdomains [3].

2. Krylov-Schwarz Algorithms

Schwarz-preconditioned Krylov solvers for linear systems, $A\Phi = g$, find the best approximation of the solution Φ in a small-dimensional subspace that is built up from successive powers of the preconditioned matrix on the initial residual. A variety of parallel preconditioners, whose inverse action we denote by B^{-1}, can be induced by decomposing the domain of the underlying PDE, finding an approximate representation of A on each subdomain, inverting locally, and combining the results. Generically, we seek to approximate the inverse of A by a sum of local inverses:

$$(9) \qquad B^{-1} = \sum_k R_k^T A_k^{-1} R_k,$$

where R_k is a restriction operator that takes vectors spanning the entire space into the smaller dimensional subspace in which A_k is defined.

The simplest domain decomposition preconditioner is block Jacobi, which can be regarded as a zero-overlap form of additive Schwarz [5]. The convergence rate of block Jacobi can be improved, at the price of a higher cost per iteration, with subdomain overlap and (for many problems) by solving an additional judiciously chosen coarse grid system. Our tests show that even a relatively small overlap of two mesh widths make preconditioner (9) comparable to the popular incomplete LU on a serial computer. In serial, the most natural form of Schwarz iteration is multiplicative, which improves the convergence rate of the algorithm, at the price of coarser parallel granularity, by enforcing sequentiality between the subdomain solves, just as Gauss-Seidel improves on Jacobi. Results are given for both additive (ASM) and multiplicative (MSM) versions of the preconditioner.

3. Numerical Results

The SISL code is built on top of the Argonne PETSc library [6]. Results below are from Sparc10 and Intel Paragon implementations. Figure 2 compares the performance of ILU, ASM, and MSM as left-preconditioners for GMRES. ILU(4) is used in the comparison, since four levels of fill led to the best ILU results. Strip subdomain problems (with overlap of $2h$) are solved exactly with a nested dissection ordering within each subdomain for the Schwarz methods. Even without a coarse grid, exact subdomain Schwarz methods performed better than ILU for this modest granularity, and improve upon ILU not only in the number of FLOPs, but also in the rate of growth with problem dimension.

P	Pois.-Dirichlet		Pois.-Neumann		SISL-with c.g.		SISL-w/o c.g.	
	add.	mult.	add.	mult.	add.	mult.	add.	mult.
4	13	6	51	16	16	7	13	7
16	11	5	30	12	17	8	15	8
64	9	4	18	9	16	8	17	8
256	8	3	15	7	15	8	18	10

TABLE 1. Number of iterations with number of subdomains P, for four problems, additive or multiplicative preconditioning.

We experimented with a two-level Schwarz preconditioner by adding a coarse grid with one degree of freedom per subdomain, with standard bilinear grid transfer

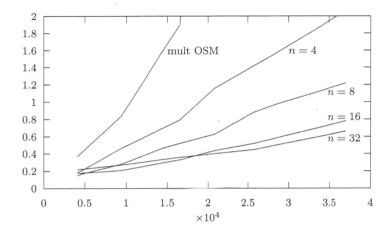

FIGURE 3. Wall-clock time (sec) vs. number of unknowns for $||r||/||r_0|| < 10^{-8}$ on the Paragon, for different numbers of sub-domains (processors), n.

operators. To verify its correct operation (convergence rate bounded independent of mesh resolution and subdomain diameter, asymptotically), the Poisson Dirichlet problem in a unit square was solved as in [2] for a range of coarse grid granularities, and the results appear in Table 1, together with convergence results for a Poisson operator with periodic BCs in x and Neumann BCs in y and potential equation (4) with explicitly integrated boundaries, with and without a coarse grid. (The Poisson Neumann problem has a null space of the constant vector, which was eliminated by employing a Dirichlet condition at one point common to both fine and coarse grids.) A box decomposition is used, for a 128×128 grid, iterated until $||r||/||r_0|| < 10^{-5}$.

The effectiveness of the coarse grid for the constant-coefficient operator in the first two pairs of columns can be seen in the decrease in the number of iterations as the granularity of the decomposition increases for both boundary types and both additive and multiplicative preconditioning. Comparison of the last two pairs of columns (for the same SISL problem) shows that the coarse grid does not make a large difference, though the difference widens for large problems. A time step of one hour is used in (4), which makes matrix A only slightly diagonally dominant, so the effect of the coarse grid is not buried in any parabolicity of the problem [1]. The ratio of the diagonal term to the biggest off-diagonal term is about 3 or 4 near equator and goes down to unity near the poles. This aspect of the problem remains under investigation.

4. PARALLEL RESULTS

Parallel implementation of the one-level Schwarz-based preconditioners was carried out on the Paragon. Vectors and matrices are distributed row-wise across processors and all subdomain solves are processed concurrently up to the number of available processors. Unfortunately, Semi-Lagrangian integration turns out to be poorly parallelizable, because the relative location of (ϕ_*, λ_*) in the processor array is solution-dependent and generally irregular. At present, it is done sequentially, which limits the scalability of the computation.

Figure 3 shows the dependence of elapsed time on the number of the unknowns N for various numbers of subdomains (processors). For zonally dominant (i.e., longitudinally oriented) flows, zonal strip decomposition gave better performance than box decomposition for the problem sizes considered. Sequential multiplicative OSM, which in our experiments has proved to be the most efficient algorithm of all tested, is used for comparison. Its high rate of growth with N for large N reflects thrashing of the memory hierarchy.

5. CONCLUSIONS

By concentrating data dependencies locally, domain decomposition preconditioners exploit the two-level memory hierarchy of high-latency distributed memory architectures. Low-communication zero or small overlaps between the preconditioner blocks are feasible with small convergence rate penalty, at least for intermediate granularities. Demonstrating the applicability of elliptic-based domain decomposition preconditioners to the shallow water equations opens the door to a variety of parallel implicit models in long-time integration geophysics applications.

REFERENCES

1. X.-C. Cai, *Some Domain Decomposition Algorithms for Nonselfadjoint Elliptic and Parabolic Partial Differential Equations*, Ph.D. thesis, Courant Institute, NYU, 1989.
2. X.-C. Cai, W. D. Gropp and D. E. Keyes, *A Comparison of Some Domain Decomposition and ILU Preconditioned Iterative Methods for Nonsymmetric Elliptic Problems*, Numer. Lin. Algebra Applics. (1994), to appear.
3. X.-C. Cai and O. B. Widlund, *Domain Decomposition Algorithms for Indefinite Elliptic Problems*, SIAM J. Sci. Stat. Comput. **13**(1992), 243–258.
4. V. Casulli, *Semi-implicit Finite Difference Method for the Two-dimensional Shallow Water Equation*, J. Comp. Phys. **86**(1990), 56–74.
5. M. Dryja and O. B. Widlund, *An Additive Variant of the Alternating Method for the Case of Many Subregions*, Courant Institute, NYU, TR 339, 1987.
6. W. D. Gropp and B. F. Smith, *Simplified Linear Equations Solvers Users Manual*, ANL-93/8, Argonne National Laboratory, 1993.
7. O. Pironneau, *On the Transport Diffusion Algorithm and Its Applications to the Navier-Stokes Equations*, Num. Math. **38**(1982), 309–322.
8. A. Robert, *A Stable Numerical Integration Scheme for the Primitive Meteorological Equations*, Atmosphere-Ocean **19**(1981), 35–46.
9. Y. Saad and M. H. Schultz, *GMRES: A Generalized Minimal Residual Algorithm for Solving Nonsymmetric Linear Systems*, SIAM J. Sci. Stat. Comp. **7**(1986), 865–869.
10. E. Turkel and G. Zwas, *Explicit Large Time-step Schemes for the Shallow Water Equations*, in Advances in Comp. Meths. for Partial Diff. Eq. III, R. Vichnevetsky and R. S. Stepleman, eds., IMACS, Brunswick, 1979, 65–69.

DEPARTMENT OF APPLIED PHYSICS, COLUMBIA UNIVERSITY, NEW YORK, NY, 10027-0029.
E-mail address: jgc@appmath.columbia.edu

DEPARTMENT OF APPLIED PHYSICS, COLUMBIA UNIVERSITY, NEW YORK, NY, 10027-0029.
E-mail address: chu@apne.columbia.edu

COMPUTER SCIENCE DEPARTMENT, OLD DOMINION UNIVERSITY, NORFOLK, VA 23529-0162
AND ICASE, MS 132C, NASA-LaRC, HAMPTON, VA 23681-0001
E-mail address: keyes@icase.edu

Contemporary Mathematics
Volume **180**, 1994

Domain Decomposition Methods
for Device Modelling

R.K. COOMER AND I.G. GRAHAM

ABSTRACT. We give an overview of some recent work [**3**] on the parallel solution of the drift diffusion equations for semiconductor device modelling. Discretization is by a variant of the finite element method. The resulting nonlinear equations are solved by Gummel's iteration, with the associated linear systems resolved by an additive Schwarz method. The algorithm has been implemented on a MasPar MP-1.

1. Drift-diffusion equations: discretization, outer iteration

In this paper we are concerned with the iterative solution of the (scaled) steady-state drift-diffusion equations for device modelling [**8**]:

$$(1.1) \qquad -\lambda^2 \Delta \psi + \delta\{\exp(\psi - v) - \exp(w - \psi)\} - d = 0,$$

$$(1.2) \qquad -\nabla.(\exp(\psi - v)\nabla v) - \sigma \rho_v r(\psi, v, w) = 0,$$

$$(1.3) \qquad -\nabla.(\exp(w - \psi)\nabla w) + \sigma \rho_w r(\psi, v, w) = 0.$$

Here ψ is the electrostatic potential and v and w are the electron and hole quasi-Fermi potentials respectively. The parameters λ, δ, σ, ρ_v and ρ_w are determined by the physics of the device, d is the (scaled) doping profile and r is the recombination/generation rate. We consider this system on a polygonal domain $\Omega \subset R^2$ with boundary $\partial\Omega$. At the *contacts* $\partial\Omega_D := \cup_i \partial\Omega_{D_i}$ (with the $\partial\Omega_{D_i}$ closed and non-empty subsets of $\partial\Omega$), we have Dirichlet conditions: $v|_{\partial\Omega_{D_i}} = w|_{\partial\Omega_{D_i}} - \alpha_i$, for each i, with the constants α_i corresponding to (scaled) applied voltages. Then ψ is specified at the contacts by requiring that the *space charge* (i.e. the zero-order term in (1.1)) should vanish there. Homogeneous Neumann conditions are imposed on $\partial\Omega \backslash \partial\Omega_D$. The function d typically has sign changes across thin transition regions or interfaces (between "p" and "n" regions

---

1991 *Mathematics Subject Classification.* Primary 65N55, 65N22; Secondary 65H10, 65F10.
This paper is in final form and no version of it will be submitted for publication elsewhere.

of the device), and as a consequence ψ, v, w have interior layers at or near these interfaces, with width determined by the small parameter λ ([**8**]).

To solve this system we first subdivide Ω into open convex pairwise disjoint quadrilateral *substructures* $\Omega^{(i)}$ such that $\overline{\Omega} = \cup_i \overline{\Omega^{(i)}}$. Subdividing each substructure into two triangles yields a "coarse grid" with diameter H which is further subdivided to give a fine triangular grid with diameter h. We assume for theoretical purposes that the refinement is *quasi-uniform* with respect to both H and h and that the fine grid is of *weakly acute type*. We also assume that the *collision points* $\partial\Omega_D \cap \overline{\partial\Omega_N}$ are vertices of substructures and hence are nodes of both the coarse and fine grids.

We discretize (1.1)-(1.3) in S_h, the space of piecewise linear finite elements with respect to the fine grid. We denote by ϕ_p the usual nodal basis functions in S_h, where p ranges over all nodes of the fine grid. Let $(\cdot, \cdot)$ denote the usual L_2 inner product, and define the corresponding discrete inner product $\langle f, g \rangle$ to be the integral of the piecewise linear interpolant of fg over Ω. These inner products can be extended to vector-valued functions f, g in the obvious way. Also, if $X \in S_h$, we define the piecewise constant function $\overline{X}$ so that $\exp(\overline{X})$ is the *harmonic average* of $\exp(X)$, i.e. for each triangle T, $\exp(\overline{X}|_T) = (\mathcal{A}(T)^{-1} \int_T \exp(-X))^{-1}$, where $\mathcal{A}(T)$ is the area of T.

We now define an iterative method for finding discrete solutions of (1.1)-(1.3). This version of the so-called *Gummel's method* consists of iterating the map $\mathcal{G} : (V, W) \longmapsto (\tilde{V}, \tilde{W})$ on $(S_h)^2$, defined as follows.

Step 1. (Fractional Step) Find $\tilde{\Psi} \in S_h$ such that for all $p \notin \partial\Omega_D$,

$$(1.4) \quad \lambda^2(\nabla\tilde{\Psi}, \nabla\phi_p) + \langle \delta\{\exp(\tilde{\Psi} - V) - \exp(W - \tilde{\Psi})\} - d, \ \phi_p \rangle = 0.$$

Step 2. Find $\tilde{V} \in S_h$ such that for all $p \notin \partial\Omega_D$,

$$(1.5) \quad (\exp(\overline{\tilde{\Psi} - V})\nabla\tilde{V}, \nabla\phi_p) - \langle \sigma\rho_v r(\tilde{\Psi}, V, W), \phi_p \rangle = 0.$$

Step 3. Find $\tilde{W} \in S_h$ such that for all $p \notin \partial\Omega_D$,

$$(1.6) \quad (\exp(\overline{W - \tilde{\Psi}})\nabla\tilde{W}, \nabla\phi_p) + \langle \sigma\rho_w r(\tilde{\Psi}, \tilde{V}, W), \phi_p \rangle = 0.$$

All iterates are assumed to satisfy the essential boundary conditions on $\partial\Omega_D$. If the iterates converge then the limit $(\Psi, V, W) \in (S_h)^3$, satisfies a finite element discretization of (1.1)-(1.3) with two modifications. Firstly the zeroth order terms have been "mass lumped" using the discrete inner product. As well as providing a simple pointwise evaluation for complicated nonlinear terms, this mass-lumping facilitates the formulation of globally convergent monotone iterative schemes for calculating the fractional step (1.4) (see §2). Secondly the exponential coefficients in the continuity equations (1.2), (1.3) are replaced by their harmonic averages. This can be interpreted as a certain generalisation to $2D$ of the classical Scharfetter-Gummel discretization for these equations. It can also be interpreted in terms of a "hybrid' mixed finite element method and, as such, ensures that the resulting (piecewise constant) approximations to the electron and hole currents ($\exp(\psi-v)\nabla v$ and $\exp(w-\psi)\nabla w$) have weak conservation properties ([2]). For more details see [3], [4].

2. Theoretical results

Since the work of Kerkhoven and Jerome ([7], [6], and the references therein) it has been known that, under appropriate assumptions, Gummel's map $\mathcal{G}$ is a contraction (on an appropriate space) for fixed h, provided the *applied bias* $\alpha := \max\{|\underline{\alpha}|, |\overline{\alpha}|\}$ is sufficiently small, where $\underline{\alpha} = \min_i \alpha_i$ and $\overline{\alpha} = \max_i \alpha_i$. Since numerical device modellers are often concerned with algorithmic complexity (i.e. with the cost in CPU time for a specified accuracy), it is also of interest to study how the Lipschitz constant of $\mathcal{G}$ changes as the mesh is refined. In [3] we have studied this question using a refinement of the techniques of Kerkhoven and Jerome. We also make use of a discrete Sobolev inequality well-known in the domain decomposition literature, namely that the uniform norm of an arbitrary element of S_h (which vanishes at least at one point of Ω) can be bounded in terms of its $H^1(\Omega)$ seminorm, where the constant of proportionality grows logarithmically with h as the mesh is refined (see, for example [1], [5]). Let B denote the set $\{(V, W) \in (S_h)^2 : \underline{\alpha} \le V, W \le \overline{\alpha}\}$, equipped with the norm $\|(V, W)\|_B = \|V\|_{H^1(\Omega)} + \|W\|_{H^1(\Omega)}$. Then we have the following result.

THEOREM 2.1. *Let $r = 0$ in (1.2),(1.3). Then, $\mathcal{G} : B \to B$ and for each $M > 0$ there exists a constant $C > 0$, independent of h, such that for all $(V^i, W^i) \in B$, $i = 1, 2$ and all $\alpha \le M$, we have*

$$\|\mathcal{G}(V^1, W^1) - \mathcal{G}(V^2, W^2)\|_B \le C\alpha(1 - \log h)^{1/2}\|(V^1, W^1) - (V^2, W^2)\|_B.$$

The assumption $r = 0$ plays a crucial role in the proof. It yields a discrete maximum principle for the solutions of (1.5), (1.6) which in turn implies that $\mathcal{G} : B \to B$. This assumption, which is made by all the other convergence analyses of Gummel's method of which we are aware, can be physically justified to some extent for devices with small current flow. However in general r is an essential part of the physical model and cannot be neglected. Thus it remains an important open question to extend the present analysis to include r.

Theorem 2.1 shows that the convergence of Gummel's method only degrades logarithmically with h as the mesh is refined, for fixed α. Inside each Gummel iterate we have to solve the semilinear problem (1.4). Since this may be regarded as singularly perturbed with respect to the small parameter λ, standard analyses of Newton's method will predict a convergence ball with radius dependent on λ (as well as h). Instead in [3], [4] we exploit the monotonicity in (1.4) to devise a monotone quasi-Newton scheme with an arbitrarily large radius of convergence. To introduce this, think of (1.4) as the problem of finding a solution $\tilde{\Psi}$ to the nonlinear problem $\boldsymbol{F}(\boldsymbol{\Psi}) = \boldsymbol{0}$, where $\boldsymbol{\Psi}$ denotes the vector of nodal values of $\Psi \in S_h$. Let $J(\boldsymbol{\Psi})$ denote the Jacobian of $\boldsymbol{F}$. Then our method is:

- Assume we have *lower* and *upper solutions* $\mathbf{\Lambda}^0, \mathbf{\Omega}^0$ which satisfy

$$\mathbf{\Lambda}^0 \leq \mathbf{\Omega}^0 \quad and \quad \boldsymbol{F}(\mathbf{\Lambda}^0) \leq \mathbf{0} \leq \boldsymbol{F}(\mathbf{\Omega}^0).$$

- Then, for $k \geq 0$, set

$$\mathbf{\Lambda}^{k+1} = \mathbf{\Lambda}^k - (J^k)^{-1}\boldsymbol{F}(\mathbf{\Lambda}^k) \quad and \quad \mathbf{\Omega}^{k+1} = \mathbf{\Omega}^k - (J^k)^{-1}\boldsymbol{F}(\mathbf{\Omega}^k),$$

where $J^k := \max\{J(\mathbf{\Lambda}^k), J(\mathbf{\Omega}^k)\}$, and the maximum is taken *elementwise*. It is shown in [**3**] that $\mathbf{\Lambda}^0$, $\mathbf{\Omega}^0$ (which are bounded independently of h) are easy to construct. For such starting vectors we prove in [**3**] the following (quadratic, mesh independent) convergence result.

THEOREM 2.2.   The sequences $\{\mathbf{\Lambda}^k\}$, $\{\mathbf{\Omega}^k\}$ converge to the same limit $\tilde{\mathbf{\Psi}}$, which is the unique solution of $\boldsymbol{F}(\mathbf{\Psi}) = \mathbf{0}$. Moreover

$$\|\mathbf{\Omega}^{k+1} - \mathbf{\Lambda}^{k+1}\|_2 \leq C\|\mathbf{\Omega}^k - \mathbf{\Lambda}^k\|_2^2, \quad k \geq 0,$$

with a constant C depending on λ and δ but independent of h and k.

The proof of this result makes use of the monotonicity of the discretization of the zero-order term in (1.4). This property is present in the undiscretized equation (1.1) and has been preserved through the use of mass-lumping in the discretization. Thus $\boldsymbol{F}$ is monotone, but unfortunately $\boldsymbol{F}$ is neither convex or concave on any domain which contains the solution, and so the results of [**9**] on monotone Newton methods cannot be used. The special quasi-Newton method defined above gets around this difficulty and the results obtained can be thought of as a generalisation of those in [**9**].

3. Inner iteration, domain decomposition

Each step of the iteration (1.4)-(1.6) (combined with the quasi-Newton method for (1.4)) amounts to the finite element approximation of a mixed boundary value problem for a symmetric linear elliptic second-order PDE, with coefficients which may suffer severe (finite) jumps across narrow interior layers. For example in a simple $p - n$ diode with no applied bias at room temperature, the potential ψ has a layer around the $p-n$ interface in which ψ changes from about -18 to $+18$ [**8**]. Consequently, for small bias, $\exp(\psi - v)$ varies between 10^{-8} and 10^8 in this layer. This jump is correspondingly present in the coefficient of the discretized equation (1.5), and the associated linear system is thus severely ill-conditioned (similarly (1.6)). It is essential (even for two-dimensional applications) to find preconditioners which mollify the effects of these jumps.

Fortunately the theory of additive Schwarz methods provides us with a reasonable solution of this problem and also yields algorithms which are readily parallelisable. We adopt here an approach analogous to that proposed in [**10**]. Each of the finite element problems is equivalent to a large sparse symmetric positive definite (SPD) linear system. To solve this we first eliminate locally the unknowns at interior nodes of substructures, yielding a new system $S\boldsymbol{x} = \boldsymbol{c}$ (where $\boldsymbol{x}$ now contains the unknown nodal values on the substructure boundaries, and S is the (SPD) Schur complement of the original system. We solve this latter system by the preconditioned conjugate gradient method (PCGM). The action of S can be computed by many local matrix-vector products plus nearest neighbour addition without assembling S explicitly.

Our preconditioner $\hat{S}$ is well-known among domain decomposition enthusiasts and goes back at least to the work of Bramble, Pasciak and Schatz [1]. It consists of the following steps: (i) For each substructure edge (except those at which essential boundary conditions are applied), invert the minor of S corresponding to interior nodes of that edge, and (ii) Invert the restriction of S to the coarse grid, with grid transfer operators defined by linear interpolation and its adjoint. Then add the results of (i) and (ii). Using a refinement of the elegant additive Schwarz analysis (e.g. [5], [10]), we can show that the condition number of $\hat{S}^{-1}S$ is bounded by $C(1 + \log(H/h))^2$, with C independent of H and h and also independent of the jumps of the coefficients of the underlying PDE across substructure boundaries.

Our implementation is on a MasPar MP-1 data parallel machine with $1K$ ($=$ 1024) processors, arranged in a 32×32 array. We assign a (small) substructure to each processor. Computation of the action of S then requires many local actions of Schur complements on substructures, followed by local addition across substructure boundaries. This fits naturally into the "massively parallel" programming model. Important questions then arise concerning the implementation of the preconditioner, especially the coarse grid problem (which is still large and may be almost as badly conditioned as S). Although this question merits further research, we have chosen in the present work to use (inner) iterations for both the local edge solves and for the coarse grid solve. After extensive experiments we concluded that approximate inner solves have a detrimental effect on the performance of the (outer) CGM and consequently we have solved the inner problems to (essentially) machine precision also using CGM. In addition we precondition the coarse grid problem by diagonal scaling. All inner iterations can be done by parallel local operations. For example one multiplication by the coarse-grid operator involves many parallel 4×4 multiplications and local addition.

At present we have experiments only for uniform grids. The unit square is divided into $m \times m$ subdomains, each of which contains a uniform mesh of triangles with $n \times n$ interior nodes. Various tests have been performed ([3], [4]) on model scalar problems which show that the number of preconditioned outer CGM iterates predicted by the theory is sufficient for convergence. More interestingly, we find that CPU time only increases very modestly with the size of coefficient jumps across substructure boundaries, and that the algorithm scales nicely with machine size: With m fixed, the solution time grows with $O(n^2)$, which is the time needed for (local) matrix-vector multiplication with the (full) Schur complements.

As a more challenging problem we have solved the semiconductor problem (1.1)-(1.3) in the case of a reverse biased $p-n$ diode. Full details of the parameter values and device geometry are in [3]. With an applied bias of 0.02 volts and with $r = 0$ we obtained the results in Table 1.

m	n	Quasi–Newton its	Gummel its	Time (s)
8	1	54	8	341
16	1	55	8	798
32	1	56	8	1666

TABLE 1

The number of quasi-Newton iterates given is those required for the solution of (1.4) in the first Gummel iterate. After that very few are required. This number remains fixed as the mesh is refined (as implied by Theorem 2.2). Moreover the number of overall Gummel iterates also appears unaffected by the mesh refinement, which is slightly better than that predicted by Theorem 2.1. Since our machine has a 32×32 array of processors, the growth in solution time can be attributed almost entirely to the cost of solution of the coarse grid problems. Qualitatively similar results were obtained with the recombination r switched on. Full details are in [3], [4].

REFERENCES

1. J. H. Bramble, J. E. Pasciak and A. L. Schatz, *The construction of preconditioners for elliptic problems by substructuring I*, Math. Comp. **47** (1986), 103–134.
2. F. Brezzi, L.D. Marini and P. Pietra, *Numerical simulation of semiconductor devices*, Comp. Meth. Appl. Mech. Engrg. **75** (1989), 493–514.
3. R K. Coomer and I. G. Graham, *Massively parallel methods for semiconductor device modelling*. Mathematics Preprint Number 93/02, University of Bath, 1993 (submitted for publication).
4. R. K. Coomer, *Parallel iterative methods in semiconductor device modelling*, Ph.D. thesis, University of Bath, 1994.
5. M. Dryja and O. B. Widlund, *Some domain decomposition algorithms for elliptic problems*, Iterative methods for large linear systems (L. Hayes and D. Kincaid, eds.), Academic Press, Orlando, 1989.
6. J. W. Jerome and T. Kerkhoven, *A finite element approximation theory for the drift diffusion semiconductor model*, SIAM J. Numer. Anal. **28** (1991), 403-422.
7. T. Kerkhoven, *Coupled and decoupled algorithms for semiconductor simulation*, Ph.D. thesis, Yale University, 1985.
8. P.A. Markowich, C.A. Ringhofer and C. Schmeiser, *Semiconductor equations*, Springer, Wien - New York, 1990.
9. J. M. Ortega and W. C. Rheinboldt, *Iterative solution of nonlinear equations in several variables*, Academic Press, New York, 1970.
10. B. F. Smith, *An optimal domain decomposition preconditioner for the finite element solution of linear elasticity problems*, SIAM J. Sci. Stat. Comput. **13** (1992), 364–378.

SCHOOL OF MATHEMATICAL SCIENCES, UNIVERSITY OF BATH, BATH BA2 7AY, UNITED KINGDOM (CORRESPONDENCE TO: I.G. GRAHAM)

E-mail address: igg@maths.bath.ac.uk

Contemporary Mathematics
Volume **180**, 1994

AN EFFICIENT COMPUTATIONAL METHOD
FOR THE FLOW PAST AN AIRFOIL

George C. Hsiao, Michael D. Marcozzi *and* Shangyou Zhang

ABSTRACT. A finite element–boundary element coupling procedure is applied to the computation of an incompressible flow past an airfoil. By utilizing a representation of the potential flow exterior to a circular auxiliary boundary, the reduced variational problem on an annular region is solved by the finite elment method alone. Two multigrid algorithms are introduced for the finite element equations. Both methods are optimal in the order of computation. The singularity at the trailing edge of the numerical solutions is corrected by the Kutta–Joukowski condition. Detailed numerical implementation is presented.

We consider a steady uniform two–dimensional fluid flow past a thin airfoil. As is customary, to arrive at a potential flow analysis, the compressibility and viscosity of the fluid are neglected, and the flow is assumed to be irrotational. The problem can then be formulated as an exterior boundary value problem for the velocity field $\mathbf{q} = (q_1, q_2)$:

(1)
$$
\begin{aligned}
\nabla \cdot \mathbf{q} &:= \frac{\partial q_1}{\partial x_1} + \frac{\partial q_2}{\partial x_2} = 0 && \text{in } \Omega^c := \mathbf{R}^2 \setminus \Omega \cup \Gamma, \\
\nabla \times \mathbf{q} &:= \frac{\partial q_2}{\partial x_1} - \frac{\partial q_1}{\partial x_2} = 0 && \text{in } \Omega^c, \\
\mathbf{q} \cdot \mathbf{n} &= 0 && \text{on } \Gamma, \\
\mathbf{q} - \mathbf{q}_\infty &= o(1) && \text{as } |x| \to \infty, \\
\lim_{\mathbf{x} \to \mathbf{T}} |\mathbf{q}(\mathbf{x})| &= |\mathbf{q}(\mathbf{T})| < \infty && \text{(Kutta–Joukowski condition).}
\end{aligned}
$$

Here Γ is the profile of the thin airfoil Ω with one corner point at the trailing edge $\mathbf{T}$ (see Figure 1), and $\mathbf{q}_\infty$ is the given free stream velocity.

Alternatively, we can introduce a stream function ψ such that $\mathbf{q} = (\nabla \psi)^\perp := \left(\frac{\partial \psi}{\partial x_2}, -\frac{\partial \psi}{\partial x_1} \right)$. Denoting by $u(\mathbf{x})$ the disturbance stream function due to the airfoil, and setting $\psi(\mathbf{x}) = \psi_\infty(\mathbf{x}) + u(\mathbf{x})$, where $\psi_\infty(x) = -\mathbf{q}_\infty^\perp(\mathbf{x}) \cdot \mathbf{x}$, we may reformulate the problem (1) as an exterior boundary value problem for $u(\mathbf{x})$:

(2)
$$
\begin{aligned}
-\Delta u &= 0 && \text{in } \Omega^c, \\
u &= -\psi_\infty && \text{on } \Gamma, \\
u + \frac{\kappa}{2\pi} \log |\mathbf{x}| &= C_0 + o(1) && \text{as } |x| \to \infty.
\end{aligned}
$$

In this formulation, however, both constants C_0 and κ are unknown. Physically, κ is the circulation around the airfoil: $\kappa = \int_\Gamma \mathbf{q} \cdot d\mathbf{x}$, which will be determined

1991 *Mathematics Subject Classification.* 65N55, 65N30, 65F10.
Key words and phrases. Finite element, boundary element, Multigrid method.
This paper is in final form and no version of it will be submitted for publication elsewhere.

by the Kutta–Joukowski condition in (1). Following the principle of the finite
element – boundary element coupling procedure in [6] and [8], we introduce a
circular boundary Γ_0 of radius a as shown in Figure 1. Let $u = u_F$ in the annular
(finite element) region Ω_F, and $u = u_B$ in the exterior (boundary element) region
Ω_B. The coupling of the two parts of u is by the following transmission conditions:

$$(3) \qquad \frac{\partial u_F^-}{\partial n} = \frac{\partial u_B^+}{\partial n} =: \sigma \qquad \text{and} \qquad u_F^- = u_B^+$$

on Γ_0. By representing $u_B(\mathbf{x}) = \int_{\Gamma_0} \frac{\partial \gamma}{\partial \mathbf{n}_y}(\mathbf{x},\mathbf{y})u_F^-(\mathbf{y})\,ds_{\mathbf{y}} - \int_{\Gamma_0} \gamma(\mathbf{x},\mathbf{y})\sigma(\mathbf{y})\,ds_{\mathbf{y}} + C_0$
in Ω_B, we obtain the following coupled differential – boundary integral system:

$$(4) \qquad \begin{cases} -\Delta u = 0 \quad \text{in } \Omega_F, \quad u_F|_\Gamma = -\psi_\infty, \quad \left.\frac{\partial u_F^-}{\partial n}\right|_{\Gamma_0} = \sigma, \\[2mm] \frac{1}{2}u_F^- - Ku_F^- + V\sigma - C_0 = 0 \quad \text{on } \Gamma_0 \\[2mm] \int_{\Gamma_0}\sigma = -\kappa, \qquad + \text{Kutta–Joukowski condition.} \end{cases}$$

Here $\gamma(\mathbf{x},\mathbf{y}) = (-1/2\pi)\log|\mathbf{x}-\mathbf{y}|$ is the fundermental solution, V and K are the
simple–layer and double–layer potential operators on Γ_0 respectively (see, e.g., [6]).

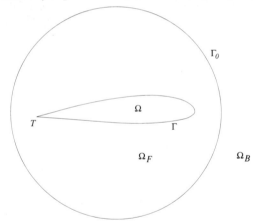

Figure 1. An airfoil and the auxiliary domains.

Because of the specially chosen auxiliary boundary Γ_0, the integral equation
in (4) can be inverted exactly at the continuous level. Therefore the problem
(4) is reduced to the following variational problem (cf. [7] and [8]): Find $u_F \in$
$H^1_{-\psi_\infty}(\Omega_F) \cap H^2(\Omega_F)$ and $\kappa \in \mathbf{R}^1$ such that

$$(5) \qquad a(u_F, v) + 2\langle V\dot{u}_F, \dot{v}\rangle = f(v) \qquad \forall v \in H^1_0(\Omega_F),$$

where $f(v) = \langle -\frac{\kappa}{2\pi a}, v\rangle$. Here, $\dot{v}$ denotes the tangential derivative (on Γ_0), $a(u,v) =$
$\int_{\Omega_F} \nabla u \cdot \nabla v\,dx$, $\langle v, \chi\rangle = \int_{\Gamma_0} v\chi\,ds$, $H^1_g(\Omega_F) = \{v \in H^1(\Omega_F) \mid v|_\Gamma = g(\mathbf{x})\}$, and
$H^m = W^{m,2}$ are standard Sobolev spaces (cf. [1]). The Kutta–Joukowski condition
ensures that the correct solution to (5) has to be a regular function even near the
trailing edge $\mathbf{T}$. Therefore by assuming $u_F \in H^2(\Omega_F)$, the problems (4) and (5)
are equivalent. For complete analysis of (4–5), we refer readers to [8].

To find the correct constant κ such that $u_F \in H^2(\Omega_F)$, we replace (5) by the
following two basic problems: Find $u_i \in H^1_0(\Omega_F)$ for $i = 0, 1$, such that

$$(6) \quad a(u_i, v) + 2\langle Vu_i, v\rangle = \begin{cases} a(\psi_\infty, v) + 2\langle V\psi_\infty, v\rangle & \text{if } i = 0, \\[2mm] \langle -1/2\pi a, v\rangle & \text{if } i = 1, \end{cases} \quad \forall v \in H^1_0(\Omega_F).$$

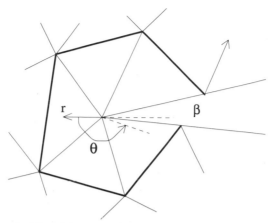

Figure 2. The polar coordinates near the trailing edge **T**.

We separate the singularity terms (at the trailing edge) from the solutions by $u_i = c_i u_s +$ (regular terms) (see [5] and [8]), where

$$(7) \qquad u_s = r^{\frac{\pi}{2\pi-\beta}} \cos \frac{\pi\theta}{2\pi - \beta}$$

where polar coordinates are used at **T** (see Figure 2). Then $u_F = u_0 + \kappa u_1$ has no singular term u_s if $\kappa = -c_1/c_0$ can be found. The stress intensity factors c_0 and c_1 are computed according to a conventional method (see, e.g., [9]) in our computation, which will be presented below.

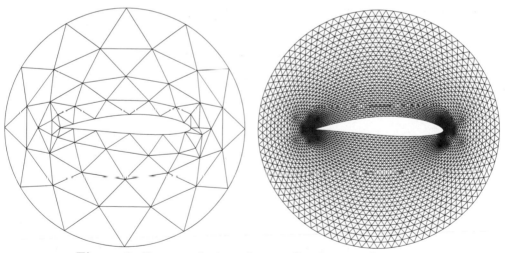

Figure 3. Nonnested triangulations (level 1 and 4) on Ω_F.

To discretize (6), we use piecewise linear finite elments (cf. [4]) on a family of nonnested triangulations (see Figure 3) $\{\mathcal{T}_k\}$, $\mathcal{T}_{k-1} \not\subset \mathcal{T}_k$ which are graded toward the trailing edge **T**. We refer readers to [3] and [10] for references on the multigrid method on nonnested grids. Defining V_k to be the intersection of the k–th level finite element space and $H_0^1(\Omega_F)$, (6) reads: Find $u_{i,k} \in V_k$, such that

$$(8) \qquad A_k u_{i,k} + B_k V \dot{u}_{i,k} = f_i,$$

where $(A_k u, v) = a(u, v)$, $(B_k u, v) = 2\langle V\dot{u}_i, \dot{v}\rangle$, $f_0 = a(\psi_\infty, v) + 2\langle V\dot{\psi}_\infty, \dot{v}\rangle$ and $f_1 = \langle -1/2\pi a, v\rangle$ for all $u, v \in V_k$. We have two multigrid algorithms (cf. [7] for details) for solving the linear system (8). In one algorithm, we apply the multigrid method to the operator $A_k + B_k$. The operator $A_k + B_k$ is shown to be symmetric and positive definite in [7].

Definition 1 (A direct multigrid method). Given w_0 approximating the solution u_i in (8), one k–th level multigrid iteration produces a new approximation w_{m+1} as follows. First, m smoothings are performed:

$$(9) \qquad w_l = w_{l-1} + \rho(C_k)^{-1}\left(f_i - C_k w_{l-1}\right), \qquad l = 1, \cdots, m,$$

where $C_k = A_k + B_k$, and $\rho(C_k)$ stands for the spectral radius of C_k or an upper bound of it. Then we solve the residual problem on the coarse level

$$(10) \qquad C_{k-1}q = I_k^\top \left(f_i - C_k w_m\right)$$

by doing $p(> 1)$ $(k - 1)$–st level multigrid iterations to get $\tilde{q}$. Finally

$$(11) \qquad w_{m+1} = \begin{cases} w_m + I_k\tilde{q} & \text{if } k > 1, \\ w_m & \text{if } k = 1. \end{cases}$$

Here the I_k in (11) is the Lagrange nodal interpolation operator, and the $I_k^\top$ in (10) is the adjoint operator of I_k under the L^2 inner–product. This operator I_k is needed since the multilevel finite element spaces are not nested: $V_{k-1} \not\subset V_k$, caused by the nonnested grids and the curved boundaries.

In the second algorithm, we apply the multigrid method only to A_k, which is a discrete Laplacian. That is, from an iterative solution $u_{i,k}^{(j)} \in V_k$, the new solution is

$$(12) \qquad u_{i,k}^{(j+1)} = u_{i,k}^{(j)} + \omega\epsilon,$$

where $\omega \leq 1$ is a relaxation parameter and $\epsilon \in V_k$ is a solution for the following residual problem:

$$(13) \qquad A_k\epsilon = f_i - A_k u_{i,k}^{(j)} - B_k u_{i,k}^{(j)}.$$

Definition 2 (A double iterative multigrid method). Given $u_{i,k}^{(j)}$ approximating the solution u_i in (8), one outer iteration produces a new approximation $u_{i,k}^{(j+1)}$ as defined in (12), where ϵ is obtained by doing $n(\geq 1)$ (inner) multigrid iteration(s) with initial guess 0. Here the (inner) multigrid iteration is defined in Definition 1 where f_i is replaced by the right hand side function in (13) and C_k replaced by A_k.

In [7], we proved that the speed of convergence for the multigrid method defined by Definition 1 is constant independent of the number of unknows in the linear system. Also in [7], we proved that the operator $A_k^{-1}(A_k + B_k)$ is well–conditioned. By the standard technique of [2], we have the following theorem of the optimal order of computation for the two multigrid methods (cf. [7]).

Theorem 1. *For both multigrid algorithms defined in Definitions 1 and 2, the number of arithmetic operations for solving the linear system (8), up to the order of truncation error, is proportional to the number of unknowns in the system.* □

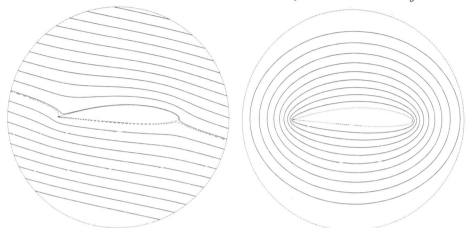

Figure 4. Contours of computed u_0 and u_1.

In our numerical test, we let $\mathbf{q}_\infty = (-1, 1/4)$ in (1) where the domain is described by a Kármán–Trefftz airfoil profile defined by the transformation

$$(14) \qquad z = -nb\frac{(c-b)^n + (c+b)^n}{(c-b)^n - (c+b)^n}.$$

This maps the outside of the unit circle to the exterior of a thin airfoil on the complex plane. For our test, in (14), $b = 1$, $n = 1.9$, $d = 0.0707$, $i = \sqrt{-1}$, and $c = bd(1+i) + b\sqrt{(b+d)^2 + d^2}e^{i\theta}$ for $0 \le \theta < 2\pi$. The auxiliary (outer) boundary is the circle of radius $a = 3.5$ centered at the origin. We emphasize again that by the specially chosen outer boundary, the boundary element discretization does not appear in the computation. We remark that the relaxation parameter ω in (12) is necessary to ensure the convergence (cf. [7]). It appears the best ω is around 0.7 which is used in this numerical test (cf. [7] for the computational results on ω).

On the 4th level we have about 5000 unknowns in the linear system (8). By the first algorithm (Definition 1), we need about 10 multigrid V–cycle ($p = 1$ in Definition 1) iterations with 8 smoothings ($m = 8$ in Definition 1). The computed stream functions (on level 4) are depicted in Figure 4. We also tested the second algorithm (Definition 2), where we apply the V–cycle iteration with 8 smoothings in the inner iteration (stops after 4 or 5 cycles, rate is about 0.1 for the V–cycle iteration). Then 5 outer iterations reduce the error (to the smooth solution) to less than 1 percent. We comment that the work for both methods is about the same because one evaluation of $B_k v_k$ ($v_k \in V_k$) is more expensive than that of $A_k v_k$ (this depends on the implementation, and can be avoided). However, both methods are very efficient due to their optimal order of computation. In our test computation on a SPARC station IPX, it takes a few seconds to solve the linear system on level 4.

From the solution u_0 in Figure 4, we can see that due to the singularity at the trailing edge of the thin airfoil, the solution u_0 needs to be corrected by u_1 to obtain a physical solution that satisfies the Kutta–Joukowski condition. Since the shape of the thin airfoil is known, we know the exact singular term at the trailing edge as specified in (7), where $\beta \approx 3.2 \approx 18^o$ is the angle for our airfoil. After we computed

u_0 and u_1, we compute (cf. [9]) the coefficient c_i of the leading singular term in $u_i = c_i w +$ (smooth terms) by the formula:

$$c_i = \int_{\Gamma^*} (\nabla u_i \cdot \mathbf{n}) w^* \, ds - \int_{\Gamma^*} (\nabla w^* \cdot \mathbf{n}) u_i \, ds$$

for $i = 0$ and 1, where Γ^* is an arc surrounding the non–convex corner and $w^* := (-1/\pi) r^{-\pi/(2\pi-\beta)} \cos(\pi\theta)/(2\pi - \beta)$ is the dual function satisfying $\Delta w^* = 0$ in the region bounded by Γ^* and Γ. In our computation, Γ^* consists of the edges (depicted by bold lines in Figure 2) around the trailing edge T. The computed $c_0 \approx 0.928348$ and $c_1 \approx 0.220257$. After we found the singular terms in u_i, we can get the correct solution $u_F = u_1 + \kappa u_0 = u_1 - (c_1/c_0) u_0$, which is plotted in Figure 5.

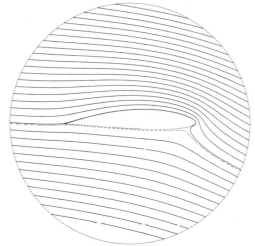

Figure 5. The computed solution u_F on the level 4 grid.

REFERENCES

1. R. A. Adams, *Sobolev spaces*, Academic Press, New York, 1975.
2. R. Bank and T. Dupont, *An optimal order process for solving finite element equations*, Math. Comp. **36** (1981), 35 – 51.
3. J. H. Bramble, J. E. Pasciak and J. Xu, *The analysis of multigrid algorithms with nonnested spaces or non–inherited quadratic forms*, Math. Comp. **56** (1991), 1 – 34.
4. P. G. Ciarlet, *The Finite Element Method for Elliptic Problems*, North–Holland, Amsterdam, New York, Oxford, 1978.
5. P. Grisvard, *Elliptic Problems in Nonsmooth Domains*, Pitman Pub. Inc., 1985.
6. G. C. Hsiao, *The coupling of boundary element and finite element methods*, ZAMM **70** (1990), 493 – 503.
7. G. C. Hsiao and S. Zhang, *Optimal order multigrid methods for solving exterior boundary value problems*, SIAM J. Num. Anal. **31-3** (to appear).
8. M. D. Marccozi, *Variational Methods for the Potential Flow Past an Airfoil*, Ph. Dissertation, Department of Mathematical Sciences, University of Delaware, 1993.
9. B. Szabo and I. Babuška, *Finite Element Analysis*, Wiley, 1991.
10. S. Zhang, *Optimal order non-nested multigrid methods for solving finite element equations I: on quasiuniform meshes*, Math. Comp. **55** (1990), 23 – 36.

DEPARTMENT OF MATHEMATICAL SCIENCES, UNIVERSITY OF DELAWARE, NEWARK, DE 19716, USA

Contemporary Mathematics
Volume **180**, 1994

Newton-Krylov-Schwarz Techniques Applied to the Two-Dimensional Incompressible Navier-Stokes and Energy Equations

P.G. JACOBS†, V.A. MOUSSEAU†, P.R. MCHUGH†, AND D.A. KNOLL†

ABSTRACT. We present research on Newton's method for the solution of the two-dimensional finite volume discretization of the incompressible Navier-Stokes and energy equations in primitive variables. Our previous research has employed a direct banded solver [6] and ILU preconditioned conjugate gradient-like algorithms (GMRES, CGS, QMRCGS, Bi-CGSTAB) [7, 8] to solve the linear systems arising on each Newton step. In this paper we show results from a preliminary investigation that uses domain decomposition to precondition the TFQMR conjugate gradient-like algorithm, showing the dependence of convergence rate on overlap, blocking strategy, and the additive/multiplicative trade-off.

1. Introduction

New numerical techniques are often tested on model problems that are over-simplified with respect to geometry, boundary conditions, and neglect of multiple scales. However, a better understanding of the processes at work, both physical and numerical, motivates solving more complicated model problems. Simulation codes for such models typically require solving very large systems of equations. Fully implicit discretization techniques that employ iterative solvers with a high degree of parallelism provide a viable mechanism for solving these more difficult problems.

In this work we present results obtained by applying a combination of numerical techniques of growing popularity for parallel computing, to model fluid flow and heat transfer problems involving both natural (free) and forced convection.

1991 *Mathematics Subject Classification.* Primary 65N55, 65F10; Secondary 76D05, 35Q30.

Work supported through the EG&G Idaho Long Term Research Initiative in Computational Mechanics under DOE Idaho Field Office Contract DE-AC07-76ID01570.

This paper is in final form and will not be published elsewhere.

The model problems addressed here include natural convection in an enclosed square cavity and internal flow past a backward facing step. The latter problem is defined on a physical domain with a large aspect ratio. Section 2 presents a description of these model problems. Section 3 presents a brief description of the numerical techniques, while results of our investigation are given in Section 4. Section 5 contains some conclusions and prospects.

2. Model Problems

The following are the governing equations for the model problems we solve:

$$\frac{\partial u}{\partial x} + \frac{\partial v}{\partial y} = 0$$

$$\frac{\partial uT}{\partial x} + \frac{\partial vT}{\partial y} = \frac{1}{Pe}\left[\frac{\partial^2 T}{\partial x^2} + \frac{\partial^2 T}{\partial y^2}\right]$$

$$\frac{\partial u^2}{\partial x} + \frac{\partial uv}{\partial y} = -\frac{\partial p}{\partial x} + \frac{1}{Re}\left[\frac{\partial^2 u}{\partial x^2} + \frac{\partial^2 u}{\partial y^2}\right]$$

$$\frac{\partial uv}{\partial x} + \frac{\partial v^2}{\partial y} = -\frac{\partial p}{\partial y} + \frac{1}{Re}\left[\frac{\partial^2 v}{\partial x^2} + \frac{\partial^2 v}{\partial y^2}\right] + Gr\, T$$

The backward facing step problem [1] is defined in the region $\Omega = [-.5, .5] \times [0, 30]$ with $Re = 100$, $Pe = Re \cdot Pr = 70$, and $Gr = 0$, and satisfies the following boundary conditions:

For $x \in [0, .5]$,
$$\begin{cases} v(x, 0) = \frac{3}{2}(4x)(2 - 4x) \\ T(x, 0) = \left[1 - (1 - 4x)^2\right]\left[1 - \frac{1}{5}(1 - 4x)^2\right] \\ u(x, 0) = 0 \end{cases}$$

For $x \in [-.5, 0]$, $u(x, 0) = v(x, 0) = \dfrac{\partial T}{\partial y}(x, 0) = 0$

For $y \in [0, 30]$,
$$\begin{cases} u(\pm.5, y) = v(\pm.5, y) = 0 \\ \frac{\partial T}{\partial x}(\pm.5, y) = \mp\frac{32}{5} \end{cases}$$

For $x \in [-.5, .5]$, $\dfrac{\partial u}{\partial y}(x, 30) = \dfrac{\partial v}{\partial y}(x, 30) = \dfrac{\partial T}{\partial y}(x, 30) = 0$

The natural convection problem is defined in the region $\Omega = [0, 1] \times [0, 1]$ with $Re = 1$, $Pe = Re \cdot Pr = 0.71$, and $Ra = Gr \cdot Pr = 10000$, and satisfies the following boundary conditions:

For $(x, y) \in \partial\Omega$, $u(x, y) = v(x, y) = 0$

For $x \in [0, 1]$, $\dfrac{\partial T}{\partial y}(x, 0) = \dfrac{\partial T}{\partial y}(x, 1) = 0$

For $y \in [0, 1]$,
$$\begin{cases} T(0, y) = 0 \\ T(1, y) = 1 \end{cases}$$

3. Numerical Techniques

The nonlinear governing equations are linearized utilizing an inexact Newton's Method [4]. The finite volume method on a staggered grid is used to discretize the model PDEs in primitive variables. Next, the Jacobian for each Newton iteration is formed by numerically evaluating the required derivatives [6]. New Newton iterates are computed until successive vector iterates change by less than 10^{-6} in Euclidean norm. This linear system is solved by the preconditioned iterative Krylov method, transpose-free quasi-minimal residual, (TFQMR) [5]. In an efficient inexact Newton approach, the Krylov solver is iterated until the scaled residual is less than 10^{-2}. For this work we use overlapping additive and multiplicative Schwarz block preconditioners [2, 3].

4. Results

The finite volume discretization of the backward facing step problem uses a uniform grid with 24 cells along the x-axis and 96 cells along the y-axis. The natural convection problem uses a 48 by 48 uniform cell structure. Thus each implicit nonlinear system has 9216 degrees of freedom. The blocking for the Schwarz preconditioners are chosen to be a uniform checkerboard pattern. In the tables, the notation $(bx \times by)$ is used to indicate the blocking strategy which has bx blocks along the x-axis and by blocks along the y-axis. The blocking strategy is given in the first column of the tables. When overlapping is used each block is "grown" uniformly to give either a three or four cell overlap. The size of the overlap region is given in the first row of the tables. The values reported in the table are the average number of TFQMR iterations per Newton step required to meet the inexact Newton convergence criteria. For comparison, a global ILU(0) preconditioner requires an average of 178 TFQMR iterations per Newton step for the backward facing step problem, and 114 for the natural convection problem.

Our initial studies are designed to obtain a better understanding of the performance of these types of iterative methods applied to models with complicated physics. We identify three major issues that should be considered when using these methods.

The first issue concerns the partitioning into subdomains. With computationally complex problems, memory requirements will often dictate the minimum number of subblocks. The choice of blocking strategy can be very important in the convergence of the iterative algorithms. For example, Table 1 compares the results for the case of six blocks (rows 4-7) for the backward step problem. Using the additive Schwarz preconditioner with no overlap, the average TFQMR iterations per Newton step is 183 for 6×1 blocking, whereas 1×6 blocking only requires 14. Similar results hold for the multiplicative Schwarz preconditioner. The natural convection problem shows a similar but less pronounced dependence on the blocking strategy, as evidenced in Table 2. These results indicate that the

Table 1: Backward facing step problem results (ILU(0) requires 178 iterations).

	Additive		Multiplicative	
Blocking	No Overlap	3 Cell Overlap	No Overlap	3 Cell Overlap
1x4	7	8	4	3
2x2	29	21	14	7
4x1	91	35	48	17
1x6	14	13	4	4
2x3	31	22	15	8
3x2	63	30	30	12
6x1	183	70	82	30
3x3	64	40	33	13
4x4	100	68	47	20
6x6	187	118	85	34
Average TFQMR iterations per Newton step				

block aspect ratio significantly influences the performance of the preconditioner.

Another issue is the use of overlap. If a poor choice in blocking strategy cannot be avoided, the use of overlap can significantly reduce the number of required TFQMR iterations. Again, consider the 6×1 blocking case; the use of a three cell overlap reduces the average TFQMR iterations from 183 to 70 for the additive Schwarz preconditioner.

The use of the multiplicative Schwarz preconditioner instead of the additive Schwarz preconditioner can also provide performance benefits. The use of the multiplicative Schwarz preconditioner generally requires less than half the iterations needed by the additive Schwarz preconditioner. For example consider the backward facing step problem with 6×6 blocking and three cell overlap; the additive Schwarz preconditioner requires 118 average TFQMR iterations per Newton step while the multiplicative Schwarz preconditioner requires only 34. We note that the serial nature of the multiplicative Schwarz preconditioner is not a large deterrent to its use in a parallel computing environment since for the checkerboard blocking the preconditioner may be realized with only four serial steps through multicoloring.

5. Summary And Future Work

Our initial work indicates the blocking strategies play an important role in the performance of the algorithms. Both cell and block aspect ratios should be considered when selecting what blocking strategy to use. Several trials may be necessary to "tune" the preconditioner. Improvements may be obtained by the use of an overlap region and the use of the multiplicative Schwarz preconditioner.

Some future work will involve: extending these algorithms to systems of convection-reaction-diffusion equations, distributing the preconditioner over a heterogeneous network using PVM, adding a coarse grid solve, and performing

Table 2: Natural convection problem results (ILU(0) requires 114 iterations).

Blocking	No Overlap	4 Cell Overlap
1x4	10	5
2x2	9	5
4x1	9	5
1x6	17	7
2x3	12	6
3x2	12	6
6x1	16	8
1x8	23	9
2x4	13	8
4x2	15	8
8x1	21	9
2x8	24	13
4x4	22	12
8x2	26	14
6x6	36	15
8x8	51	19
Average TFQMR iterations per Newton step		

numerical eigenvalue analysis to study preconditioner effectiveness.

REFERENCES

1. B. Blackwell and D. W. Pepper, editors. *Benchmark Problems for Heat Transfer Codes*, The American Society of Mechanical Engineers, United Engineering Center, 345 East 47th Street, New York, N.Y. 10017, November 8-13 1992. The Heat Transfer Division, ASME HTD-Vol. 222, American Society of Mechanical Engineers.
2. X.-C. Cai, W. D. Gropp, and D. E. Keyes. A comparison of some domain decomposition and ILU preconditioned iterative methods for non-symmetric elliptic problems. *J. Numerical Lin. Alg. Applics.*, to appear 1994.
3. X.-C. Cai and Y. Saad. Overlapping domain decomposition algorithms for general sparse matrices. Technical Report Preprint 93-027, Army High Performance Computing Research Center, University of Minnesota, 1993.
4. R. S. Dembo, S. C. Eisenstat, and T. Steihaug. Inexact Newton methods. *SIAM J. Numer. Anal.*, 19:400–408, 1982.
5. R.W. Freund. A transpose-free quasi-minimal residual algorithm for non-hermitian linear systems. *SIAM J. Sci. Comput.*, 14:470–482, 1993.
6. D. A. Knoll and P. R. McHugh. A fully implicit direct Newton solver for the Navier-Stokes equations. *Internat. J. Numer. Methods Fluids*, 17:449–461, 1993.
7. P. R. McHugh and D. A. Knoll. Inexact Newton's method solutions to the incompressible Navier-Stokes and energy equations using standard and matrix-free implementations. In *Proceedings of the 11th AIAA Computational Fluid Dynamics Conference: Part 1*, page 385, Orlando, July 1993.
8. P. R. McHugh and D. A. Knoll. Fully coupled finite volume solutions of the incompressible Navier-Stokes and energy equations using inexact Newton's method. *Internat. J. Numer. Methods Fluids*, (in press).

· †IDAHO NATIONAL ENGINEERING LABORATORY, P.O. BOX 1625, IDAHO FALLS, IDAHO 83415-3730

Contemporary Mathematics
Volume **180**, 1994

Direct Numerical Simulation of Jet Flow via a Multi-block Technique

H. C. Ku, H. E. Gilreath, R. Raul and J. C. Sommerer

Abstract

A multi-grid domain decomposition approach by the pseudospectral element method is used to simulate a two-dimensional jet emanating from the nozzle. The solution technique is to implement the Schwarz alternating procedure for exchanging data among subdomains, where the coarse-grid correction is used to remove the high frequency error.

Numerical results of jet flow not only provide the possible mechanism of turbulence formation, but also quantitatively capture all the phenomena for flow transition from the laminar to the turbulent structure.

1 Introduction

Turbulence, a phenomenon related to but distinct from chaos, has been increasingly in the focus of physics research in a variety of flows for the last two decades. Although there is no unique mathematical model that encompasses all flow environments, it is possible to gain an insight into turbulence through direct numerical simulation (low Reynolds numbers). In order to model all the features of turbulence, one needs to resolve the smallest length scale, i.e., the Kolmogorov length scale at which the turbulent energy carried from the large length scales is dissipated into heat by the molecular viscosity. Based on the Kolmogorov dissipation scale, the ratio of length scales in one dimension is estimated as the reciprocal of $Re^{\frac{3}{4}}$. Thus, in the Kolmogorov theory of three-dimensional turbulent flow, there are at least on the order $Re^{\frac{3}{4}}$ dynamically active degrees of freedom for a given volume.

A special device is used to simulate the two-dimensional turbulence of jets by imposing a strong stratification along the vertical direction, so that the resulting

1991 Mathematics Subjet Classification. Primary 65N35, 68Q10; Secondary 76D05

This work was partially supported by the SPAWAR under the Contract N00039-91-C-0001.

This paper is in final form and no version of it will be submitted for publication elsewhere.

flow occurs mainly in a horizontal plane. This allows the direct numerical simulation to be performed on the HP9000/735 work station machine.

To address the goal of the direct numerical simulation of turbulent flow, the desired features of numerical algorithms are: (1) applicability to a variety of geometrical shapes; (2) high resolution in steep gradient areas (multi-grid or single-grid technique); (3) minimal working space (domain decomposition); and (4) low running time of computation (multiple processors). A novel pseudospectral element method [1] that contains the above mentioned features is ideally suitable for the proposed work.

2   Navier-Stokes Equations

For an incompressible flow, the time-dependent Navier-Stokes equations are:

$$\frac{\partial \mathbf{u}}{\partial t} + \mathbf{u} \cdot \nabla \mathbf{u} = -\nabla p + \frac{1}{Re}\nabla^2 \mathbf{u}, \tag{1a}$$

$$\nabla \cdot \mathbf{u} = 0. \tag{1b}$$

Here $\mathbf{u}$ is the velocity vector, p the pressure, $Re = UL/\nu$ the Reynolds number (U, L the characteristic velocity and length, respectively), and ν the kinematic viscosity.

To simplify the notation while explaining the basic ideas, we write the equations as if we could compute exact spatial derivatives in the curvilinear coordinates. The method utilized to solve the Navier-Stokes equations is fourth-order Runge-Kutta time integration scheme based on the Chorin's [2] splitting technique. According to this scheme, the equations of motion read

$$\frac{\partial u_i}{\partial t} + \frac{\partial p}{\partial x_i} = F_i \tag{2}$$

where $F_i = -u_j\,\partial u_i/\partial x_j + 1/Re\partial^2 u_i/\partial x_j^2$.

At each stage, the first step is to split the velocity into a sum of predicted and corrected values. The predicted velocity is determined by time integration of momentum equations without the pressure term and the second step develops pressure and corrected velocity fields that satisfy the continuity equation.

1st stage:
$$\bar{u}_i^1 = u_i^n + \frac{\Delta t}{2}F_i(u^n) \tag{3a}$$

$$u_i^1 = \bar{u}_i^1 - \frac{\Delta t}{2}\frac{\partial p}{\partial x_i}, \quad \frac{\partial u_i^1}{\partial x_i} = 0; \tag{3b}$$

2nd stage:
$$\bar{u}_i^2 = u_i^n + \frac{\Delta t}{2}F_i(u_i^1) \tag{4a}$$

$$u_i^2 = \bar{u}_i^2 - \frac{\Delta t}{2}\frac{\partial p}{\partial x_i}, \quad \frac{\partial u_i^2}{\partial x_i} = 0; \tag{4b}$$

3rd stage:
$$\bar{u}_i^3 = u_i^n + \Delta t F_i(u_i^2) \tag{5a}$$

$$u_i^3 = \bar{u}_i^3 - \Delta t\frac{\partial p}{\partial x_i}, \quad \frac{\partial u_i^3}{\partial x_i} = 0; \quad \text{and} \tag{5b}$$

4th stage:
$$\bar{u}_i^{n+1} = u_i^n + \Delta t\left\{\frac{F_i(u_i^n)}{6} + \frac{F_i(u_i^1)}{3} + \frac{F_i(u_i^2)}{3} + \frac{F_i(u_i^3)}{6}\right\} \tag{6a}$$

$$u_i^{n+1} = \bar{u}_i^{n+1} - \Delta t\frac{\partial p}{\partial x_i}, \quad \frac{\partial u_i^{n+1}}{\partial x_i} = 0. \tag{6b}$$

The main features of this method include: (i) for a given accuracy the time step size is larger than that of the first-order scheme and (ii) the most promising time integration scheme conserves the total energy during the evolution of inviscid flow [1]. The approach is very effective for flow at high Reynolds numbers because the gain in time step size offsets more than the cost of four pressure solvers.

At each stage, the pressure Poisson equation can be generated by taking the divergence operator of the corrected velocity. In the Cartesian coordinates a direct solution of pressure equation can be obtained by the eigenfunction expansion technique [3], while in the curvilinear coordinates the pressure solution is governed by the iterative preconditioned minimal residual method [1].

3   Domain Decomposition with Multi-Grid SAP

The SAP iterative scheme has been successfully applied to those configurations where the overlapped grids coincide with each other [3]. Under this condition, which we call the single-grid SAP, no data interpolation error occurs. The success of the single-grid SAP lies in the exclusive use of the continuity equation as the pressure boundary condition in the overlapping area, and the velocity difference in the overlapping area is reduced by one order of magnitude after each SAP iteration. Under some circumstances, the overlapped grid positions may not coincide with each other due to the complexity of the geometrical configuration, as in a submarine or automobile, where there arises a need for a possible layout of mixed grids or the application of adaptive fine grids in one subdomain to resolve steep changes of variables. But simply exchanging the data through interpolation in the inter-overlapping areas will cause high-frequency error and pollute the results throughout the whole computational domain [4].

The multi-grid technique, which has long been advocated by finite-difference users [5], employs a sequence of grids to accelerate the convergence of iterative methods. The work rests on "standard coarsening," i.e., doubling the mesh in each direction from one grid to the next coarsest grid. The problem is solved on the coarse grid, and the coarse-grid correction is recursively transferred back to the fine grid to obtain rapid convergence. It can apply to the overlapping area as well [5].

In addition to the Lagrangian constraint between the pressure and velocity field, the noncoinciding overlapped grids (nonequal-spaced collocation points) in the inter-overlapping areas enhance the difficulty of applying the multi-grid technique. However, the idea of "coarse-grid correction" is still effective in reducing high-frequency error. The strategy behind the coarse-grid correction process is to adopt the idea proposed by Thompson and Ferziger [6], modified as

$$\nabla_c \cdot \mathbf{u}_c - \nabla_c \cdot (I_c^f \mathbf{u}_f) = I_c^f (r_f - \nabla_f \cdot \mathbf{u}_f). \tag{7}$$

Here $\nabla_c\cdot$ represents the operator of divergence on the coarse-grid subdomain. I_c^f is an interpolation operator from the fine-grid subdomain "f" to coarse grid subdomain "c", and $\mathbf{u}$ is the velocity. r_f is simply the result of the divergence of the velocity field which should be set to zero. The left-hand side of Eq. (7) is the difference between the coarse-grid operator acting on the coarse-grid subdomain and the coarse-grid operator acting on the interpolated fine-grid subdomain (which is held fixed). When substituting the coarse-grid velocity in terms of pressure gradient (Eqs. (3 - 6)), the first term on the-left hand side of Eq. (7) becomes the pressure equation acting on the coarse-grid subdomain, while the right-hand side of Eq. (7)

is the interpolated residual of $\nabla \cdot \mathbf{u}$ from the fine-grid subdomain. It is apparent that once the solution of the fine-grid subdomain has been found the residual will be zero (exactly satisfy the pressure Poisson equation), which implies

$$\mathbf{u}_c = I_c^f \mathbf{u}_f. \tag{8}$$

When the residual is non-zero, Eq. (7) acts as a forcing term for the coarse-grid correction and transfers the correction of $\mathbf{u}$ back to the fine-grid subdomain, i.e.,

$$\mathbf{u}_f^{new} = \mathbf{u}_f^{old} + I_f^c(\mathbf{u}_c - I_c^f \mathbf{u}_f^{old}). \tag{9}$$

This is vital for the success of the scheme. Changes in the velocity field are transferred back to the fine-grid subdomain rather than the velocity field itself. Meanwhile, the error index, ℓ_2 norm of $\omega = \parallel \mathbf{u}_c - I_c^f \mathbf{u}_f \parallel$, provides a good monitor for the evolution of the flow field. Any unexpected increase in ω reflects that an extremely steep change of flow field occurs in the overlapping area, otherwise, the error index ω should remain in the same range of order of magnitude. Notice that when the overlapped grids in the overlapping areas are coinciding with each other, the interpolation operator I_c^f becomes a unitary matrix.

The multi-grid domain decomposition technique for the direct numerical simulation of jet flow sketched in Fig. 1 is summarized by the following algorithm:

1. First set $\mathbf{u}^{n+1}$ on $\overline{AB}, \overline{CD}, \overline{EF}, \overline{GH}$. Usually $\mathbf{u}^n$ will be a good initial guess.

2. Solve the fine-grid domain III employing the boundary conditions derived from the continuity equation on $\overline{AB}, \overline{CD}$, where the pressure solution is directly obtained by the eigenfunction expansion technique. While in domain I with the same type of boundary condition for the pressure on $\overline{EF}, \overline{GH}$, the iterative preconditioned method is used to get the pressure solution.

3. With the interpolated $\mathbf{u}^{n+1}$ from step (2) on domain III $\cap$ IV, solve the coarse-grid domain IV to update $\mathbf{u}^{n+1}$ on domain III $\cap$ IV by the coarse-grid correction process. With the velocity $\mathbf{u}^{n+1}$ along $\overline{IJ}, \overline{KL}$ from domain I & III, a single grid method is to give the pressure solution for domain II.

4. Repeat steps (2) & (3) until the error index ω among the overlapping areas meets the convergence criterion.

4   Results and discussion

In the interest of brevity, no effort is made hereafter in this paper to discuss the turbulent quantities, the fluctuation velocity (u_i') and turbulent stresses $(\overline{u_i'u_j'})$. However, they can be derived from the mean velocity (temporal average based on every time interval), i.e., $u_i' = u_i - \bar{u}_i$, and $\overline{u_i'u_j'} = \overline{u_i u_j} - \bar{u}_i \bar{u}_j$.

In order to make a comparison with the results obtained by the experiment, a realistic experimental scale of will be used by the direct numerical simulation. A slit of 0.1 inch wide is designed as the narrow part of a round nozzle whose diameter is 0.125 inch. A jet flow is discharged into a stratified tank with 1.5 feet wide and 1.5 feet long, and the upstream of a nozzle is connected by a 6 inch wide reservoir in which the fluid is driven by a constant moving piston.

As illustrated in Fig. 1, the computational domain is decomposed into four subdomains with overlapping areas: the upstream reservoir where a constant moving piston is used to drive the flow, the convergent nozzle where the incoming flow from the reservoir is developed to gain a high speed, the immediate downstream from the entrance of the nozzle where the high speed jet is discharged into the tank, and the far downstream area where a well-developed turbulent flow can be traced. In order to resolve the Kolmogorov length scale in the interesting area (indicated by the fine grid distribution), for the case of $Re = 500$, the number of points (element layout as plotted in Fig. 2) applied to each subdomain is 61 × 79 for domain I, 61 × 31 for domain II, 115 × 151 for domain III, and 97 × 139 for domain IV, separately.

Figs. 3 depict the streamline plot of jet flow at Re = 500. During the time evolution of jet flow, the symmetry of jet front will not be distorted at the early stage (laminar flow as seen in Fig. 3a) until the phase speed of the vortex shedding (due to flow instability) travels faster than that of jet front. Fig. 3b shows the onset of vortex shedding and Fig. 3c demonstrates that the jet front is not symmetric any more. A pair of vortices adjacent to the jet front represents the extrusion of the jet into the ambient fluid. Once the jet front is caught up by the incoming travelling waves, the energy transferred by the vortex shedding, in a cascade process from the highest at the nozzle exit (high shedding frequency) to the lowest at the jet front (low shedding frequency), splits into two parts: one for the jet front pushing against the ambient viscous fluid, and another travelling back, causing a wave-wave interaction. Initially, the wave-wave interaction starts close to the jet front and gradually propagates backward toward the nozzle exit. This process constitutes a complete mechanism to account for the turbulent formation. The longer the elapsed time, the more unstable the flow becomes. Fig. 3d gives a clear picture of flow development (nearly turbulent) at a longer time. A few distinct pairs of vortices always exist within 1 to 5 inches of the nozzle exit, the appearance of which are also confirmed by an APL Fluid Dynamics Laboratory experiment.

References

[1] H. C. Ku, A. P. Rosenberg and T. D. Taylor, *in the 12th Intl. Conference on numerical Methods in Fluid Dynamics*, Proceedings, Oxford, 1990, Lecture Notes in Physics, Springer-Verlag, 223-227.

[2] A. J. Chorin, "Numerical Solution of Navier-Stokes Equations," *Math. Comp.* **22** (1968), 745-762.

[3] H. C. Ku, R. S. Hirsh, T. D. Taylor and A. P. Rosenberg, "A Pseudospectral Matrix Element Method for Solution of Three-Dimensional Incompressible Flows and Its Parallel Implementation," *J. Comput. Phys.* **83** (1989), 260-291.

[4] H. C. Ku and B. Ramaswamy, *in the 6th Copper Mountain Conference on Multigrid Methods*, NASA Conference Publication (edited by N. D. Melson et al.), Colorado (1993), 293-304.

[5] W. Hackbusch, "Multi-Grid Methods and Applications," Springer-Verlag, Berlin, (1985).

[6] M. C. Thompson and J. H. Ferziger, "An Adaptive Multigrid Technique for the Incompressible Navier-Stokes Equations," **82** (1989), 94-121.

Johns Hopkins University Applied Physics Laboratory, Laurel, Maryland 20723

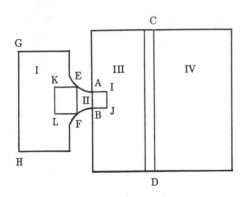

Figure 1: Configuration of domain decomposition for jet flow

Figure 2: Element layout of jet flow (6 × 6 points per element)

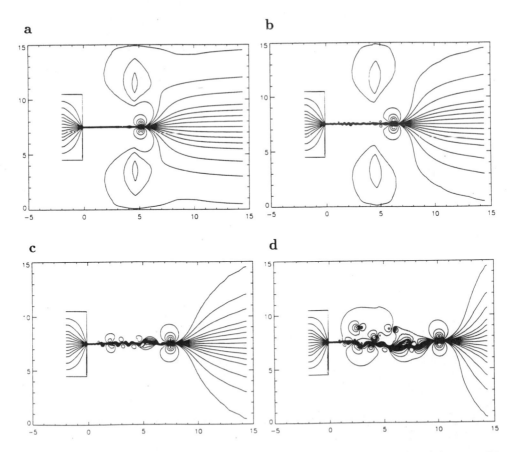

Figure 3: Streamline plots for Re = 500 at time (a) t = 120, (b) t = 150, (c) t = 180, and (d) t = 255.

Contemporary Mathematics
Volume **180**, 1994

Balancing Domain Decomposition for Plates

PATRICK LE TALLEC, JAN MANDEL, AND MARINA VIDRASCU

ABSTRACT. We show that the Neumann-Neumann preconditioner with a coarse problem can be applied to the solution of a system of linear equations arising from the thin plate problem discretized by the HCT and DKT elements. The condition number is asymptotically bounded by $\log^2(H/h)$, with H the subdomain size and h the element size. The bound is independent of coefficient jumps of arbitrary size between subdomains. Numerical results are presented.

1. Introduction

This note presents an application of the Balancing Domain Decomposition (BDD) method to the solution of linear systems of equations arising from the finite element discretization of thin plate problems. The BDD method was developed from the *Neumann-Neumann preconditioner* of De Roeck and Le Tallec [5] by Mandel [9], who has modified the algorithm by adding a *coarse problem* with few unknowns per subdomain. Solving the coarse problem in each iteration coordinates the solution between the subdomains and prevents any slow-down with an increasing number of subdomains.

The coarse problem, as introduced in [9], is composed of the rigid body modes of the substructures. Other modes can be added to the coarse problem to remove troublesome modes from the iterative process; in effect, these modes are resolved directly in every iteration. While the possibility of adding such modes has been known, it was not clear how to do that efficiently. This paper presents the first such example: for thin plates, these modes are the subdomain solutions for point loads applied at crosspoints (i.e., at subdomain corners).

1991 *Mathematics Subject Classification.* 65N55.
Key words and phrases. Iterative methods, plates, biharmonic equation, boundary value problems, finite elements.
Prepared for Proceedings of the 7th International Symposium on Domain Decomposition Methods, Penn State, November 1993. This research was supported by NSF grants ASC-9121431 and ASC-9217394. This paper is in final form and no version of it will be submitted for publication elsewhere.

For the reduced Hsieh-Clough-Tocher (HCT) and Discrete Kirchoff Triangle (DKT) elements, the condition number of the algorithm is proved to grow at most as fast as $\log^2(H/h)$, where H and h are the characteristic subdomain size and element size. Numerical results confirm that the fast growth of the condition number for decreasing h is indeed prevented by the additional coarse functions. Using a result of Mandel and Brezina [10], it is shown that the bound does not depend on the jumps of elasticity coefficients between subdomains. We omit the proofs of technical lemmas, but the principal argument is complete.

For related work on the Neumann-Neumann preconditioner, we refer to Glowinski and Wheeler [8]. For a somehow different formulation of the Neumann-Neumann problem with similar bounds for second order problems we refer to Dryja and Widlund [7]. The BDD method was also applied to mixed problems by Cowsar, Mandel, and Wheeler [4].

2. Finite Element Plate Model

Let $\Omega \subset \mathbb{R}^2$ be a bounded polygonal domain decomposed into nonoverlapping subdomains $\Omega_1, \ldots, \Omega_k$. There is given a conforming triangulation $\{T\}$ of Ω such that each Ω_i is the union of some triangles (elements) from $\{T\}$. The subdomains Ω_i and the elements $\{T\}$ are shape regular. The characteristic subdomain size is H and the characteristic element size is h. Throughout the paper, C and c are generic constants that do not depend on H and h but may depend on the shape regularity of the triangulation and subdomain decomposition. The union of all subdomain boundaries is $\Gamma = \cup_{i=1}^{k} \partial\Omega_i$, and it consists of *edges* and *crosspoints* at the junctures of the edges. We assume that each subdomain Ω_i has at least three points that are crosspoints or in the part of the boundary $\partial\Omega_s$ where the plate is simply supported, and the three points form a triangle with angles bounded below by $1/C$.

Spectral equivalence of quadratic forms is defined by

$$a(u, u) \sim b(u, u) \iff \exists C \; \forall u : \frac{1}{C} a(u, u) \le b(u, u) \le C a(u, u).$$

The domain of the forms will be always clear from the context. The Sobolev seminorms in $W^{m,p}(\Sigma)$ are denoted as usual by $|u|_{m,p,\Sigma}$, $|\cdot|$ is the Euclidean norm, and $P_p(\Sigma)$ is the space of all polynomial functions of order p on Σ.

We solve the problem of finding the displacement of a thin plate occupying the domain Ω, clamped on $\Omega_c \subset \partial\Omega$ and simply supported on $\Omega_s \subset \partial\Omega$. Plate elements used in engineering practice have typically three degrees of freedom per node, corresponding to the transversal displacement u and the rotations $\vec{\theta} = (\theta_\alpha)$, $\alpha = 1, 2$. Under the Kirchoff hypothesis

(1) $$\vec{\theta} = \nabla u,$$

the problem to be solved is to find the transversal displacement u so that

(2) $$u \in \mathbb{H}(\Omega) : \quad a(u, v) = L(v), \quad \forall v \in \mathbb{H}(\Omega),$$

where

$$a(u,v) \;=\; \int_{\Omega} \varepsilon(\vec{\theta}(u)) : K : \varepsilon(\vec{\theta}(v)),$$

$$L(v) \;=\; \int_{\Omega} fv + \int_{\partial\Omega - \partial\Omega_c} m_g \partial_n v + \int_{\partial\Omega - \partial\Omega_s} gv,$$

$$\mathbb{H}(\Omega) \;=\; \{v \in H^2(\Omega), v = 0 \text{ on } \partial\Omega_s, \quad \partial_\nu v = 0 \text{ on } \partial\Omega_c\},$$

where ε is the linearized strain tensor. The plate flexural stiffness tensor K is assumed to be symmetric, measurable, and, on each subdomain Ω_i, uniformly positive definite and bounded:

$$\varepsilon : K(x) : \varepsilon \sim \rho_i \varepsilon : \varepsilon, \quad \rho_i > 0, \qquad \forall x \in \Omega_i.$$

Then one has the spectral equivalence

$$(3) \qquad\qquad a(u,u) \sim \sum_{i=1}^{k} \rho_i \, |u|^2_{2,2,\Omega_i},$$

which will be the starting point of our investigations. The numbers ρ_i have interpretation as the *relative stiffnesses* of the subplates Ω_i.

We first consider a discretization of the problem (2) by the reduced Hsieh-Clough-Tocher (HCT) triangle. Here we only note that the HCT element is C^1 continuous, satisfies $u \in P_3(\sigma)$, $\partial_\nu u \in P_1(\sigma)$ on each side σ of the element, and is composed of three Γ_3 subtriangles; for more details on the HCT element, see [3]. Denote by I_{HCT} the interpolation operator associated with the HCT elements. The finite element discrete problem is obtained by replacing the space $\mathbb{H}$ in (2) by the finite element space $\mathbb{H}_h = \mathbb{H}_h(\Omega) = \mathbb{H} \cap \operatorname{Im} I_{\mathrm{HCT}}$. The value of $I_{\mathrm{HCT}} U$ on a side of T depends on the degrees of freedom on the side. Hence, by abuse of notation, we also note

$$I_{\mathrm{HCT}} : V_i \to \mathbb{H}_h(\partial\Omega_i), \qquad I_{\mathrm{HCT}} : V \to \mathbb{H}_h(\Gamma).$$

Here V_i and V are the spaces of vectors of degrees of freedom on $\partial\Omega_i$ and Γ, respectively. $\mathbb{H}_h(\partial\Omega_i)$ and $\mathbb{H}_h(\Gamma)$ are the spaces of traces of functions from $\mathbb{H}_h$ on $\partial\Omega_i$ and Γ, respectively.

The local stiffness matrices, defined by

$$X^t A_T Y = \int_T \varepsilon(\vec{\theta}(I_{\mathrm{HCT}} X)) : K : \varepsilon(\vec{\theta}(I_{\mathrm{HCT}} Y)),$$

satisfy the spectral equivalence property

$$(4) \qquad\qquad U^t A_T U \sim \rho_i \|\nabla I_{\mathrm{HCT}} U\|^2_{1,T}.$$

The theory presented in this paper applies to the HCT element and to any element with the degrees of freedom u, θ_1, θ_2 at each vertex satisfying (4). The Discrete Kirchoff Triangle (DKT) element is an example which enforces (1) only along each side of the element T [1]. The proof of (4) for the DKT element as well as for stabilized Reissner-Mindlin elements will be presented elsewhere. Quadrilateral elements may be treated as two triangles for which (4) holds.

3. Formulation of the Algorithm

We recall the algorithm following [**9, 10**]. The local stiffness matrix corresponding to subdomain Ω_i is A_i and U_i is the corresponding vector of degrees of freedom. Let $\bar{N}_i$ denote the matrix with entries 0 or 1 mapping the degrees of freedom U_i into global degrees of freedom: $U_i = \bar{N}_i^t U$. Write

$$A_i = \begin{pmatrix} \bar{A}_i & B_i \\ B_i^t & \dot{A}_i \end{pmatrix}, \qquad \bar{N}_i = (\bar{N}_i, \dot{\bar{N}}_i),$$

where the first block corresponds to degrees of freedom on Γ. Eliminating the remaining degrees of freedom, one obtains the reduced system

(5) $SX = B,$

for unknown values X of the degrees of freedom on Γ, posed in the space V. The matrix S is the Schur complement, defined by

$$S = \sum_{i=1}^{k} \bar{N}_i S_i \bar{N}_i^t, \qquad S_i = A_i - B_i \dot{A}_i^{-1} B_i^t \ .$$

The reduced system (5) is solved by a preconditioned conjugate gradient algorithm. To define the preconditioner, we need auxiliary matrices D_i and Z_i such that

$$\sum_{i=1}^{k} \bar{N}_i D_i \bar{N}_i^t = I, \quad \operatorname{Ker} S_i \subset \operatorname{Im} Z_i.$$

The choice of D_i and Z_i will be specified later. Define the *coarse space*

$$W = \{v \in V \ : \ v = \sum_{i=1}^{k} \bar{N}_i D_i u_i, u_i \in \operatorname{Im} Z_i\}.$$

Our algorithm is :

ALGORITHM 1. *Given $R \in V$, compute $U \in V$ as follows:*
   (i) *Find λ_j so that $Z_i^t D_i^t \bar{N}_i^t \left(R - S \sum_{j=1}^{k} \bar{N}_j D_j Z_j \lambda_j \right) = 0, i = 1, \dots, k$.*
   (ii) *Set $R_i = D_i^t \bar{N}_i^t (R - S \sum_{j=1}^{k} \bar{N}_j D_j Z_j \lambda_j)$.*
   (iii) *Find a solution U_i for each of the local problems $S_i U_i = R_i, i = 1, \dots, k$.*
   (iv) *Find μ_i so that $Z_i^t D_i^t \bar{N}_i^t \left(R - S \sum_{j=1}^{k} \bar{N}_j D_j (U_j + Z_j \mu_j) \right) = 0, i = 1, \dots, k$.*
   (v) *The output is $U = \sum_{i=1}^{k} \bar{N}_i D_i (U_i + Z_i \mu_i)$.*

The solution of the auxiliary problem in step 1 can be omitted by choosing a suitable starting vector to guarantee that $\lambda_j = 0$ in every step. Mandel [9] has shown that the right-hand sides of the singular problems in step 3 are consistent, the output z from Algorithm 1 is independent of the choice of a solution in step 3, and the following condition number estimate holds.

THEOREM 1. *Algorithm* 1 *returns* $U = M^{-1}R$, *where* M *is symmetric positive definite and*

$$\kappa(M^{-1}S) = \lambda_{max}(M^{-1}S)/\lambda_{min}(M^{-1}S)$$

$$(6) \qquad \leq \quad \sup\left\{\frac{\sum_{j=1}^{k}\|\bar{N}_j^t\sum_{i=1}^{k}\bar{N}_iD_iU_i\|_{S_j}^2}{\sum_{i=1}^{k}\|U_i\|_{S_i}^2} : U_i \perp Ker(S_i),\ S_iU_i \perp Im\,Z_i\right\}.$$

The main trick in this paper is now to use the flexibility in the choice of the matrices Z_i to enforce that the supremum in (6) is taken only over vectors U_i such that the normal displacement $I_{\mathrm{HCT}}U_i$ is zero at all crosspoints. For this purpose, choose

$$(7) \qquad\qquad Z_i = [X_1, \ldots, X_{n_i}, Y_{i1}, \ldots Y_{im_i}]$$

where $\{X_1, \ldots, X_{n_i}\}$, $n_i \leq 3$, is a basis of $Ker\,S_i$, and for each crosspoint $j = 1, \ldots, m_j$ of Ω_i, Y_{ij} is a solution of the problems $S_iY_{ij} = E_{ij}$, with E_{ij} the vector corresponding to a unit normal load applied at crosspoint j. Indeed, since S_i is symmetric, $S_iU_i \perp Y_{ij}$ implies that $U_i \perp S_iY_{ij} = E_{ij}$, so $I_{\mathrm{HCT}}U_i$ is zero at all crosspoints.

It remains to construct the weight matrices D_i. If G is an edge or a crosspoint of Γ, define $E_G : V \to V$ as follows : $E_G(U)$ is the vector with the same values of the degrees of freedom as U on G, and zero values of all the other degrees of freedom. Here, an edge does not contain its end crosspoints, so $\sum_G E_G = I$. Now set, with $\beta \geq 1/2$,

$$(8) \qquad D_i = \sum_{G \subset \partial\Omega_i} d(i, G)\bar{N}_i^t E_G \bar{N}_i, \qquad d(i, G) = \frac{\rho_i^\beta}{\displaystyle\sum_{j\,:\,G\cap\partial\Omega_j \neq \emptyset} \rho_j^\beta}.$$

That is, the weight matrices D_i are diagonal, with the diagonal entry equal to the ratio of ρ_i^β to the sum of ρ_j^β for all subdomains sharing that degree of freedom. In our computations, we choose $\beta = 1$ as in [5].

4. Condition Number Estimate

The following theorem follows immediately from Theorem 3.3 in Mandel and Brezina [10].

THEOREM 2. *Let the weight matrices D_i be constructed as in (8) with $\beta \geq 1/2$, and for all subdomains crosspoints or edges $G \subset \partial\Omega_i \cap \partial\Omega_j$,*

$$(9) \qquad \frac{1}{\rho_j}\|\bar{N}_j^t E_G \bar{N}_i U_i\|_{S_j}^2 \leq \frac{1}{\rho_i}R\|U_i\|_{S_i}^2, \quad \forall U_i \perp Ker\, S_i,\, S_i U_i \perp Im\, Z_i \,.$$

Then the condition number from (6) satisfies $\kappa \leq 9(K+1)^2 R$, where K is the maximal number of adjacent subdomains to any subdomain Ω_i.

Define continuous analogues of the projection operators E_G via the interpolation mapping I_{HCT},

$$\mathcal{E}_G : \mathbb{H}_h(\Gamma) \to \mathbb{H}_h(\Gamma), \qquad \mathcal{E}_G I_{HCT} U = I_{HCT} \mathcal{E}_G U, \qquad \forall U \in V.$$

Verification of the bound (9) will be based on estimates in the trace norm of the operators $\mathcal{E}_G$. For this purpose, we first state several technical results concerning the trace norm.

An extension lemma can be proved by similar arguments as in Widlund [11].

LEMMA 1. *For any $u \in \mathbb{H}_h(\partial\Omega_i)$, there is a $v \in \mathbb{H}_h(\Omega_i)$ so that $v|_{\partial\Omega_i} = u$, and $|\nabla v|_{1,\Omega_i} \leq C|\nabla u|_{1/2,2,\partial\Omega_i}$.*

From Lemma 1 and the trace inequality follows the equivalence of seminorms

$$(10) \qquad \frac{1}{\rho_i}|U|_{S_i}^2 \sim |\nabla I_{HCT} U|_{1/2,2,\partial\Omega_i}^2.$$

The following estimate of the trace norm of the extension by zero is proved as in Bramble, Pasciak, and Schatz [2, Lemma 3.5].

LEMMA 2. *There exists a constant C such that if the support of u is contained in a segment σ of $\partial\Omega_j$ of length τ, and $|\frac{\partial u}{\partial s}|_{0,\infty,\sigma} \leq \frac{c}{h}|u|_{0,\infty,\sigma}$, then*

$$|u|_{1/2,2,\partial\Omega_j}^2 \leq |u|_{1/2,2,\sigma}^2 + C\left(1 + \log\frac{\tau}{h}\right)|u|_{0,\infty,\sigma}^2 \,.$$

We will also need an extension of the discrete Sobolev inequality of Dryja [6] to piecewise polynomial functions of order $p > 1$.

LEMMA 3. *Let $p \geq 1$. Then there exists a constant $C = C(p)$ such that for every u continuous on $\partial\Omega_i$ such that $u \in P_p$ on the side of every triangle T,*

$$|\nabla u|_{0,\infty,\partial\Omega_i}^2 \leq C\left(1 + \log\frac{H}{h}\right)\left(|\nabla u|_{1/2,2,\partial\Omega_i}^2 + \frac{1}{H}|\nabla u|_{0,2,\partial\Omega_i}^2\right).$$

We are now ready for the main estimate.

LEMMA 4. *There exists a constant C such that if G is a crosspoint or an edge of Ω_i, then it holds for all $u \in \mathbb{H}_h(\Gamma)$, such that $u = 0$ on all crosspoints of Ω_i, that*

$$|\nabla \mathcal{E}_G u|_{1/2,2,\partial\Omega_j}^2 \leq C\left(1 + \log^\beta\frac{H}{h}\right)\left(|\nabla u|_{1/2,2,\partial\Omega_j}^2 + \frac{1}{H}|\nabla u|_{0,2,\partial\Omega_j}^2\right),$$

with $\beta = 1$ if G is a crosspoint and $\beta = 2$ if G is an edge.

PROOF. Assume $u \in \mathbb{H}_h(\Gamma)$ and $u = 0$ on all crosspoints of Ω_i. Let $F \in \partial\Omega_i$ be a crosspoint. The shape function $\phi_{\alpha,F}$ associated with the degree of freedom $\hat{c}_\alpha u(F)$ satisfies

(11) $|\nabla\phi_{\alpha,F}|_{1,2,\Omega_i} \leq C,$ $|\nabla\phi_{\alpha,F}|_{0,2,\Omega_i} \leq Ch,$ $|\nabla\phi_{\alpha,F}|_{0,\infty,\Omega_i} \leq C.$

From (11), the trace theorem, and Lemma 2 with $\tau = Ch$, it follows that

(12) $|\nabla\phi_{\alpha,F}|_{1/2,2,\partial\Omega_i} \leq C.$

Since $u(F) = 0$, we have $\mathcal{E}_F u = \sum_\alpha \phi_{\alpha,F} \partial_\alpha u(F)$, and the proposition with $G = F$ follows using Lemma 3 and (12).

Let F_1, F_2 be crosspoints at the ends of an edge G. Since

$$\mathcal{E}_G u|_G = (u - \mathcal{E}_{F_1} u - \mathcal{E}_{F_2} u)|_G ,$$

it follows using the inequality $\|a + b\|^2 \leq 2\left(\|a\|^2 + \|b\|^2\right)$, from the already proved estimate for the case of crosspoint, from inequalities (11), (12), and from Lemma 3, that

$$|\nabla\mathcal{E}_G u|_{1/2,2,G}^2 \leq C\left(1 + \log\frac{H}{h}\right)\left(|\nabla u|_{1/2,2,\partial\Omega_i}^2 + \frac{1}{H}|\nabla u|_{0,2,\partial\Omega_i}^2\right)$$

$$|\nabla\mathcal{E}_G u|_{0,\infty,G}^2 \leq C\left(1 + \log\frac{H}{h}\right)\left(|\nabla u|_{1/2,2,\partial\Omega_i}^2 + \frac{1}{H}|\nabla u|_{0,2,\partial\Omega_i}^2\right)$$

$$|\nabla\mathcal{E}_G u|_{0,2,G}^2 \leq |\nabla u|_{0,2,\partial\Omega_i}^2 + Ch^2\left(1 + \log\frac{H}{h}\right)\left(|\nabla u|_{1/2,2,\partial\Omega_i}^2 + \frac{1}{H}|\nabla u|_{0,2,\partial\Omega_i}^2\right).$$

Since $\mathcal{E}_G u = 0$ and $\nabla\mathcal{E}_G u = 0$ at F_1 and F_1, it remains only to apply Lemma 2 to $\partial_\alpha\mathcal{E}_G u$, $\alpha = 1, 2$.   □

The desired bound on the condition number follows.

THEOREM 3. *Suppose that the assumptions made in Section 2 hold, that Z_i are defined by (7), and D_i are defined by (8). Then the condition number of Algorithm 1 satisfies*

$$\kappa \leq C\left(1 + \log^2 \frac{H}{h}\right),$$

with the constant C independent of H, h, and of the coefficients $\rho_i > 0$.

PROOF. The assumption (9) of Theorem 2 follows from Lemma 4, the equivalence of seminorms (10), and the inequality

$$|\nabla u|_{0,2,\partial\Omega_i}^2 \leq CH|\nabla u|_{1/2,2,\partial\Omega_i}^2$$

for all $u \in \mathbb{H}_h(\partial\Omega_i)$ that are zero at all crosspoints.   □

TABLE 1. Results for a Rectangular Plate

	h	iter	condition
no corners	h	25	350
	$h/2$	29	1430
	$h/4$	32	5764
corners	h	11	5.4
	$h/2$	13	8.2
	$h/4$	14	11.8

2×8 substructures, regular decomposition
$\Omega = [-2, 2] \times [0, 20]$, 8×64 HCT elements for initial h

TABLE 2. Results for oval plate with 24 subdomains (Fig. 1)

				CPU CRAY 2 sec	
	h	iter	cond	setup	iter
HCT element	h	43	153	14.8	3.6
no corners	$h/2$	59	588	25.8	7.6
	$h/4$	75	1981	53.9	27.7
HCT element	h	16	7.8	15.8	2.9
corners	$h/2$	23	22.2	26.0	4.3
	h/4	33	76.0	57.8	13.2
DKT element	h	33	62	14.9	3.2
no corners	$h/2$	49	239	25.1	6.5
	$h/4$	65	898	51.3	23.8
DKT element	h	12	3.3	15.4	2.6
corners	$h/2$	17	7.4	25.7	3.8
	$h/4$	25	25.1	56.6	10.8

5. Computational Results

In all tests, "corners" refers to the case when Z_i are defined by (7), and "no corners" is the case when the point load solutions Y_{ij} ("corner functions") are omitted from the columns of Z_i. The plate was clamped on the whole boundary. All experiments show that adding the corner functions improved the condition number considerably. The condition numbers were estimated from Ritz values in the Krylov space generated by conjugate gradients. The stopping criterion was the ratio of the ℓ^2 norm of the residual and the right hand side less than $\varepsilon = 10^{-6}$. In all experiments, the domain and the subdomains remain the same, and the elements are uniformly refined, so H is fixed. The condition number appears to grow about as $|\log^2 h|$ with the added corner functions, and about as $1/h^2$ without.

The purpose of the first test was to confirm the theory and demonstrate the effect of adding corner functions on the condition numbers (Tab. 1). Then to determine if adding the corner functions results in an improvement for

FIGURE 1. Oval plate

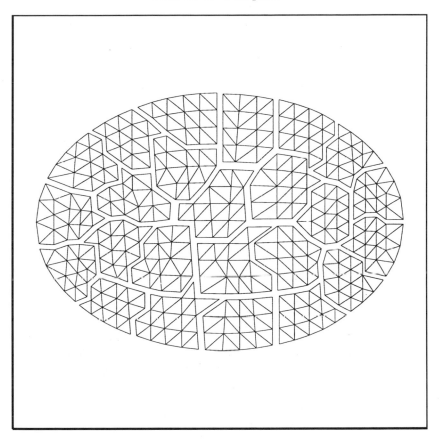

a realistic problem, we considered an oval plate discretized by an irregular mesh decomposed in 24 subdomains (Tab. 2, Fig. 1). The tests have shown that the improvement in the CPU times for the iterations outweigh the increase in the setup time due to the larger dimension of the coarse space.

6. Acknowledgements

The computations were run on the CRAY2 at CCVR, Palaiscau, France. The authors would like to thank Professors Leo Franca and Doug Arnold for useful discussions, and Mr. Marian Brezina for reading a preliminary version of this paper.

REFERENCES

1. J. L. Batoz, K. J. Bathe, and W. H. Ho, *A study of three-node triangular bending element*, Int. J. Numer. Methods Engrg. **15** (1980), 1771–1812.
2. James H. Bramble, Joseph E. Pasciak, and Alfred H. Schatz, *The construction of preconditioners for elliptic problems by substructuring, I*, Math. Comp. **47** (1986), no. 175,

103–134.

3. P. G. Ciarlet, *Basic error estimates for elliptic problems*, Handbook of Numerical Analysis (Amsterdam) (P.G. Ciarlet and J. L. Lions, eds.), vol. II, North-Holland, Amsterdam, 1989, pp. 17–352.

4. Lawrence Cowsar, Jan Mandel, and Mary F. Wheeler, *Balancing domain decomposition for mixed finite elements*, Math. Comp., to appear.

5. Yann-Hervé De Roeck and Patrick Le Tallec, *Analysis and test of a local domain decomposition preconditioner*, Fourth International Symposium on Domain Decomposition Methods for Partial Differential Equations (Roland Glowinski, Yuri Kuznetsov, Gérard Meurant, Jacques Périaux, and Olof Widlund, eds.), SIAM, Philadelphia, PA, 1991.

6. Maksymilian Dryja, *A method of domain decomposition for 3-D finite element problems*, First International Symposium on Domain Decomposition Methods for Partial Differential Equations (Roland Glowinski, Gene H. Golub, Gérard A. Meurant, and Jacques Périaux, eds.), SIAM, Philadelphia, PA, 1988.

7. Maksymilian Dryja and Olof B. Widlund, *Schwarz methods of Neumann-Neumann type for three-dimensional elliptic finite element problems*, Tech. Report 626, Department of Computer Science, Courant Institute, March 1993, submitted to Comm. Pure Appl. Math.

8. Roland Glowinski and Mary F. Wheeler, *Domain decomposition and mixed finite element methods for elliptic problems*, First International Symposium on Domain Decomposition Methods for Partial Differential Equations (Roland Glowinski, Gene H. Golub, Gérard A. Meurant, and Jacques Périaux, eds.), SIAM, Philadelphia, PA, 1988.

9. Jan Mandel, *Balancing domain decomposition*, Comm. in Numerical Methods in Engrg. **9** (1993), 233–241.

10. Jan Mandel and Marian Brezina, *Balancing domain decomposition: Theory and computations in two and three dimensions*, submitted.

11. Olof B. Widlund, *An extension theorem for finite element spaces with three applications*, Numerical Techniques in Continuum Mechanics (Braunschweig/Wiesbaden) (Wolfgang Hackbusch and Kristian Witsch, eds.), Notes on Numerical Fluid Mechanics, v. 16, Friedr. Vieweg und Sohn, 1987, Proceedings of the Second GAMM-Seminar, Kiel, January, 1986, pp. 110–122.

UNIVERSITÉ DE PARIS-DAUPHINE, 75775 PARIS CEDEX 16, FRANCE
E-mail address: Patrick.Le_Tallec@inria.fr

CENTER FOR COMPUTATIONAL MATHEMATICS, UNIVERSITY OF COLORADO AT DENVER, DENVER, CO 80217-3364
E-mail address: jmandel@colorado.edu

INRIA, DOMAIN DE VOLUCEAU, 78153 LE CHESNAY CEDEX, FRANCE
E-mail address: Marina.Vidrascu@inria.fr

Contemporary Mathematics
Volume **180**, 1994

Nonlinear Block Iterative Solution of Semiconductor Device Equations by a Domain Decomposition Method[1]

S.Micheletti, A.Quarteroni and R.Sacco

Abstract

A block nonlinear Gauss-Seidel procedure is employed to decouple the full system arising from the finite element approximation of the steady-state semiconductor device equations. At each iteration, a Neumann-Neumann domain decomposition method is applied to solve the linearized equations yielding electric potential and free carrier densities. Numerical results for one dimensional realistic test problems are given.

1 Introduction

The aim of this paper is to study a typical test device in semiconductor modeling, namely, a one-dimensional p-n diode. In section 2 we introduce the well-known *drift-diffusion* equations (see, e.g., [4]) that describe charge flow in a semiconductor device at steady-state conditions. The mathematical problem consists of a set of three highly nonlinearly coupled equations in the unknowns (ψ, n, p), which are respectively electric potential and carrier concentrations (electrons and holes). In order to reduce the computational effort, a block nonlinear Gauss-Seidel algorithm known in semiconductor literature as Gummel's map [3] is considered in section 3 to decouple the full system. The three resulting linearized equations are suitably formulated and then successively solved by a Neumann-Neumann domain decomposition method [1]. Concerning the spatial discretization, we respectively employ piecewise linear and exponentially fitted (*à la* Scharfetter-Gummel [7]) finite elements to handle the electric potential equation and the convection-diffusion equations for both electron and hole densities (see also [2] and [6]). In the concluding section 4 we discuss several numerical results relative to the study of the p-n diode at some working conditions of noteworthy interest.

[1]1991 *Mathematics Subject Classification.* Primary 78A30, 65M55; Secondary 65N30.
This research has been supported in part by the Italian Ministery of Research (Fondi MURST 40 %) and partly by the Sardinian Regional Authorities.
This paper is in final form and no version of it will be submitted for publication elsewhere.

2   The drift-diffusion model for semiconductors

Steady-state charge flow throughout a semiconductor device in the one dimensional case is commonly modeled by the following elliptic boundary value problem (see, e.g., [4])

$$(1) \quad -\psi'' = \rho \qquad\qquad\qquad (4) \quad \rho = \rho(\psi, n, p) = (p - n + C)$$

$$(2) \quad J'_n = R(\psi, n, p) \qquad\qquad (5) \quad J_n = \mu_n(n' - n\psi')$$

$$(3) \quad J'_p = -R(\psi, n, p) \qquad\quad (6) \quad J_p = -\mu_p(p' + p\psi')$$

The equations above represent the *drift-diffusion* model for a one dimensional semiconductor device and can be solved in any open set $\Omega = (0, L)$ provided by suitable boundary conditions. The unknowns (ψ, n, p) are respectively electric potential and free carrier densities (electrons and holes), so that (1) is a Poisson equation and (2)-(3) are two continuity equations for electron and hole current densities J_n and J_p. These latter are, for a given electric field $E = -\psi'$, convection-diffusion equations with a (possibly) highly dominating transport in thin regions across the so called *p-n junctions*. The functions $\rho(x)$, $C(x)$, $\mu_{n,p}$, and R are respectively the space charge density, the doping profile, the carrier mobilities, and the net recombination/generation rate. For more details on the physical model, see [4] and the references therein.

Maxwell-Boltzmann statistics is assumed to hold, and therefore

$$(7) \quad n = \rho_n e^{\psi} \qquad\qquad\qquad (8) \quad p = \rho_p e^{-\psi}$$

where $\rho_{n,p}$ are usually known as *Slotboom* variables [8].

The choice of such a model problem makes the algorithmic effort relatively easy while retaining a wide range of generality, since the basic physical properties of the solutions in more complex and realistic geometries are well reproduced by the one dimensional approximation. A more detailed description of the mathematical model in the two dimensional case and of the numerical aspects is given in [4] and [5].

3   The solution algorithm

When facing the numerical approximation of (1)-(3), the most critical aspects are:

- the need of finding an effective linearization procedure that allows the decoupling of the three equations;

- the presence of sharp interior layers that demands use of domain decomposition methods and upwind finite elements.

On the ground of the numerical experiments performed, we propose the following solution algorithm, which is long established in semiconductor device modeling and is commonly known as *Gummel's map* [3]. It reduces the over-all computational effort by forcing a decoupling in system (1)-(3) and leading to the successive solution of one *nonlinear* Poisson problem and two *linearized* convection-diffusion equations.

Gummel's map reads as follows.

Outer iteration: Loop on k until convergence

- Construct Newton iterates $\{\psi_m^k\}$ for the nonlinear elliptic boundary value problem

$$(9) \quad \begin{cases} -\psi''^k = (\rho_p^{k-1}e^{-\psi^k} - \rho_n^{k-1}e^{\psi^k} + C(x)) = \rho(x, \psi^k(x)), & 0 < x < L \\ \psi^k(0) = \psi_0, \qquad \psi^k(L) = \psi_L \end{cases}$$

such that

$$\lim_{m \to \infty} \psi_m^k = \psi^k$$

Inner iteration:

- for each m solve the linearized problem by the Neumann-Neumann (NN) multidomain method proposed in [1].

- Solve the two linear drift-diffusion problems

$$(10) \quad \begin{cases} -(\mu_n^k(n'^k - n^k\psi'^k))' = -R(\psi^k, n^{k-1}, p^{k-1}), & 0 < x < L \\ n^k(0) = n_0, \qquad n^k(L) = n_L \end{cases}$$

$$(11) \quad \begin{cases} -(\mu_p^k(p'^k + p^k\psi'^k))' = -R(\psi^k, n^{k-1}, p^{k-1}), & 0 < x < L \\ p^k(0) = p_0, \qquad p^k(L) = p_L \end{cases}$$

for n^k and p^k by the NN multidomain method (upon "symmetrizing" the convection-diffusion problems)

The intermediate nonlinear step (9) provides a new electric potential ψ^{k+1} which is plugged into the two convection-diffusion equations to be solved for the carrier concentrations n^{k+1} and p^{k+1} respectively. The procedure is stopped as soon as the variations of electric potential and carrier concentrations between two consecutive iterations (for a suitable norm) fall below a fixed tolerance.

The finite element method is used for the spatial approximation, where piece-wise linear and exponentially fitted (*à la* Scharfetter-Gummel (SG) [7]) shape functions are respectively employed for electric potential and carrier densities (see also [2] and [6]).

Use of a Neumann-Neumann domain decomposition method is motivated by the strongly varying nature of the solutions of realistic semiconductor device problems. Indeed, they typically exhibit very sharp interior layers across the p-n junctions where convection dominates and whose position may be easily deter-mined *a priori*, while they behave smoothly in the rest of the device domain. The Neumann-Neumann domain decomposition method allows the solution of self-adjoint boundary-value problems in regions partitioned into subdomains through an iterative procedure among subdomains. At each step the updating is achieved

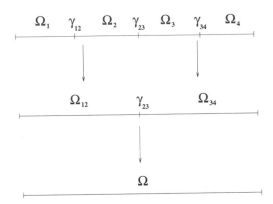

Figure 1: Recursive partitioning algorithm to reduce to single interface problems

by solving independent subproblems with Dirichlet conditions; this phase is followed by a correction yielding subproblems with Neumann conditions at the interfaces among subdomains. The method is described in [1], where an acceleration procedure relying on Conjugate Gradient (CG) iterations for the so called Steklov-Poincaré's interface operator is also proposed.

We may interpret this algorithm at the continuous level by saying that, starting from an initial guess u^0, the NN method generates a sequence of approximants $\{u^n\}$ of the exact solution; such functions are continuous over the device domain Ω but, in general, will have discontinuous derivatives at the interfaces. The Neumann step serves to smooth these irregularities by smearing the jumps at each interface of the electric field and of the discrete current densities all over the subdomains.

As for the numerical treatment of current continuity equations, there are two conflicting needs to be satisfied, namely, the self-adjointness of the operators and the computation of exponential terms arising from the change of variables (7)-(8) that make these equations self-adjoint. The successful strategy (*cf.* [5]) consists in modifying the standard Neumann-Neumann method in such a way as to solve a series of differential problems having *just one* interface each, and, consequently, only one degree of freedom. This kind of subproblem may be in fact easily solved in a *single* iteration by the original domain decomposition procedure.

Actually, the starting coupled problem partitioned in N subdomains is *recursively* led to $N/2$ subdomains problems by an algorithm that systematically eliminates the interfaces until *single-interface* problems are reached and eventually solved. This strategy is illustrated in figure 1 in the case of $N = 4$.

4   Numerical results

In this section we discuss some numerical examples relative to the simulation of a one dimensional p-n diode with an abrupt doping profile $C(x)$ and subject to some different values of the biasing potential $V_a = V_{ap} - V_{an}$, where V_{ap}, V_{an} are respectively the external applied potentials at $x = 0, L$. We assume the device length $L = 10\mu m$ (junction at $x = L/2$) and the piecewise constant doping profile

$$(12) \quad C(x) = \begin{cases} -10^{17} cm^{-3} & 0 \leq x \leq \frac{L}{2} \\ 10^{17} cm^{-3} & \frac{L}{2} < x \leq L \end{cases}$$

We show in figure 2 the numerical results relative to the simulation of the p-n diode at the *reverse bias* $V_a = -15$ Volt (i.e., $V_a < 0$). Sharp interior layers across the junction are clearly exhibited by both electric potential and carrier distributions; the former are due to the discontinuity in the doping profile $C(x)$, while the highly dominating transport $E = -\psi'$ around the p-n junction is responsible for the latter. The main parameters of the simulation are sketched in the headings, where IDOM is the number of subdomains, NP is the total number of internal nodes and ITGLOB denotes the number of iterations on k needed to achieve convergence of Gummel's map. Notice the highly nonuniform distribution of the mesh nodes over the device domain; the grid spacing has been taken constant within each subdomain, being quite coarse in the lateral quasi neutral regions and much finer in the depleted zone across the junction, where the maximum variations of the solutions are expected.

We stress that the solution by the NN domain decomposition method of each subproblem has always required a *number of iterations as low as the number of interfaces*. We also remark the effectiveness of the SG finite elements in reproducing the sharp interior layers exhibited by the carrier distributions.

Other examples of simulations of the p-n diode test device under different biasing conditions are reported in [5], with special emphasis on the case of a *high reverse voltage* V_a. The results show the quick convergence of the non linear block iterative procedure, which turns out to be reasonably independent of the applied voltage V_a; further checks on the discrete distributions of electric potential and carrier concentrations have also proved the accuracy of the method measured in the sup-norm.

It must be pointed out that convergence of the Gummel's map quickly worsens as the reverse bias assumes increasing negative values, until it definitevely stops as V_a approaches the breakdown voltage (see, e.g. [9]). The numerical solution of the drift-diffusion system in this physical situation is extensively addressed in [5], where a very effective variant of Gummel's map based on BI-CGSTAB [10] preconditioned iterations is proposed.

References

[1] J.F. Bourgat, R. Glowinski, P. Le Tallec, M. Vidrascu, '*Variational Formulation and Algorithm for Trace Operator in Domain Decomposition Calculations*', in '*Domain Decomposition Methods for Partial Differential Equations*', T. F. Chan *et al.* (eds.), SIAM Philadelphia, 1990, 3—16.

[2] F. Brezzi, L.D. Marini, P. Pietra, '*Two Dimensional Exponential Fitting and Application to Drift-Diffusion Models*', SIAM J. Numer. Anal., **26** (1989), 1347—1355.

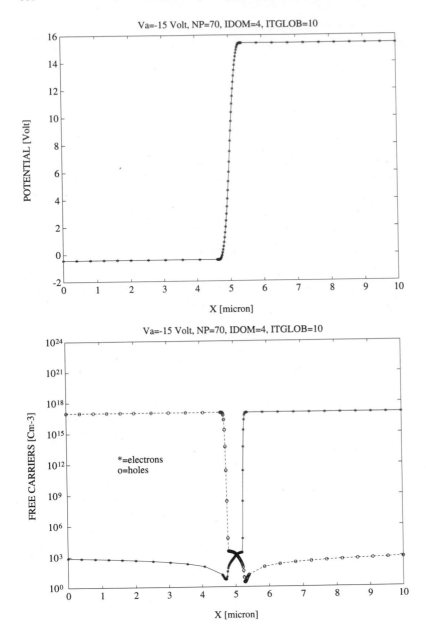

Figure 2: Potential and carrier distributions at $V_a = -15$ Volt

[3] H. K. Gummel, 'A Self-Consistent Iterative Scheme for One-Dimensional Steady-State Transistor Calculations', IEEE Trans. El. Dev., **ED-11** (1964), 455—465.

[4] P. Markowich, 'The Stationary Semiconductor Device Equations', Springer-Verlag, Wien-New York, 1986.

[5] S. Micheletti, A. Quarteroni, R. Sacco, 'Current-Voltage Characteristics Simulation of Semiconductor Devices using Domain Decomposition', Technical Report, Department of Mathematics, Polytechnic of Milan (1994).

[6] R. Sacco, 'Convergence of a Second-Order Accurate Petrov-Galerkin Scheme for Convection-Diffusion Problems in Semiconductors', Appl. Num. Math., **11** (1993), 517—528.

[7] D. L. Scharfetter, H. K. Gummel, 'Large-Signal Analysis of a Silicon Read Diode Oscillator', IEEE Trans. Electr. Dev., **ED-16** (1969), 64—77.

[8] J. W. Slotboom, 'Computer-Aided Two-Dimensional Analysis of Bipolar Transistors', IEEE Trans. Electr. Dev., **ED-20** (1973), 669—679.

[9] S. M. Sze, 'Physics of Semiconductor Devices', 2nd Ed., John Wiley & Sons, New York, 1981.

[10] H. Van Der Vorst, 'BI-CGSTAB : A Fast and Smoothly Converging Variant of BI-CG for the Solution of Nonsymmetric Linear Systems', SIAM J. Sci. Stat. Comput., **13**, No. 2, (1992), 631—644.

Authors' current address: Department of Mathematics, Technical University of Milan, Via Bonardi 9, 20133, Milan, Italy.
Additional address for A. Quarteroni is: CRS4, Via N. Sauro 10, 09123, Cagliari, Italy.

Contemporary Mathematics
Volume 180, 1994

A Direct Chebyshev Multidomain Method for Flow Computation with Application to Rotating Systems

I. Raspo, J. Ouazzani and R. Peyret

ABSTRACT. This paper presents a spectral multidomain method for solving Navier-Stokes equations in the vorticity - stream function formulation. Numerical results are reported and compared with spectral monodomain solutions to show the advantage of the domain decomposition for some problems with singular solution.

1. Introduction

Spectral methods are very efficient for calculating smooth solutions in rectangular domains. On the other hand, when the solution exhibits a large gradient or a singularity inside the domain, the efficiency of the spectral methods is lost. One way to remove this difficulty is to use a domain decomposition in order to isolate the singularity at a corner boundary of subdomains. The multidomain method presented here is based on an extensive use of the influence matrix technique ([1], [2]). The aim of this approach is to obtain in a direct way, i. e. without iterative process, the values of the variables at the interface between two adjacent subdomains insuring the continuity of their normal derivatives.

The method is applied to the computation of the crystal growth by the Czochralski process. In such a configuration ([3]), the vorticity and the azimuthal velocity derivative are singular at the junction between the crystal and the free surface of the melt, where the type of boundary conditions changes. In order to easily describe the method and also to avoid supplementary numerical difficulties associated with the axis of rotation, we consider, in the following sections, a plane mathematical model. Results concerning the axisymmetric Czochralski process will be presented in the last section.

2. Mathematical model

We consider the geometrical configuration of figure 1 (see on the next page). The governing equations are the 2D Navier Stokes equations in the vorticity ω - stream function ψ formulation ([3]). The other variable is the temperature T determined by a

---

1991 *Mathematics Subject Classification.* Primary 65M55, 65M70; Secondary 76U05.
The first author was supported by a DRET contract.
The detailed version of this paper will be submitted for publication elsewhere.

transport - diffusion equation. The equations are solved in the domain $\Omega=[0,1] \times [0,\alpha]$, the characteristic length being R_c and the characteristic velocity υ/R_c ; $\alpha=H/R_c$ is the gap ratio and $\gamma=R_x/R_c$ is the radius ratio by reference to the Czochralski configuration. The boundary conditions are given on figure 1. The dimensionless parameters are $Pr = \upsilon/\chi$ and $Gr = \beta\,\delta T\,g\,R_c{}^3/\upsilon^2$, where g is the gravity, υ the kinematic viscosity, χ the thermal diffusivity, β the thermal volume expansion coefficient and δT the temperature difference between the crucible of radius R_c and the crystal of radius R_x.

The time discretization is done through the second - order finite difference backward Euler - Adams Bashforth scheme ([2]). Therefore, at each time step, we have to solve a Helmholtz problem for the temperature and a Stokes - type problem for (ω,ψ), using a collocation Chebyshev method.

3. Multidomain method for Stokes-type problem

We describe the method for solving the Stokes-type problem for (ω,ψ). The method is the same, with obvious simplifications, for solving the Helmholtz problem for the temperature. More details are given in [3].

The problem to solve, formulated in cartesian coordinates, is of general form:

(1)
$$\Delta\omega - \sigma\omega = F, \quad \Delta\psi - \omega = 0 \qquad \text{in } \Omega$$
$$\psi = g, \; \frac{\partial\psi}{\partial n} = h \qquad \text{on } \Gamma_2\,U\,\Gamma_3 U\,\Gamma_4$$
$$\psi = g, \; \omega = f \qquad \text{on } \Gamma''_2\,U\,\Gamma_1$$

where $\Gamma_1 = \{\, x = 0\,, 0 \le z \le \alpha \,\}; \Gamma'_2 = \{\, 0 \le x \le \gamma, z = \alpha \,\}; \Gamma''_2 = \{\, \gamma \le x \le 1\,, z = \alpha \,\},$
   $\Gamma_3 = \{\, x = 1\,, 0 \le z \le \alpha \,\}; \Gamma_4 = \{\, 0 \le x \le 1, z = 0 \,\}.$
$\partial/\partial n$ denotes the normal derivative.

The computational domain Ω is divided into 2 subdomains Ω_i, i=1, 2. Then, the

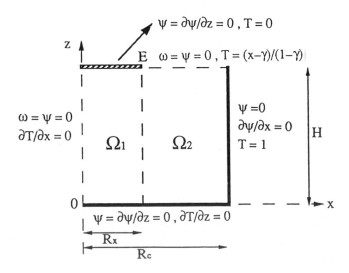

Figure 1 : Geometrical configuration.

global problem (1) is replaced by a set of two problems solved respectively in the corresponding subdomains.

Let us symbolically note $B\psi_i = S_i$ and $B'\omega_i = S'_i$ the above boundary conditions on $\Gamma^{(i)}$, the physical boundary of Ω_i. More precisely, we have:

$$\Gamma^{(1)} = \Gamma_1 \cup \Gamma_2 \cup \{ (x,0) , 0 \le x \le \gamma \} \text{ and } \Gamma^{(2)} = \Gamma''_2 \cup \Gamma_3 \cup \{ (x,0) , \gamma \le x \le 1 \}.$$

At the interface γ_{12} between the two adjacent subdomains Ω_1 and Ω_2, we impose the following conditions of continuity, for $\phi = \omega$, ψ:

(2)
$$\phi_1 = \phi_2, \quad \frac{\partial\phi_1}{\partial x} = \frac{\partial\phi_2}{\partial x}.$$

These conditions of continuity are enforced through the influence matrix technique. More precisely, the solution in Ω_i is sought in the form:

(3)
$$\begin{pmatrix} \omega_i \\ \psi_i \end{pmatrix} = \begin{pmatrix} \tilde{\omega}_i \\ \tilde{\psi}_i \end{pmatrix} + \sum_{k=1}^{K} \lambda_{ik} \begin{pmatrix} \omega'_{ik} \\ \psi'_{ik} \end{pmatrix} + \sum_{k=1}^{K} \mu_{ik} \begin{pmatrix} \omega''_{ik} \\ \psi''_{ik} \end{pmatrix},$$

where $(\tilde{\omega}_i, \tilde{\psi}_i)$ is solution of:

(4)
$$\begin{aligned}
&\Delta\tilde{\omega}_i - \sigma\tilde{\omega}_i = F_i, \quad \Delta\tilde{\psi}_i - \tilde{\omega}_i = 0 && \text{in } \Omega_i , \\
&B\tilde{\psi}_i = S_i , \, B'\tilde{\omega}_i = S'_i && \text{on } \Gamma^{(i)} , \\
&\tilde{\omega}_i = 0 , \, \tilde{\psi}_i = 0 && \text{on } \gamma_{12} ,
\end{aligned}$$

$(\omega'_{ik}, \psi'_{ik})$, for $k = 1, .., K$, is solution of :

(5)
$$\begin{aligned}
&\Delta\omega'_{ik} - \sigma\omega'_{ik} = 0, \quad \Delta\psi'_{ik} - \omega'_{ik} = 0 && \text{in } \Omega_i , \\
&B\psi'_{ik} = 0 , B'\omega'_{ik} = 0 && \text{on } \Gamma^{(i)} , \\
&\omega'_{ik}(\eta_m) = \delta_{km} , \, \psi'_{ik} = 0 && \text{on } \gamma_{12} ,
\end{aligned}$$

(where η_m, $m = 1, .., K$, refers to the collocation points on γ_{12} and δ_{km} is the Krönecker symbol) and :

$(\omega''_{ik}, \psi''_{ik})$, for $k = 1, .., K$, is solution of:

(6)
$$\begin{aligned}
&\Delta\omega''_{ik} - \sigma\omega''_{ik} = 0, \quad \Delta\psi''_{ik} - \omega''_{ik} = 0 && \text{in } \Omega_i , \\
&B\psi''_{ik} = 0 , B'\omega''_{ik} = 0 && \text{on } \Gamma^{(i)} , \\
&\omega''_{ik} = 0 , \, \psi''_{ik}(\eta_m) = \delta_{km} && \text{on } \gamma_{12} .
\end{aligned}$$

Each Stokes problem (4), (5) and (6) is solved using the influence matrix technique ([1], [2]). The conditions (2) with the decomposition (3) give an algebraic system to determine the constants λ_{ik} and μ_{ik} which are the values at the collocation points on γ_{12} of ω and ψ, respectively. The matrix of this system is called the continuity influence matrix. It is important to note that this matrix, as well as the boundary influence matrices of the Stokes problems (4), (5) and (6), has some eigenvalues equal to zero. So we have to remove some points of the boundary to make these matrices invertible ([1], [6]).

4. Numerical results

As already said, the vorticity ω exhibits a singularity at point E (see figure 1 on page 2). For the Stokes problem, ω behaves like $\rho^{-1/2}$ where ρ is the distance to the singular point E. The spectral coefficients of a function presenting such a singular behaviour do not decrease at all, as it can be seen on figure 2.a) which shows these coefficients for the function $f(x) = \{(\varepsilon - 2x)^{-1/2}$, for $-1 \leq x \leq 0; 0$, for $0 \leq x \leq 1\}$. This function is infinite at x=0 if ε=0. In order to avoid infinite values, we chose $\varepsilon = 2\pi/N^2$, N being the number of collocation points. When using a two-domain method, the singularity is located at a corner and the convergence of the spectral coefficients is much better. As an example, Figure 2b) shows the Chebyshev coefficients of the function $f(x) = (1 - x + \varepsilon)^{-1/2}$, for $-1 \leq x \leq 1$, for which the singularity is at x=1 (if ε=0). It is important to note that, nevertheless, we have not the spectral accuracy but we can see clearly the advantage of the domain decomposition on the horizontal vorticity profile at the first line of collocation points under the boundary containing the singular point E (see figure 3). This vorticity distribution is solution of the problem described in section 2. For the resolution NxM = 55x73 (N and M are the numbers of collocation points respectively in the x - direction and in the z - direction), the monodomain solution exhibits large oscillations whereas the multidomain solution does not ($N_1=N_2=(N+1)/2$). Let us note that these profiles are obtained from a Chebyshev polynomial interpolation on a regular mesh 201 x 201.

5. Application to the Czochralski melt configuration

Let us now consider the physical problem which induced the mathematical model described in section 2. For this axisymmetric problem, we have not only the singularity of the vorticity but also the singularity of the azimuthal velocity derivative because of a discontinuity of boundary conditions at the crystal - free surface junction. Indeed, in z=α, the boundary conditions for the azimuthal velocity v are :

(7) $$v = (Re_x - Re_c)\, r \text{ , for } 0 \leq r \leq \gamma; \frac{\partial v}{\partial z} = 0 \text{ , for } \gamma \leq r \leq 1$$

Figure 2 : Spectral coefficients A_k for a resolution N=65 : a) of a singular function f(x) in x=0 ; b) of a singular function f(x) in x=1.

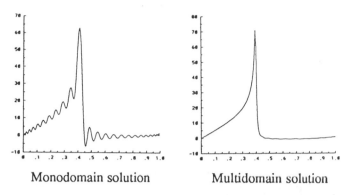

Monodomain solution   Multidomain solution

Figure 3 : Vorticity profile under the boundary containing the point E , for the resolution NxM=55x73.

where Re_x is the rotation Reynolds number of the crystal of radius R_x and Re_c is the rotation Reynolds number of the crucible of radius R_c. The radial coordinate r replaces the cartesian coordinate x. A coordinate transformation in the radial direction has been done in the 2 domain solution in order to put the calculation points away from the axis. On the other boundaries, we have the boundary condition v=0.

The results presented here were obtained with a Prandtl number Pr=0.05, a gap ratio α=1, and a radius ratio γ=0.4. The multidomain solution is compared with a spectral monodomain solution. Figure 4 shows the configuration of the flow for Re_x=2500,

Monodomain solution

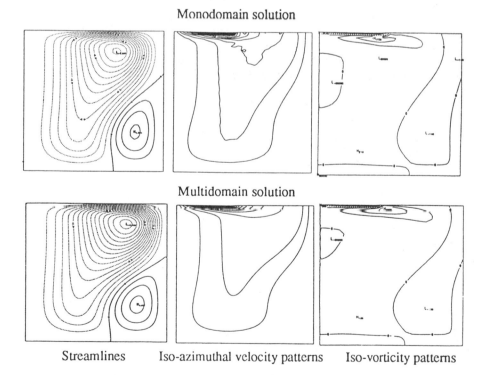

Multidomain solution

Streamlines   Iso-azimuthal velocity patterns   Iso-vorticity patterns

Figure 4 : Flow configuration for Re_x=2500, Re_c=0 and Gr=10^5 and for the resolution NxM=41x41.

$Gr=10^5$ and $Re_c=0$. The advantage of the multidomain method for this physical problem can be seen clearly on the iso-azimuthal velocity patterns drawn using an interpolation on a regular mesh 101 x 101: the monodomain solution exhibits large Gibbs oscillations under the crystal, because of the discontinuity, whereas there are no oscillations on the multidomain solution. These oscillations are less visible on the iso - vorticity lines because no interpolation has been used: the isolines are drawn on the collocation points themselves.

6. Conclusion

We have presented a direct multidomain technique which allows to use efficiently spectral methods for problems whose solution is not regular. We have shown that this method gives a good accuracy for such a solution and can be used in an efficient way for complex physical problems such as the Czochralski melt configuration. It is important to note that this technique can also allow to use spectral methods in non rectangular domains. Indeed, we are now using this multidomain method with a decomposition in four subdomains for the flow computation in a rotating cavity with a T - shape. The results obtained are satisfying and will be published in a forth coming paper.

References

1. U. EHRENSTEIN and R. PEYRET, 'A Chebyshev collocation method for the Navier - Stokes equations with application to double - diffusive convection', Int. J. Numer. Methods in Fluids, **9** (1989), 427-452.
2. J.M. VANEL, R. PEYRET and P. BONTOUX, 'A pseudospectral solution of vorticity-stream function equations using the influence matrix technique', Numer. Meth. for fluid dynamics II, K. W. Morton and M. J. Baines, Eds., Clarendon Press ,Oxford, (1986), 463-475.
3. I. RASPO, J. OUAZZANI and R. PEYRET, 'A spectral multidomain technique for the computation of the Czochralski melt configuration', to appear.
4. J.P. PULICANI, 'A spectral multidomain method for the solution of 1D-Helmholtz and Stokes-type equations', Computers and Fluids, **16** (1988), 207-215.
5. R. PEYRET, 'The Chebyshev multidomain approach to stiff problems in fluid mechanics', Comp. Meth. Appl. Mech. Eng., **80** (1990), 129-145.
6. G. R. BWEMBA and R. PASQUETTI, 'About the influence matrices used in the 2D spectral solution of the Stokes problem (vorticity - stream function formulation)', to appear.

IMFM - UM 34 C.N.R.S.,1, rue Honnorat, 13003 Marseille, France
E-mail address : raspo@imtcray.imt-mrs.fr

I.M.T., Technopôle de Château Gombert, 13451 Marseille, France

Département de Mathématiques, C.N.R.S. - URA 168
Université de Nice - Sophia Antipolis, 06000 Nice, France

Contemporary Mathematics
Volume **180**, 1994

Multi-domain Fourier Algorithms for Parallel Solution of the Navier-Stokes Equations *

L. Vozovoi, M. Israeli, A. Averbuch

Abstract

We present a parallel multi-domain algorithm for the solution of the incompressible Navier-Stokes equations. A high-order scheme is employed for the time discretization. The discretization in space is based on Fourier methods. For the interior subdomains we use an overlapping Local Fourier Basis technique, for the boundary subdomains - a non-overlapping Fourier-Gegenbauer method.

The matching of the local solutions is performed via a direct point-wise procedure on the interfaces, using properly weighted interface Green's functions. The unknown coefficients are found explicitly in terms of the jumps of the solution and its first derivatives at the interfaces.

The localization properties of the interface Green's functions are exploited in order to simplify the matching relations so that communication is reduced mostly to local data exchanges between neighboring subdomains. In effect, the parallel algorithm becomes highly scalable with the percentage of the global communications decreasing as the resolution requirements of the problem increase.

1 Introduction

Parallel multiprocessor computers are becoming indispensable for large-scale scientific computing, in particular, in computational fluid dynamics (CFD) where direct numerical simulations at high Reynolds numbers are one of the principle means of research.

The high performance of parallel machines can be realized only if effective parallel algorithms are supplied. An efficient parallel strategy for CFD problems is domain decomposition. The geometric domain decomposition (DD) is intrinsically suited for the purposes of parallelization since neighboring pieces of space (subdomains) can be allocated to processors with short communication

---

*AMS(MOS) Subject Classification 65P05

This research is supported partly by a grant from the French-Israeli Binational Foundation for 1992-1994

This paper is in final form and no version of it will be submitted for publication elsewhere.

links. Such a decomposition is in agreement with the natural data dependencies of elliptic and time-parabolic problems. The original problem is solved independently in each processor, then some patching procedure is employed to enforce the continuity conditions on the interfaces. Such a patching step requires communication between processors. The objective of any DD method is to minimize the interprocessor communication and the amount of data to be transfered in order to avoid communication and synchronization bottlenecks.

Numerical simulation of fluid flows at high Re numbers requires high resolution in time and space. Spectral methods, using series expansions in terms of polynomial or trigonometric functions, are most appropriate for such problems. For smooth flow fields these methods converge exponentially fast as the number of modes increases. However, all spectral methods are inherently global as they couple all variables in the domain for the computation of any local quantity. The implication of this fact for parallel processing is that parallel algorithms using spectral methods are expected to become inefficient in the case of massively parallel processor due to communication bottlenecks (but see [2] for the spectral element approach).

In [5, 6, 9] a low communication multi-domain approach is developed. A notable feature of this approach is that it makes use of Fourier methods for space discretization. A modified Local Fourier Basis (LFB) technique [1] is employed for the smooth decomposition of the original problem into subproblems. The use of the Fourier basis leads to a great reduction of the parallel complexity as it makes it possible to match independently each separate harmonic in the spectral space. Another important feature of this approach is that it takes advantage of the local behavior of the Green's functions employed to impose continuity conditions at the interfaces. As a result, the influence of remote domains on the processing at a particular location becomes negligible and the interprocessor communication is confined to local data exchange between neighboring units.

This paper is a further development of the local Fourier methods for the solution of PDEs in multi-domain regions. A novel feature of the present algorithm is the combination of the LFB technique, applied in the interior subdomains, and the Fourier-Gegenbauer (FG) method of [3] in the boundary subdomains. The use of a non-overlapping FG method enables us to treat non-periodic problems. The method is applied here to the incompressible Navier-Stokes equation for regions decomposed into parallel strips or rectangular boxes.

2   Formulation of the Problem and Numerical Schemes

We are interested in the numerical solution of the unsteady incompressible Navier- Stokes equations that govern viscous flows with constant properties:

$$\frac{\partial \mathbf{v}}{\partial t} = Re^{-1}\nabla^2 \mathbf{v} + \mathbf{N}(\mathbf{v}) - \nabla\Pi \quad \text{in } \Omega \subset R^2. \tag{2.1}$$

Here $\mathbf{v}(\mathbf{x}, t) = (u, v, w)$ is the velocity, subject to the incompressibility constraint

$$\nabla \cdot \mathbf{v} = 0 \qquad \text{in } \Omega, \tag{2.2}$$

Π is the total pressure, and Re is the Reynolds number. The nonlinear term is written in the rotational form

$$\mathbf{N}(\mathbf{v}) = \mathbf{v} \times (\nabla \times \mathbf{v}). \tag{2.3}$$

The numerical solution of the problem (2.1)-(2.3) with specified boundary conditions requires discretization in both time and space.

2.1   Discretization in Time

The discretization in time is performed via the third-order splitting algorithm of [7]:

$$\frac{\hat{\mathbf{v}} - \sum_{q=0}^{2} \alpha_q \mathbf{v}^{n-q}}{\Delta t} = \sum_{q=0}^{2} \beta_q \mathbf{N}(\mathbf{v}^{n-q}), \tag{2.4}$$

$$\frac{\hat{\hat{\mathbf{v}}} - \hat{\mathbf{v}}}{\Delta t} = -\nabla \Pi^{n+1}, \quad \nabla^2 \Pi^{n+1} = \frac{1}{\Delta t} \nabla \cdot \hat{\mathbf{v}}, \tag{2.5}$$

$$\frac{\gamma_0 \mathbf{v}^{n+1} - \hat{\hat{\mathbf{v}}}}{\Delta t} = Re^{-1} \nabla^2 \mathbf{v}^{n+1}. \tag{2.6}$$

It consists of an explicit advection step (2.4), a global pressure adjustment for incompressibility (2.5) and an implicit viscous step (2.6).

The high-order approximation used for the pressure boundary conditions is:

$$\frac{\partial \Pi^{n+1}}{\partial \nu} = \nu \cdot \left[\sum_{q=0}^{2} \beta_q \mathbf{N}(\mathbf{v}^{n-q}) + Re^{-1} \sum_{q=0}^{2} \beta_q (-\nabla \times (\nabla \times \mathbf{v})^{n-q}) \right] \quad \text{on } \partial\Omega \tag{2.7}$$

where ν is the direction normal to the boundary.

Semi-implicit schemes of this type have much less severe restriction on the time step than fully explicit schemes. However the parallelization of such schemes is considerably more difficult because the boundary values at interfaces are not known explicitly and thus the local problems in subdomains are globally coupled.

2.2   Discretization in Space

The splitting procedure in time results in two types of elliptic equations, of the Helmholtz type

$$\nabla^2 u - \lambda^2 u = f(x, y), \tag{2.8}$$

and of the Poisson type

$$\nabla^2 u = f(x, y), \tag{2.9}$$

which have to be solved repeatedly (for each time step). The parameter λ in (2.8) is related to the time-stepping increment, $\lambda \propto 1/\sqrt{\Delta t}$.

The solution of Eq. (2.8), (2.9) is based on spectral methods with a Fourier basis. The Fourier method is the most efficient for the evaluation of spatial derivatives since the differential operators are represented in the transform space by diagonal matrices so that harmonics with different wave numbers remain uncoupled. Another advantage of the Fourier method, when compared to Chebyshev or Legendre based methods, is that it has uniform resolution and is thus most appropriate for turbulence computations.

However it is well known that exponential (spectral) convergence of the Fourier series takes place only if it approximates a continuous periodic function. For a continuous but nonperiodic function, which has a discontinuous periodic extension, the truncated Fourier series does not converge uniformly near the boundaries giving rise to spurious oscillations of order $O(1)$ (Gibbs phenomenon). Therefore, when the Fourier method is applied to the solution of nonperiodic problems, like the elemental problems in subdomains, the key question is how to remove the Gibbs phenomenon.

We use two approaches in order to preserve spectral accuracy: a modified Local Fourier Basis (LFB) method of [1], implemented in the interior subdomains, and the Fourier-Gegenbauer (FG) method of [3]. In the first approach a smooth decomposition of the source function $f(x, y)$ is performed by using a system of overlapping bell functions. Then the local Fourier method is applied within each subdomain (see [5, 6] for more details).

The second approach makes use of re-expansion of the (local) Fourier partial sums into rapidly convergent Gegenbauer series. This technique is implemented in the subdomains adjacent to the boundaries where overlapping of two contiguous subdomains is not possible in the case of non-periodicity (e.g. for Dirichlet or Neumann boundary conditions).

The rapid convergence of the Gegenbauer series is related to the fact that the Gegenbauer polynomials $G_l^\lambda(x)$ are the solutions of singular Sturm-Liouville problems (l is the order of the polynomial; λ is a parameter appearing in the weight function $(1 - x^2)^{\lambda-1/2}$). Unlike the Chebyshev or the Legendre polynomials, the Gegenbauer polynomials constitute a two-parameter family. It is proven in [3] that in some parametric region λ, $l \propto N$ (N is the number of the Fourier coefficients, representing an analytic and nonperiodic function) the Gegenbauer series converges exponentially. The application of the FG method for the solution of nonperiodic PDEs is described in [10].

The FG method is a good choice for our purpose because it operates inside the interval where the function is defined, so that it does not require overlapping of subdomains. Also, the FG method has uniform resolution like the LFB method, so that both Fourier techniques are easily combined.

However, the resolution properties of the FG method are much worse than those of the Fourier method or the LFB method. Therefore the FG method by itself is too expensive to implement in the whole domain. The combination of both the LFB and the FG techniques meets the requirement of high efficiency.

3 The DD Technique in 1-D

We describe our Multi-domain Local Fourier (MDLF) approach as applied to a modified Helmholtz equation in 1-D:

$$\frac{d^2u}{dx^2} - \lambda^2 u = f(x), \quad x \in [0, L]. \tag{3.1}$$

The computational interval L is divided into P pieces (subdomains) of arbitrary size l_n, $n = 1, 2, ..., P$.

The algorithm consists of two steps:

Construction of the elemental particular solutions. We decompose (3.1) into local subproblems for $u_p^{(n)}(x)$, $x \in [\bar{x}_{n-1}, \bar{x}_n]$ and solve them independently in each subdomain with *arbitrary* boundary conditions on the interfaces. The LFB technique is employed in the interior subdomains and the FG technique in the boundary subdomains.

Matching step. The interface conditions impose matching of adjacent local solutions at $x = \bar{x}_n$: $u_p^{(n)} = u_p^{(n+1)}$, $\frac{d}{dx} u_p^{(n)} = \frac{d}{dx} u_p^{(n+1)}$. The matching procedure makes use of the properly weighted interface Green's functions $h_{\pm}^{(n)}(x)$ which satisfy the homogeneous form of (3.1). For each interface $\bar{x}_n$, these are two exponential functions decaying away on each side.

The smooth global solution is a linear combination

$$u = \bigcup_{n=1}^{P} u^{(n)}, \quad u^{(n)} = u_p^{(n)} + A_n h_+^{(n)} + B_n h_-^{(n)} \tag{3.2}$$

The unknown coefficients A_n, B_n can be found explicitly in terms of the jumps of $u_p^{(n)}$, $du_p^{(n)}/dx$ at the interfaces $x = \bar{x}_n$ (see [6] for more details).

This two-step algorithm can be viewed as a reduction of the full matrix, representing the elliptic operator in (3.1), to a block-diagonal form (see [4]). It results in a great reduction of the parallel complexity. Instead of a global coupling of the collocation points in the whole domain (which necessitates a global data transfer), the interaction is confined mostly to the neighborhood of the subdomain of interest. However, all interface points remain coupled globally.

An important feature of the MDLF approach is that it takes advantage of the local behavior of the interface Green's functions in order to decouple the matching relations. For small enough time steps Δt the functions $h_{\pm}$ decay rapidly away from the interfaces so that the influence of remote interfaces become negligible. In effect, only local communication between neighboring subdomains (processors) is important. Note that the localization property of the modified Helmholtz operator, resulting from an implicit time discretization procedure, reflects the locality of the diffusive linear operator in the evolution problem (2.1).

On the contrary, the Poisson equation (2.8) describes equilibrium processes with global interactions. Therefore the solution of this equation requires global communication between subdomains. Nevertheless, we will show that in two (or more) dimensions the necessary global communications constitute only a small percentage of the required communication.

4   Extension to 2-D

The previous MDLF technique can be extended to two dimensions without loosing the property of locality. We consider a computational region $\Omega = (0, L_x) \times (0, L_y)$ divided into parallel strips or rectangular cells.

Case of strips. After applying the FFT in the periodic direction y along the strips we get a set of *uncoupled* 1-D ODEs for the Fourier coefficients $\hat{u}_k(x)$:

$$\frac{d^2 \hat{u}_k^{(n)}}{dx^2} - \lambda_k^2 \hat{u}_k^{(n)} = \hat{f}_k^{(n)}(x) \qquad (4.1)$$

where $\lambda_k^2 = \lambda^2 + k^2$ for the modified Helmholtz equation and $\lambda_k^2 = k^2$ for the Poisson equation. These problems are solved by using the 1-D routine, described in section 3.

An important observation is that even in the case $\lambda = 0$, $\lambda_k \neq 0$ for $k \neq 0$, the homogeneous solutions of (4.1) decay exponentially as functions of x. Thus, a global matching procedure is required only for the long waves, $k \leq k_*$, whereas the short waves, $k \geq k_*$, can be treated by using local matching on the interfaces. The cut-off wave number k_* should be chosen in accordance with the prescribed accuracy.

Let us denote the number of collocation points in the x and y directions as N_x and N_y, and define $N_y/N_x = P^\gamma$ where P is the number of subdomains. Then $\eta = 2k_*/N_y$ will be the relative amount of equations (modes) treated locally. It can be shown that for $\gamma = 0$, that is $N_x = N_y$, the relative amount of local communication $\eta \propto \sqrt{P}$. Another limiting case, $\gamma = 1$, corresponds to $N_x = const$, $N_y \propto P$ (the resolution is changed only in the direction of strips, e.g. the case of a long channel). In this case η is independent of P, i.e. the algorithm is fully scalable.

Case of cells. In this case the direct point-wise matching at all interface points (in 2-D), using the corresponding two-dimensional interface Green's functions, results in a large linear system. The use of the LFB technique allow us to perform matching in the Fourier space for each Fourier harmonic independently. The procedure consists of several matching steps alternately in x and y directions. The maximum precision is attained after $2 - 3$ iterations (for more details see [9]). The analysis of scalability in this case is similar to the case of strips.

5   Results and Conclusion

To demonstrate the accuracy of the MDLF method we consider the Kovaznay flow [8], which is an exact solution of the Navier-Stokes system (2.1):

$$u = 1 - e^{\rho x} \cos(2\pi y),$$

$$v = \frac{\rho}{2\pi} e^{\rho x} \sin(2\pi y)$$

where $\rho = Re/2 - (Re^2/4 + 4\pi^2)^{1/2}$. The domain decomposition into strips is considered. The computational parameters are: $P = 4$, $N_x = 128$, $N_y =$

32, $Re = 40$. Fig.1 shows the error (in the maximum norm) as a function of Δt^3. The linear dependence, which is in agreement with the third-order scheme (2.4)-(2.7), gives an evidence that the spatial errors are below the temporal ones in this range of Δt.

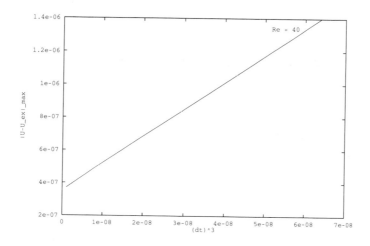

Figure 1:

To summarize, the MDLF approach, based on the local Fourier methods, overcomes most of the global coupling, inherent both in the use of a spectral method in space and an implicit discretization in time. It presents a low-communication, highly scalable parallel algorithm.

References

[1] R. Coifman, Y. Meyer, *Remarques sur l'analyse de Fourier à fenêtre, séric*, C.R.Acad.Sci., Paris **312** (1991), 259–261.

[2] P.F. Fisher, A.T. Patera, *Parallel spectral element solution of the Stokes problem*, J. Comput. Phys., **92**, 380–342.

[3] D. Gottlieb, Chi-Wang Shu, A. Solomonoff, H. Vandeven, *On the Gibbs phenomenon I: Recovering exponential accuracy from the Fourier partial sum of a non-periodic analytic function*, J.Comput.Appl.Math. **43**, (1992), 81–98.

[4] D. Gottlieb, R.S. Hirsh, *Parallel pseudo-spectral domain decomposition techniques*, J.Scient.Comput., **4**, (1989), 309–325.

[5] M. Israeli, L. Vozovoi, A. Averbuch, *Spectral multi-domain technique with Local Fourier Basis*, J.Scient.Comput., **8**, (1993), 181–195.

[6] M. Israeli, L. Vozovoi, A. Averbuch, *Parallelizing implicit algorithms for time-dependent problems by parabolic domain decomposition*, J.Scient.Comput., **8**, (1993), 197–212.

[7] G. E. Karniadakis, M. Israeli, S. A. Orszag, *High order splitting methods for the incompressible Navier-Stokes Equations*, J.Comput.Phys., **97**, (1991), 414–443.

[8] L. I. G. Kovaznay - *Laminar flow behind a two-dimensional grid*, Proc. Cambridge Philos. Soc., (1948), 48.

[9] L. Vozovoi, M. Israeli, A. Averbuch, *Spectral multi-domain technique with Local Fourier Basis II: decomposition into cells*, SIAM, J.Scient.Comp., to appear.

[10] A. Weill, L. Vozovoi, M. Israeli, *Spectrally accurate solution of non-periodic differential equations by the Fourier-Gegenbauer method*, SIAM, J.Numer.Anal., to appear.

Faculty of Computer Science, Technion, Haifa 32000, Israel
E-mail address: vozovoi@cs.technion.ac.il
                israeli@cs.technion.ac.il
School of Mathematical Sciences, Tel Aviv University, Tel Aviv 69978, Israel
E-mail address: amir@math.tau.ac.il

Recent Titles in This Series

(See the AMS catalog for earlier titles)